Modern Techniques for Crop Management

NIPA® GENX ELECTRONIC RESOURCES & SOLUTIONS P. LTD.
New Delhi-110 034

About the Author

Dr U S Walia, Sr. Agronomist-cum-Head (Retd.,), Department of Agronomy, Punjab Agricultural University, Ludhiana has teaching, research and extension experience for about 43 years. He handled 9 adhoc research projects funded by national and international agencies especially in the field of weed science. His publications include 6 books, 6 book chapters, 5 research bulletins, 4 extension bulletins, 13 practical manuals, 325 publications (research and extension). Apart from these 93 recommendations (as co-worker) have been included in the package of practices for *rabi* and *kharif* season crops for the benefit of Punjab farmers. He has guided 25 M.Sc. and 9 Ph.D. students. His contributions have been recognized through various awards viz., FAI Silver Jubilee Award-1984 for outstanding Doctoral Research in Fertilizer Usage, Fellow as well as Gold Medal of India Society of Weed Science during 2004 and 2007 respectively.

Modern Techniques for Crop Management

U.S. Walia
(Retd.) Senior Agronomist cum Head
Department of Agronomy
Punjab Agricultural University
Ludhiana, Punjab

NIPA® GENX ELECTRONIC RESOURCES & SOLUTIONS P. LTD.
New Delhi-110 034

NIPA® GENX ELECTRONIC RESOURCES & SOLUTIONS P. LTD.

101,103, Vikas Surya Plaza, CU Block
L.S.C. Market, Pitam Pura, New Delhi-110 034
Ph : +91 11 27341616, 27341717, 27341718
E-mail: newindiapublishingagency@gmail.com
Website: www.nipabooks.com

For customer assistance, please contact

Phone: + 91-11-27 34 17 17
Fax: + 91-11-27 34 16 16
E-Mail: feedbacks@nipabooks.com

ISBN: 978-81-19103-00-3

Composed and Designed by NIPA®.

Dedicated to
My Late Parents

Sardar Sarwan Singh

(November, 1916 to January, 1971)
and

Sardarni Parsin Kaur

(April 1921 to January, 2005)
With respect and love

Acknowledgement

I am grateful to the almighty who blessed me with patience and fortitude for the completion of this noble job. Secondly, I feel privileged to express my heartful gratitude and indebtedness to late Dr. H.S. Gill, Sr. Agronomist (weed control) cum Head (Retd.) Department of Agronomy, PAU, Ludhiana and major advisor during my M.Sc. and Ph.D. programmes, who shaped my mind set and existence among weed science scientists to serve farmers, teachers, researches and students. Also, I express my sincere regards and thankfulness to Dr. L.S. Brar Sr. Agronomist (weed control) cum Head (Retd.), Department of Agronomy, PAU, Ludhiana, who inspired me to undertake an arduous job. It is beyond my access to acknowledge in words, the over encouraging moral support and co-operation which I enjoyed from Sardarni Parmjit Kaur Walia (wife) Er. Gurpreet Singh Walia and Er. Gaganpreet Singh Walia (sons) as well as Neena Walia and Rakhi Walia (daughter-in-law) throughout the completion of this manuscript. Above all, adorable grandsons, (Aarin, Manvik and Mehransh), whose cheerful presence has provided a perfect milieu to my writing.

U.S. Walia

Preface

Agriculture nowadays is facing a complex problem. Alternative framing methods have become increasingly popular in recent years as arable land is turning to non-arable land rapidly in the country. The income of farmers is declining day by day due to increasing cost of inputs including wages of farm labour. The net income can be increased by reducing cost of inputs and increasing the yield of current cropping systems with the adoption of recommended practices. There is an urgent need for vertical increase in agricultural productivity to feed the increasing population. Due to changing climate conditions, there is a need for reorientation of presently practised agricultural technologies. At the same time there is necessity to save/conserve the natural resources. Hence to increase income of farmers, there is a need to judiciously use production and protection technologies required for crop cultivation.

The book covers comprehensive information on highly profitable and soil less farming techniques i.e. hydroponic - a high income technique. It also includes the modern techniques like precision farming, conservation agriculture, irrigation saving devices, remote sensing, nanotechnology, leaf colour chart for economical use of nitrogen, laser land levelling, value addition in farm produce, precision weed management, climate smart agriculture, organic farming, global warming, crop residues management techniques, bioherbicides and application of agricultural biotechnology in weed management etc. Latest practices for the cultivation of cereal crops, oilseed crops, pulses and commercial crops have been discussed in detail. Apart from these topics highly profitable systems i.e. integrated farming system and allied enterprises, important uses of medicinal, aromatic and spice crops are also covered. The book also contains information on dry land agriculture, role of plant nutrition in agriculture, crop management in problematic soils, hints for achieving higher yield of crops as well as judicious use of herbicides for controlling problematic weeds. So, this book will be highly helpful for graduates and post graduate students of agriculture particularly Agronomy throughout India as it covers nearly the syllabus as per ICAR guidelines.

Author

Contents

1

History and Role of Agronomy in Agriculture

Agriculture, also known as farming, is the production of food, fiber, animal feed, and other goods by means of growing and harvesting plants and animals. Agriculture is practiced throughout the world. Many agricultural products are used every day, from the clothes we put on in the morning to the sheets we sleep under at night. When you think of agriculture, think of the five F's: food, fabric, forestry, farming, and flowers.

Agriculture is the science, art and industry of managing the growth of plants and animals to make use by human beings. In general, agriculture includes soil cultivation, growing and harvesting field crops, raising and breeding livestock, dairy and forest trees etc. The word agriculture is the English adaptation of Latin word i.e. agricultūra, from ager, "a field" and cultura, "cultivation" in the strict sense of "tillage of the soil". Thus, a literal reading of the word yields "tillage of a field/ fields".

Economic Survey

As per Economic Survey 2021-22 (Government of India), the agricultural sector is a central pillar of the Indian economy, employing 60 percent of the nation's workforce and contributing to about 17 percent of its GDP. India is the largest producer as well as the largest exporter of cereal products in the world. India's export of cereals stood at $12.87 bn during the year 2021-22. Rice (including Basmati and Non-Basmati) occupy the major share in India's total cereals export with 75% (in value terms) during the same period. Whereas, other cereals including wheat represent only a 25 % share of total cereals exported from India during this period.

India produced around 3.5 million tonnes (2020-21) of certified organic products which includes all varieties of food products namely Oil Seeds, Fibre, Sugar cane, Cereals & Millets, Cotton, Pulses, Aromatic & Medicinal Plants, Tea, Coffee, Fruits, Spices, Dry Fruits, Vegetables, Processed foods etc. The production is not limited to the edible sector but also produces organic cotton fiber, functional food products etc.

As per Fourth Advance Estimates for 2020-21, total food grain production in the country is estimated at a record 308.65 million tonnes which is 11.15 million tonnes higher than that during 2019-20. The production of rice, wheat and coarse cereals has increased at compound annual growth rates (CAGR) of 2.7, 2.9 and 4.8 per cent respectively during last six years i.e. 2015-16 to 2020-21. The CAGR for pulses, oilseeds and cotton has been 7.9, 6.1 and 2.8 per cent, respectively during the same period. As per the First Advance Estimates for 2021-22 (*kharif* only), total food grain production in the country is estimated at a record level of 150.50 million tonnes which is higher by 0.94 million tonnes than *kharif* food grain production of 2020-21.

India is one of the major oilseeds growing country. The oilseed production in India has steadily increased since 2016-17 onward after showing a fluctuating trend prior to that. The oilseed production in India has grown by almost 43 per cent from 2015-16 to 2020-21. The oil production in India has however lagged behind its consumption necessitating import of edible oils. India is the world's second largest consumer and number one importer of vegetable oil. As urbanisation increases in developing countries, dietary habits and traditional meal patterns are expected to shift towards processed foods that have a high content of vegetable oil. Vegetable oil consumption in India is, therefore, expected to remain high due to high population growth and consequent urbanisation. As per the OECD-FAO Agricultural Outlook 2021-2030, India is projected to maintain a high per capita vegetable oil consumption growth of 2.6 per cent per annum reaching 14 kg/capita by 2030 necessitating a high import growth of 3.4 per cent per annum

Significance of sugarcane and sugar industry for India's economy can be gauged from the fact that it is the country's second largest agro-based industry, next to cotton. It impacts the livelihood of over 5 crore farmers and their dependents. India is the largest consumer and the second-largest producer of sugar in the world. Average annual production of sugarcane is around 35.5 crore tonnes which is used to produce around 3 crore tonnes of sugar. The domestic consumption is estimated to be around 2.6 crore tonnes in 2020-21. Over the years, India has become a sugar surplus nation as reflected from the trend of sugar production and consumption. Since 2010-11, production has outstripped consumption except in 2016-17

Water is a critical input for agriculture which accounts for about 80 per cent of the current water use in the country. The share of net irrigated area accounts for about 49 per cent of the total net sown area in the country and out of the net irrigated area, about 40 per cent is irrigated through canal systems and 60 per cent through groundwater.

Livestock Sector is an important subsector of agriculture in the Indian economy. It grew at a CAGR of 8.15 per cent during 2014-15 to 2019-20 (at constant prices). As per the estimates of National Accounts Statistics (NAS) 2020 for sector wise GVA of agriculture and allied sectors, the contribution of livestock in total agriculture and allied sector GVA (at constant prices) has increased from 24.32 per cent (2014-15) to 29.35 per cent (2019-20). Livestock sector contributed 4.35 per cent of total GVA in 2019-20. Development of livestock sector has led to improvement in per capita availability of milk, eggs and meat.

Dairy is the single largest agricultural commodity contributing 5 per cent of the national economy and employing more than 8 crore farmers directly. India is ranked 1st in milk production contributing 23 per cent of global milk production. Milk production in the country has grown at a compound annual growth rate of about 6.2 per cent to reach 209.96 million tonnes in 2020-21 from 146.31 million tonnes in 2014-15.

According to FAOSTAT production data (2020), India ranks 3rd in egg production and 8th in meat production in the world. Egg production in the country has increased from 78.48 billion in 2014-15 to 122.11 billion in 2020-21 (provisional). The per capita availability of egg is at 91 eggs per annum in 2020-21 (provisional). Meat production in the country has increased from 6.69 million tonnes in 2014-15 to 8.80 million tonnes in 2020-21 (provisional)

India is the second largest fish producing country in the world accounting for 7.56 per cent of global production. It contributes about 1.24 per cent to the country's GVA and over 7.28 per cent to the agricultural GVA. Fisheries sector has demonstrated an outstanding double-digit average annual growth of 10.87 per cent since 2014-15 with record fish production of 145 lakh tons in FY 2020-21 (provisional). In terms of employment, the sector supports the livelihood of over 28 million people in India especially the marginalized and vulnerable communities. Export earnings from the fisheries sector was ₹ 46,662.85 crore during 2019-20.

1.1. History of Agriculture

The history of agriculture begins in the Fertile Crescent. This area of Western Asia comprises the regions of Mesopotamia and Levant while being confined by the Syrian Desert to the south and the Anatolian Plateau to the north. In the early 1900s, University of Chicago archaeologist James Henry Breasted coined the term «Fertile Crescent» to describe this location›s role as the birthplace of agriculture. It has often been called the "Cradle of Civilization" as well, since both the wheel and writing first appeared there. Modern Turkey, Iran, Iraq, Syria, Lebanon, Israel, Jordan, and the Palestinian territories all include some land within the Fertile Crescent.

Humans invented agriculture between 7,000 and 10,000 years ago, during the Neolithic era, or the New Stone Age. There were eight Neolithic crops: emmer wheat, einkorn wheat, peas, lentils, bitter vetch, hulled barley, chickpeas, and flax. The Neolithic era ended with the development of metal tools.

Evidence suggests that irrigation first appeared in Egypt and Mesopotamia in the fourth millennium B.C. Floods caused by the yearly inundation of the Nile would have had disastrous effects for ancient farmers, washing away dikes and swamping fields. Conversely, when the waters were low, the land dried up, killing crops. The oldest method of irrigation made use of man-made underground streams, called qanats, and is still being used in parts of the Middle East. Various groups of people began digging and repairing older, more primitive canal networks to help regulate the flow of rivers, such as the Nile River in Egypt. The canal networks eventually developed into sophisticated irrigation systems.

During the Bronze Age and the eras that followed, civilizations all over the world gradually invented or acquired advanced metalworking techniques, creating ever-stronger farming implements. Humans continued domesticating animals and plants to serve as food sources or sources of other useful products. During the Middle Ages, European farmers began using complex irrigation systems such as dams, reservoirs, and water-raising machines. They also developed a three-field system of crop rotation and the mouldboard plough. These inventions greatly increased the efficiency of agriculture.

Between the 17th and 19th centuries, Britain experienced a dramatic increase in agricultural productivity known as the British Agricultural Revolution. This "revolution" consisted of a variety of improvements to agricultural methods which more or less took place simultaneously. Farmers developed new methods of crop rotation, began cultivating land that had been marshy or forested, and planted new crops such as the turnip.

The technology of agriculture has continued to evolve over the years. Ploughs and other farming implements improved, and the mechanical combine harvester—a machine that harvests grain—was invented in the 1830s. In the early 1900s, the horse-drawn plough was a catalyst for introducing more machinery on the farm. The first tractors were steam-powered engines designed to haul agricultural equipment and were too expensive for most farmers. The gasoline-powered tractor was invented in 1892.

The last century has seen a host of changes in the way farming is carried out and in the way farm produce shows up on our tables. In 1938, a trucking executive complained during a round of golf about how the summer heat was spoiling much of the food shipped by his company. His companion, Joseph Numero, jokingly suggested refrigerating the trailers. The idea stuck, and Numero went on to start a refrigerated transport company with the help of inventor Frederick Jones.

Meanwhile, the development of hybrid seed, and particularly hybrid corn, revolutionized agriculture. Hybrids improve the results of planting, and often produce plants and fruits that are hardier and more uniform. Hybrid seeds contributed to the increased agricultural output of the second half of the 20th century.

Despite these great improvements, other changes arrived with controversy. During the past few years, a huge controversy has raged over the use of chemical pesticides and fertilizers on farms as well as genetically modified organisms. Growing awareness of how chemicals impact human health and the environment has led to calls for increased government regulation of the agricultural industry.

The Farming Revolution

Taking root around 12,000 years ago, agriculture triggered such a change in society and the way in which people lived that its development has been dubbed the "Neolithic Revolution." Traditional hunter-gatherer lifestyles, followed by humans since their evolution, were swept aside in favour of permanent settlements and a reliable food supply. Out of agriculture, cities and civilizations grew, and because crops and animals could now be farmed to meet demand, the global population rocketed—from some five million people 10,000 years ago, to more than seven billion today.

There was no single factor, or combination of factors, that led people to take up farming in different parts of the world. In the Near East, for example, it's thought that climatic changes at the end of the last ice age brought seasonal conditions that favoured annual plants like wild cereals. Elsewhere, such as in East Asia, increased pressure on natural food resources may have forced people to find homegrown solutions. But whatever the reasons for its independent origins, farming sowed the seeds for the modern age.

Plant Domestication

The wild progenitors of crops including wheat, barley, and peas are traced to the Near East region. Cereals were grown in Syria as long as 9,000 years ago, while figs were cultivated even earlier; prehistoric seedless fruits discovered in the Jordan Valley suggest fig trees were being planted some 11,300 years ago. Though the transition from wild harvesting was gradual, the switch from a nomadic to a settled way of life is marked by the appearance of early Neolithic villages with homes equipped with grinding stones for processing grain. The origins of rice and millet farming date to the same Neolithic period in China. The world's oldest known rice paddy fields, discovered in eastern China in 2007, reveal evidence of ancient cultivation techniques such as flood and fire control.

In Mexico, squash cultivation began around 10,000 years ago, but corn (maize) had to wait for natural genetic mutations to be selected for in its wild ancestor, teosinte. While maize-like plants derived from teosinte appear to have been cultivated at least 9,000 years ago, the first directly dated corn cob dates only to around 5,500 years ago. Corn later reached North America, where cultivated sunflowers also started to bloom some 5,000 years ago. This is also when potato growing in the Andes region of South America began.

Farmed Animals

Cattle, goats, sheep, and pigs all have their origins as farmed animals in the so-called Fertile Crescent, a region covering eastern Turkey, Iraq, and southwestern Iran. This region kick-started the Neolithic Revolution. Dates for the domestication of these animals range from between 13,000 to 10,000 years ago.

Genetic studies show that goats and other livestock accompanied the westward spread of agriculture into Europe, helping to revolutionize Stone Age society. While the extent to which farmers themselves migrated west remains a subject of debate, the dramatic impact of dairy farming on Europeans is clearly stamped in their DNA. Prior to the arrival of domestic cattle in Europe, prehistoric populations weren't able to stomach raw cow milk. But at some point during the spread of farming into south eastern Europe, a mutation occurred for lactose tolerance that increased in frequency through natural selection thanks to the nourishing benefits of milk. Judging from the prevalence of the milk-drinking gene in Europeans today—as high as 90 percent in populations of northern countries such as Sweden—the vast majority are descended from cow herders.

1.2. Agronomy

Agronomy is the science and technology of producing and using plants by agriculture for food, fuel, fiber, chemicals, recreation, or land conservation. Agronomy has come to include research of plant genetics, plant physiology, meteorology, and soil science. It is the application of a combination of sciences such as biology, chemistry, economics, ecology, earth science, and genetics. Professionals of Agronomy are termed agronomists.

'Agronomy' has been derived from the Greek derivatives 'Agros' and 'nomos' which respectively mean 'field' and 'management'. The foundation of the American Society of Agronomy in 1907 led to a vigorous development of the subject; but in India, the field management of crops has been advocated from very early days of agriculture.

'Krishi Parashar' is a text of ancient agriculture in Sanskrit. Although there is no documentary evidence, it is argued that the author sage 'Parashar' in the 4th century AD paid great attention to the management of agriculture.

This is evident from his following verses

- "Farms yield gold if properly managed but lead to poverty if neglected".
- "Management of one's harem may be entrusted to one's father, that of the kitchen to one's mother and that of cattle to someone equal in status. But farm should never be left to the care of anyone other than oneself."

- "Agriculture, cattle, business, women and royal families if left unattended even for a short while perish in no time".
- "Only the capable, motivated by the welfare of people should undertake farming. An incapable farmer lands himself in poverty".

The above statements of sage 'Parashar' are true for research in Agronomy as well as for agricultural administrators. Very meticulously planned field experiment will give good result only if it is supervised and executed well. The messages of Parashar are very relevant on macro-scale also. It is ironical therefore, that during the last 50 years of the existence of our Society, the Agronomists and the policy makers of Indian agriculture have not bothered much about the good management of agriculture and field experiments. Agriculture has been paid lip service. Government policies have not been based on sound management principles but have been ad-hoc and sometimes even faulty. It is believed that only high-yielding cultivars and hybrid seeds are enough to increase production. Irrigation facilities have been created but without proper attention to drainage. Application of inorganic fertilizers has been promoted without equal insistence on organic manures like farmyard manure, compost and residue management. Indiscriminate use of pesticides has created problems of toxic vegetables, fruits, and grains and residues of the pesticides in the soil. Now the sustainability issues have come to forefront and we are worried with low factor productivity. The message of Parashar for good management is an integral part of good agronomy and management of entire agriculture.

In 1903, Department of Agronomy was formed at South Dakota State University. In 1904, a Division of Experimental Agronomy was established in the College of Agriculture of the University of California, Berkeley. From the beginning, the division's major emphasis was on cereal crops. By 1906, it was clear that a Berkeley location for experimental work could not meet requirements of the agriculture that was developing in the Central Valley, and the division was largely shifted to the University Farm at Davis. The two-year non-degree program included practical work in agronomy, while experiment station research focused on developing improved cereal and range crop varieties

Purdue University was organized in 1869 and classes began in 1874. Departments in the College of Agriculture derived from "courses of study" listed in class catalogues. The word agronomy appeared in the 1905-1906 catalogue. In the 1907-1908 catalogue courses in soils and crops were grouped under a main heading of "agronomy," which was the first indication of an independent agronomy curriculum. In 1908, Agronomy was established as a department in the Agricultural Experiment Station.

At North Carolina University, during the early 1900s, research and extension activities focused on crop production and soil management for North Carolina crops such as forages, cotton, tobacco and upland rice. Classes in crops and soils were first offered in 1914, and two student clubs were organized, including the present-day Agronomy Club.

In the 20th century, Agronomy first developed as a science with a typical American urge to grow “bigger and better crops.” The foundation of the American Society of Agronomy in 1907 led to a vigorous development of the subject as a field science. Sir R.G. Staplendon made a comment in the Herbage Abstract in 1938. He stated: “The major aim of agronomical research, which is essentially field research, is to study all the factors which are operative at once and together, and in their natural interplay, for “nature is a theatre for the interrelations of activities”. Such a procedure, it may be said, is impossible, and if it is unscientific it will yet remain agronomical and many of the problems of agriculture are more likely to be solved, by agronomical investigations than by scientific research, while nearly all the results of scientific research have to pass through the sieve of an immense amount of agronomical investigations before they can be other than positively dangerous to the practitioner. Well-being every item of human food and clothing, has somewhere along the line from its source to its final synthesis, challenged the interest of an Agronomist. The Agronomist not only occupies a position that is fundamentally important in agriculture, but in industry as well. Most of the industrial production is dependent upon successful crop production. Problems of land use adjustment and soil conservation, which affect both urban and rural people, require the guiding hand of an Agronomist to help in their solution

The beginning of the Indian Society of Agronomy can be traced in a letter issued in 1955 from Dr. H.R Arakeri, (who served as the first Patron of the Society from 1974 to 1979), then Sugarcane Specialist, Padegaon, then in Bombay presidency and then in Maharashtra state in continuation of the one that was circulated earlier from Gwalior then in Madhya Bharat and now in Madhya Pradesh. The activities in the formative stage were shifted to the Division of Agronomy at the Indian Agricultural Research Institute, New Delhi. It was registered as a Society with its headquarters at the IARI, New Delhi under Societies Registration Act XXI of 1860 vide No. S 1109 dated 23rd July 1957.

Responsibility of Agronomist

An Agronomist deals with investigating the problems related to crop production and thereby suggesting/implementing suitable practices for better production of crops and efficient management of soil, in order to maximize both the income and the yield.

- Agronomists work towards maximizing production by utilizing the basic knowledge of science, with minimum expenditure.

- If we talk in a broader sense then agronomists are people who aim towards producing food and fiber, in order to meet the demands of the ever-growing population.
- Agronomists come up with efficient and feasible methods of field preparation for different seasons and the respective crop grown, for example, flatbeds, ridges, and furrows.
- He also selects a suitable variety of crops which would go well with the ongoing season at that particular time and for a particular kind of soil.
- The most efficient method of cultivation (broadcasting, transplantation, or dibbling etc.) which would help in better crop establishment is also selected by the agronomists.
- The nutrient requirement of the crops, including the time at which they are required and the method of application which is to be used, is studied by the agronomist itself.
- Agronomists must choose the best weed management method. Either using mechanical means or using physical labour or by chemical methods such as herbicides or weedicides, or by cultural means. Nowadays, integrated weed management method plays an important role in weed management.
- Choosing the suitable irrigation method, irrigation timing, and quantity according to the crops to be irrigated, whether to irrigate in one go or take breaks in between and the quantity of water to be applied are calculated by agronomy science so as to use water efficiently.
- Appropriate crop sequence (crop planning) is also decided by agronomists. They decide the type of crop, the pattern of cropping, the sequence of cropping, etc.
- Agronomists are also responsible for deciding the harvesting time, harvesting method, etc., so as to minimize the loss of yield.
- At last, agronomists are accountable for every decision related to the farm and management of the farm.

1.3. Relationship of Agronomy with other Disciplines

Agronomy is synthesis of several disciplines like soil science, agricultural chemistry, crop physiology, crop ecology, biochemistry, economics, agricultural meteorology, plant breeding, entomology, plant pathology, animal science, extension education, food technology etc.

Soil Science: Agronomists study sustainable ways to make soils more productive and profitable. They classify soils and reproduce them to determine whether they contain substances vital to plant growth such as compounds of nitrogen, phosphorus, and potassium. If a certain soil is deficient in these substances, need based fertilizers may be provided. Soil science also involves investigation of the movement of nutrients through the soil, the amount of nutrients absorbed by the plant roots, and the development of roots and their relation to the soil. In addition, agronomists develop methods to preserve the soil and to decrease the effects of erosion by wind and water. For example, a technique called contour ploughing may be used to prevent soil erosion and conserve rainfall. Researchers in Agronomy also seek ways to use the soil more effectively in solving other problems. Such problems include the disposal of human and animal wastes; water pollution; and the build-up of pesticides in the soil. Role of different agronomic practices on physical properties of soil are also studied.

Plant Breeding: This involves selective breeding of plants to produce the high yielding and high quality varieties under various conditions and under different cropping systems. Plant breeding has increased crop yields and has improved the nutritional value of numerous crops, including corn, soybean, wheat, rice etc. It also has led to the development of new types of plants. For example, a hybrid grain called triticale was produced by cross breeding rye and wheat. Triticale contains more usable protein than does either rye or wheat. Agronomists and plant breeder have close associations as agronomists are to work out the agronomic requirements of newly evolved varieties under different environments.

Biotechnology: Biotechnology is often a lab activity requiring field testing of the new crop varieties that are developed under controlled conditions. In addition to increasing crop yields, biotechnology is increasingly being applied for novel uses other than food. For example, oilseed is at present used mainly for margarine and other food oils, but it can be modified to produce fatty acids for detergents, substitute fuels and petrochemicals. The success of plants developed by biotechnological approaches are studied under field conditions by the agronomists.

Agroecology: Agroecology is the management of agricultural systems with an emphasis on ecological and environmental perspectives. This area is closely associated with work in the areas of sustainable agriculture, organic farming, alternative food systems and the development of alternative cropping systems.

Plant Protection: Plant protection includes Entomology and Plant Pathology. These disciplines prevent the crop plants from insect-pests and diseases so ultimately prevent yield losses. Time of sowing as well as crop vigour also decide attack of insect, pest and diseases.

Economics: The economist takes into account each and every input spent on raising crops. The economic analysis also determines the benefit cost ratio, net profit fetched.

Agrometeorology: Agrometeorology is a branch of meteorology which deals with the response of crops with the physical environment. The various climatic factors play important role in crop production. Its knowledge is very important in crop production. Agronomists can modify crop cultivation technologies as per changing environmental conditions suggested by the agrometeorologists i.e. agronomic practices can be adjusted as per weather forecasting.

Animal Science: Animal Science deals with scientific rearing of livestock. These enterprises are combined with cropping systems to form farming system. Integrated farming system approach can be adopted for higher benefits.

Agricultural Chemistry: Agricultural Chemistry is associated with some important farming concerns like fertilizer application, insecticides or pest control herbicides, fungicides, soil composition, agricultural products analysis, and nutritional requirements of farm animals.

Biochemistry: It is the branch of science which deals to study the composition of plant residues, effect of different herbicides, their residues in grain, straw, soil and drinking water etc. The nutrient content of plant parts viz., roots, shoots, grain and straw and soil is determined using different analytical methods.

Extension Education: Extension Education plays an important role in transfer of the advanced technology to the farmers. The extension workers act as a bridge between the farmers problems and scientists as the farmers problems are brought to the scientists and new advanced technologies developed by agronomists are given to the farmers by extension workers.

Horticulture: It deals with raising of fruit plants. It is also called as pomology. During initial stages, intercropping of field crops in the fruit plants can be done for higher profits and their suitability is worked out by the agronomists.

Vegetable Crops: It deals with raising of vegetable plants. It is also called as olericulture. For higher economics, intercropping of vegetables in the field crops can be done.

Floriculture: It deals with raising of ornamental flower and plants. Agronomist can study the selective herbicide for weed control in ornamental plants and their suitability as intercrop in different field crops can be worked out.

Forestry: Which deals with the raising of sole tree plantations and along with crop plants. Agronomist play very important role for evaluation of field crops which are best suited in forestry plantations of variable ages.

Plant Physiology: It is very closely related with Agronomy as growth regulators or growth retardants can be used to modify plant growth for higher yields. Moreover, to solve the problems of seed germination, excessive growth, flower/fruit drop, plant physiologist plays very important role.

Food Science Technology: Post harvest technologies for value additions for enhancing income of the farmers with the adoption of any value addition techniques. Food Science has also revolutionized the way we pack, process and market agricultural products. For example, dehydration and quick-freezing methods have increased farm products markets (food processing, preservation, meat packing etc). The refrigerated carriers are used to transport the perishable agricultural products to airport and then these high value crops and vegetables are transported to the other countries.

1.4. ICAR Institutions

Following are the ICAR Deemed Universities, National Research Centres, National Bureaux & Directorate/Project Directorates

Deemed Universities – 4

1. ICAR-Indian Agricultural Research Institute, New Delhi
2. ICAR-National Dairy Research Institute, Karnal
3. ICAR-Indian Veterinary Research Institute, Izatnagar
4. ICAR-Central Institute on Fisheries Education, Mumbai

Institutions – 65

1. ICAR-Central Island Agricultural Research Institute , Port Blair
2. ICAR-Central Arid Zone Research Institute, Jodhpur
3. ICAR-Central Avian Research Institute, Izatnagar
4. ICAR-Central Inland Fisheries Research Institute, Barrackpore
5. ICAR-Central Institute Brackishwater Aquaculture, Chennai
6. ICAR-Central Institute for Research on Buffaloes, Hissar
7. ICAR-Central Institute for Research on Goats, Makhdoom
8. ICAR-Central Institute of Agricultural Engineering, Bhopal
9. ICAR-Central Institute for Arid Horticulture, Bikaner
10. ICAR-Central Institute of Cotton Research, Nagpur
11. ICAR-Central Institute of Fisheries Technology, Cochin
12. ICAR-Central Institute of Freshwater Aquaculture, Bhubneshwar
13. ICAR-Central Institute of Research on Cotton Technology, Mumbai
14. ICAR-Central Institute of Sub Tropical Horticulture, Lucknow

15. ICAR-Central Institute of Temperate Horticulture, Srinagar
16. ICAR-Central Institute on Post harvest Engineering and Technology, Ludhiana
17. ICAR-Central Marine Fisheries Research Institute, Kochi
18. ICAR-Central Plantation Crops Research Institute, Kasargod
19. ICAR-Central Potato Research Institute, Shimla
20. ICAR-Central Research Institute for Jute and Allied Fibres, Barrackpore
21. ICAR-Central Research Institute of Dryland Agriculture, Hyderabad
22. ICAR-National Rice Research Institute, Cuttack
23. ICAR-Central Sheep and Wool Research Institute, Avikanagar, Rajasthan
24. ICAR- Indian Institute of Soil and Water Conservation, Dehradun
25. ICAR-Central Soil Salinity Research Institute, Karnal
26. ICAR-Central Tobacco Research Institute, Rajahmundry
27. ICAR-Central Tuber Crops Research Institute, Trivandrum
28. ICAR-ICAR Research Complex for Eastern Region, Patna
29. ICAR-ICAR Research Complex for NEH Region, Barapani
30. ICAR-Central Coastal Agricultural Research Institute, Ela, Old Goa, Goa
31. ICAR-Indian Agricultural Statistics Research Institute, New Delhi
32. ICAR-Indian Grassland and Fodder Research Institute, Jhansi
33. ICAR-Indian Institute of Agricultural Biotechnology, Ranchi
34. ICAR-Indian Institute of Horticultural Research, Bengaluru
35. ICAR-Indian Institute of Natural Resins and Gums, Ranchi
36. ICAR-Indian Institute of Pulses Research, Kanpur
37. ICAR-Indian Institute of Soil Sciences, Bhopal
38. ICAR-Indian Institute of Spices Research, Calicut
39. ICAR-Indian Institute of Sugarcane Research, Lucknow
40. ICAR-Indian Institute of Vegetable Research, Varanasi
41. ICAR-National Academy of Agricultural Research & Management, Hyderabad

42. ICAR-National Institute of Biotic Stresses Management, Raipur
43. ICAR-National Institute of Abiotic Stress Management, Malegaon, Maharashtra
44. ICAR-National Institute of Animal Nutrition and Physiology, Bengaluru
45. ICAR-National Institute of Natural Fibre Engineering and Technology, Kolkata, Kolkata
46. ICAR-National Institute of Veterinary Epidemiology and Disease Informatics, Hebbal, Bengaluru
47. ICAR-Sugarcane Breeding Institute, Coimbatore
48. ICAR-Vivekananda Parvatiya Krishi Anusandhan Sansthan, Almora
49. ICAR-Central Institute for Research on Cattle, Meerut, Uttar Pradesh
50. ICAR-National Institute of High Security Animal Diseases, Bhopal
51. ICAR-Indian Institute of Maize Research, Ludhiana
52. ICAR- Central Agroforestry Research Institute , Jhansi
53. ICAR-National Institute of Agricultural Economics and Policy Research, New Delhi
54. ICAR- Indian Institute of Wheat and Barley Research, Karnal
55. ICAR- Indian Institute of Farming Systems Research, Modipuram
56. ICAR- Indian Institute of Millets Research, Hyderabad
57. ICAR- Indian Institute of Oilseeds Research, Hyderabad
58. ICAR- Indian Institute of Oil Palm Research, Pedavegi, West Godawari
59. ICAR- Indian Institute of Water Management, Bhubaneshwar
60. ICAR-Indian Institute of Rice Research, Hyderabad
61. ICAR- Central Institute for Women in Agriculture, Bhubaneshwar
62. ICAR-Central Citrus Research Institute, Nagpur
63. ICAR-Indian Institute of Seed Science, Mau
64. ICAR-Indian Agricultural Research Institute, Jharkhand
65. ICAR-National Institute for Plant Biotechnology, New Delhi

National Research Centres – 14

1. ICAR-National Research Centre for Banana, Trichi
2. ICAR-National Research Centre for Grapes, Pune
3. ICAR-National Research Centre for Litchi, Muzaffarpur
4. ICAR-National Research Centre for Pomegranate, Solapur
5. ICAR-National Research Centre on Camel, Bikaner
6. ICAR-National Research Centre on Equines, Hisar
7. ICAR-National Research Centre on Meat, Hyderabad
8. ICAR-National Research Centre on Mithun, Medziphema, Nagaland
9. ICAR-National Research Centre on Orchids, Pakyong, Sikkim
10. ICAR-National Research Centre on Pig, Guwahati
11. ICAR-National Research Centre on Seed Spices, Ajmer
12. ICAR-National Research Centre on Yak, West Kemang
13. ICAR-National Centre for Integrated Pest Management, New Delhi
14. Mahatma Gandhi Integrated Farming Research Institute ,Motihari

National Bureaux – 6

1. ICAR-National Bureau of Plant Genetics Resources, New Delhi
2. ICAR-National Bureau of Agriculturally Important Micro-organisms, Mau, Uttar Pradesh
3. ICAR-National Bureau of Agricultural Insect Resources, Bengaluru
4. ICAR-National Bureau of Soil Survey and Land Use Planning, Nagpur
5. ICAR-National Bureau of Animal Genetic Resources, Karnal
6. ICAR-National Bureau of Fish Genetic Resources, Lucknow

Directorates/Project Directorates – 13

1. ICAR-Directorate of Groundnut Research, Junagarh
2. ICAR-Directorate of Soybean Research, Indore
3. ICAR-Directorate of Rapeseed & Mustard Research, Bharatpur
4. ICAR-Directorate of Mushroom Research, Solan
5. ICAR-Directorate on Onion and Garlic Research, Pune

6. ICAR-Directorate of Cashew Research, Puttur
7. ICAR-Directorate of Medicinal and Aromatic Plants Research, Anand
8. ICAR-Directorate of Floricultural Research, Pune, Maharashtra
9. ICAR-Directorate of Weed Research, Jabalpur
10. ICAR-Project Directorate on Foot & Mouth Disease, Mukteshwar
11. ICAR-Directorate of Poultry Research, Hyderabad
12. ICAR-Directorate of Knowledge Management in Agriculture (DKMA), New Delhi
13. ICAR-Directorate of Cold Water Fisheries Research, Bhimtal, Nainital

2

Hydroponics: A Profitable Technique in Agriculture

Agriculture nowadays is facing a complex problem. It's the rate at which we are turning arable land into non-arable. To solve this problem, we cannot simply stop farming and wait for the Earth to return to its natural state. Hence, alternative farming methods have become increasingly popular in recent years. Hydroponics is one of these options that can help produce crops without soil and other resources needed for crops to grow.

2.1. Definitions

Hydroponics is the term for growing plants without soil and in a water-based solution, the roots may be physically supported by an inert medium such as perlite, gravel, or other substrates. While it can be used to grow all kinds of crops, hydroponics is particularly effective for growing leafy vegetables, herbs and salad greens that have high water requirements, as well as for growing plants that have very shallow root systems. With hydroponic systems less space is needed, arable land is not needed and nutrients are easily controlled because they are added directly to the nutrient solution. The nutrient solution is continuously drained and replaced by fresh solution in the hydroponic systems. The nutrients used in hydroponic systems can come from many different sources, including fish excrement, duck manure, purchased chemical fertilizers, or artificial nutrient solutions.

Plants grow faster and produce more crops when they are hydroponically grown compared to those grown in soil because they increase access to oxygen, and provide easier ways of controlling their growth environment.

Hydroponics, or the cultivation of plants in a soilless medium, is a growing global industry with new applications for food, medicine and other products requiring superior quality. The innovative use of hydroponics in agriculture is expected to increase yields up to 40% while lowering operating costs. Hydroponics allows a farmer to control every aspect of their plants environment as well as increasing the yield per plant.

Plants sustain themselves by a process called photosynthesis. Plants capture sunlight with chlorophyll (a green pigment present in their leaves). They use the light's energy to split water molecules they absorbed via their root system. The hydrogen molecules combine with carbon dioxide to produce carbohydrates, which plants use to nourish themselves. Oxygen is then released into the atmosphere, a crucial factor in preserving our planet's habitability. Plants do not need soil to photosynthesize. They need the soil to supply them with water and nutrients. When nutrients are dissolved in water, they can be applied directly to the plant's root system by flooding, misting, or immersion. Hydroponic innovations have proven direct exposure to nutrient-filled water can be a more effective and versatile method of growth than traditional irrigation.

Hydroponics is an ancient practice that has been used since Roman times and may have been developed even earlier. Early contributors to experimental hydroponic technology include William Farrar who in 1808 wrote a book on the "artificial culture of vegetables without humous earth" by using infusions of acid, alkaline salts and metallic oxides.

"Hydroponics, or the growing of plants in a nutrient solution rather than soil, is a growing trend in agriculture. The need to meet increasing demand while preserving our environment and natural resources has spurred the development of hydroponics as a way to expand production.

In Hydroponic method, the plants are grown in a custom-made solution, from which nutrients and oxygen are supplied to the plant through a water flow system. Hydroponic systems can be set up at any size, depending on how much crop produce an individual wants to grow. The right system will also depend on whether one operates at home or in a professional setting.

Hydroponics is a type of horticulture and a subset of hydroculture which involves growing plants, usually crops, without soil, by using water-based mineral nutrient solutions in aqueous solvents. Terrestrial or aquatic plants may grow with their roots exposed to the nutritious liquid or in addition, the roots may be physically supported by an inert medium such as perlite, gravel, or other substrates. Despite inert media, roots can cause changes of the rhizosphere pH and root exudates can affect rhizosphere biology and physiological balance of the nutrient solution by secondary metabolites. (Walker et al. 2003)

Need for Hydroponic Farming

Our current agricultural system is facing a severe problem: by 2050, we will need to increase food production by about 70% in order to meet the caloric needs of a global population of 9.8 billion people—68% of whom are projected to live in urban areas. If we were to project linear growth in yield from our agricultural output from the past five decades, we would be nowhere near achieving this kind of growth by 2050.

The amount of resources used by traditional agriculture is astronomical. With most crop production already pushed to its ceiling both genetically and chemically (a significant increase in fertilizer or pesticide use will not sufficiently increase yields), intensification and the expansion of land used for food production have been seen as the only viable options to meet these growing food demands. Globally, 70% of water usage goes towards agricultural production, largely due to unsustainable irrigation practices. At present, 38% of earth's non-frozen land is used for growing food. This percentage will continue to rise: by 2050, 593 million hectares of land will need to be transformed into agricultural land to meet the projected calorie needs of the global population if we continue with business as usual. This needed land is equal to roughly double the size of India. This outlook is placing many essential ecosystems at risk of being completely destroyed, especially those that are key to maintaining an already disturbed balance of carbon dioxide in our atmosphere. Hydroponic farming offers a solution to many of the detriments of our world's current agricultural problems.

2.2. History

The word "hydroponic" was first used in 1937, coined by Dr. William Frederick Gericke, aptly known as the "father of hydroponics" (he made news when he was able to grow tomato vines more than 7-meters long just using mineral-nutrients solutions, right in his backyard!). The word is a combination of the Greek words "hydro" (meaning water) and "ponos" (meaning labour). This means water-work, literally. After leaving his University position, Gericke wrote and published 'Complete Guide to Soil-less Gardening' in 1940. Then came into the picture Dennis R. Hoagland, who created a nutrient formula that provided the plants with all the nutrients essential for growth and catered to a large diversity of plant species. The Hoagland solution still serves as the basis for most of the nutrient solutions that are being utilized even today. Alongside Daniel I. Aaron, Hoagland published '*The Water Culture Method for Growing Plants without Soil*', which is considered the most important text in hydroponics culture. There are accounts of the US Army using hydroponics to grow fresh food for the troops stationed in the infertile Pacific Islands, during World War II.

As the hydroponics technology picked up the attention across the world and the usage grew, its inevitable commercialization took place. Lawrence Brooks founded 'General Hydroponics' in 1976. Now, even after long years, GH remains one of the leaders of continuously expanding the market of hydroponic mineral-nutrient solutions. In 1982, a new area called 'The Land Pavilion' was opened in Disneyworld's EPCOT Center, where a ride goes through "gardens of tomorrow", which are grown using hydroponic technology. When you think about hydroponics, aside from the typical applications of the practice, you may also think about its applications in space and space travel. NASA is actually quite heavily invested in the practice of hydroponics. A research department completely dedicated to hydroponics and bio-farming was developed by NASA where they researched the growing of plants in extreme environments and hydroponic technology was a big player in the research.

As per a recent Hydroponics Market Research Report, it is projected that the hydroponics market will be worth $27.29 billion in the year 2022. Currently, almost 90 percent of lettuce and tomatoes that are being grown in the US and Canada are being grown using hydroponics. As the population of the world keeps growing, hydroponics becomes a more viable substitute for traditional farming. Hydroponically grown crops take only 25 percent of the area used by traditional soil growing, for the same amount of crops. We all know the harrowing threat of water scarcity in the coming years. With hydroponics, water usage can be cut down by 90 percent. This is just another reason why hydroponics is booming. Zero pesticide and herbicide involvement, growth rate twice as fast as traditional agriculture, nutrient solutions with perfect pH eliminating soil worries and the list is never-ending, which is why hydroponics is need for the present. Countries have already started dedicating acres of lands for commercial hydroponics, including 10,000 acres in Holland, Israel reserving 30,000 acres, Australia and New Zealand having 8,00 acres between them and Canada with hundreds of acres of land for large-scale commercial greenhouses, mainly growing tomatoes, peppers, and cucumber. In India, hydroponics is still a pretty new technology but initiatives are being taken. Many startups have taken it upon them to make hydroponics more popular, affordable and accessible. These startups use new technologies like IoT (Internet of Things), data analysis, etc. to perfect the procedure and make it as efficient as possible.

2.3. Advantages

Hydroponics is the fastest, easiest, most productive and cost-effective way to grow plants. The main advantages include

1. **An extended growing season:** Cold climates with chilly winter temperatures and shorter day lengths prohibit plant growth. But with a hydroponics system, plants can be grown hydroponically year-round because the grower controls the temperature, light, and nutrient-supply.

2. **Improved growth and yield:** Hydroponics systems typically result in faster-growing, higher-yielding plants. This is likely due to the increased oxygen levels found in the nutrient solution and the carefully controlled environmental factors. By increasing a plant's oxygen levels, you stimulate root growth and enhance nutrient uptake. These optimal growing conditions equate to less stress on plants and a more bountiful harvest.

3. **Higher plant density:** Plants grown in soil have rigid spacing guidelines that must be followed to allow each plant equal access to the soil's somewhat limited supply of water and nutrients. Because hydroponics systems deliver a more nutrient-charged solution to the root zone, plants can be grown closer together without competing for root space.
4. **Plants can grow anywhere:** Unlike traditional gardens that require outdoor space for plants, hydroponics systems are easily incorporated into many homes, regardless of their size or location. Under controlled conditions any plant can be grown at any time depending upon its need.
5. **Less water consumption:** Even though hydroponic systems depend primarily on water to grow plants, they use between 80 to 90% less water than plants grown in the ground. In traditional gardening, a large amount of water is applied to the soil to allow adequate moisture to reach the root zone. When moving through the soil, the water evaporates and only a percentage of it reaches the roots. In hydroponics, the water immediately reaches the roots, with little lost to evaporation. In many systems, the nutrient solution is also recirculated multiple times before becoming unusable and discarded, further improving water efficiency.
6. **Fewer pest problems:** Because hydroponics systems are indoors, pests aren't as prevalent and have controlled entrances. Insects find it more challenging to infiltrate the system and attack plants. Plus, fewer pest problems mean little to no need for pesticides.
7. **Easier to harvest mature plants:** Plants grown in hydroponics systems are typically grown on counters, benches, tables, etc., which puts them at waist height for most growers. At this height, mature plants are easier to harvest since there's no need to bend down or kneel to reach the plants. This is an important advantage for growers with limited mobility or physical ailments that prevent them from gardening at ground-level.

2.4. Disadvantages

It's easy to see the advantages of hydroponics. But some distinct disadvantages exist. As with anything, it's essential to understand the drawbacks to avoid surprises.

1. **Expensive to set up:** Compared to a traditional garden, a hydroponics system is more expensive to acquire and build. Costs range depending upon the type and size of the system purchased, and whether or not it's prefabricated or built with individual components to create a customized design.

2. **Vulnerable to power outages:** Both passive and active hydroponics systems depend on electricity to power the different components such as grow lights, water pumps, aerators, fans, etc. Therefore, a power outage will affect the entire system. In active systems, a loss of power can be detrimental to plants if it goes unnoticed by the grower.
3. **Requires constant monitoring and maintenance:** Hydroponics requires a higher level of monitoring and micro-managing than growing plants traditionally. To maintain a carefully controlled growing environment, all system components need constant vigilance—lights, temperature, and many aspects of the nutrient solution such as pH and electrical conductivity. The nutrient solution also needs to be flushed and replaced regularly, and the system parts cleaned often to prevent buildup and clogging.
4. **Waterborne diseases:** Because hydroponically grown plants are grown in water instead of soil, waterborne diseases are considerably higher. With the water circulating continuously through the system, infections can spread quickly throughout the growing system as a whole, affecting the whole collection of plants. In extreme cases, a waterborne disease can kill all the plants in a hydroponics system within hours.
5. **Problems affect plants quicker:** Soil protects the roots from extreme temperature changes, slows diseases and pests from attacking, and regularly releases and absorbs nutrients. Without soil to act as a buffer, plants grown in hydroponics systems react negatively to problems like nutrient deficiencies and disease much quicker.

2.5. Types of Hydroponic Systems

There are many hydroponic methods, but all of them are a modification or combination of following basic hydroponic systems.

1. Deep Water Culture Systems (DWC)

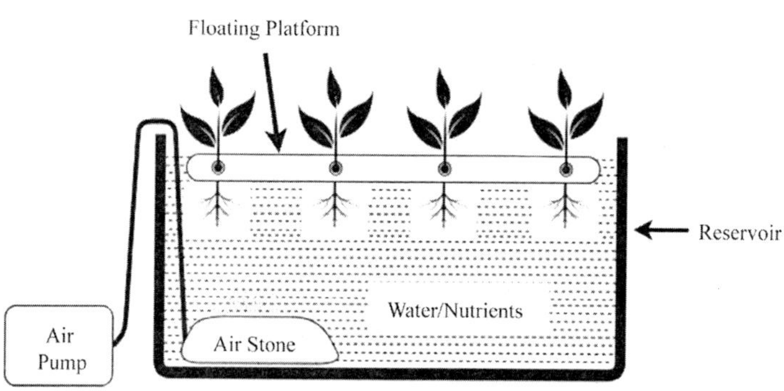

Deep water culture hydroponics are simply plants suspended in aerated water. Deep water culture systems, also known as a DWC system, are one of the easiest and most popular methods of hydroponics on the market. A DWC system dangles net pots holding plants over a deep reservoir of oxygen-rich nutrient solution. The plant's roots are submerged in the solution, providing it with perpetual access to nutrition, water, and oxygen. Deep water culture is considered by some to be the purest form of hydroponics.

Since the root system is suspended in water at all times, proper water oxygenation is vital to the plant's survival. If there is not enough oxygen supplied to the plant's roots, the plant will drown in the solution. Add an air stone connected to an air pump at the bottom of the reservoir to supply oxygenation to the entire system. The bubbles from the air stone will also help circulate the nutrient solution.

It is very easy to assemble a deep-water culture system at home or in a classroom without needing expensive hydroponics equipment. You can use a clean bucket or old aquarium to hold the solution and place a floating surface like Styrofoam on top to house the net pots. Plants in DWC systems should only have their roots submerged in the solution. No part of the stem or vegetation should be underwater. You can even leave about an inch and a half of the roots above the waterline. The air stone bubbles will pop out of the surface and splash onto the exposed roots, so they will not be at risk of drying out.

Advantages

Low maintenance: Once a DWC system is set up, there's very little maintenance required. Just replenish the nutrient solution when needed and make sure your pump is running oxygen to the air stone. The nutrient solution typically only needs replenishing every 2-3 weeks, but this does depend on the size of your plants.

DIY appeal: Unlike many hydroponic systems, deep water culture systems can be made cheaply and easily at home, with a quick run to your pet store and local nursery to pick up the air pump and nutrients.

Disadvantages

Limitations : Deep water culture systems are adept at growing herbs and lettuce but they struggle with larger and more slow-growing plants. DWC systems are not ideal for anything that flowers. However, with some extra work, you can grow plants like tomatoes, bell peppers, and squash in a DWC system.

Temperature control: It's important that your water solution does not exceed 68°F and does not go below 60°F. In a DWC system, the water is static and not recirculating, so it can be more difficult to regulate temperature.

2. Wick Systems

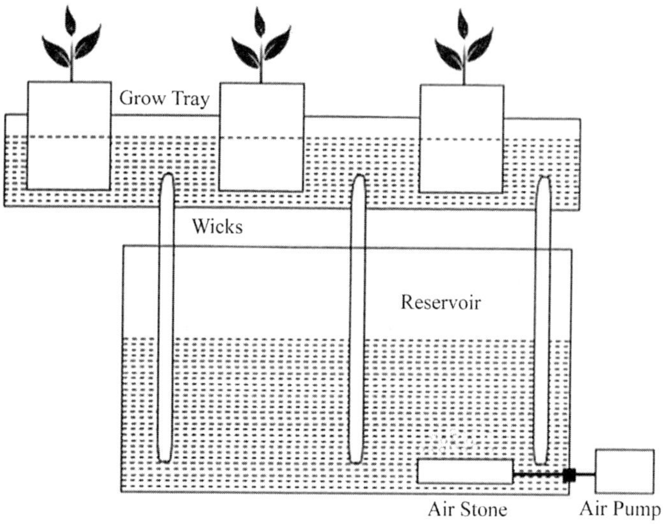

In a wick system, plants are nestled in growing media on a tray that sits on top of a reservoir. This reservoir houses a water solution with dissolved nutrients. Wicks travel from the reservoir to the growing tray. Water and nutrients flow up the wick and saturate the growing media around the root systems of the plants. These wicks can be made of material as simple as rope, string, or felt. Wick systems are by far the simplest form of hydroponics. Wick systems are passive hydroponics - meaning they don't require mechanical parts like pumps to function. This makes it ideal for situations where electricity is either unreliable or unavailable.

Wicks systems work by a process called capillary action. The wick absorbs the water it's immersed in like a sponge, and when it comes in contact with the porous growing media, it transfers the nutrient solution. Wick system hydroponics only work if accompanied by growing media that is able to facilitate nutrient and water transference. Coco coir (fibers from the outer husks of coconuts) have excellent moisture retention and the added benefit of being pH neutral. Perlite is also pH neutral and extremely porous, making it ideal for wicking systems. Vermiculite is also very porous, and also possesses a high cation-exchange capacity. This means it can store nutrients for later use. These three growing media are the most suitable for hydroponic wick systems.

Wick systems work quite slowly compared to other hydroponic systems, which does limit what is practical to grow with them. You want to make sure for every plant in the growing tray you have at least one wick running from the reservoir. These wicks should be placed close to the root system of the plant. Though capable of functioning with aeration, many people do choose to add an air stone and air pump to the wick system's reservoir. This adds extra oxygenation to the hydroponic system.

Advantages

Simplicity: A wick system can be set-up by anyone and does not demand excessive attention after it is running. The wicks will constantly supply your plants with water, so there is no risk of them drying out. Furthermore, plants like lettuce will flourish in a wick system, providing a great return on your hands-free investment

Space-efficient: Wick systems are unobtrusive and can be installed anywhere, seeing as they do not need electricity to run. It is a perfect system for educators, beginners or anyone interested in exploring hydroponics.

Disadvantages

Limitations : Lettuce and herbs like rosemary, mint, and basil are fast-growing and do not demand large quantities of water. Tomatoes, on the other hand, will struggle to thrive in a wick system because of their high demand for nutrients and hydration. Other plants cannot thrive in an environment that is perpetually moist. Root vegetables like carrots and turnips will not succeed in a wick system.

Susceptible to rot: A hydroponic wick system is always humid and damp. This creates the risk that fungal outbreaks and rot can develop in the organic growing media and on the roots of your plants.

3. Nutrient Film Technique System (NFT system)

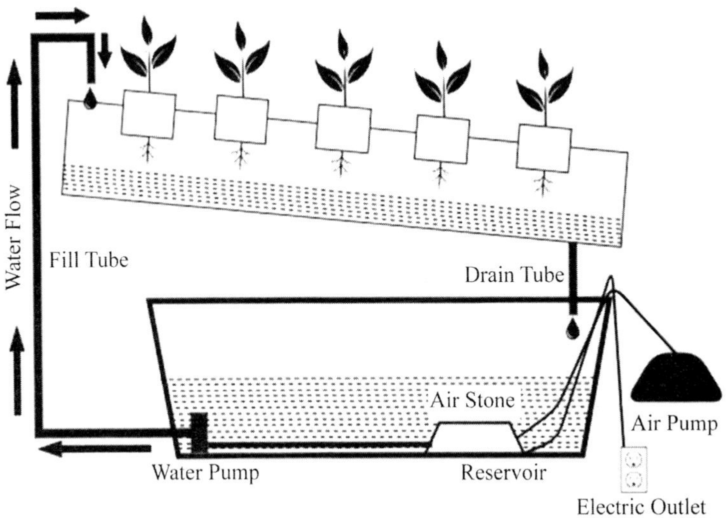

Nutrient film technique (NFT) systems suspend plants above a stream of continuously flowing nutrient solution that washes over the ends of the plant's root systems. The channels holding the plants are tilted, allowing water to run down the length of the grow tray before draining into the reservoir below. The water in the reservoir is then aerated via air stone. A submersible pump then pumps the nutrient-rich water out of the reservoir and back to the top of the channel. The nutrient film techniquc is a recirculating hydroponic system

Unlike with deep water culture hydroponics, the roots of the plants in an NFT system are not immersed in water. Instead, the stream (or "film") only flows over the ends of their roots. The roots' tips will wick the moisture up into the plant, while the exposed root system is given plenty of access to oxygen. The bottoms of the channels are grooved, so the shallow film can pass over the root tips with ease. This also prevents water from pooling or damming up against the root systems.

Even though nutrient film technique systems are constantly recycling water, it is wise to drain the reservoir and replenish the nutrient solution every week or so. This ensures your plants are being delivered ample nutrition. NFT channels must be angled at a gradual slope. If it's too steep, the water will rush down the channel without properly nourishing the plants. If too much water is being pumped through the channel, the system will overflow and the plants can drown. NFT hydroponics are popular commercial systems, as they can support several plants per channel and can easily be mass-produced. Nutrient film technique systems are best suited for lightweight plants, like mustard greens, kale, lettuce, spinach as well as fruits like strawberries. Heavier fruiting plants like tomatoes and cucumbers will require trellises to support the excess weight.

Advantages

Low consumption: Since NFT hydroponics recirculate the water they do not demand large quantities of water or nutrients to function. The constant flow also makes it harder for salts to accumulate on the plant's roots. Nutrient film technique systems also don't require growing media, so you are saved the expense of purchasing media and the hassle of replacing it.

Modular design: Nutrient film technique systems are perfect for large-scale and commercial endeavors. Once you have one channel set up and functioning, it is very easy to expand. You can fill your greenhouse with multiple channels supporting different crops. It's a good idea to feed each channel with a separate reservoir. This way, if pump failure occurs or disease spreads in the water, you will not lose your entire operation.

Disadvantages

Pump failure: If the pump fails and the channel is no longer circulating the nutrient film, your plants will dry out. In a matter of hours, your entire crop can perish if it is not being supplied with water. Maintaining an NFT hydroponic system does require vigilance. You will want to diligently observe the performance of your pump.

Overcrowding: If the plants are spaced too close together or the root growth is too proliferated, the channel can become clogged. If the channel is obstructed by roots, water will be unable to flow and your plants will starve. This is especially true of the plants at the bottom of the channel. If the plants at the end ever seem to be underperforming compared to the rest of the channel, consider removing some plants or switching to a smaller unit.

4. Ebb and Flow Systems

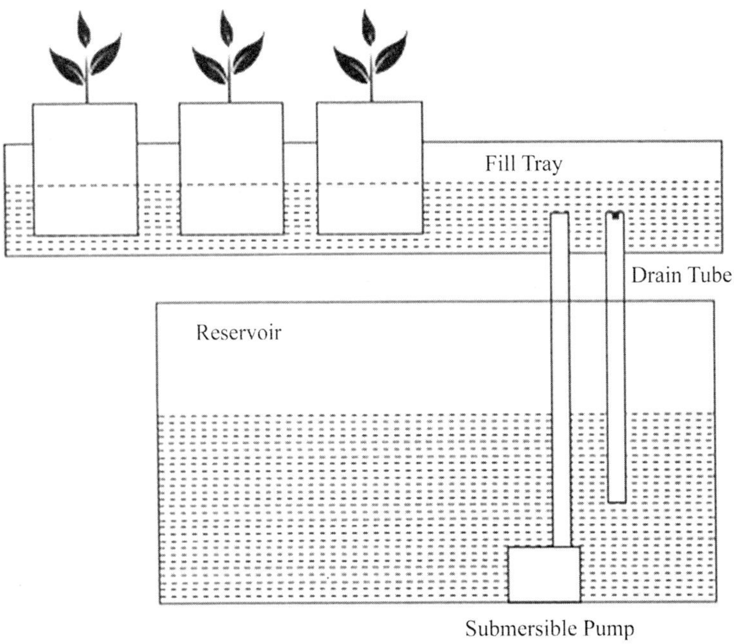

Ebb and flow hydroponics work by flooding a grow bed with a nutrient solution from a reservoir below. The submersible pump in the reservoir is equipped with a timer. When the timer starts, the pump fills the grow bed with the water and nutrients. When the timer stops, gravity slowly drains the water out of the grow bed and flushes it back into the reservoir. The system is equipped with an overflow tube to ensure flooding doesn't surpass a certain level and damage the stalks and fruits of the plants. Unlike the previous systems mentioned, the plants in an ebb and flow system are not constantly exposed to water. While the grow bed is flooded, the plants drink up the nutrient solution through their root systems. When the water ebbs and the grow bed empties, the roots dry out. The dry roots then oxygenate in the interval before the next flood. The length of time between floods is dictated by the size of your grow bed and the size of your plants.

Ebb and flow systems (also called flood and drain systems) are one of the most popular hydroponic growing methods. The plants are supplied with abundance of oxygen and nutrition which encourages quick and vigorous growth. The ebb and flow system are easily customizable and versatile. The grow bed can be filled with an assortment of net pots and a variety of fruits and vegetables. Perhaps more than any other hydroponic system, the ebb and flow system allow you to experiment with your plants and media.

Ebb and flow systems can accommodate almost any type of vegetation. Your primary limitation is the size and depth of your grow tray. Root vegetables will require a much deeper bed than lettuce or strawberries. Tomatoes, peas, beans, cucumbers, carrots, and peppers are all popular ebb and flow crops. In fact, you can even attach trellises directly to the grow bed. "Grow rocks" and expanded clay pebbles (hydroton) are some of the most popular growing media in ebb and flow hydroponics. These are cleanable and reusable, lightweight, and while they do retain moisture they will also drain. This is an important quality in ebb and flow systems.

Advantages

Versatilste: With an ebb and flow system, you can grow much larger plants than in most other hydroponic systems. Fruits, flowers, and vegetables alike respond very well to ebb and flow hydroponics. If you've taken care to provide your plants with the appropriately sized grow bed and nutrition, you will see bountiful yield.

DIY appeal: There are hundreds of ways to construct your own ebb and flow hydroponic system at home. A visit to the hardware store and pet store will provide you with all the supplies you need to construct an ebb and flow system. Though more expensive to set up than other DIY systems like wick and deep-water culture, ebb and flow systems sustain a much broader scope of plant life than they can.

Disadvantages

Pump failure: Like any hydroponic system reliant on a pump, if the pump ceases to work, your plants will die. You do have to monitor your ebb and flow system to ensure that the system's performance is not compromising the health of your plants. If the water is rushing in and out too fast, your plants will not receive an adequate amount of water and nutrients.

Rot & disease: Sanitation and maintenance are essential to an ebb and flow system. If the bed is not draining properly, root diseases and rot can set in. A dirty ebb and flow system can grow mold and attract insects. If you neglect cleanliness, your crops will suffer. Additionally, some plants do not respond well to the rapid pH change that occurs as a result of the flooding and draining extremes.

5. Drip Systems

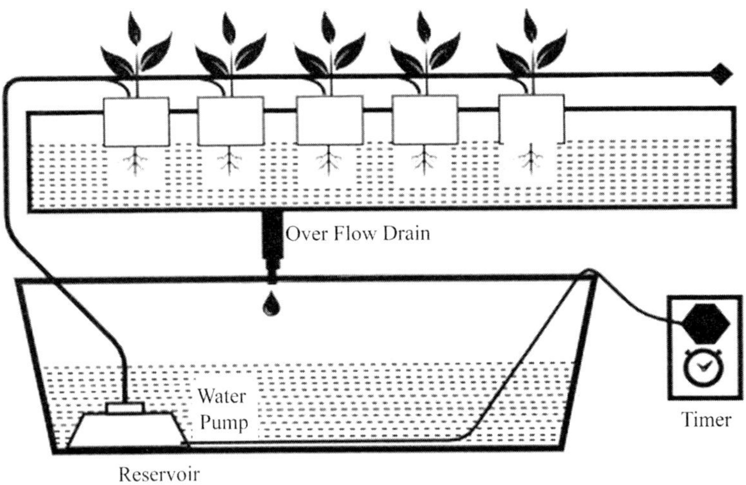

In a hydroponic drip system, the aerated and nutrient-rich reservoir pumps solution through a network of tubes to individual plants. This solution is dripped slowly into the growing media surrounding the root system, keeping the plants moist and well-nourished. Drip systems are the most popular and widespread method of hydroponics, especially among commercial growers. Drip systems can be individual plants or massive irrigation operations.

There are two configurations of drip system hydroponics: recovery and non-recovery. In recovery systems, more popular with smaller, at-home growers, the excess water is drained from the grow bed back into the reservoir to be recirculated during the next drip cycle. In non-recovery systems, the excess water drains out of the growing media and runs to waste. This method is more popular among commercial growers. Though non-recovery drip systems can sound wasteful, large-scale growers are very conservative with water usage. These drip systems are designed only to deliver precisely the amount of solution required to keep the growing media around the plant dampened. Non-recovery drip systems employ elaborate timers and feeding schedules to keep waste to a minimum.

If you are growing plants in a recovery drip system, you will need to be attuned to the fluctuations in the pH of the nutrient solution. This is true of any system where wastewater re-circulates into the reservoir. Plants will deplete the nutrient content of the solution as well as alter the pH balance, so the grower will need to monitor and adjust the solution reservoir more than they would need to in a non-recovery system. Growing media can also become oversaturated with nutrients, so they will need to be washed and replaced periodically.

Advantages

Variety of plant options: A drip system can support much larger plants than most other hydroponic systems. This is one of the reasons it is so appealing to commercial growers. Melons, pumpkins, onions, and zucchinis can all be amply supported by a properly sized drip system. Drip systems hold greater quantities of growing media than other systems, allowing them to support the larger root systems of these plants. Drip systems work best with slow draining media, like rockwool, coco coir, and peat moss.

Scale: Drip systems can easily support large-scale hydroponics operations. If a grower desires to add more plants, new tubing can be connected to a reservoir and divert solution to the new vegetation. New crops can be introduced to an existing drip system, as additional reservoirs can be added with differing timer schedules tailored to fit the needs of the new plants. This is another factor that makes drip systems popular commercial hydroponics

Disadvantages

Maintenance : If you a growing plants using a non-recovery drip system at home, there is a significant amount of maintenance involved. You'll need to consistently monitor pH and nutrient levels in your solution, draining and replacing if necessary. Recovery systems lines can also become clogged by debris and plant matter, so you will need to regularly wash and flush delivery lines.

Complexity: Drip systems can easily become elaborate and complex undertakings. This matters less for professional hydroponics, but it is not the most ideal system for home growers. There are many much simpler systems, like ebb and flow, that lend themselves better to at-home hydroponics.

6. Aeroponics

Aeroponics systems suspend plants in the air and expose the naked roots to a nutrient-filled mist. Aeroponics systems are enclosed frameworks, like cubes or towers, that can hold a multitude of plants at once. Water and nutrients are stored in a reservoir, and then pumped to a nozzle that atomizes the solution and distributes it as a fine mist. The mist is usually released from the top of the tower, allowing it to cascade down the chamber. Some aeroponics continuously mist the plant's roots, much like NFT systems expose the roots to the nutrient film at all times. Others function more like the ebb and flow system, spraying the roots with mist in intervals. Aeroponics do not need substrate media to survive. The root's constant exposure to air allows them to drink in oxygen and grow at an accelerated rate.

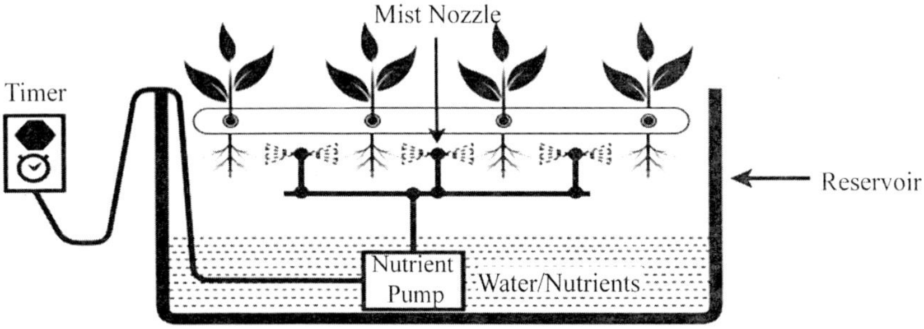

Aeroponics systems use less water than any other form of hydroponics. In fact, it takes 95% less water to grow a crop aeroponically than in an irrigated field. Their vertical structure is designed to take up minimal room and allows for numerous towers to be housed in a single location. With aeroponics, great yields can be produced even in confined spaces. Furthermore, because of their maximized exposure to oxygen, aeroponic plants grow faster than other hydroponically grown plants.

Aeroponics allow for simple harvesting year-round. Vine plants and nightshades like tomatoes, bell peppers, and eggplants all perform well in an aeroponic environment. Lettuce, baby greens, herbs, watermelons, strawberries, and ginger all also flourish. However, fruiting trees are too large and heavy to be grown aeroponically, and underground plants with extensive root systems like carrots and potatoes cannot be grown.

Advantages

Oxygen surplus : The surplus of oxygen taken in by the bare roots supercharges the plant's growth. Aeroponics are not only the eco-friendliest hydroponic system, they are also among the highest performing. They are versatile, customizable systems that reliably produce high-quality results.

Mobility: Aeroponic towers and trays can be easily transported from one location to another without disrupting the plant's growth. During the transportation process, you will want to hand-mist the roots to prevent them from drying out. Additionally, aeroponic systems are designed to be ergonomic and maximize space. Aeroponics allow you to grow plants in greater densities than other hydroponics systems.

Disadvantages

Expensive: Aeroponics do have a higher initial cost than other hydroponic systems. To set up a fully functional system complete with reservoirs, timers, and pumps can cost thousands of dollars. It is possible to construct a DIY aeroponics system for much less, but it is a much harder undertaking than a DIY deep water culture or wick system.

Maintenance Aeroponics systems maintain a delicate balance and, if disrupted, the results are disastrous for your plants. If your timer doesn't go off or a pump fails, you risk losing your entire crop unless you mist the roots by hand. You will need to regularly clean the root chamber to prevent root disease from compromising your plants. Generally speaking, aeroponic systems do require more technique to succeed than other systems.

7. Rotary Hydroponics

A rotary hydroponic garden is a style of commercial hydroponics created within a circular frame which rotates continuously during the entire growth cycle of whatever plant is being grown.

While system specifics vary, systems typically rotate once per hour, giving a plant 24 full turns within the circle each 24-hour period. Within the center of each rotary hydroponic garden can be a high intensity grow light, designed to simulate sunlight, often with the assistance of a mechanized timer.

Each day, as the plants rotate, they are periodically watered with a hydroponic growth solution to provide all nutrients necessary for robust growth. Due to the plants continuous fight against gravity, plants typically mature much more quickly than when grown in soil or other traditional hydroponic growing systems (Al-Kodmany 2018). Because rotary hydroponic systems have a small size, they allow for more plant material to be grown per area of floor space than other traditional hydroponic systems.

Rotary hydroponic systems should be avoided in most circumstances, mainly because of their experimental nature and their high costs for finding, buying, operating, and maintaining them. (Manos and Xydis 2019)

8. Fogponics

Fogponics is a derivation of aeroponics wherein the nutrient solution is aerosolized by a diaphragm vibrating at ultrasonic frequencies. Solution droplets produced by this method tend to be 5–10 μm in diameter, smaller than those produced by forcing a nutrient solution through pressurized nozzles, as in aeroponics. The smaller size of the droplets allows them to diffuse through the air more easily, and deliver nutrients to the roots without limiting their access to oxygen. (Rakib and Suliaman, 2021)

2.6. Growing Media for Hydroponics

1. Perlite

Perlite is a very common growing medium that has been around for years, usually used by traditional gardeners since from a long time to add aeration to soil mixes. As a mined material, a form of volcanic glass, perlite is created under intense and rapid heat. Then countless tiny bubbles pop outs like popcorn. So, as you can guess, it is very lightweight and porous.

Perlite is a good choice for the wick-type hydroponic system since it makes an excellent standing wicking action. But because of its porous and easy-to-flow nature, it is not recommended using this medium for quick and strong watering systems like the ebb & flow. It can be washed away fast & easily.

Able to hold air very well, having neutral pH, but easy to flow, Perlite is rarely used alone. People often mix it with other growing media like vermicule, coco coir or soil – A common combination is with vermicule in equal quantity (50-50)

Benefits

- Reasonably inexpensive
- Lightweight
- High air retention.
- Reusable

Downsides

- Lightweight, not suitable for certain system kinds.
- Dust from the medium – effects on the environment and health

Can be used in drip systems, aeroponic systems.

2. Coconut Coir

Coconut coir (also known as "Coco-tek," "Cocopeat" and "Ultrapeat" is essentially a byproduct of the coconut industry. People make it from the brown husks surrounding the coconut shell. Ultrapeat" is essentially a byproduct of the coconut industry. It is made from the brown husks surrounding the coconut shell.

What makes this one of the most efficient growing media for hydroponics is because it is totally organic, highly inert, and holds water very well. But it has a really good air to water ratio, saving plants from drowning. Coconut coir is friendly to the environment and renewable. The unused material can go to disposal places of or be composted. These make coconut become a so commonly used material for Hydroponics recently.

Coconut coir can be used alone or mixed with other growing media like perlite, expanded clay pellets. A common formula used by growers is 50% of the coconut fiber & 50% clay pellets.

Benefits

- Excellent water retention and aeration
- Organic material
- Environmentally friendly

Downsides

- Uncompressed after several uses.
- Doesn't drain well, so often mixed with other media.

Can be used in drip systems, aquaponic systems, ebb and flow system.

3. Vermiculite

Vermiculite is a form of hydrated laminar minerals which resemble mica. Just like perlite, vermiculite is processed by exposing the material to extreme heat to expand them into small clean, odorless pellets.

Vermiculite is an excellent soilless planting medium. It is non-toxic, sterile, moist-resistant and has a nearly neutral pH. What's more, the material is very lightweight and able to hold water very well, which is quite contrasting to perlite. Yet, it does not keep aeration as well as perlite. This growing media also has a reasonably great cation-exchange capacity, which helps keep the nutrients for later use.

There are different vermiculite types, so bear that in mind and choose the correct material for the hydroponic purpose. And because of its ability to retain much water (approximately 200%-300% of its weight), there's a risk of suffocating the plants. So, it is usually used to mix with other media. The most familiar one would be perlite as the two media complement each other quite well – perlite drains so fast whereas vermiculite holds moisture. They are often conjuncted in a 50/50 formula, which keeps the mix from being washed away in ebb and flow systems. Can be used in the drip system, aeroponic system.

Benefits

- Water and nutrient retention

Downsides

- Poor drainage capacity
- Danger of suffocating plants

4. Rockwool

Hydroponic growers would not find rockwool that strange as this material has been used widely in recent years, largely on commercial farms. And this material has been proven as quite effective, and hence its popularity.

This sterile, porous medium is primarily made up of granite and/or limestone rocks which are heated until melting and then are spun into super thin and long fibers. After that, people compress these fibers into cubes, bricks of preferred sizes.

Rockwool owns many benefits of an ideal growing material like microbe immunity, good water & air retention. This protects your plants from dehydrating while giving plant roots continuous available amounts of oxygen. However, there is one word of caution about the pH level when using rockwool. The natural pH of the material is usually high, which can alter the nutrient solution pH. Prevent this by soaking this media into pH balanced water before using.

Another drawback is that rockwool is non-degradable and not sustainable. The unused fibers of rockwool are almost unable to be disposed of. Also, the dust giving off from the medium can cause irritation. This is not good for the lungs and eyes, so it's a good practice to wash it into the water once when using.

Benefits

- Great water absorption and aeration capacity

Downsides

- Not environmentally friendly
- The dust is not good for the health.
- Disturb the pH of the nutrient solution.

Can be used in drip systems, ebb and flow systems, deep water cultures systems, nutrient-film technique systems. Almost any system except aquaponics.

5. Expanded Clay Pellets or Leca

Expanded clay pellets or hydroton (also referred as LECA (lightweight expanded clay aggregate) are small round balls, about the size of marbles. They are created by heating and expanding clays to form thousands of small bubble-shaped balls. Overall, this is a great and effective growing medium for hydroponic use.

As a form of rocks, but clay pellets are lightweight because of its porous nature. And yet they are still heavy enough to give strong support for the plants and are quite excellent in wicking up the nutrients to plant roots.

The round and porous form makes clay pellets an incredibly balanced water to oxygen medium. This neutral pH medium is reusable. People can clean, sterilize and reuse it. Clay pellets are not a good water holding material because of the spaces between each pellet. They will drain and dry out fast. So be sure to get the material watered enough. Also, the pellets are heavier and more expensive than other media.

Benefits

- Reusable, sustainable.
- Effective water drainage, and air retention.

Downsides

- Poor moisture retention capacity
- More expensive than other growing media

Can be used in drip systems, ebb and flow systems, aquaponic systems.

6. Oasis Cubes

Oasis cubes are made from floral foam which is designed like a sheet form. Each of the medium's individual cell (which looks alike to cubes) contains the right amount of nutrients as well as air and water for plants' growing. Growers use oasis cubes primarily as a starting environment for seedlings or plants' cuttings, not as a full growing medium.

The medium is pH neutral. Its cells absorb water and air fairly well, which is very necessary for seeds or cuttings. Also, the roots can easily grow and expand within the medium's open cell structure.

Benefits

- Inexpensive
- Good water and air holding capacity

Downsides

- Not organic
- Not sustainable
- Used for germination and seedling growing phases.

7. Starter Plugs

Starter plugs are another effective product to start plant propagation or seed germination. Hydroponic growers who care about the environment can find this product a good material for consideration. They can choose starter plugs that are composted of organic materials.

To make it useful for the starting phase of planting, starter plugs are often created from materials that keep moisture well, are not easily waterlogged, and at the same time allows the roots to expand and drive through the loose bottom.

Benefits

- Excellent for seedlings, and propagation
- Sustainable (depending on the material used)

Downsides

- Suitable for seed starting, or cloning.
- Relatively expensive

8. Rice Hulls

The normal byproduct of rice crops and usually thrown away material can make good use for soilless growing. This medium is not pH neutral and the final composts usually range from 5.8 to 7.2. Keep in mind not to use fresh rice hulls because they are not sterilized, and there are risks of microorganisms, decaying bugs, and weed seeds. Better to use parboiled rice hulls (PRH), which has been stemmed, rice-free, dried and clean.

Rice hulls are able to drain water well. One disadvantage is that the material is decomposed after some time, so should be replaced very often.

Benefit

- Totally organic

Downsides

- Break down over time
- Not pH neutral

9. Pumice

Pumice is a mined mineral just like perlite and is formed by a super-heated and highly pressured volcano. It can be white, pale gray or pale yellow, depending on the area's minerals. But usually, pumice is darker than perlite.

This is a lightweight stone, which is porous, slow to break down and holds the air well because of spaces between each particle. This makes the medium tend to float on water, but not to the extent of perlite. Pumice has good water retention, but not as good as vermiculite, which is also a mined mineral.

Benefits

- Lightweight.
- Excellent air holding capacity

Downsides

- Lightweight for some hydroponic system

10. Growstones

Growstones are made from recycled glass from landfills or glass collecting & processing places. Some people may worry that they can get cut from this medium because they are made from glass. Don't get worried about that because this recycled glass has been hardly crushed, melt, sometimes mixed in some calcium carbonates.

This medium is very lightweight, highly porous and above all, it has great air aeration, and average moisture retention to the root system. One should start by washing and rinse the medium thoroughly to remove small particles, and dust. The medium is ideal for using alone or for mixing into peat, coco coir and other growing media.

Benefits

- Lightweight
- Good air to water ratio
- Sustainable

Downsides

- Cling to some roots, which can cause root damage for some plant types.
- A little bit of dust.
- More expensive than other media.

11. Sawdust

Sawdust is a by-product of sawmills and retail hardware stores. It is quite inexpensive, and even you can find or ask for free of cost. This material is lightweight, yet retains water well so be sure not to water it. Sawdust is biodegradable but will decompose over time. This medium is not pH neutral. So extra care of pH monitoring for your system needs taking.

One word of caution is that you should know what kinds of wood your sawdust is made; it should not be chemically treated and contaminated. This is absolutely unhealthy if you are growing edible food on the contaminated medium. So double check about that, and if it has chemical treatment, make sure it is sterilized before use.

Many hydroponic growers don't like sawdust because it compacts quickly, pH-altering, and can be contaminated. However, sawdust has some limited successes in Africa for soilless planting, and for hydroponic tomato growing in Australia.

Benefits

- Organic
- Sustainable.

Downsides

- pH changes frequently
- Rot over time, and can cause bacteria.
- May not be sterile

12. Wood Chips/Fibers

Just like sawdust, wood chips or fibers are from woods. So, they are totally organic, a great option for growers who have an eco-friendly mindset. And again, make sure that wood chips are not used from contaminated wood, or should be sterilized before using. Wood chips hold water well. But it is not so compacted and waterlogged as sawdust.

Benefit

- Totally organic
- Able to retain water well.

Downsides

- Biodegradable
- Can contain chemicals.
- May bring fungi, pests.

13. Peat Moss

Soil gardening lovers would find this material quite familiar as an excellent moisture nutrient retaining, and organic soil-softening durable material. But peat moss also makes good use in hydroponics whether alone, or in conjunction with other material.

Peat moss is a dead fibrous material that develops in a wet, cold, acidic, and lack-of-air environment called peat bogs. In that environment, sphagnum moss and other living things decompose into a dense dark brown compact. This is the peat moss. This process takes place so long, even a few thousands of years. So, peat moss is not considered a renewable medium, and not friendly to the environment.

This is a good medium for soil and hydroponic planting because of its ability to retain water and hold nutrients well. Peat moss can get wet then rehydrate quick and doesn't compact or break down easily. These keep the material's life cycle of several years of use. The medium doesn't contain adverse weed seeds or microorganisms, unlike other organic compost.

One can blend peat moss with perlite, vermiculite, or styrofoam particles to add aeration and adjust the PH of this medium.

Benefits

- Good water, and nutrient holding.
- Doesn't compact
- Doesn't comprise of harmful bacteria or weeds

Downsides:

- Not renewable.
- Low pH, acidic.
- Relatively expensive

14. Sand

Soil and hydroponic growers would not find sand any new as it is one of the most plentiful materials on earth and are in great use for construction, road building, glass industry, etc. as well as for growing of crops. That's the reason why sand is respectively cheap or available for free at places like the beach.

Sand is composed of super small finely divided rocks. Their particle size is really tiny, so water doesn't dry out fast. Sand can be used as a great starting medium and then one can try with other media as this media is quite heavy, and has poor aeration qualities, and must usually be cleaned. One can mix it with vermiculite, perlite, or coconut to increase the medium's aeration.

Benefits

- Cheap (or free).

Downsides

- Heavy
- Low aeration
- The tiny size which can block some system types.

15. Gravel

This growing medium has been in use quite early with great success. The popular one is the gravel-based ebb & flow system. As fragmented media from hard-wearing rocks like sandstone, limestone, or basalt., spaces between each particle are quite large. This helps give a plentiful supply of air to the roots. But at the same time, that means the medium doesn't hold water well, which can cause plants' roots to dry out quickly.

Because gravel is rocks, its weight makes a gravel-based system, especially large one hard to carry. But that comes with an advantage that is durable and can be reusable. As long as you wash and sterilize it, you can start using for first time growing or reusing. Just like sand, this is a good cheap starter medium for soilless growing, then people can move on to other growing materials.

Benefit

- Cheap (or free).

Downsides

- Poor water retention, not suitable for heavy plant roots.
- Heavy

16. Air

Strangely as it sounds, yet hydroponic plants can live without any growing media. In some hydroponic systems like aeroponics, plant roots are hung freely in the air and are periodically moisturized with water and nutrient through sprays.

So, it means that plants can grow in the air with no issues as long as you provide it with enough air, water, and nutrient. Plants grown in the air have a big advantage that their root can get access to plenty of oxygens (which totally comes at no cost).

On the other hand, this makes plants totally depend their life on the system. Suppose there is a power cut, a pump or timer failure, you can guess that the roots will dry out quickly, and the death of plants can be expected.

Benefit

- Plenty of oxygen.

Downside

- Plant danger in the event of-of power, pump, timer failure

Can be used in nutrient film technique systems, deep water culture systems.

2.7. Nutrients Required

Basic Nutrients

For survival, plants need to respire, get moisture and photosynthesize. And they are in the form of oxygen, carbon, hydrogen, and nitrogen. O, C, H, and N are all readily available in the air, water. And plants have the ability to get these elements from the environment. Meanwhile, light supply plants with energy to make food.

And before providing plants with necessary nutrients, you must take care of these organic composts first. Without them, plants are bound to die. Meanwhile, without nutrients, plants can still live but will not develop properly.

Macro Nutrients

These are nutrients that plants need to absorb in large quantities. They are the most vital nutrient minerals you must take care first.

Nitrogen

Without doubts, Nitrogen is considered the most important elements among all the nutrients. It is mainly responsible for

- The vegetative growth of plants – leaves, stems and their colors.
- Formation of chlorophyll, amino acids, co-enzymes, and proteins used in new cell walls.

People usually use much of Nitrogen for the growth periods of plants before they start bearing fruits or flowering. Because plants need nitrogen so quickly at some periods, nitrogen is used as part of the supplementary materials – fertilizers in soil, and nutrient solutions in hydroponics.

Plants that lack nitrogen show symptoms with their yellow leaves. Older and lower leaves will have the effects first, and drop off fast. When plants have an excess of nitrogen, the symptoms are harder to recognize. Your plants may look green and vibrant, but their ability to bear fruits and flower are greatly decreased. That is because plants are spending all of its energy producing foliage.

Phosphorus

As a vital nutrient for plants like Nitrogen, Phosphorus is the essential component of DNA, the genetic memory unit of plants. That is very important for cell formation and cell division. Phosphorus plays a critical role in the development of

- Flowering
- Fruits
- Seeds
- Roots

Your plants will require a large amount of phosphorus at the early phase of seedling, germination and flowering stage, but it's also necessary during the whole plants' life cycle.

Plants deficient of phosphorus show indication of shorter plant growth – abnormal weak leaves, flowers, and roots. Meanwhile, the excessive of phosphorus affect the plants by preventing it from absorbing other elements like calcium, copper, iron, magnesium and zinc. So the deficiencies of these elements can be seen.

Potassium (K)

Another indispensable plant nutrient that is required in a large adequate amount for the effective development and reproduction of plants. Potassium does not form compounds in plants like other nutrients, but it does help the plant of crucial processes including photosynthesis, starch formation, protein synthesis, and enzyme activation

Plants that are in need of potassium usually show yellow leaves first. When in high of potassium, like phosphorus, plants will be unable to interact with other nutrients such as zinc, iron, magnesium.

Calcium

Necessary to cell formation and development. Too little of calcium, leaf tips, and edges will turn brown and can die. Too much calcium at the younger phase can stunt plant growth.

Sulfur

A structural component of two of the 21 amino acid that creates protein. Also, helps activate and form certain enzymes and vitamins.

Magnesium

One of the chemical components of chlorophyll. Magnesium helps create the oxygen through photosynthesis and is often used in large amounts in fast-growing plants.

Among these, the three macronutrients (N, P, K) described above are the most crucial nutrients for plant's development.

Micro Nutrients

Micro nutrients are still important in plant development, but they are required in smaller quantities.

Zinc

Zinc is very important in the formation of chlorophyll and other engines, and nitrogen metabolism.

Boron

Boron is used with calcium in synthesizing the structure and functions of cell membranes. Also, help with pollination and seed production.

Iron

A component of many enzymes associated with energy provision, nitrogen fixation. Help form chlorophyll, and is used in photosynthesis.

Manganese (Mn)

Catalyze the growth process, and help form oxygen in photosynthesis.

Apart from these, plants also need Chlorine, Nickel, Copper and Molybdenum for their growth.

2.8. Hydroponically Grown Crops

1. **Lettuce:** Perfect for your salads and sandwiches, Lettuce is probably one of the most common vegetables that are grown in the hydroponic system. They have incredibly fast growth and caring for them is equally simple.

   You can grow this vegetable in almost any kind of hydroponic system. Since lettuce is so easy to care for, this might also be one of your first choices if you planning to try hydroponic farming for the first time. Lettuce usually requires cool temperatures and the pH level should be between 6 to 7. In most instances, they will grow within 30 days or less.

2. **Strawberries:** Harvesting strawberries is incredibly simple and convenient. The best part: you will have them throughout the year. Unlike Lettuce that works well in almost any kind of hydroponics systems, strawberries seem to do relatively better when in the Ebb-and-Flow system. You can also try deep water culturing with a nutrient film. Strawberries take around two months or 60 days to grow and they seem to do well at a pH level of 5.5 to 6.2. One should avoid purchasing strawberry seeds as they won't be ready with berries for a couple of years. Switch seeds with cold-stored options that are already at a growing stage.

3. **Cucumbers:** These are an extremely common variant of vining plants that are primarily grown in commercial greenhouses. Luckily, with the hydroponics system in place, you can also grow them in the comfort of your home.

   Cucumbers grow rapidly and they also end up producing excellent yields. Ideally, you can choose from the thick American slicers, the elongated and seedless variants of European cucumbers, or the impeccably smooth Lebanese cucumbers. Each of these plants will thrive well in your hydroponic system. Since cucumber is known to be a warm plant make sure you supply it with ample light and high temperatures. These plants thrive best at a pH level of 5.5 to 6.0.

4. **Spinach:** Spinach is yet another fast-growing plant that can be a suitable fit for your hydroponics system. While they thrive well in almost all types of hydroponics systems, the Nutrient Film technique works best. If you aren't comfortable with this option, choose any other technique that will keep the soil nutrient-rich and duly oxygenated.

   As with every other form of hydroponics farming, with spinach, one will need significantly less water when compared to your home garden. Besides, you can grow them from scratch right when they are seeds. In most instances, spinach is known to grow within one month or 40 days. Make sure the pH is between 6.0 to 7.5 and the temperature is duly regulated. For growing spinach, the temperature ranges must be between 65 degrees to 72 degrees F. The plant may take longer to grow if the temperature is slightly lower than the mentioned options.

5. **Beans:** hese are yet another productive and extremely low-maintenance options for your hydroponic garden. Since beans are used in most meals, growing them at home will ensure that you can duly source the veggie as and when you need to. If you are planning to grow the pole or string beans in your hydroponic garden, you will need a trellis. This will ensure that the plants are duly supported when they need it the most. Germination will take anything from three to eight days and harvesting can start around six to

eight weeks after that. Beans thrive best in warm temperatures and the ideal pH level will range from 6.0 to 6.3.

6. **Bell peppers:** Bell peppers are certainly an excellent plant for hydroponic garden. While growing them, you need to strictly ensure that they do not achieve their full height. To avoid this, make sure the plant is pruned and pinched at about seven or eight inches. This will spur their growth and also make way for bigger and fuller vegetables. In terms of technique, you should try the Deep-Water Culture while growing Bell Peppers. If this is not possible, the Ebb and Flow technique are equally good.

   Most Bell peppers take around 90 days for achieving full growth and their ideal pH level ranges from 6.0 to 6.5. While growing Bell peppers, make sure they get plenty of full light. Ideally, they need around 18 hours of complete light at a stretch. You can also raise the light rack as your bell peppers start growing. While regulating the light also make sure it is at a distance from six inches from the light.

7. **Chives:** Herbs like chives, are not just easy to grow but they are also equally easy to maintain. You can source them from your local stores and then plant them in a standard condition. Ideally, it takes around six or eight weeks for the Chives to completely mature. You should also harvest it almost daily.

   While Chives do not need plenty of water, they do need plenty of light. Their ideal temperature is warm, and they grow best upon receiving full light at a stretch of 12 to 14 hours. For pH, a level between 6.0 to 6.1 is sufficient. Note that after harvesting the Chives, they will need around three to four weeks to regrow again. So, be patient and allow them this time for regrowth.

8. **Blueberries:** Rich in vitamins, Blueberries are yet another top choice for hydroponics. Unlike strawberries, Blueberries take a fairly long time to grow. In some cases, it may take months to bear the first fruit. Blueberries thrive best in NFT systems with a warm temperature and a pH level between 4.5 and 6.0. Since growing them from seeds is not the best option, we would recommend transplants.

9. **Tomatoes:** Tomatoes (both the regular and the cherry variants) can be grown in your Hydroponic garden. So, if you are tired of consuming commercially grown tomatoes, consider growing them soilless right at the comfort of your home. While growing these veggies, make sure you provide ample light and hot temperatures. If you plan to grow them indoors purchase some grow lights in advance. The ideal pH level ranges from 5.5 to 6.5.

10. **Kale:** If you are a health freak, Kale might be one of the best options for your hydroponic system. Nutritious and yet incredibly delectable, this is an

excellent vegetable for anyone who's looking to stay healthy all way round. Since Kale has been grown in hydroponic systems for years, you can be completely assured that they will thrive well. Kale is easy to maintain that they grow well in moderate temperatures. The pH level should ideally range from 5.5 to 6.5.

11. **Basil:** Like Chives, Basil is yet another herb that thrives really well in a soilless system. NFT or the drip techniques will give the best result. After the plant is fully mature, you can start harvesting it almost daily. It is recommended to trim it instead of pinching the leaves for better and more consistent growth. Basils require plenty of lights and they are likely to undergo poor growth when they do not have ample lighting. Ideally, they should get around 11 hours of uninterrupted lighting at a warm temperature. The recommended pH level will range from 5.5 to 6.5.

12. **Grapes:** Grapes are vine-based fruits that require plenty of care and maintenance. So, if you are a newbie who is just getting started with hydroponic farming, grapes as your first option is not recommended. Grapes work best in a bucket system and being vine-based they will need a trellis for added support. Make sure the root doesn't rot and the pH is duly balanced. The ideal pH level would range from 5.5 to 6.0. While growing grapes, also make sure they get plenty of water.

13. **Celery:** This is an extremely healthy veggie with plenty of flavors and excellent texture. You can grow them in an Ebb and flow hydroponic system. Make sure the stalks are wet and the roots aren't submerged. The ideal pH level would range from 5.7 to 6.0.

14. **Sage:** Sage is yet another vital herb for most chefs. Growing them is equally simple. While sage thrives best in an NFT systems you can also try other techniques especially if you are planning to grow it with other herbs. The ideal pH level would range from 5.5. to 6.5.

15. **Cantaloupe/Melon:** Bearing plenty of similarities with the watermelon, Cantaloupes are often termed as netted melons for their net-like skin. They work best in the Ebb and Flow System and the ideal pH would range from 6.0 to 6.8. As with melons, you may have to use nets to support the Cantaloupes.

2.9. Scope in India

Growing demand for sustainable and protected farming practices and higher yield of crops grown under hydroponic facility over traditional farming are the driving factors for the growth of the market. According to the study published by V. Chandrshekaran a founder of hydroponics technology company, the average yield of lettuce per acre yearly was recorded between 9-10 tons in soil-based agriculture and 300-400 tons in hydroponics agriculture. Increased rate of investment in hydroponics over the traditional soil based agricultural process through increased yield, minimal water cultivation might further boost the agricultural communities towards adoption of the hydroponics technologies during the forecast period.

However, lack of awareness among the farming communities regarding hydroponics and high initial investment and maintenance cost associated with hydroponics are the key restraints hampering the market growth globally. The major farming communities located across the country have minimum knowledge over the hydroponics and they are majorly into competitive farming. The lack of spread of word of mouth regarding hydroponics technologies from one farming communities to other led to low understanding of technology by farmers across the globe.

By produce type, the India hydroponics market is segmented into fruits & vegetables and flowers. The fruits & vegetables segment has a market value of USD 2544.15 thousand in 2018. The market is estimated to grow at a CAGR of 13.53% in the forecasted period 2019-2027. The fruits & vegetables segments is further segmented to tomatoes, strawberries, cucumber, leafy vegetables, bell pepper and others. These products are produced majorly through nutrient film technique, deep water culture and Dutch bucket technique.

The consumer markets for hydroponically produced fruits and vegetables includes retail and hotels, fast food chains, railway catering, foreign food service companies, defense establishments and NGO's. Leafy vegetables dominate the fruits and vegetables in India owing to growing demand for high quality fresh lettuce by the commercial institutions, food chains and restaurants to meet the daily requirements are boost the market for hydroponically produced leafy greens such as lettuce. Currently, India is importing 85% of exotic vegetables from neighboring countries and it is expected to grow at a growth rate of 15-20% per year.

The key strategies adopted by the prominent players in India hydroponics market includes establishment of new commercial plants in tier 1 cites in collaboration between regional producers and global technology providers to meet the growing demand for exotic and organic foods.

Over the next few years, market might experience huge competition through major food chains across the globe establishing their hydroponics farms in India.

According to a report by DataM Intelligence, though still nascent, Indian hydroponic market is expected to grow at a Compound Annual Growth Rate (CAGR) of 13.53 per cent between 2020-2027. In comparison, the growth in the global market is estimated at just half of it at 6.8 per cent.

Vegetable Type	Soil agriculture (Avg per Acre)	Hydroponics Agriculture (Avg per acre)
Lettuce	9-10 tons	300-400 tons
Strawberries	20-25 tons	50 tons
Cucumber	15-20 tons	200 tons
Tomato	10-12 tons	180-200 tons
Bell Pepper	10-12 tons	120-140 tons
Cabbage	6-7 tons	10-12 tons

2.10. Success Stories

1. **Aqua Farms, Chennai:** Rahul Dhoka, the founder of Green Rusk Organics and Hydroponics Farming Consultancy Aqua Farms grows everything like Italian basil, mint, spinach, lettuce by using planter made of PVC pipes.
2. **Letcetra Agritech, Goa:** Ajay Naik, software engineer-turned- hydroponics farmer, grows chemical-free vegetables like lettuce, salad greens, cherry tomato, bell peppers etc.
3. **Urban Kissan, Hyderabad:** Vihari Kanukollu, Dr. Sairam and Srinivas Chaganti establish Urban Kissan jointly, where they produce lettuce, herbs, greens, exotic vegetables throughout the year.
4. **Future Farms, Chennai:** Sriram Gopal, who earlier ran an IT firm, establishes the Future Farm to develop hydroponics farming, which spread over 10 states growing leafy vegetables.

Rohit Nagdewani began researching a small kitchen garden and heard about hydroponics. With its major USP being the ability to grow anything, anywhere, it didn't take a lot of time for me to venture into it commercially. He started experimenting with small imported setups but could not scale due to the lack of information and equipment. A lot of research and calls later, he found two companies that help set up hydroponic farms – Urban Farmer and Valens Arbor in Mumbai. With their help he set up my first commercial farm. The challenge for hydroponic farmers like him is the fact that the technology is at a very nascent stage, so getting access to the right people, equipment and process is not easy. No government permissions were needed.

He started with a small 300 sqft indoor farm that cost me Rs 4 lakh. Now he has expanded to a 6,600 sqft indoor farm and one 6,000 sqft outdoor farm. All this cost me Rs 45 lakh. Both the capex and the opex of a hydroponic farm is higher than soil grown produce, which results in a higher selling price. In the case of commodity crops like wheat and rice, the price will be four times more that soil-grown crops.

Last year India grew 15 crores tones of greens. There is a market for it and producing greens costs marginally more than traditional crops, about 2-3%. With the nutritional benefits there are people who will pay more. He is having 500 customers who are subscribers and a few restaurants who buy regularly.

They had a great idea and we were certain that we wanted to execute it at some point in time. The question was - would it be possible to run an actual business? They did a lot of research and even zeroed in on one international firm that specializes in vertical farms, but their estimate was about 10 times the cost I finally paid to a domestic company.

Extensive market research revealed that people don't like to eat salads every day, but prefer it twice or thrice a week. With the rising trend of veganism, people like their salads to be vegetarian. We on-boarded a chef and a nutritionist and developed gourmet salad recipes that we are currently testing in the market.

He invested Rs 15 lakh to set up a 500 sqft farm and that changed the game for me. Today, their farm can churn our 1,500 plates of salad per week. They are also planning to set up another farm of 4000 sqft. Their primary challenge is to educate our potential customers on the benefits of our product. This topic is gaining momentum on social media and more people are aware of the benefits of eating hydroponic food. Their product – greens and herbs - is something that traditional farmers do not grow or rather cannot grow due to the weather conditions, so there is no issue with that at all. We only grow varieties that you don't usually get at the local farmers' market.

Future

We are in the final stages of validating our product and have planned on expanding to a farm that is almost 10 times the current size next year. Our marketing strategy is also in place, but we are waiting to execute everything at the right time as we don't want a bandwidth, we are not able to manage.

MarketsandMarkets, a market research company, the global hydroponics market is estimated to grow from $8.1 billion in 2019 to $16.0 billion by 2025, registering a CAGR of 12.1%. The primary drivers for this handsome growth include growing population and the need to ensure food security through alternative high-yield farming techniques as arable land and water have been depleting.

Hydroponics globally has grown because of the high yield (20-25%) and 2-5 times productivity. This system eliminates the use of artificial ripening agents and pesticides creating nutritionally superior vegetables. That said. India's overall fruit and vegetables (F&V) market was estimated at around Rs 5,00,000 crore in 2015. Even if we assume a meagre 0.5% of this overall market, it translates into a humongous opportunity of around Rs 2,500 crore for hydroponic produce, which may take at least five years to materialize. Renowned business groups such as DS Group and Patanjali are also considering hydroponics on a commercial scale.

3

Advanced Topics in Agronomy

The income of farmers is declining day by day due to increasing cost of inputs including wages of farm labor. The net income can only be increased by reducing cost of inputs and by increasing yield of currently adopted cropping systems. The income of farm can be increased by growing horticultural crops (vegetables, flowers, fruits etc.) as well as medicinal, aromatic or spice crops. However, these crops cannot be grown on large scale due to their market problems and involvement of high labor. So, in order to improve profit from ongoing cropping systems, some latest crop production practices are being discussed in this chapter.

3.1. Precision Agriculture

Precision agriculture (PA) is an approach to farm management that uses information technology (IT) to ensure that crops and soil receive exactly what they need for optimum health and productivity. The goal of PA is to ensure profitability, sustainability and protection of the environment. PA is also known as satellite agriculture, as-needed farming and site-specific crop management (SSCM). Precision agriculture relies upon specialized equipment, software and IT services. The approach includes accessing real-time data about the conditions of the crops, soil and ambient air, along with other relevant information such as hyper-local weather predictions, labor costs and equipment availability.

Predictive analytics software uses the data to provide farmers with guidance about crop rotation, optimal planting times, harvesting times and soil management. Sensors in fields measure the moisture content and temperature of the soil and surrounding air. Satellites and robotic drones provide farmers with real-time images of individual plants. Information from those images can be processed and integrated with sensor and other data to yield guidance for immediate and future decisions, such as precisely what fields to water and when or where to plant a particular crop.

Agricultural control centers integrate sensor data and imaging input with other data, providing farmers with the ability to identify fields that require treatment and determine the optimum amount of water, fertilizers and pesticides to apply. This helps the farmer avoid wasting resources and prevent run-off, ensuring that the soil has just the right amount of additives for optimum health, while also reducing costs and controlling the farm's environmental impact. In the past, precision agriculture was limited to larger operations which could support the IT infrastructure and other technology resources required to fully implement and benefit from the benefits of precision agriculture. Today, however, mobile apps, smart sensors, drones and cloud computing makes precision agriculture possible for farming cooperatives and even small family farms.

Challenges

Research suggest educational and economic challenges as these are important in the application of precision agriculture. Among the variables that contribute to educational challenges, lack of local experts, funds, knowledgeable research and extension personnel have more of an impact compared to others. PA and initial costs have more of an impact among the economic challenges compared to the other issues.

Need for Precision Agriculture

- To increase agriculture productivity
- Prevents soil degradation
- Reduction of chemical application in crop production
- Efficient use of water resources
- Dissemination of modern farm practices to improve quality, quantity and reduced cost of production
- Developing favorable attitudes
- Precision farming changing the socio-economic status of farmers

Advantages

- Agronomical perspective
- Technical perspective
- Environmental perspective
- Economic perspective
- A refined set of cultivation practices and choice of crops based on suitability of land
- Elimination of volatility and risk
- Waste management
- Reduced production costs
- Minimum environmental impact

- Optimized use of fertilizers
- Water management

Drawbacks

- High cost
- Lack of technical expertise knowledge and technology
- Not applicable or difficult/costly for small land holdings
- Heterogeneity of cropping systems and market imperfections

Tools and Equipment

Precision farming is a combination of application of different technologies. All these combinations are mutually inter related and responsible for developments. The same are discussed below:

I. **Global Positioning System (GPS):** It is a set of 24 satellites in the Earth orbit. It sends out radio signals that can be processed by a ground receiver to determine the geographic position on earth. It has a 95% probability that the given position on the earth will be within 10-15 meters of the actual position. GPS allows precise mapping of the farms and together with appropriate software informs the farmer about the status of his crop and which part of the farm requires what input such as water or fertilizer and/or pesticides etc.

II. **Geographic Information System (GIS):** It is software that imports, exports and processes spatially and temporally geographically distributed data.

III. **Grid Sampling:** It is a method of breaking a field into grids of about 0.5-5 hectares. Sampling soil within the grids is useful to determine the appropriate rate of application of fertilizers. Several samples are taken from each grid, mixed and sent to the laboratory for analysis.

IV. **Variable Rate Technology (VRT):** The existing field machinery with added Electronic Control Unit (ECU) and onboard GPS can fulfill the variable rate requirement of input. Spray booms, the spinning disc applicator with ECU and GPS have been used effectively for patch spraying. During the creation of nutrient requirement map for VRT, profit maximizing fertilizer rate should be considered more rather than yield maximizing fertilizer rate.

V. **Yield Maps:** Yield maps are produced by processing data from adapted combine harvester that is equipped with a GPS, i.e. integrated with a yield recording system. Yield mapping involves the recording of the grain flow through the combine harvester, while recording the actual location in the field at the same time.

VI. **Remote Sensors:** These are generally categories of aerial or satellite sensors. They can indicate variations in the colours of the field that corresponds to changes in soil type, crop development, field boundaries, roads, water, etc. Arial and satellite imagery can be processed to provide vegetative indices, which reflect the health of the plant.

VII. **Proximate Sensors:** These sensors can be used to measure soil parameters such as N status and soil pH) and crop properties as the sensor attached tractor passes over the field.

VIII. **Computer Hardware and Software:** In order to analyze the data collected by other precision agriculture technology components and to make it available in usable formats such as maps, graphs, charts or reports, computer support is essential along with specific software support.

IX. **Precision Irrigation Systems:** Recent developments are being released for commercial use in sprinkler irrigation by controlling the irrigation machines motion with GPS based controllers. Wireless communication and sensor technologies are being developed to monitor soil and ambient conditions, along with operation parameters of the irrigation machines (i.e. flow and pressure) to achieve higher water use efficiency.

X. **Precision Farming on Arable Land:** The use of PA techniques on arable land is the most widely used and most advanced amongst farmers. CTF (Controlled Traffic Farming) is a whole farm approach that aims at avoiding unnecessary crop damage and soil compaction by heavy machinery, reducing costs imposed by standard methods. Controlled traffic methods involve confining all field vehicles to the minimal area of permanent traffic lanes with the aid of decision support systems. Another important application of precision agriculture in arable land is to optimize the use of fertilizers especially, Nitrogen, Phosphorus and Potassium.

Precision Farming vs. Digital Farming vs. Smart Farming

Technology has become a critical piece of business for every farmer, agriculture retailer, and agronomist. The increasing technology adoption rate in agriculture shouldn't be surprising to anyone. Farming is highly land- and labor-intensive. Farmers use technology to increase efficiency and manage costs. The evolution of digital technology and the modernization of agriculture have caused new concepts to emerge. These terms include precision agriculture or farming, digital farming, and smart farming. These terms are often being used interchangeably but have a subtle difference in meaning.

Precision Agriculture or Precision Farming

Precision agriculture is also referred to as precision farming or precision agriculture. Perhaps the easiest way to understand precision agriculture is to think of it as everything that makes the practice of farming more accurate, optimized, and controlled when it comes to the growing of crops and raising of livestock. A vital component of this is the use of information technology and a wide array of items such as:

- automated hardware and software,
- autonomous vehicles,
- drones,
- GPS guidance,
- robotics,
- sensors,
- soil sampling, and
- telematics.

The critical point here is optimization. For example, instead of applying an equal amount of fertilizers over a field, precision agriculture involves measuring the within-field soil variations and adapting the fertilizer strategy accordingly. This leads to optimized fertilizer usage, saving costs, and reduced environmental impact.

Precision agriculture was initially born with John Deere's introduction of GPS guidance for tractors in the early 1990s. Their goal with this technology was to automate steering, reducing errors, and increasing crop yield by not wasting seed. Adopting this GPS guidance is so widespread that it's probably the most commonly-used example of precision agriculture today. Plus, it gave way to a full-fledged industry: precision agriculture!

While precision farming principles have been around for more than twenty-five years, it's only been over the past decade that they became mainstream. This is due to technological advancements and the adoption of other, broader technologies. The adoption of mobile devices, access to high-speed internet, low-cost and reliable satellite communications – for positioning and imagery – and farm equipment optimized for precision agriculture by the manufacturer are vital technologies characterizing the trend for precision agriculture.

Digital Farming

The essence of digital farming lies in creating value from data. Digital farming means to go beyond the presence and availability of data to develop actionable intelligence and meaningful added value from such data. Digital agriculture is integrating both concepts – precision farming and smart farming.

For farmers, digital agriculture allows for the opportunity to increase their farm's production, save costs in the long-term, and eliminate risks. Many view digital agriculture as the future of the agricultural industry.

Smart Farming

Basically, "smart farming" is applying information and data technologies for optimizing complex farming systems. The focus is on access to data and how farmers can use the collected information intelligently. The goal is to increase the quality and quantity of the products while optimizing human labor production. Or, more simply, producing more food with less investment and the same amount of land.

The technology used in smart farming range from IoT and robotics to drones and AI. With these tools, farmers can monitor field conditions without going to the field. This enables them to make decisions for the whole farm, a lot, or even a single plant.

Smart farming is not only for large agricultural corporations. It's also able to boost family farms, organic farms, and other smaller operations. The entire process of smart farming is software-managed and sensor-monitored, reducing overall prices, rising overall yield, increasing quality of the availability, and ultimately the customer experience. Automation has enabled giant steps forward in production efficiency, quality improvements, and sustainability.

New Technologies in Agriculture

Among the technologies that are revolutionizing the present, and that will define the future of agriculture, the most notable are:

Drones

Drones simplify supervision tasks for farms by being able to cover hundreds of acres in one flight, gathering, thanks to infrared technology, multispectral images and a wide variety of information about the condition of the land, irrigation needs, crop growth, the existence of pathogens, and, in the case of cattle, the number of animals, their weight and possible anomalies such as lameness or unusual movements.

IoT

The Internet of Things makes it possible to optimize the monitoring of farms, mainly through smart sensors capable of measuring everything from solar radiation to leaf moisture and stem diameter, or the temperature of each animal in the case of livestock, making it easier to make all sorts of management decisions.

Big data

Thanks to the ability of big data to analyze massive amounts of data, farmers can manage all the information obtained from drones, the Internet of Things and other measuring instruments and integrate it both with historical information for the farm and with weather data, in order to optimize all stages of the production process.

Blockchain

Blockchain makes it possible to monitor crops and cattle from growth until handover to suppliers, improving, for example, the traceability of the supply chain. By using this technology, if an imported vegetable poisons consumer the source of the outbreak can be easily traced and only the affected products withdrawn, instead of prohibiting imports of vegetables from the entire country of origin.

Robots

Self-steering tractors have existed for some time now, as John Deere equipment works like a plane on autopilot. The tractor does most of the work, with the farmer stepping in for emergencies. Technology is advancing towards driverless machinery programmed by GPS to spread fertilizer or plow land. Autonomy of technology is driven by the demanding need of diagnoses, often difficult to accomplish solely by hands-on farmer-operated machinery. In many instances of high rates of production, manual adjustments cannot sustain. Other innovations include a solar powered machine that identifies weeds and precisely kills them with a dose of herbicide or lasers. Agricultural robots, also known as AgBots, already exist, but advanced harvesting robots are being developed to identify ripe fruits, adjust to their shape and size, and carefully pluck them from branches.

Smartphone applications

A possible configuration of a smartphone-integrated precision agriculture system. Smartphone and tablet applications are becoming increasingly popular in precision agriculture. Smartphones come with many useful applications already installed, including the camera, microphone, GPS, and accelerometer. There are also applications made dedicated to various agriculture applications such as field mapping, tracking animals, obtaining weather and crop information, and more. They are easily portable, affordable, and have high computing power.

Machine learning

Machine learning is commonly used in conjunction with drones, robots, and internet of things devices. It allows for the input of data from each of these sources. The computer then processes this information and sends the appropriate actions back to these devices. This allows for robots to deliver the perfect amount of fertilizer or for IoT devices to provide the perfect quantity of water directly to the soil. Machine learning may also provide predictions to farmers at the point of need, such as the contents of plant-available nitrogen in soil, to guide fertilization planning. As more agriculture becomes ever more digital, machine learning will underpin efficient and precise farming with less manual labour.

Satellite imagery

Satellite technology are used in precision farming. Aerial photography from light aircraft can be combined with data from satellite records to predict future yields based on the current level of field biomass. Aggregated images can create contour maps to track where water flows, determine variable-rate seeding, and create yield maps of areas that were more or less productive.

3.2. Conservation Agriculture

Conservation Agriculture (CA) is a farming system that can prevent losses of arable land while regenerating degraded lands. It promotes maintenance of a permanent soil cover, minimum soil disturbance, and diversification of plant species. It enhances biodiversity and natural biological processes above and below the ground surface, which contribute to increased water and nutrient use efficiency and to improved and sustained crop production.

CA principles are universally applicable to all agricultural landscapes and land uses with locally adapted practices. Soil interventions such as mechanical soil disturbance is reduced to an absolute minimum or avoided, and external inputs such as agrochemicals and plant nutrients of mineral or organic origin are applied optimally and in ways and quantities that do not interfere with, or disrupt, the biological processes.

CA facilitates good agronomy, such as timely operations, and improves overall land husbandry for rainfed and irrigated production. Complemented by other known good practices, including the use of quality seeds, and integrated pest nutrient, weed and water management, etc., CA is a base for sustainable agricultural production intensification. It opens increased options for integration of production sectors, such as crop-livestock integration and the integration of trees and pastures into agricultural landscapes.

Some distinguishing features of conventional and conservation agriculture systems

Conventional agriculture	Conservation agriculture
Cultivating land, using science and technology to dominate nature	Least interference with natural processes
Excessive mechanical tillage and soil erosion	No-till or drastically reduced tillage (biological tillage)
High wind and soil erosion	Low wind and soil erosion
Residue burning or removal (bare surface)	Surface retention of residues (permanently covered)
Water infiltration is low	Infiltration rate of water is high
Use of ex-situ FYM/composts	Use of in-situ organics/composts
Green manuring (incorporated)	Brown manuring/cover crops (surface retention)
Kills established weeds but also stimulates more weed seeds to germinate	Weeds are a problem in the early stages of adoption but decrease with time
Free-wheeling of farm machinery, increased soil compaction	Controlled traffic, compaction in tramline, no compaction in crop area
Mono cropping/culture, less efficient rotations	Diversified and more efficient rotations
Heavy reliance on manual labor, uncertainty of operations	Mechanized operations, ensure timeliness of operations
Poor adaptation to stresses, yield losses greater under stress conditions	More resilience to stresses, yield losses are less under stress conditions
Productivity gains in long-run are in declining order	Productivity gains in long-run are in incremental order

Source: Sharma et al., 2012.

Conservation Agriculture Principles

Issues	Conventional Agriculture (CT)	Conservation Agriculture (CA)
Practice	Disturbs the soil and leaves a bare surface	Minimal soil disturbance and soil surface is permanently covered
Cropping system	Monocropping/culture, less efficient rotation	Diversified farming and more efficient rotations
Residue management	Residue burning or removal	Residue retention on the soil surface
Erosion	Wind and water erosion is maximum	Wind and water erosion is the least
Soil physical health	Very poor	Comparatively good
Soil biological health	Poor biological health owing to frequent disturbance	More diverse and healthy biological properties and populations
Water infiltration	Lowest after soil pores clogged	Better water infiltration

Issues	Conventional Agriculture (CT)	Conservation Agriculture (CA)
Soil organic matter	Low due to oxidation of organic matter and residue removal	Soil organic carbon build-up
Weeds	Controls weeds and also causes more weed seeds to germinate	Weeds are a problem especially in the early stages of adoption, but problems are reduced with time and residues can help suppress weed growth
Diesel use and costs	Diesel use high	Diesel use much reduced
Production costs	High	Low
Timeliness	Operations can be delayed	Timeliness of operations more optimal
Yield	Can be lower where planting delayed	Yields same as CT but can be higher if planting done more timely

Source: Hobb et al. (2008), Sharma et al., 2012.

Minimum soil disturbance refers to low disturbance no-tillage and direct seeding. The disturbed area must be less than 15 cm wide or less than 25% of the cropped area (whichever is lower). There should be no periodic tillage that disturbs a greater area than the aforementioned limits. Strip tillage is allowed if the disturbed area is less than the set limits.

Permanent soil organic cover

Three categories are distinguished: 30-60%, 60-90% and >90% ground cover, measured immediately after the direct seeding operation. Area with less than 30% cover is not considered as CA.

Species diversification

Rotation/association should involve at least 3 different crop species

Advantages

Conservation agriculture is generally a "win-win" situation for both farmers and the environment. Conservation agriculture (CA) increases the productivity of:

- **Land** - Conservation agriculture improves soil structure and protects the soil against erosion and nutrient losses by maintaining a permanent soil cover and minimizing soil disturbance. Furthermore, CA practices enhance soil organic matter (SOM) levels and nutrient availability by utilizing the previous crop residues or growing green manure/ cover crops (GMCC's) and keeping these residues as a surface mulch rather than burning. Thus, arable land under CA is more productive for much longer periods of time.
- **Labor** - Because land under no-till is not cleared before planting and involves less weeding and pest problems following the establishment of

permanent soil cover/crop rotations, farmers in Ghana reported a 22% savings in labor associated with maize production. Similar reductions in labor requirements have been reported with no-till rice-wheat systems in South Asia and various CA technologies in South America. Much of the reduced labor comes from the absence of tillage operations under CA, which use up valuable labor days during the planting season.

- **Water** - Conservation agriculture requires significantly less water use due to increased infiltration and enhanced water holding capacity from crop residues left on the soil surface. Mulches also protect the soil surface from extreme temperatures and greatly reduce surface evaporation, which is particularly important in tropical and sub-tropical climates. In Sub-Saharan Africa, as with other dryland regions, the benefits of conservation agriculture are most salient during drought years, when the risk of total crop failure is significantly reduced due to enhanced water use efficiency.
- **Nutrients** - Soil nutrient supplies and cycling are enhanced by the biochemical decomposition of organic crop residues at the soil surface that are also vital for feeding the soil microbes. While much of the nitrogen needs of primary food crops can be achieved by planting nitrogen-fixing legume species, other plant essential nutrients often must be supplemented by additional chemical and/or organic fertilizer inputs. In general, soil fertility is built up over time under conservation agriculture, and fewer fertilizer amendments are required to achieve optimal yields over time.
- **Soil biota** - Insect pests and other disease-causing organisms are held in check by an abundant and diverse community of beneficial soil organisms, including predatory wasps, spiders, nematodes, springtails, mites and beneficial bacteria and fungi, among other species. Furthermore, the burrowing activity of earthworms and other fauna create tiny channels or pores in the soil that facilitate the exchange of water and gases and loosen the soil for enhanced root penetration.
- **Economic benefits** - Farmers using CA technologies typically report higher yields (up to 45-48% higher) with fewer water, fertilizer and labor inputs, thereby resulting in higher overall farm profits. In Paraguay, net farm income of no-till (NT) farming on large-scale commercial farms increased from \$23,467 to \$32,608 more than farms using conventional tillage over a 10-year period. The economic benefits of NT and other conservation agriculture technologies, more than any other factor, has led to widespread adoption among both large- and small-scale farmers throughout the world.
- **Environmental benefits** - Conservation agriculture represents an environmentally-friendly set of technologies. Because it uses resources more efficiently than conventional agriculture, these resources become

available for other uses, including conserving them for future generations. The significant reduction in fossil fuel use under no-till agriculture results in fewer greenhouse gases being emitted into the atmosphere and cleaner air in general. Reduced applications of agrochemicals under CA also significantly lessens pollution levels in air, soil and water.

- **Equity considerations** -Conservation agriculture also has the benefit of being accessible to many small-scale farmers who need to obtain the highest possible yields with limited land area and inputs. Perhaps the biggest obstacle thus far for the technology spreading to more small-scale farmers worldwide has been limited access in certain areas to certain specialized equipment and machinery, such as no-till planters. This problem can be remedied by available service providers renting equipment or undertaking conservation agriculture operations for farmers who would not otherwise have access to the needed equipment. Formulating policies that promote adoption of CA are also needed. As more and more small-farmers gain access to CA technologies, the system becomes much more "scale neutral".

Disadvantages

Despite the apparent conservation tillage environmental benefits, these methods also have disadvantages. First of all, they are associated with creating special conditions that are not suitable for all manufacturers. There is also a specific negative impact on the environment. Consider some drawbacks:

Conservation tillage requires expensive equipment or a lot of manual labor if we are talking about small producers;

Unlike traditional methods of weed and pest control, it demands more pesticides and herbicides;

The first positive results of conservation tillage are visible only in long-run;

The method contributes to greenhouse gases emission (CH_4 and N_2O).

Case Study by FAO (Indo-Gangetic Plains)

Resource-conserving technologies produce high wheat yields while reducing farmers' costs by 20 percent. A shift to conservation agriculture in rice would create positive synergies in the production of other crops in rotation.

In South Asia farmers practice zero-tillage to reduce costs and grow more wheat. Alternate wetting and drying of rice fields helps cut water consumption by up to 50 percent. Yields of both cereals improve after laser-assisted land-levelling. Farmers save on fertilizer with 'needs-based' nitrogen management and use legumes that also suppress weeds.

For rice, the Rice-Wheat Consortium, an eco-regional initiative of national agricultural research systems and the CGIAR, promoted the substitution of long-season cultivars with short-seasons ones, and direct dry-seeding which, by eliminating the need for transplanting, reduces water use, energy costs and labour requirements. In dry-seeding, fields are prepared in June and a short-season rice crop is sown after irrigation to establish it before the onset of the monsoon in July.

During crop growth, various approaches are being promoted to help farmers increase rice output with the same amount of water, or use less water without reducing yields. One approach is alternate wetting and drying, in which the paddy is flooded and the water is allowed to dry out before re-flooding. Another is aerobic rice, where seeds are sown directly into the dry soil, then irrigated. Both approaches result in water savings of 30 to 50 percent. Raised-bed planting also produces significantly higher rice yields.

Another resource-conserving technology introduced to the Indo-Gangetic Plains is laser-assisted land-levelling. Many fields have uneven surfaces, which lead to wasted water, sub-optimal germination and lower yields. Traditionally, farmers have levelled their fields using scrapers and wooden boards. Now, laser-guided tractors, operated by private contractors, offer more precise levelling of fields at prices smallholders can afford. Recent studies in northwest India found that the technology is far more efficient than traditional levelling, reducing water applications by as much as 40 percent, improving the efficiency of fertilizer, and boosting rice and wheat yields by from 5 to 10 percent. It is equally profitable on all farm sizes.

Farmers have also introduced new crop rotations that disrupt the life cycles of insect pests and weeds, and promote soil health. In Pakistan's Punjab province, smallholder farmers rotate rice with berseem clover, a fodder crop that improves soil fertility and suppresses weeds that might otherwise infest subsequent cereal crops. On the eastern plains, where fields generally remain fallow for 80 days after the wheat harvest, a summer mungbean crop planted on zero-tilled soil produces 1.45 tons per ha, worth US$745. Mungbeans also add nitrogen to the soil through biological nitrogen fixation.

To reduce the wasteful use of fertilizer, the Rice-Wheat Consortium promoted 'needs-based' nitrogen management by introducing a leaf color chart indicating the best times for fertilization. The charts were originally designed for rice, but were spontaneously adapted to wheat by farmers. Using the charts, farmers have reduced fertilizer applications by up to 25 percent with no reduction in yield.

Village surveys conducted across the plains in 2009 found that one in three farm households had adopted at least one resource-conserving technology, with the highest rates – of almost 50 percent – in the northwest. Farmers learned about the technologies from a variety of sources, including other farmers and equipment manufacturers, and most had integrated them into their traditional crop management practices. In northwest India, zero-tillage seed drills were the most common item of agricultural machinery after tractors. Their high adoption rate was made possible by the ready availability of seed drills developed by the private sector, with strong support from state and local governments.

3.3. Irrigation Saving Devices

Drip irrigation

Drip irrigation is one of the most common localized irrigation systems. Sometimes also known as a trickle or micro-irrigation system, it consists of a network of pipelines and valves. The valves facilitate the dripping water directly to the plant root zone. This helps to reduce water loss through evaporation and leaking.

The size of the valve, pipe diameter, and flow rate are determined after considering the water requirements at particular times. Then, following regular intervals or depending on the yard's design, there are openings in the irrigation pipe that allow water to flow. This keeps the base of the plant damp and moisturized. And because only the root area is watered, weeds don't get enough water to flourish.

When used on farms, the soluble fertilizer, different chemicals such as pesticides, and cleaning agents are also supplied through this arrangement to the crop by dissolving in irrigated water. You can pre-estimate the amount of water and fertilizer to minimize waste.

In areas with water scarcity, drip irrigation is mainly used to prevent the spread of disease caused by contact with water. The irrigation pipe is usually surrounded by bare earth or stone pavers, so no plantings are crushed under the drip irrigation system. It is widely used in commercial agricultural systems such as greenhouses, containerized plants, coconut cultivation, and landscapes.

Advantages

- Fertilizer and nutrient loss will be minimized due to a localized application of water and reduced leaching.
- Water application efficiency is very high if managed correctly.
- Field leveling is not necessary and fields with irregular shapes are easily accommodated.
- Soil type plays a less important role in the frequency of irrigation.
- Soil erosion and weed growth are lessened.
- Water distribution is highly uniform and controlled by the output of each nozzle.

- Labor cost is less in drip irrigation than other irrigation methods.
- Variation in supply can be regulated by regulating the valves and the drippers.
- Fertigation can be easily included with minimal waste of fertilizers.
- Usually operated at a lower pressure than other kinds of pressurized irrigation, reducing energy costs.
- Saves water to 70% compared to flood irrigation and more land can be irrigated with the water thus saved.
- Early maturity effects in higher and faster returns on investment and fertilizer use efficiency increases by 30%.
- Undulating terrains, saline, waterlogged, sandy, and hilly lands can also be brought under productive cultivation.

Disadvantages

- The initial cost is high.
- It can cause clogging if water is not filtered correctly.
- There are problems in moisture distribution and salinity.

Crops suitable for drip irrigation

Orchard Crops	Grapes, Banana, Pomegranate, Orange, Citrus, Mango, Lemon, Custard Apple, Sapota, Guava, Pineapple, Coconut, Cashewnut, Papaya, Aonla, Litchi, Watermelon, Muskmelon etc.
Vegetables	Tomato, Chilly, Capsicum, Cabbage, Cauliflower, Onion, Okra, Brinjal, Bitter Gourd, Ridge Gourd, Cucumber, Peas, Spinach, Pumpkin ete.
Cash Crops	Sugarcane, Cotton. Arecanut, Strawberry, etc.
Flowers	Rose, Carnation, Gerbera, Anthurium, Orchids, Jasmine, Dahilia, Marigold, etc.
Plantation	Tea, Rubber, Coffee, Coconut, etc.
Spices	Turmeric, Cloves, Mint, etc,
Oil Seed	Sunflower, Oil palm, Groundnut, etc.
Forest Crops	Teak Wood, Bamboo, etc.

Table 3.1 shows that in drip irrigation system, the increase in yield may vary from 20 to 514 percent and water saving from 14 to 79 percent as compared to surface irrigation. The resulting efficiency, of course, depends on climatic, soil and crop characteristics and farmer practices. Thus, water saving ratio can vary depending on these conditions. The advantages of drip irrigation could be specified as water saving (30–50%), higher crop yield, maximum utilization of available water, no water being available for weeds, high efficiency in the use of fertilizers, less weed growth, lower labour, no soil erosion, possible sophisticated automatic control,

no runoff, no leaching of fertilizers into ground water, and less evaporation losses compared to surface irrigation.

Table 3.1: Efficiency of drip irrigation when compared to surface irrigation in different crops

Crop	Yield (q/ha)			Irrigation		
	Surface	Drip	% Increase	Surface	Drip	% Saving
Beet root	5.7	8.8	54	86	18	79
Bitter gourd	32	43	34	76	33	57
Brinjal	91	148	63	168	64	62
Broccoli	140	195	39	70	60	14
Chilli	171	274	60	27	18	33
Cucumber	42.3	60.9	44	109	41.7	62
Okra	155	255	45	54	24	56
Onion	284	342	20	52	26	50
Potato	172	291	69	60	27.5	54
Radish	10.5	11.9	13	46	11	76
Sweet potato	42.4	58.9	39	63	25	60
Tomato	61.8	88.7	44	49.8	10.7	79
Banana	575	875	52	176	97	45
Grapes	264	325	23	53	28	47
Papaya	130	230	77	228	73	68
Pomegranate	34	67	97	21	16	24
Watermelon	82.1	504	514	72	25	65

Source: Shah, 2011

Case Study

Kumara and Palanisami (2010) worked on Impact of Drip Irrigation on Farming System: Evidence from Southern India, while working at Department of Agricultural Economics, Tamil Nadu Agricultural University, Coimbatore, Tamil Nadu and International Water Management Institute, Patancheru, Andhra Pradesh reported as follows

The adoption of drip irrigation has significant positive impact on the cost of cultivation and cost of production and returns of the farmers. The economics of banana cultivation revealed that the cost of labour was significantly lower under the drip method (Rs 9761/ha), which was 69 per cent less than in the control villages (Rs 31487/ha) (Table 3.2). The drip method significantly saves the human labour involved in crop production activities.

Also, data presented in Table 3.2 indicated that there is saving of irrigation water by 29.2 percent and increase in yield per unit of water (kg/m3) was 51.0 percent in banana irrigated with drip irrigation as compared to control villages.

Table 3.2: Economics of crop production for banana in sample farms in Tamil Nadu

	(Per Hectare)	
Particulars	Drip villages	Control villages
Quantity of water applied (M^3)	8979*	12669
Quantity of energy consumed (kWh)	2219*	8294
Cost of labour (Rs)	9761*	31487
Capital (Rs)	80369*	104351
Yield (tonnes)	60.34*	57.79
Gross income (Rs)	280602*	267400
Gross margin (Rs)	200232*	163048
Yield per unit of water (kg/M^3)	74*	4.9
Yield per unit of energy (kg/kWh)	28.6*	7.2
Returns per unit of water (Rs/M^3)	23.8*	13.3
Returns per unit of energy (Rs/kWh)	92.3*	19.8

Source: Field survey during 2007-08. *Note*: *indicates that values are significantly different at 1 per cent level from the corresponding values of control village

It also saves irrigation labour and weeding labour. On an average, the human labour days used for weeding banana were 17 labour days / ha under drip method and 60 labour days/ha under flood method of irrigation. The drip method saved nearly 71 per cent of weeding labour when compared to flood method of irrigation. The irrigation labour has been worked out to be 168 labour days /ha under flood method and 18 labour days/ha under drip method of irrigation. Due to this, the cost of cultivation was significantly less under drip over the flood method.

The reduction in cost on human labour has a significant bearing on the cost of cultivation. Though, the cost of installation of drip equipment and maintenance is incurred by the drip farmers, the cost of cultivation per hectare has been worked out to be Rs 80396/ha in drip farms, which is around 23 per cent less than in the control villages (Rs 109685/ha). The gross margin per hectare has been found as Rs 200232/ ha in drip and Rs 163048/ha in control farms. It clearly shows that drip method of irrigation has resulted in an increase of 22 per cent of gross margin over the control. As the adoption of drip irrigation saves considerable water and energy, the water and energy productivity are significantly more in drip farms than the control villages where the flood irrigation is followed. For instance, the water productivity has been worked out to be 7.4 kg/ M3 of water in drip farms and 4.9 kg/M3 of water in control villages. Significant difference in energy productivity has also been noticed. The returns per unit of water and energy have shown that drip farms have significantly higher returns over the control. Thus, one could conclude that the drip adoption would be a viable technology with significant bearing on the private profits.

They also worked on drip irrigation in Grapes and Coconut (Table 3.3). The economics of coconut cultivation in drip and control villages has revealed that the cost saving due to reduction in labour was 69 per cent. Similarly, the cost of cultivation was considerably lower under the drip method, registering a reduction of 15.5 per cent. The impact of drip irrigation on resource saving and productivity enhancing was highly significant in grapes. Since grape cultivation is sensitive to water stress and involves huge labour for irrigation, weeding, training and pruning, the drip could result in significant savings in water and labour, leading to reduction in cost of cultivation. In grape cultivation, the cost incurred on human labour was Rs 17324/ha in drip farms and Rs 29433/ha in control farms with an average reduction of 41 per cent. Also, there was a reduction in the cost of cultivation by 15.6 per cent in drip farms over control farms. The gross margin across farms indicated that the drip farms achieved relatively higher returns with a given price of output when compared to control farms mainly due to difference in yield. The physical productivity of water and energy was significantly higher in drip than control farms. The analysis of economics of crop cultivation under drip and flood methods of irrigation has revealed that the former has a significant impact on resources saving, cost of cultivation, yield of crops and farm profitability. The physical water and energy productivity were significantly high in drip than flood method of irrigation. One could conclude that the drip has a significant bearing on the private costs and hence on profit of farmers

Table 3.3: Economics of crop production for coconut and grapes in sample farms in Tamil Nadu (per hectare)

Particulars	Drip villages	Control villages
Quantity of water applied (M3)	8979*	12669
Quantity of energy consumed (kWh)	2219*	8294
Cost of labour (Rs)	9761*	31487
Capital (Rs)	80369*	104351
Yield (tonnes)	60.34*	57.79
Gross income (Rs)	280602*	267400
Gross margin (Rs)	200232*	163048
Yield per unit of water (kg/M3)	74*	4.9
Yield per unit of energy (kg/kWh)	28.6*	7.2
Returns per unit of water (Rs/M3)	23.8*	13.3
Returns per unit of energy (Rs/kWh)	92.3*	19.8

Source: Field survey during 2007-08

Notes: *indicates that values are significantly different at 1 per cent level from the corresponding values of control village

Sprinkler irrigation system

A sprinkler irrigation system is widely used as a localized method of dispensing water to plants and crops. So, you can install the lawn sprinkler systems to water your lawn and backyard grass and crop plants. Use it as a cooling system or to stave off airborne dust. It consists of pipelines, spray guns, and spray nozzles. The gun will rotate in a circular pattern by using the power of the spraying water.

As it is a localized irrigation method, the water loss is less than surface irrigation but higher than drip irrigation. However, spraying water all over the field may cause the spread of some plant diseases and increase the pest population. But you can automate the water flow as per your requirement to avoid excess water use and disease spread.

It requires far less maintenance than other irrigation methods. As a result, it is affordable, dependable, inexpensive, and hassle-free.

Advantages

- The spray gun can wet a large circular area in less time.
- It serves as a cooling system and clears off the dust from plants.
- It can water every part of plants.
- It is beneficial for all different types of soil.
- You can program your sprinkler system to run on a timer.
- You can add additives like pesticides and fertilizers to some systems
- It is affordable and easy to install.

Disadvantages

- The evaporative loss of water is higher.
- The water can drift if there is a lot of wind.
- A stable water supply is required

Case Study

Indian national committee on irrigation and drainage, New Delhi worked out water saving percentage and yield increase of different field crops of sprinkler irrigation as compared to surface irrigation (Table 3.4). Among different field crops, highest saving of irrigation water (69%) was for Gram and lowest (16%) in Lucerne under sprinkler system. Also, highest increase in yield (57%) was in Gram and lowest in Cabbage (3%) and Cowpea (3%) with sprinkler irrigation system as compared to surface irrigation.

Table 3.4. Response of different crops to sprinkler irrigation

Crops	Water saving, %	Yield Increase, %
Bajra	56	19
Barley	56	16
Bhindi	28	23
Cabbage	40	3
Cauliflower	35	12
Chillies	33	24
Cotton	36	5
Cowpea	19	3
Fenugreek	29	35
Garlic	28	6
Gram	69	57
Groundnut	20	40
Jowar	55	34
Lucerne	16	27
Maize	14	36
Onion	33	23
Potato	46	4
Sunflower	33	20
Wheat	35	24

3.4. Remote Sensing in Agriculture

According to Ohio State University, "Remote sensed imagery can be used for mapping soil properties, classification of crop species, detection of crop water stress, monitoring of weeds and crop diseases, and mapping of crop yield. Use of remote sensing is influenced by the type of platforms (satellite, air or ground) used for data collection; number and width of spectral bands captured by the sensor (multi versus hyperspectral); and spatial (high, medium and low), temporal (hourly, daily and weekly) and radiometric (8-, 12- and 16-bit) resolutions at which sensors collect data."

Remote sensing fundamentally involves the checking of land from advanced equipment from far distances. The checking includes the physical characteristics of a piece of land. The process works by estimating the produced and reflected radiation from the land, which helps structure a picture of the land and assists specialists with passing judgment on specific parts of the land. It can be used for:

- **Crop production forecasting:** Remote sensing is used to forecast the expected crop production and yield over a given area and determine how much of the crop will be harvested under specific conditions. Researchers can be able to predict the quantity of crop that will be produced in a given farmland over a given period of time.

- **Assessment of crop damage and crop progress:** In the event of crop damage or crop progress, remote sensing technology can be used to penetrate the farmland and determine exactly how much of a given crop has been damaged and the progress of the remaining crop in the farm.
- **Cropping systems analysis:** Remote sensing technology has also been instrumental in the analysis of various crop planting systems. This technology has mainly been in use in the horticulture industry where flower growth patterns can be analyzed and a prediction made out of the analysis.
- **Crop identification:** Remote sensing has also played an important role in crop identification especially in cases where the crop under observation is mysterious or shows some mysterious characteristics. The data from the crop is collected and taken to the labs where various aspects of the crop including the crop culture are studied.
- **Crop acreage estimation:** Remote sensing has also played a very important role in the estimation of the farmland on which a crop has been planted. This is usually a cumbersome procedure if it is carried out manually because of the vast sizes of the lands being estimated.
- **Crop condition assessment and stress detection:** Remote sensing technology plays an important role in the assessment of the health condition of each crop and the extent to which the crop has withstood stress. This data is then used to determine the quality of the crop.
- **Identification of planting and harvesting dates:** Because of the predictive nature of the remote sensing technology, farmers can now use remote sensing to observe a variety of factors including the weather patterns and the soil types to predict the planting and harvesting seasons of each crop.
- **Crop yield modelling and estimation:** Remote sensing also allows farmers and experts to predict the expected crop yield from a given farmland by estimating the quality of the crop and the extent of the farmland. This is then used to determine the overall expected yield of the crop.
- **Identification of pests and disease infestation:** Remote sensing technology also plays a significant role in the identification of pests in farmland and gives data on the right pests control mechanism to be used to get rid of the pests and diseases on the farm.
- **Soil moisture estimation :** Soil moisture can be difficult to measure without the help of remote sensing technology. Remote sensing gives the soil moisture data and helps in determining the quantity of moisture in the soil and hence the type of crop that can be grown in the soil.

- **Irrigation monitoring and management:** Remote sensing gives information on the moisture quantity of soils. This information is used to determine whether a particular soil is moisture deficient or not and helps in planning the irrigation needs of the soil.
- **Soil mapping:** Soil mapping is one of the most common yet most important uses of remote sensing. Through soil mapping, farmers are able to tell what soils are ideal for which crops and what soil require irrigation and which ones do not. This information helps in precision agriculture.
- **Monitoring of droughts:** Remote sensing technology is used to monitor the weather patterns including the drought patterns over a given area. The information can be used to predict the rainfall patterns of an area and also tell the time difference between the current rainfall and the next rainfall which helps to keep track of the drought.
- **Land cover and land degradation mapping:** Remote sensing has been used by experts to map out the land cover of a given area. Experts can now tell what areas of the land have been degraded and which areas are still intact. This also helps them in implementing measures to curb land degradation.
- **Identification of problematic soils:** Remote sensing has also played a very important role in the identification of problematic soils that have a problem in sustaining optimum crop yield throughout a planting season.
- **Crop nutrient deficiency detection:** Remote sensing technology has also helped farmers and other agricultural experts to determine the extent of crop nutrients deficiency and come up with remedies that would increase the nutrients level in crops hence increasing the overall crop yield.
- **Reflectance modeling:** Remote sensing technology is just about the only technology that can provide data on crop reflectance. Crop reflectance will depend on the amount of moisture in the soil and the nutrients in the crop which may also have a significant impact on the overall crop yield.
- **Determination of water content of field crops:** Apart from determining the soil moisture content, remote sensing also plays an important role in the estimation of the water content in the field crops.
- **Crop yield forecasting:** Remote sensing technology can give accurate estimates of the expected crop yield in a planting season using various crop information such as the crop quality, the moisture level in the soil and in the crop and the crop cover of the land. When all of this data is combined it gives almost accurate estimates of the crop yield.

- **Flood mapping and monitoring:** Using remote sensing technology, farmers and agricultural experts can be able to map out the areas that are likely to be hit by floods and the areas that lack proper drainage. This data can then be used to avert any flood disaster in future.
- **Collection of past and current weather data:** Remote sensing technology is ideal for collection and storing of past and current weather data which can be used for future decision making and prediction.
- **Crop intensification:** Remote sensing can be used for crop intensification that includes collection of important crop data such as the cropping pattern, crop rotation needs and crop diversity over a given soil.
- **Water resources mapping:** Remote sensing is instrumental in the mapping of water resources that can be used for agriculture over a given farmland. Through remote sensing, farmers can tell what water resources are available for use over a given land and whether the resources are adequate.
- **Precision farming:** Remote sensing has played a very vital role in precision agriculture. Precision agriculture has resulted in the cultivation of healthy crops that guarantees farmers optimum harvests over a given period of time.
- **Climate change monitoring**: Remote sensing technology is important in monitoring of climate change and keeping track of the climatic conditions which play an important role in the determination of what crops can be grown where.
- **Compliance monitoring:** For the agricultural experts and other farmers, remote sensing is important in keeping track of the farming practices by all farmers and ensuring compliance by all farmers. This helps in ensuring that all farmers follow the correct procedures when planting and when harvesting crops.
- **Soil management practices:** Remote sensing technology is important in the determination of soil management practices based on the data collected from the farms.
- **Air moisture estimation :** Remote sensing technology is used in the estimation of air moisture which determines the humidity of the area. The level of humidity determines the type of crops to be grown within the area.
- **Crop health analysis:** Remote sensing technology plays an important role in the analysis of crop health which determines the overall crop yield.
- **Land mapping:** Remote sensing helps in mapping land for use for various purposes such as crop growing and landscaping. The mapping technology used helps in precision agriculture where specific land soils are used for specific purposes.

3.5. Nano Technology in Agriculture

Nanotechnology has gained intense attention in recent years due to its wide applications in several areas like medicine, medical drugs, catalysis, energy and materials. Those nanoparticles with small size to large surface area (1–100 nm) have potential medical, industrial and agricultural applications. Scientists have carried out significant efforts toward the synthesis of nanoparticles by different means, including physical, chemical and biological methods. (Ghidan and Antary, 2019)

These methods have many disadvantages due to the difficulty of scale-up of the process, separation and purification of nanoparticles from the micro-emulsion (oil, surfactant, co-surfactant and aqueous phase) and consuming large amount of surfactants. Green methods for synthesizing nanoparticles with plant extracts are advantageous as it is simple, convenient, eco-friendly and require less reaction time. Nanomaterials prepared by eco-friendly and green methods could increase agriculture potential for improving the fertilization process, plant growth regulators and pesticides. In addition, they minimize the amount of harmful chemicals that pollutes the environment. Hence, this technology helps in reducing the environmental pollutants, and nanotechnology has recently gained attention due to its wide applications in different fields such as in medicine, environment and agriculture. Particularly, the large surface area offered by the tiny nanoparticles, which have high surface area, makes them attractive to address challenges not met by physical, chemical pesticides and biological control methods.

Nanotechnology provides new agrochemical agents and new delivery mechanisms to improve crop productivity, and it promises to reduce pesticide applications. Nanotechnology can increase agricultural production, and its applications include:

- Nano formulations of agrochemicals for applying pesticides and fertilizers for crop improvement.
- The application of Nano sensors in crop protection for the identification of diseases and residues of agrochemicals.
- Nanodevices for the genetic engineering of plants.
- Plant disease diagnostics.
- Animal health, animal breeding, poultry production.
- Postharvest management.

Precision farming techniques might be used to further improve the crop yields but not damage soil and water. In addition, it can reduce nitrogen loss due to leaching and emissions, and soil microorganisms. Nanotechnology applications include nanoparticle-mediated gene or DNA transfer in plants for the development of insect-resistant varieties, food processing and storage and increased product shelf life. Nanotechnology may increase the development of biomass-to-fuel production. Experts feel that the potential benefits of nanotechnology for agriculture, food, fisheries and aquaculture need to be balanced against concerns for the soil, water and environment and the occupational health of workers. Nanotechnology uses are currently being researched, tested and in some cases already applied in food technology. Nanomaterials are considered with specific chemical, physical and mechanical properties.

In recent years, agricultural waste products have attracted attention as source of renewable raw materials to be processed in substitution of several different applications as well as a raw material for nonmaterial production. Insecticide resistance is one of the best examples of evolution occurring on an ecological time scale. The study of insecticide resistance is needed, both because it leads to understanding mechanisms operating in real time and because of its economic importance. It has become in insects an increasing problem for agriculture and public health. Agricultural practices could include wide range of selective regimes. Nanotechnology applications are being tested in food technology and agriculture. The applications of nanomaterials in agriculture aim to reduce spraying of plant protection products and to increase plant yields. Nanotechnology means like Nano capsules, and nanoparticles are examples of uses for the detection and treatment of diseases. Nanotechnology derived devices are also explored in the field of plant breeding and genetic transformation. The potential of nanotechnology in agriculture is large, but a few issues are still to be addressed as the risk assessment. In this respect, some nanoparticle attractants are derived from biopolymers such as proteins and carbohydrates with low effect on human health and the environment. Nanotechnology has many uses in all stages of production, processing, storing, packaging and transport of agricultural products. Nanotechnology will revolutionize agriculture and food industry such as in case of farming techniques, enhancing the ability of plants to absorb nutrients, disease detection and control pests.

Nanotechnology in pesticides and fertilizers

These days, sustainable agriculture is needed. It may be understood to present a good approach of ecosystem for long run. Practices that can cause long-term damage to soil include excessive tilling of the soil which leads to erosion and irrigation without needed drainage. This will lead to salinization. This is to satisfy human being food, animal feed and fiber needs.

Long-term experiments are required to show the effect of different practices on soil properties which are essential to sustainability and to provide important data on this objective. In the United States, a federal agency, the development of Nano-chemicals has appeared as promising agents for the plant growth and pest control.

The fertilizers are required in plants growth. Nanomaterials act as fertilizers might have the properties such as crop improvement and with less eco-toxicity. Plants can give an important way for their bioaccumulation into the food chain. The recent developments in agriculture cover the applications of Nano Pesticides (NPs) for more effective and safe use of chemicals for plants. The effects of different NPs on plant growth and phytotoxicity were reported by several workers including magnetite (Fe□O□) nanoparticles and plant growth, alumina, zinc, and zinc oxide on seed germination and root growth of five higher plant species; radish, rape, lettuce, corn, and cucumber, silver nanoparticles and seedling growth in wheat, sulfur nanoparticles on tomato, zinc oxide in mungbean, nanoparticles of AlO, CuO, FeO, MnO, NiO, and ZnO. Silver nanoparticles can stimulate wheat growth and yield. Soil applied 25 ppm SNPs had highly favorable growth promoting effects on wheat growth and yield.

Zinc has been considered as an essential micronutrient for metabolic activities in plants although it is required in trace amounts in plants. It was found that zinc has an important role in management of reactive oxygen species and protection of plant cells against oxidative stresses. Zinc has important functions in the synthesis of auxin or indoleacetic acid (IAA) from tryptophan as well as in biochemical reactions required for formation of chlorophyll and carbohydrates. The crop yield and quality of produce can be affected by deficiency of Zn. The development of insecticide resistance in pest insects has been an increasing problem for agriculture and public health.

Magnesium oxide (MgO) is important inorganic materials with many uses such as adsorbents, fire retardants, advanced ceramics, toxic waste remediation, and photo electronic materials. Therefore, various techniques and routes for synthesis of MgOHNPs have been reported. MgOH was synthesized by green methods using nontoxic neem leaves extract, Citrus lemon leaves extract, acacia gum.

Control of plant pests

Fusarium wilt is a destructive disease of tomato and lettuce in several countries due to its severe production loss, prolonged survival of fungus in soil and generation of resistant races. The disease can be reduced to some extent with the use of resistant cultivars and chemicals. However, the occurrence and development of new pathogenic races is a continuing problem, and the use of chemicals is expensive and not always effective. In recent years, the use of nanomaterials has been considered as an alternative solution to control plant pathogens. Ghidan et al. (2018) has synthesized nanoparticles of magnesium oxide (MgO) and tested the effect of different concentrations on the green peach aphid (GPA) under the greenhouse conditions. The synthesis of nanomaterials of copper oxide (CuO), zinc oxide (ZnO), magnesium hydroxide (MgOH) and magnesium oxide (MgO) has been carried out successfully by using aqueous extracts of *Punica granatum* peels, *Olea europaea* leaves and *Chamaemelum nobile* flowers. The screening of synthesized bio-nanoparticles revealed that these nanoparticles were effective in increasing the mortality percent of green peach aphid.

After the glasshouse experiments, the metal oxide nanoparticle accumulations were analyzed in the fruits and leaves of green sweet pepper. The results showed that there was no any metal accumulation in any of the plant fruits. Foliar spray by synthesized of MgOH nanoparticles for green pepper leaves revealed that the foliar spraying leaves with 100–800 ppm metal nanoparticles are very beneficial to plant growth and produced healthy plants with greener leaves and high fruit quality compared to the control. Researchers made significant efforts toward the synthesis of nanoparticles by various means, including physical, chemical and biological methods. Green methods for synthesizing nanoparticles with plant extracts are advantageous as it is simple, convenient, environment friendly and require less reaction time. Nanomaterials prepared by eco-friendly and green methods may increase agriculture potential for improving the fertilization process, plant growth and pesticides. In addition, this technology minimizes the amount of harmful chemicals that pollutes the environment. The green peach aphid is considered as a key pest on peach and globally important pest of a broad range of arable and horticultural crops, including Jordan. The pest is categorized as of the most important agricultural pest in the world. This devastated pest combats organophosphorus and carbamate insecticides by overproducing insecticide-degrading carboxyl esterases. Moreover, control of such a pest is becoming increasingly difficult, because the overproduction of resistance for aphid individuals when using chemical insecticides such as carbamates, organophosphates and pyrethroids.

Nanomaterials such as copper oxide (CuONPs), zinc oxide (ZnONPs), magnesium hydroxide (MgOHNPs) and magnesium oxide (MgONPs) were synthesized by different physical and chemical methods. With the growing needs to minimize the use of environmental-risk substances, such as insecticides, the biosynthesis of nanoparticles as an emerging highlight of the intersection of nanotechnology and biotechnology has received increasing attention. The rate of reduction of metal ions using plants has been found to be much faster as compared to microorganisms and stable formation of nanoparticles has been reported.

Nano-insecticidal potential

Copper oxide nanoparticles (CuONPs) are synthesized through different methods such as precipitation and chemical reduction. Many plant aqueous extracts have been reported such as Citrus lemon juice and carob leaves. Applications led many researchers to develop different ways to synthesis ZnONPs such as chemical route, precipitation method, hydrolyzed in polar organic solvents and microwave synthesis. Different plant extracts have been reported in the open literature for green synthesis of ZnONPs such as *Olea europaea, Solanum nigrum* leaves and *Azadirachta indica.* Different methods for synthesis MgOHNPs and MgONPs have been reported such as hydrothermal route, water-in-oil microemulsion and microwave reaction. MgOH was synthesized by green methods using nontoxic and eco-friendly such as Neem leaves extract, *Citrus lemon* leaves extract, acacia gum, *Brassica oleracea* and *Punica granatum* peels. In agriculture sector, there are several uses available like nanotech based pesticides and fertilizers with effective impact on plant growth and molecular farming with the help of nano vectors which is hoping to take the place of viral vectors.

Antimicrobial activity

Several nanomaterials are used as antimicrobial agents in food packing in which silver nanoparticles are in great interest. This is because of its extended use. Some other nanoparticles currently used are titanium dioxide (TiO_2), zinc oxide (ZnO), silicon oxide (SiO_2), magnesium oxide (MgO), gold and silver. All of them have specific characteristics and functions, for example, zinc nanocrystal shows antimicrobial and antifungal activity. Silver was a disinfectant and sterilizing agent used by NASA and Russian Space station for water, silver zeolite and silver. Gold has high temperature stability and low volatility and good antifungal and antimicrobial effects against 150 different bacteria. FDA in 2009 approves the direct use of silver as disinfectant in commercial water, since with effective result against microorganisms. The antimicrobial effect of these are *E. coli, L. monocytogenes* and *Staphylococcus aureus*, and nano silver particles coated with cellulose acetate phthalate also provided similar results. Some nanoparticles have shown their antifungal activity. These fungi include *Candida albicans, Aspergillus niger* and yeast. AgNPs are also found to be effective against methicillin resistant *Staphylococcus aureus.* Other nanoparticles besides silver are also found to have antimicrobial characteristics like titanium oxide (TiO_2). Its antimicrobial activity in UV light was obvious. Zinc oxide is reported to have antimicrobial activity in packaging material.

Zinc oxide nanoparticles synthesized using *Punica granatum* peel aqueous extract has shown effectiveness as antibacterial agents against standard strains of Gram-positive *Staphylococcus aureus* and Gram-negative *Escherichia coli.*

Nanotechnology application as nano fungicides

The use of nano silver has been studied recently against phytopathogen *Colletotrichum gloeosporioides*. Other nanoparticles (Fe, Cu, Si, Al, Zn, ZnO, TiO_2, CeO_2, Al_2O_3 and carbon nanotubes) have been reported to have some adverse effects on plant growth apart from the antimicrobial properties. Sometimes, nanoparticles also have an effect on the growth of useful soil bacteria, such as *Pseudomonas putida* KT2440. Various research groups focused their interest on the usage of eco-friendly pesticides. Similar to chemical pesticides, nanoparticle-based pesticides and herbicides are being explored for the application of the antimicrobial agents to protect crops from various diseases. Extensive studies on nanoparticle-based systems may eliminate the intensive use of pesticides in the agricultural sector. The antifungal properties of nanoparticles can help to formulate nanoparticle-based pesticides. Among the different inorganic nanoparticle-based antimicrobial agents, silver has been extensively studied by many researchers because of its several advantages over other nanoparticles such as copper, zinc, gold, ZnO, Al_2O_3 and TiO_2.

Effect of nickel nanoparticles on fungal mycelial growth

It was observed that both the concentrations of nickel nanoparticles (50 and 100 ppm) inhibited the fungal mycelial growth on solid media, and the inhibitions were significant ($p \leq 0.05$) over control. Nickel nanoparticles at 100 ppm concentration inhibited the mycelial growth of *F. oxysporum f. sp. lactucae and F. oxysporum f. sp. lycopersici* by 60.23 and 59.77%, respectively, over control.

The inhibitory effects of nickel nanoparticles were also assessed in liquid medium, and the results were similar with solid media. In the liquid media, the fresh mycelial weight of the tested fungal pathogens decreased significantly and more than 50% reduction was recorded with the use of nickel nanoparticles at the concentration of 100 ppm. The results revealed that mycelial growth of tested pathogens was inhibited in a concentration dependent manner. These results suggest that using Ni nanoparticle solution can significantly increase the surface areas acting on the mycelia of Fusarium and mycelial growth.

Nickel nanoparticles at the concentration of 100 ppm decreased the number of spore development by 81.40 and 74.60% in *F. oxysporum f.* sp. lactucae *and F. oxysporum f.* sp. lycopersici, respectively. The conidial germination was negatively affected by nickel nanoparticles.

The inhibitory effect of Ni nanoparticles on spore germination could be due to their fungicidal effect. These results agreed with the results obtained from other workers on antifungal effects of different metal nanoparticles against some pathogenic fungi such as silver nanoparticles and zinc nanoparticles against copper nanoparticles. Inhibitory effect of Ni nanoparticles could be due to producing of extracellular enzymes from fungi as survival agents caused by stress of toxic materials or could be due to large surface areas and small sizes to penetrate into the cell membrane of pathogen and work in the cytosols.

Nanotechnology for controlling plant virus

Plant virus particularly spherical virus is considered to be the naturally occurring nanomaterials. The smallest plant viruses known till date are satellite tobacco necrosis virus measuring only 18 nm in diameter. Plant viruses are made up of single or double stranded RNA/DNA as genome which is encapsulated by a protein coat. Their ability to infect, deliver nucleic acid genome to a specific site in host cell, replicate, package nucleic acid and come out of host cell precisely in an orderly manner have necessitated them to be used in nanotechnology.

3.6. Leaf Color Chart For N Applications

Nitrogen (N) fertilizer is important in rice production. Apply N fertilizer several times during the growing season to ensure that the crop's nitrogen need is supplied, particularly at critical growth stages.

The Leaf Color Chart (LCC) is used to determine the N fertilizer needs of rice crops. LCC has four green strips, with color ranging from yellow green to dark green. It determines the greenness of the rice leaf, which indicates its N content.

Using the Leaf Color Chart (LCC) for N Management in Rice

The leaf color chart (LCC) is an easy-to-use and inexpensive diagnostic tool for monitoring the relative greenness of a rice leaf as an indicator of the plant N status. Leaf N status of rice is closely related to photosynthetic rate and biomass production, and it is a sensitive indicator of changes in crop N demand within a growing season. A tool to rapidly assess leaf N status and thereby guide the application of fertilizer N to maintain an optimal leaf N content can consequently be vital for achieving high rice yield with effective N management. The leaf color chart (LCC) is a plastic, ruler-shaped strip containing four panels that range in color from yellowish green to dark green. The LCC is used to monitor leaf N status from tillering to panicle initiation or later, by either of two equally effective options. The decision on which option to use can be based on farmers' preferences and location specific factors, such as frequency of visits by farmers to their fields and their knowledge of critical growth stages for N application. The fixed-time/adjustable-dose option saves time, and is thus preferred by farmers who have gainful alternative activities. The real-time option is generally preferred when farmers lack sufficient understanding of the critical stages for optimal timing of fertilizer N.

Fixed time/adjustable-dose N management option

With this option, farmers measure leaf color before applying N at active tillering and panicle initiation. If mean leaf color is intermediate between 3 and 4, apply a standard rate of fertilizer N. If the mean leaf color is higher (for example $\geq$4), apply less fertilizer N than the recommended rate. If the mean leaf color is lower (for example $\leq$3), apply more fertilizer N than the recommended rate. Such adjustments in N doses at active tillering and panicle initiation ensure application of more N in fields and years with high plant demand for N, and application of less N in fields and years with low demand for N.

Real-time N management option

With this more complicated option, farmers monitor the rice leaf color at 7- to 10-day intervals from tillering to about 5−10 days after panicle initiation. Farmers apply fertilizer N whenever the leaves become more yellowish-green than a critical threshold value indicated on the LCC. Leaf color is typically monitored a total of four to six times per cycle. The effective use of real-time N management requires the selection of an N dose and a critical threshold LCC color that ensure 2−3 N applications in an average yielding field or year. In fields and years with above average growth and crop N demand, rice leaves will turn yellow more rapidly, resulting in more N applications and hence more fertilizer N use. In fields and years with below average growth and crop N demand, the rice will require less N and leaves will remain greener longer, resulting in fewer N applications and less fertilizer N use. The critical threshold value can be adjusted for cultivars and crop establishment method. Thresholds for cultivars with inherently yellowish leaves should be more yellowish green than for cultivars with inherently dark green leaves.

How to use the LCC

- Randomly select at least 10 disease-free rice plants or hills in a field with uniform plant population.
- Select the topmost fully expanded leaf from each hill or plant. Place the middle part of the leaf on a chart and compare the leaf color with the color panels of the LCC. Do not detach or destroy the leaf.
- Measure the leaf color under the shade of your body, because direct sunlight affects leaf color readings. If possible, the same person should take LCC readings at the same time of the day every time.
- Determine the average LCC reading for the selected leaves.

Plants without N application are yellowish. Nitrogen deficiency is confirmed when the LCC reading is between panels 2 and 3. At lower fertilizer N rates in photos the plants look better, but the low LCC reading still indicates N deficiency. At higher fertilizer N rates, the plants look well developed and the canopy is closed. The LCC reading is between panels 3 and 4, which is the critical range for most transplanted rice. In plants with a high N rate, leaves are dark green. Leaf color is darker than the LCC panel no. 4 indicating a surplus of fertilizer N.

Case Study

Ali et.al. (2009) conducted field experiment during *kharif* at the Crop Research Farm, Department of Agronomy, Naini Agricultural Institute, Sam Higginbottom University of Agriculture, Technology & Sciences, Allahabad, Uttar Pradesh,

India to study the effect of different spilt doses of nitrogen (N), Leaf Colour Chart (LCC) based nitrogen management on growth and yield of transplanted rice (Table 3.5). Nitrogen management practices were consisted of application of 150 kg N ha^{-1} in different split doses (Two equal splits – 1/2 at basal, 1/2 at active tillering, Three splits – 1/2 at basal, 1/4 at active tillering, 1/4 at panicle initiation, Three splits 1/3 at basal, 1/3 at active tillering, 1/3 at panicle initiation, Four splits – 1/4 at basal, 1/4 at active tillering, 1/4 at maximum tillering, 1/4 at panicle initiation, and Four splits – 1/3 at basal, 1/3 at active tillering, 1/6 at maximum tillering, 1/6 at panicle initiation), LCC 3 and LCC 4 with 20 kg N ha^{-1} as basal with 23 and 28 kg N ha^{-1} respectively based on weekly reading. The results revealed that nitrogen management at LCC value of 4 (4 splits; 132 kg N ha^{-1}) produced significantly higher plant height (113.4 cm), crop and relative growth rates 21.84 g m^{-2} day^{-1} and 0.072 g g^{-1} day^{-1} respectively, effective tillers $hill^{-1}$ (14.60), grains $panicle^{-1}$ (162.13) and grain yield (60.70 q ha^{-1}) than LCC value of 3 (4 splits; 112 kg N ha^{-1}) that produced grain yield of (42.30 q ha^{-1}) as well as other nitrogen management treatments (N at 150 kg ha^{-1}). Thus, considering the influence of LCC based nitrogen application on productivity and profitability of rice and saving of fertilizer N. It is concluded that the LCC offers gigantic opportunities to increase N use efficiency, rice yield and net return for farmers in Allahabad region of Eastern Uttar Pradesh.

Table 3.5: Effect of Nitrogen management treatments on grain and stover yield of transplanted rice

Treatment	Number of splits	Total N applied (kg ha^{-1})	Grain yield (q ha^{-1})	Straw yield (q ha^{-1})	Harvest Index (%)	B:C Ratio
Nl- 1/2+1/2	2	75+ 75 = 150	55.00	77.00	39.60	1.80
N2- 1/2+1/4+1/4	3	75 + 37.5 + 37.5 = 150	50.60	77,00	39.70	1.81
N3- 1/3+1/3+1/3	3	50+ 50 + 50 = 150	48.20	73.90	39.50	1.67
N4- 1/4+1/4+1/4+1/4	4	37.5+37.5+37.5+ 375 = 150	51.00	78,00	39.60	1.78
N5- 1/3+1/3+1/6+1/6	4	50 + 50+ 25+25 = 150	43.50	67.40	39.20	1.52
N6- LOC3-20 kg basal + 23 kg N per ha based on weekly reading	4	23+23+23+23=112	42.30	65.90	39.10	1.50
N7- LCC4-20 kg basal + 28 kg N per ha based on weekly reading	4	28+28+28+28=132	60.70	86.20	41.30	1.99
LSD(P=.05)	-	-	8.10	NS	-	-

3.7. Laser Land Levelling

A laser land leveller- a machine equipped with a laser guided drag bucket- is much more effective and quicker in ensuring a flat table-top like surface. An even land means irrigation water reaches every part of the field with minimal waste from run-off or water-logging. (Jeetendra and Jat, 2015, CIMMYT)

While mechanization is good news to farmers, climate change and climate variability pose unprecedented challenges to agriculture. We need climate-smart agriculture practices and technologies that save on scarce resources like water and energy but increase yields, incomes and reduce environmental footprints. When practiced together, a portfolio of climate-smart practices can equip farmers to adapt to changing weather patterns amidst depleting natural resources. For instance, groundwater in north-western India has been declining at alarming rates because of the overuse of electric pumps, largely by subsidized electricity, and inadequate recharge from erratic rainfall. Recent studies predict that there would be at least a 10 percent increase in irrigation water demand with a 1 °C rise in temperature in arid and semi-arid regions of Asia. Irrigation is the biggest user of groundwater in north-western India and unless steps are taken to revert this trend, farmers are looking at a water-scarce future.

A study, Impacts of Laser Land Leveling in Rice-Wheat Systems of the North–Western Indo-Gangetic Plains of India, by researchers from the International Maize and Wheat Improvement Centre (CIMMYT), Borlaug Institute for South Asia (BISA) and the CGIAR Research Program on Climate Change, Agriculture and Food Security (CCAFS) looks into the impact of laser land levelling in rice-wheat systems in north India. The study aims to assess if and how laser land levelling helps farm communities by improving farm productivity, saving on water and energy, reducing greenhouse gas emissions and increasing incomes.

Using household surveys and comparing data from traditional levelling, the study comes up with key research findings that show the benefits behind laser land levelling but also the policy implications of promoting new climate-smart technologies among farmers.

Key Findings

- Laser land levelling considerably lowers irrigation time for rice by 47-69 hours per hectare per season and for wheat by 10-12 hours per hectare per season.
- It increases yields by an average 8 percent for both crops.
- It saves electricity about 755 kWh per hectare per year for rice-wheat systems.

- It's cost effective. As demand increases, service providers rent out equipment and farmers collectively share the costs.
- A study showed that it reduces greenhouse gas emissions from saving on energy, reducing cultivation time and increasing input efficiency.
- Higher yields mean higher incomes for farmers.

Benefits of Laser Levelling

Irrigation time: Laser-levelling in rice fields reduced irrigation time by 47-69 hours per hectare per season and improved yield by approximately 7 percent compared with traditionally levelled fields. In wheat, irrigation time was reduced by 10-12 hours per hectare every season and yield increased by 7-9 percent in laser levelled fields.

Food security: Based on analysis from the data, the study computes that if 50 percent of the area under ricc-wheat systems in Haryana and Punjab States were laser levelled, it would result in an additional production of 699 million kg of rice and 987 million kg of wheat which amounts to USD 385 million a year extra. Not only does this translate to higher incomes for farmers but it also increases food security of the region given that the Indo-Gangetic Plains are a bread basket and rice bowl in South Asia.

Energy: Less time spent on irrigation means less energy spent on irrigation. The study shows that laser land levelling saves electricity amounting to about 755 kWh per hectare per year for rice-wheat systems.

Costs: Another perception that the study challenges are that only large-holder or rich farmers can afford and benefit from laser-levellers. In fact, laser land levelling is scale neutral and not particularly biased towards large-holdings. It is common for many small-holder farmers to rent the equipment or form a cooperative to share the costs. The rental cost is about USD10 for an hour. It takes about 4 to 5 hours to laser level a hectare of land and the farmers need to do it once in 3 years.

Water: Laser land levelling is essentially a water-saving technology as it uses scarce groundwater optimally by ensuring even coverage. Compared to traditionally levelled land, a laser levelled farm minimizes run-off and water-logging ensuring that farmers use just as much water they need in the best possible way.

Less greenhouse gas emissions: A related study (Jat et al 2015) reported that use of laser land leveller over traditional land levellers reduces emission of greenhouse gases through decreased water pumping time, decreased cultivation time and better use of fertilizers.

Income: The increased yields and the money saved on water and energy means farmers benefited by an additional USD 143.5 per hectare a year from growing rice and wheat.

Policy Implications

Laser land levelling is just one among numerous farming activities that contribute to sustainable agriculture. When used in combination with other resource-saving practices and technologies like solar energy-based irrigation, precision nutrient application, direct-seeded rice, zero tillage and proper residue management, the gains can be multiplied for each farmer and the community as a whole.

Governments need to assess how they can support farming communities take up new technologies that deliver multiple benefits. The aim should also be to ensure that technology uptake is inclusive and does not discriminate against marginalized farmers.

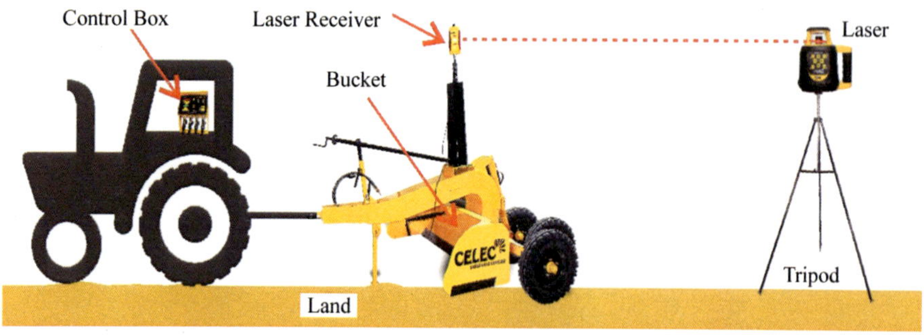

How does a laser level work?

A laser transmitter transmits a laser beam, which is intercepted by the laser receiver mounted on the leveling bucket. The control panel mounted on the tractor interprets the signal from the receiver and opens or closes the hydraulic control valve, which will raise or lower the bucket. Loose soils are picked up by the bucket and released lower in the field. (Rickman, 2002 and Jat 2006)

4-wheel tractor

A 4wheel tractor is required to drag the leveling bucket. The size of the tractor can -vary from 30-500 hp depending on the time restraints and field sizes, though a 4-wheel drive with power shift transmission is preferable.

Drag bucket

The leveling bucket can be either 3- point linkage mounted or pulled by the tractor's drawbar. Pull type systems are preferred as it is easier to connect the tractor's hydraulic system to an external hydraulic ram.

Laser transmitter

The laser transmitter mounts on a tripod which allows the laser beam to sweep above the tractor unobstructed. With the plane of light above the field, several tractors can work from one transmitter.

Laser receiver

The laser receiver detects the position of the laser reference plane and transmits these signals to the control box. The receiver mounts on a mast attached to the drag bucket.

Control panel

The control box accepts and processes signals from the machine-mounted receiver. It then indicates the drag bucket's position relative to the finished grade.

Hydraulic control system

The hydraulic system supplies oil to raise and lower the leveling bucket. A pressure relief valve is needed in the system to return the excess oil to the tractor reservoir. If this relief valve is not large enough or it malfunctions, damage can be caused to the tractor's hydraulic pump.

Procedure for laser leveling

Laser leveling requires soil to be shifted from the high points of the field to the low points in the most cost-effective way. In most of the situation's fields will need to be plowed and a topographic survey undertaken before leveling commences.

Step 1: Plowing the field

The field should be preferably plowed from the center of the field outwards. Plow when the soil is moist, because if it is plowed dry a significant increase in tractor power is required and large clod sizes may result. Cut up or remove surface residues to aid soil flow from the bucket.

Step 2: Conducting a topographic survey

Once the field is plowed, a topographic survey should be conducted to record the high and low spots in the field. The mean height of the field can be calculated by taking the sum of all the readings and dividing by the number of readings taken. Then, using a field diagram and the mean height of the field, determine how to effectively move soil from the high to low areas.

Step 3: Leveling the field

Leveling a field involves the following steps:

- The laser-controlled bucket should be positioned at a point that represents the mean height of the field.
- The cutting blade should be set slightly above ground level (1-2 cm). The tractor should then be driven in a circular direction from the high areas to the lower areas in the field.
- To maximize working efficiency, as soon as the bucket is near filled with soil the operator should turn and drive towards thc lower area. Similarly, when the bucket is near empty the tractor should be turned and driven back to the higher areas.
- When the whole field has been covered, the tractor and bucket should then do a final leveling pass in long runs from the high end of the field to the lower end.
- Re-survey to make sure that the desired level of precision has been attained.
- The fields should not require further major leveling works for at least 8 years.

Different types of laser levellers

Manual leveling lasers

Set-up of a laser leveling instrument requires the operator to manually level the unit by using the units' screws and bubble vials. These lasers can achieve a maximum accuracy of 1 cm at 100m.

Semi self-leveling lasers

These lasers adjust themselves automatically within a range, and can achieve accuracy of at least 1 cm at 100 m. They are equipped with either a circular bubble with a bull's eye or electronic lights that turn green when you reach the self-leveling range.

Fully self-leveling lasers

These lasers automatically find and maintain level within a specified range. They are the easiest to use and can achieve accuracy of up to 2.5 mm at 100 m. Split-beam lasers - These lasers emit simultaneous horizontal and vertical beams to establish both level and plumb reference lines.

Case Study

Sidhu et.al. 2008 worked at Punjab Agricultural University Ludhiana and they reported that the grain yield of rice on leveled fields was nearly 11 % higher and there was 25.37 % saving in irrigation water than that on traditional leveled fields (Table 3.6) on the basis of mean of three locations.

Table 3.6: Yield and irrigation water saving for laser leveled and traditionally level plots for rice crop under replicated experiments

Sr. No.	Yield (t/ha)		% age increase over conventional	% Saving in irrigation under laser levelled field
	Levelled	Conventional		
Site 1	8.78+0.33	7.73+0.21	13.6	26.15
Site 2	8.30+0.46	7.53+0.39	10.30	...
Site 3	7.60+0.21	7.00+0.25	8.57	25.00
Mean			10.82	25.57

3.8. Value Addition in Agriculture

Value-added agricultural products are characterized by one or more of the following criteria: a change in the physical state or form of the product; the production of a product in a manner that enhances its value, as demonstrated through a business plan; the physical segregation of a commodity or product in a manner that results in the enhancement of the value of that commodity.

Value addition is a process in which for the same volume of a primary product, a high price is realized by means of processing, packing, upgrading the quality or other such methods.

Value-added agriculture refers most generally to manufacturing process that increases the value of primary agricultural commodities. Value-added agriculture may also refer to increasing the economic value of a commodity through particular production process, e.g., organic produce, or through regionally branded products that increase consumer appeal and willingness to pay a premium over similar but differentiated products. Value-added agriculture is regarded by some, a significant rural development strategy.

Small scale processing unit, organic food processing, non-traditional crop production, agritourism and bio-fuels development are examples of various value-added projects that have created new jobs in some rural areas.

Need for value addition

- To improve the profitability of farmers
- To empower the farmers and other weaker sections of society especially women through gainful employment opportunities and revitalize rural communities.
- To provide better quality, safe and branded foods to the consumers.
- To emphasize primary and secondary processing.
- To reduce post-harvest losses.
- Reduction of import and meeting export demands.
- Way of increased foreign exchange.
- Encourage growth of subsidiary industries.
- Reduce the economic risk of marketing.
- Increase opportunities for smaller farms and companies through the development of markets.
- Diversify the economic base of rural communities.
- Overall, increase farmers' financial stability.

Value Addition in Field Crops

Selling the agricultural produce as such in the market, may fetch low price, whereas selling their value-added products (through processing) may improve the net income of the farmers. Few examples of value-added product in agricultural crops are as follows

Value addition - Rice

As per NIIR Project Consultancy Services (2018), rice is the most important cereal food crop of India, occupying one-fourth of the gross cropped area of the country. As the basic food crop, rice is cultivated comfortably in hot and humid climate. In the country, the crop is mainly grown as a *Kharif* crop in rainfed areas that receive heavy annual rainfall.

India is the second largest rice producer in the world after China, with 21% of the global production share. The rice production has increased by 3.5 times in the last 60 years. The country's productivity is higher than Thailand and Pakistan at 2.2 tons/ha. Major Rice producing states in India are West Bengal, Uttar Pradesh, Andhra Pradesh, Punjab, Tamil Nadu, Odisha, and Bihar. India has been the top exporter in global rice trade, accounting for 25% of the export in the last four years. Indian rice caters to the Middle East and Africa for non-Basmati, and the EU and the US for Basmati variety.

Rice and rice-based products derived from rice grain and rice flour include parboiled rice; quick-cooking rice and ready-to-eat convenience foods; rice flours; rice starch; cakes and puddings; baked breads and crackers; breakfast cereals and expanded rice products; extrusion-cooked and puffed-rice snacks; noodles, paper, and pasta; baby/weaning foods; fermented foods and beverages; pet foods; and bran products.

Value addition also enhances the profitability of rice production. A wide range of product development like processed and canned, ready-to-eat products, vitamin, iron or calcium enriched flaked or puffed rice, flavored rice, starch extraction from broken rice and so on are nowadays getting popular. Value-added products from organic rice and therapeutic value medicinal rice varieties have good niche in domestic and export markets.

By-products from the rice milling process have high amounts of nutrients when compared to white rice itself. Rice straw, rice hull, broken rice, rice germ, rice bran, rice bran oil and wax are the by-products from the rice industry. These by-products usually have basic applications in their original form, but now can be used as raw materials for different value-added research or in food applications with functional properties. Rice by-products not only contain various types of functional components, but also contain dietary fiber. The fiber can be mostly found in rice hull and the types of fiber present include cellulose, hemicellulose, lignin and hydrated silica. Because of the high fiber content in rice hull and rice bran, they are used as ingredients by the bakery industries to increase the fiber content and improve the nutrition of bakery products.

Rice milling unit

Rice grain, rice husk and straw are used for many purposes as indicated below.

Rice mill (Parboiled rice)

Parboiled rice, popularly known as "ushna" rice in boiled form. India contributes about one-third of the world acreage under rice. Rice is available in over 5000 varieties, of which Izong rice of Assam occupies a important position on account of its superfine grains, pleasant, fine cooking quality, sweet taste, soft texture. Parboiling is preliminary to hulling in which the rough paddy is first soaked, then steamed and dried before removing the hulls.

Rice powder

Rice flour (also rice powder) is a form of flour made from finely milled rice. It is distinct from rice starch, which is usually produced by steeping rice in lye.

Rice bran oil

Rice bran is the main source of rice oil. The majority of available bran continues to be used for animal feeds without being extracted for the oil. Rice oil, also called rice bran oil, has been used extensively in Asian countries. Rice oil is a minor constituent of rough rice when compared with the carbohydrate and protein content. Two major classes of lipids are present: those internal within the endosperm and those associated with the bran. The internal lipids contribute to the nutritional, functional, and sensory qualities of rice.

Rice beer

Rice beer is an alcoholic drink generally made from rice. Those who consume moderate amounts of beer (one to two a day at the most) have a 30-40% lower rate of coronary heart disease compared to those who do not drink. Beer contains a similar amount of polyphenols (antioxidants) as red wine and 4-5 times as many polyphenols as white wine. Alcohol has also been attributed of its ability to increase the amount of good cholesterol (HDL) into the bloodstream as well as help to decrease blood clots.

Rice flakes

Rice flakes are tasty flakes that are created using rice grains. The process for creating rice flakes involves parboiling the rice, then flattening the grains to product a solid flake. There are a number of rice flake recipes used in Asian cuisine. Often in western countries, rice flakes are used to create cereals and different types of snacks. Once the rice is tender, the cooked grains are rolled, then flattened. The thickness of the flattened rice will depend on the amount of pressure that is applied.

Rice bran based solvent extraction

Rice has been and continues to be the largest source of human nutrition. Rice bran is a by-product of the rice milling process. Rice bran is the most important source of edible oil among the unconventional sources. Rice bran is the brown coating around the white starchy rice kernel, which is obtained by dehusking paddy and polishing the rice. While white rice holds little nutritional value, the bran that is removed contains 65 percent of the rice kernel's nutrients and boasts a bounty of healthful benefits.

Fructose syrup from broken rice (HFS 90%)

Glucose syrup is a popular substitute for sugar. It is more commonly referred to as corn syrup because its main ingredient is usually cornstarch. From a strictly chemical perspective, glucose syrup is the proper term for any liquid starch consisting of carbohydrates. In addition to cornstarch, it can be made from any form of starch, including wheat, rice or potatoes. Because of its successful and easy bonding with dry or solid substances, this syrup is the preferred sweetener for candy, chewing gum, jams, jellies and canned fruit.

Puffed rice (Muri)

Puffed rice is a commonly consumed commodity as a pastime snack. It can be used in combination with nuts such as groundnut or roasted and salted cashews; with fried gram; with Jaggery and coconut gratings, or dusted with salt and spices after enrobing with oil. Since the product is easily digested and assimilated, it finds a wide acceptance among a cross section of the households. It is a versatile product with an excellent market potential. Today the snack food is one of the most important areas of the food industry.

Liquid glucose from broken rice

Liquid glucose is usually manufactured by subjecting starch to high temperature in the presence of acid. However, liquid glucose of same dextrose equivalent can be manufactured by enzymatic conversion but the spectrum of saccharides will differ and so also the properties, in comparison to acid converted liquid glucose. Liquid glucose is an aqueous solution of nutritive saccharide obtained by starch hydrolysis, by using corn and rice as raw material, which is purified and concentrated to required solids.

Rice flakes from broken rice

Rice flakes are tasty flakes that are created using rice grains. Rice flakes are used as breakfast food and are famous by the names like poha/chiwda when cooked as breakfast meal. Rice flakes can be consumed with milk as cereal. Wide application of rice flakes is in the application of beer industry to produce beer. Rice flakes can also be mixed with other dried grain flakes to create a tasty breakfast cereal. Rice flakes from broken rice used in beer industries, which is cheap convenient and is an innovated concept.

Uses of rice husk (hull)

As an industrial fuel

Ethanol from rice husk

Preparation of activated carbon

Rice husk as a fertilizer and substrate

As pet food fiber

Substrate for silica and silicon compound

Used for making bricks

Rice husk as fireworks

Used as pillow stuffing

As an admixture in low cost concrete block manufacturing

Manufacturing refractory bricks

Control of insect pests in stored food stuffs

Activated carbon from rice husk

Cement from rice husk ash

Packing & forwarding

Electricity from rice husk

Hardboard from rice husk

Oxalic acid from rice husk

Paper from rice husk

Particle board from rice husk

Rice husk briquettes

Rice husk pellet (RHP)

Rice straw

Rice straw is used as feed for ruminants and for many other uses like manure, thatching, paper pulp, alcohol, mats, poultry litter and mushroom production. Rice straw in principle can be processed and used in agriculture for purposes such as soil improvement through carbonization and composting, in bio-energy production, and to produce materials for industrial uses such as silica and bio-fiber.

Rice husk contains about 30–50% of organic carbon and have high heat value of 13–16 MJ per kg. It can be used to generate fuel, heat, or electricity through thermal, chemical, or bioprocesses.

Value Addition - Maize

As per Saritha et.al. 2020 following byproducts can be made from maize.

Corn oil

Its main use is in cooking, where its high smoke point makes refined corn oil to a valuable frying oil. It is also a feedstock used for biodiesel. Other industrial uses for corn oil include soap, paint, rust proofing for metal surfaces, inks, textiles, nitroglycerin and insecticides.

Corn syrup

Corn syrup is a food syrup which is made from the starch of corn and composed mainly of glucose. Corn syrup is used in foods to soften texture, add volume, prevent crystallization of sugar, and enhance flavor.

Corn flakes

Corn flakes being one of most nutritious foods and is consumed as breakfast food not only in India but elsewhere in the world. Besides the good taste, crispy nature corn flakes are also popular because of their friable texture blend flavor and above all the ease with which it can be prepared for consumption. There is a good scope to develop this important agro-based food processing industry especially in the maize growing states to cater to the increasing demand of the metropolitan and industrialized cities.

Corn pops

Popcorn is a popular snack food at sporting events and in movie theatres. Traditions differ as to whether popcorn is consumed as a hearty snack food with salt (predominating in the United States) or as a sweet snack food with caramelized sugar (predominating in Germany). Popcorn is naturally high in dietary fiber and antioxidants low in calories, fat, free of sugar and sodium. This can make it an attractive snack to people with dietary restrictions on the intake of calories, fat or sodium. For the sake of flavor, however, large amounts of fat, sugar, and sodium are often added to prepared popcorn, which can quickly convert it to a very poor choice for those on restricted diets.

Roasted corn

Masala Butta or Spiced fire roasted corn on the cob is a favorite street snack in India when corn is in season. Street vendors roast the corn on coal fire or open grill, use a lime to rub in the Indian spices and serve it hot. People can be seen walking around with the corn in hand.

Corn starch

Corn starch, maize starch, or corn flour is the starch derived from corn (maize) grain. The starch is obtained from the endosperm of the kernel. Corn starch is a common food ingredient, often used to thicken sauces or soups, and to make corn syrup and other sugars. Corn starch is versatile, easily modified, and finds many uses in industry such as adhesives, in paper products, as an anti-sticking agent, and textile manufacturing. It has medical uses as well, such as to supply glucose for people with glycogen storage disease

Corn sauce

Corn sauce or fermented corn sauce is produced by fermentation using corn starch as the primary substrate. It is used as a food condiment and ingredient, both in paste and in powder form. Corn sauce, like soy sauce, has a characteristic savory taste. It is used to flavor dishes including soups, broths, and gravies.

Cornick

Cornick is a Filipino deep-fried crunchy puffed corn nut snack. It is most commonly garlic-flavored but can also come in a variety of other flavors. It is traditionally made with glutinous corn. Cornick is made by soaking corn kernels in water for three days, changing the water used for soaking daily. The corn used is traditionally glutinous corn, but other types of corn can also be used, including popcorn. After soaking, the kernels are drained and dried thoroughly. It is then deep-fried in oil at about 120 to 130 °C (248 to 266 °F), to ensure that the kernels do not pop. It is cooked for around two to three minutes then drained on paper towels

Cornmeal

Cornmeal is a meal (coarse flour) ground from dried corn. It is a common staple food, and is ground to coarse, medium, and fine consistencies, but not as fine as wheat flour can be. In Mexico, very finely ground cornmeal is referred to as corn flour. When fine cornmeal is made from maize that has been soaked in an alkaline solution, e.g., limewater (a process known as nixtamalization), it is called masa harina (or masa flour), which is used for making arepas, tamales and tortillas. Boiled cornmeal is called polenta in Italy and is also a traditional dish and bread substitute in Romania

Corn stover

Corn stover is made up of the stalk, leaves, husks and tassels left in the field after harvesting the grain with a combine. This stover can be used to make advanced biofuels or be used as a low quality, emergency livestock feed. Maize stover (or straw) normally consists of variable proportions of the leaves, stalks, and cobs of maize plants left in a field after harvest. It can be a valuable addition in ruminant feeding systems (dairy, beef, heifers, as well as sheep and goats) as it has the highest feeding value of all cereal straws.

Value Addition - Wheat

Wheat germ oil

Wheat germ oil is extracted from the germ of the wheat kernel, which makes up only 2- 3% by weight of the kernel. It is a refined vegetable oil that contains a high level of natural vitamin E which is a natural antioxidant, and high levels of unsaponifiable fraction. Its fatty acid composition supplies a high content of essential fatty acid that promotes the regeneration of cells. Wheat germ oil is particularly high in octacosanol - a long- chain saturated primary alcohol found in a number of different vegetable waxes.

Wheat starch and wheat gluten

Wheat starch is produced from wheat species such as *Triticum aestivum*, which is also known as bread wheat. Starch is one of the main carbohydrates in wheat which comprises of around 60-75% of grain and around 70-80% of flour.

Wheat gluten is a food made from gluten, the main protein of wheat. It is made by washing wheat flour dough with water until all the starch granules have been removed, leaving the sticky insoluble gluten as an elastic mass which is then cooked before being eaten.

Atta, maida, suji & wheat bran

Around 800 large flour mills in the country convert about 10.5 Million Tons of wheat into wheat products i.e., coarse flour, flour, semolina, bran & wheat germ. The installed capacity of flour mills is more than 21 million metric tons. Roller flour milling sector processes around 12 – 15 per cent of the total wheat consumed in the country. Most wheat is consumed in the form of baked goods, mainly bread; therefore, wheat grains must be milled to produce flour prior to consumption. Wheat is also used as an ingredient in compound feedstuffs, starch production and as a feed stock in ethanol production.

Biscuits making

Biscuit industry in India in the organized sector produces around 60% of the total production, the balance 40% being contributed by the unorganized bakeries. The industry consists of two large scale manufacturers, around 50 medium scale brands and small-scale units ranging up to 2500 units in the country, as at 2000-01. The unorganized sector is estimated to have approximately 30,000 small & tiny bakeries across the country. Biscuit can he broadly categorized into the following segments: glucose 44% marie 13% cream 10% crackers 13% milk 12% others 8%.

Atta

Atta is one of the staple and basic food to Indians not only that it is basic food ingredients for all the men lives in the world. Atta is manufactured from wheat which is cultivated in the winter season or just before it. Through there is development of tissue culture base seeds which can be cultivated throughout the year. Ata (wheat flour) is basically used for preparation of chappati for the human basic needs of food in every classes of people. It is also used for the preparation of adhesives.

Bread making

In modern days bread is now becoming one of the most essential food items in human diet due to its readymade availability and high nutritive value. It is the most consumable wheat-based bakery product. Wheat flour, yeast, sugar, salt, water and shortening agent are required as raw materials to manufacture bread. The plant and machinery and the technology required to manufacturing the bread, are completely available in India. Since the consumption of bread is increasing rapidly day by day, the demand also is increasing enormously.

Grain based alcohol distillery

The alcohol industry is the commercial industry involved in the manufacturing, distribution, and sale of alcoholic beverages. Alcohol has assumed very important place in the country's economy. It is a vital raw material for a number of chemicals and become the source of large amount of revenue for government as excise duty. Rapid increase in urban population is one of the key factors that is boosting the India alcohol market. In India about 350 distillery units of 4.0 billion liters & 100 distillery units of 1.8 billion liters are molasses & grain based respectively.

Functional food-based bakery products (bread, cookies and biscuits)

Biscuits Sweet biscuits are commonly eaten as a snack food, and are, in general, made with wheat flour or oats, and sweetened with sugar or honey. Varieties may contain chocolate, fruit, jam, nuts, or even be used to sandwich other fillings. Savory biscuits or crackers (such as cream crackers, water biscuits, oatcakes, or crisp breads) are usually plainer and commonly eaten with cheese following a meal. Other savory biscuits include the Jewish biscuits known as matzos. Many savory biscuits also contain additional ingredients for flavor or texture, such as poppy seeds, onion or onion seeds, cheese (such as cheese melts), and olives.

Khakra-Ready to Eat

Khakra is an Indian traditional ready to eat snack food based on wheat. It is known for its crisp texture and baked flavour. It can be consumed as breakfast/snack food and can also fit into the main meal. It is a traditional Gujarat food and is quite familiar in Gujarat and other states in India. It is a food, which is liked by people of all states and is well within the reach of common people. It is consumed at breakfast, lunch and dinner. It is very good in taste. Due to its good nutritious value and taste its demand is increasing day-by-day. The product at present is made on cottage scale.

Wheat straw plastic products

Wheat straw is very versatile and can be used for many different products. Since plastic made from wheat straw is very durable and heat-resistant (it can resist temperatures up to 200 degrees Fahrenheit), it's a great material to make mugs and cups from like this travel mug. Wheat straw plastic is also being used to create plates, cutlery and food-containers. It's also a great way to produce single-use products since there is no need to destroy forests.

Paper and paperboard

Chemical components and anatomy of wheat straw, pulping process and pulping properties, paper and paperboard products from wheat straw pulps, as well as environmental protection issues. Wheat straw is a kind of annual vascular-bundled herbal arthrophyte containing cellulose fibers that are acceptable for pulping and papermaking. Wheat straw can be widely used for the preparation of paper as well as paper boards.

Value addition - Pulses

Following value added products can be prepared from different pulses

Puffed chickpea and peas	Pakoda mix
Parching of pulses	Preparation of instant adai mix
Papad	Instant vada and bajji mixes
Soy milk	Preparation of idli podi
Tofu (Soy paneer)	Preparation of noodles
Soya sauce	Preparation of idiappam
Soya protein isolates Supplement	Shrikhand
Khakhra	Soy biscuit
Soy chocolate	Soy wafers

Value Addition - Oil Seeds

India holds unique position in the world in terms of rich diversity of vegetable oil sources and oil crops followed by cereal crops. Mostly nine annual oilseeds crops are growing in India, namely, groundnut, rapeseed-mustard, sesame, safflower, niger, soybean and sunflower, forming the edible group and linseed and castor, forming the non-edible group. Linseed due to its unique nutritional value has high potential to be an edible oil crop as well. In addition, cotton seed, rice bran and seeds of some tree species are being exploited for edible oil. Oilcakes and meals are majorly used in animal feeds and as manures

For groundnut, there are opportunities to make peanut butter, fat-free peanuts, roasted peanuts and salted peanuts, specially coated peanuts, etc. Castor oil is a versatile vegetable oil with varied industrial applications covering soaps, detergents, paints, varnish, cosmetics, plasticizers, lubricant and many more.

Sesame seeds

Sesame seeds are used in baking, to top breads, buns and bagels, in crackers and in cakes. The ground seeds are used in East African cuisine in soups and fish dishes. They are also used in sweets, including a product similar to peanut brittle.

Sesame oil

Commercially, sesame oil comes in two types. One type of sesame oil is a pale-yellow liquid and has a pleasant grain-like odor and somewhat nutty taste. This oil is high in polyunsaturated fats, ranking fourth behind safflower, soybean and corn oil. It is excellent for use as a frying oil, in cosmetics and in food preparations. The other type of oil is amber-colored and aromatic, made from pressed and toasted sesame seeds. This popular ingredient in ethnic cooking is not used as a cooking oil, however, because the flavor is too intense and it burns quite easily. Instead, sesame oil is normally added as a flavoring agent in the final stages of cooking.

Sesame oil is also used in paints, soaps, cosmetics, perfumes, bath oils, insecticides and pharmaceuticals (vehicle for drug delivery).

Pharmaceutical and nutraceutical applications

Sesame seed oil has been used as a healing oil for thousands of years. It is naturally antibacterial and effective against common skin pathogens as well as common skin fungi including the athlete's foot fungus. It is naturally antiviral and is a natural anti-inflammatory agent. Many "natural" cosmetics now include sesame oil because of its antioxidant properties.

Value Addition - Sugarcane

The juice of sugarcane is widely used for the preparation of sugar crystals, jaggery, sugarcane juice concentrate, powder jaggery, rab, vinegar etc.

Multiple bagasse products: - Tableware, paper, packaging, bottles, MDF board, plastic composite resin, disposable plate, bowl, tray, cutlery, photocopier paper, kraft, printing, tissue, toilet, wall paper, bags, cardboard and packing products. Electric power can be generated from bagasse, by installing high pressure boilers and special type of steam turbine. The cellulosic residue of sugarcane crop viz., bagasse could also be used for ethanol production by simultaneous saccharification-cum-fermentation (SSF) route using enzymatic or acid hydrolysis technology

Molasses: It is one of the most valued by-products of the sugar industry which is being used primarily by the distillation industry in India. This edible molasses has wide application ranging from bakery product to enhancing flavoring properties in meat, herbs, chocolate, spices etc. also help in fortifying sweets, savory or spice flavors. Bio-ethanol is most abundant bio-fuel for automobile transportation.

Molasses constitutes the main raw material for distilling industries. Molasses can be used in other fermentation industries also. Vinegar and acetic acid are two important products from molasses. Citric acid is another important fermentation product of molasses.

Filter cake: It is largely being utilized as bio-fertilizers and is used as compost in sugarcane fields. An emerging trend to utilize the press cake for production of Bio-gas or Bio-CNG is gaining momentum in coming times. The press cake since contains appreciable proportion of biodegradable organic matter which has very good potential for the production of biogas.

Value Addition - Vegetables

Most of the vegetables are highly perishable and have limited life, which need to be marketed immediately or processed into varied value-added products. Diversification of vegetable products for export and domestic consumption reduces the risk of loss due to price fluctuation. This increased demand for value-added products is due to the change in the market behavior, changes in consumer preferences and emergences of supermarkets, which have resulted in the usages of more value-added, ready-to-use vegetable products and vegetable in consumer packs. These products also have a high demand in the defense sector of the different country. Among the different vegetable value-added products, dehydrated potato, peas, carrot, cauliflower, tomato-based processed products, pickles and chutney from different vegetables are the most important. Apart from this lycopene from tomato and watermelon, tomato seed oil, frozen vegetables, organic vegetables, minimal processed vegetables, consumer-packed vegetables, etc., are also important.

Fresh frozen vegetables

There are thousands of frozen foods available. Before freezing, it is necessary to blend the product to destroy enzymes, which would result on off flavors and consequent quality deterioration. As frozen food preserves the vegetables, they don't get spoiled even after months once they are kept in cold storage

Potato powder, granules and flakes

Potato is one of the important tuber vegetables, which is consumed throughout the year. Indian vegetable basket is incomplete without mentioning the king of vegetables-potato-a sustaining force and a culinary delight. The tantalizing taste of nutrient-rich potatoes makes it an essential part of every breakfast, lunch and dinner worldwide

Garlic oil and powder

Garlic is one of the important bulb crops grown and used as spice and condiment throughout India. It possesses a high nutritive value and medicinal property. Allicin, the principle of garlic has antibiotic properties.

Ginger oil

Ginger, one of the most important and oldest of spices used in every kinds of food preparation. The rhizomes known in the trade as hand or races reach the spice trade either, with the outer cortical layers intact (coated unscraped ginger) or with the outer coating partially or completely removed

Ginger powder

The rhizomes of ginger are pale brower in colour and fibrous with peel not entirely removed washed and dried in the sun. The material may be garbled by removing pieces that are too lighter and it may also be lime bleached

Bitter gourd (Karela) powder

Karela is a vegetable which is grown in every part of India. Karela is especially grown in India in from April to August. In the session it is abundantly available. It is generally used as vegetable throughout India. It has also good medicinal value. It is generally bitter in taste. Karela powder is now special product which can be produced by spray drying process

Potato chips

India is one of the largest producers of potato. Besides being used as a daily food item in various vegetable preparations, potato today increasingly finds use in the form of chips or wafers as snacks food. The potato chips and wafers are popular processed food items that give considerable value addition to potatoes

Chilli powder

Chili is an important cash crop in India. Its annual production is 8.4 lakh tones out of which only 10 per cent is being exported to other nations. The primary chemical constituents of chili fruit are (1) colour (2) pungency which are chiefly responsible for export demand.

Value Addition - Fruits

- Horticulture deals a large group of crops. Therefore, cultivation of crops which belong to us and possess great medicinal, nutritional, health promoting values.

- India as second largest producer of fruits and vegetables, only 10 per cent of that horticultural produce is processed, but other developed and developing countries where 40-80 per cent produce is value added.
- Horticultural crops provide varied type of components, which can be effectively and gainfully utilized for value addition like pigment, amino acids, oleoresins, antioxidants, flavors, aroma etc.
- Post-harvest losses in horticultural produce are 5 to 30 per cent which amounts to more than 8000 crore rupees per annum. If we subject our produce to value addition the losses can be checked.
- Horticultural crops are right material for value addition because they are more profitable, has high degree of process ability and richness in health promoting compounds and higher potential for export.

Therefore, horticultural crops are right material for value addition in the present context of agricultural scenario and we should go for new product development to be unique and novel

Fruits are processed into a variety of products such as juices and concentrates, pulp, canned and dehydrated products, jams and jellies, pickles and chutneys etc.

Fruit juice is a naturally contained liquid in fruits. Juice is prepared by mechanically squeezing or macerating fresh fruits or vegetables flesh without the application of heat or solvents. Many commercial juices are filtered to remove fiber or pulp, but high-pulp fresh orange juice is a popular beverage

Mango pickles

Mango is one of the best fruits in India. There are number of products produced from mango like mango juice, mango pulp, mango flavour, mango kernel oil, mango pickles and powder etc. which are well accepted throughout India and in aboard. There are some manufacturers in organized sector and some are in unorganized sector

Coconut

Coconut is one of the important fruit having large uses starting from bakery, confectionery, pharmaceutical industries to beverage industry and many more industries. Coconut oil is extensively used for edible and industrial purposes. Desiccated coconut powder is a foremost edible product and is used largely as the basic ingredients in the preparation of many South Indian dishes

Strawberry Jam

Strawberry is one of the delicious fruits. It contains vitamins minerals, enzymes, fruit juice etc. Our daily need of food is incomplete unless we take sugar. We can intake sugar by many means, one as fruit jam. Strawberry is one the fruits having high perishable character for making jam

Amla (gooseberry)

It is very commonly found in India, Ceylon, Malaya and China. It is very important fruit from medicinal point of view. It has characteristics to maintain good health. It has capability to establish fighting system against diseases related to stomach. It is helpful in the growth of hairs and also good for skin

Lychee Juice

Lychee is one of the tropical fruits which is mainly available in Bihar, and Eastern part of India. It is one of the healthy fruit juices. It will help to grow the health. It is sweet tasty with good flavour

3.9. Organic Farming

FAO suggested that "Organic agriculture is a unique production management system which promotes and enhances agro-ecosystem health, including biodiversity, biological cycles and soil biological activity, and this is accomplished by using on-farm agronomic, biological and mechanical methods in exclusion of all synthetic off-farm inputs".

More specifically, organic farming entails:

- Use of cover crops, green manures, animal manures and crop rotations to fertilize the soil, maximize biological activity and maintain long-term soil health.
- Use of biological control, crop rotations and other techniques to manage weeds, insects and diseases.
- An emphasis on biodiversity of the agricultural system and the surrounding environment.
- Using rotational grazing and mixed forage pastures for livestock operations and alternative health care for animal wellbeing.
- Reduction of external and off-farm inputs and elimination of synthetic pesticides and fertilizers and other materials, such as hormones and antibiotics.
- A focus on renewable resources, soil and water conservation, and management practices that restore, maintain and enhance ecological balance.

Organic Farming Scenario in Indian States

Since the early days, India has had a close relationship with organic farming–agricultural methods that align with nature, causing lesser harm to the environment and people. While India is still known for organic farming practices, over the years, it has seen some challenges. According to research reports, only 2% of the net sown area in the country is under organic cultivation.

The country is blessed with a rich, vast land that is divided into many states. And it appears that certain regions are doing better than the others in terms of organic farming. While larger states like Madhya Pradesh, Rajasthan, and Maharashtra have around half of the areas under organic cultivation, smaller states like Sikkim, Kerala, Mizoram, Nagaland are more progressive in their mission to becoming natural-farming states.

Let's take a look at the organic farming scenario in some of the states of India–

• Madhya Pradesh

Madhya Pradesh accounts for 40% of India's organic farming share and has over 100,000 registered organic farmers. Their government envisions leading the nation in organic farming to ensure sustainable livelihoods and healthier food for the community.

• Rajasthan

Rajasthan has about 60,000 hectares of land under organic farming, with 97.3% farmers agreeing to adopt the agricultural method. The state government has started identifying blocks in 11 districts to convert them into fully organic farming areas.

• Maharashtra

Maharashtra has about 8-10 lakh hectares of land under organic farming, promoted across 84 villages. The state has around 4 million organic farmers coming together to form a volunteer organization called the Maharashtra Organic Farming Federation (MOFF).

• North-East India

- **Sikkim:** The North-East region in India is largely untouched by industrial, agricultural practices, making it pro-organic farming. Among the eight states in NE India, all of Sikkim's farmlands are officially organic certified—making it the only 100% certified organic State in the world. The transition has benefitted more than 66,000 farming families and boosted the state's economy. Meanwhile, other states like Meghalaya and Assam are also showing tremendous progress in embracing organic farming.

- **Assam:** The demand for organic farming is increasing in Assam—conscious organic farming is fast gaining traction in the hills of the State. Farmers are rediscovering the holistic benefits of organic farming and living a sustainable livelihood that does not harm the environment.
- **Meghalaya:** Meghalaya government launched the Mission Organic with a vision to convert at least 2 lakh hectares into organic farmland in 2015. The state is also taking initiatives to further encourage sustainable and organic farming with 'Organic Meghalaya'.

Case Studies on Organic Agriculture in India

Nitiparna R presented case studies for different states of India as below.

1. North India

The Shivalik Hills stretch into five states- Jammu and Kashmir, Punjab, Himachal Pradesh, Haryana, and Uttaranchal. These foothills of the Himalayan range have been identified as one of the degraded agro-ecosystems of India facing acute shortages of drinking water and deforestation to meet fodder and fuel requirements. An Integrated Watershed Development Project (IWDP) was launched by the government with World Bank support to improve the production potential of the area by evolving the watershed management technologies and encouraging community participation. The entire project covers 835 villages with an area of 103652 ha. Farmers throughout are small and typically marginal with land holdings 0.2 to 1 ha. Many plots are in higher elevations and on steep slopes throughout the watershed area. Most of the farmers use traditional methods with the more recent advent of conventional components such as chemical fertilizers and pesticides, with the latter having become particularly common.

A variety of crops such as ginger, peas, capsicum, wheat, paddy and seasonal vegetables are cultivated in the region. Since the land holdings are small, many farmers have very little marketable surplus. Most of the production, remains in the region, being sold in the local market and going to the towns and urban centres. In few villages, farmers pool their produce and hire a truck and sell the produce about 100 kilometres away from the village to get a better price.

In Uttarakhand, Self-help groups and village level organisations play some role in the development and dissemination of organics, but public agencies, i.e. extension services are still predominant. In several cases, these have integrated with specialized NGOs to help improve the uptake of improved compost and other organic methods.

Marketing efforts have been focused primarily at the domestic level. Through direct contact and participation in trade fairs and exhibitions, modest sales have been generated. In some cases, these sales are for the domestic market and in other cases, traders export them. The state has planned to develop 33 marketing centres for organic products and one has already opened for business. Thirteen tons of organic rice has also been directly exported.

2. West India

Much of Maharashtra's Aurangabad region is considered to be very poor. Agriculture is the main source of income and the area depends on modest rainfall that is concentrated in the summer months. The area is multicultural with a sizable muslim minority comprising approximately 35% of the total population.

The Institute for Integrated Rural Development (IIRD) is a civic organization that has targeted women, and particularly destitute women, for training needs and rural development activities. As a result, 60% of its beneficiaries are women. Accordingly, it is also women who facilitate and organize local groups. Food security was a predominant concern for a number of years and the focus crops included cereals, legumes, oilseeds, and spiccs. More recently, as food security has improved, marketing has emerged as a prime concern. The main organic products grown are wheat, sorghum, cotton, and pearl millet.

IIRD has established a weekly organic bazaar in the city of Aurangabad to foster more direct linkages between producers and consumers as well as providing a consistent platform for the exchange of products and services related to organic farming. The bazaar now sells approximately 40% of farmers' marketable surplus. The rest is sold to local traders and markets.

3. South India

The Idukki District is part of Kerala's Western Ghats region, recognized as one of the world's 25 bio-diversity hot spots. This hilly region receives adequate rainfall and has maintained a considerable amount of forest cover despite increasing threats from agriculture and timber interests. Three systems of production have dominated the project area. On steep slopes small farmers cultivate multiple crops.

Peermade Development Society (PDS) emerged as an NGO in 1980 and as a social service wing of a Christian diocese in response to extensive agrochemical contamination in the area's drinking water. This resulted in acute toxicity of farmers in the region leading to their hospitalization. It has focused on tribal and marginalized farmers and contributed to the development of farmer led organisations for the dissemination of organic practices and to effect quality control and standards compliance.

PDS has established processing facilities for the farmers to capture more value and its Export Division is one of several functional marketing units that export primarily pepper and bananas. Most of the production is destined for the domestic market and an integral part of PDS' success has been its entrepreneurial experience. It helped to develop the local medicinal plants industry that integrated with pharmaceutical processing and national as well as overseas marketing.

Spices and tea are the primary crops but several other varieties of nuts and fruits are also produced. In keeping with the project biodiversity commitment, no cereals are cultivated and diverse tree fruits are encouraged. These include jackfruit, banana, plantains, coconut, and guava. Most are for self-consumption as are the few vegetables and greens cultivated by many households.

The Eco-Agri Research Foundation (EARF) — as the central organizing body — is a registered trust that since 1994 serves as a 50-ha model farm demonstrating organic farming and biodynamic practices. In Karnataka, the concept demonstrates a complex system of eco-farming with- animal husbandry; conservation of soil and water through water harvesting structures; vegetative soil erosion checks; and production of high value crops like vanilla, pepper, and banana.

The major organic products generated are jaggery sugar (54 mt), rice (25 mt), vanilla (0.8mt) and banana (10 mt); much of which is sold by the EARF in domestic markets at nearby Bangalore and Mysore and to market agents in other major cities. The Foundation pays farmers a substantial premium, even during the transition. It is also procuring organically grown vanilla, ginger and pepper from other parts of Karnataka as well as from the neighboring state of Kerala to combine with its own and improve export efficiencies to the USA and European countries.

4. Central India

In Madhya Pradesh during 1993, the Maikaal organic cotton initiative was started by a major Swiss yarn trading company, together with Maikaal Fibres Ltd., an Indian spinning mill. The experiment developed into a commercial project, which has grown and is now run by an independent company called Maikaal bioRe (India) Ltd, employing 36 persons. Farmers are both suppliers (raw cotton) to the company and its customers for support services such as training, consulting, crop monitoring, inputs, etc.

The project focuses on biodynamic, certified organic cotton for the export market. It demonstrates how strong corporate leadership can create mutually profitable initiatives that address the environmental needs of farming communities. The project has a strong market orientation and has helped farmers to efficiently apply state-of the- art organic technology and methods.

Training is an integral part of participation and the company provides all necessary inputs. This results in considerable efficiencies. For example, labour utilization is less in the organic systems than in the conventional system. Production costs are lower and yields are higher than in similar conventional systems.

3.10. Precision Weed Management

Precision weed management practices

Precision weed management is a technique where it involves identification and discrimination of variance in weed population at sub field level. It manages the weeds effectively by using the advanced technologies like sensors, GPS, robotics, MAP-based approaches, laser thermal beam, microwaves, mechanical arms etc to optimize agricultural inputs and improve crop yields. The recent advancement in agriculture and precision management made the possibility of site-specific management of weeds with limited inputs like herbicides. The target oriented and variable rate based application of herbicides is an effective way to minimize herbicide usage, input cost, increase herbicide usage potency and weed control efficiency with reduction in environmental waste.

The Site-Specific Weed Management (SSWM) is the component of precision agriculture, which includes sensor-based machinery embcdded technique, where it observes, records and executes the controlling operations according to the predefined weed management plan. The main purpose of SSWM is to apply the herbicide at right dose at specific place and time. In case of mechanical weed management, the role of SSWM is to actuate the weeding tool at right time and at right place. Through SSWM, one can save 30-90 % of operating cost and 50-80 % of herbicide over conventional practices. The SSWM mainly has four components through which it executes the different weed controlling strategies. Under precision weed management practices, the weed control measures have to be executed in following ways:

Real-time or online method: involves real time detection of weeds, decision making to adjust the dose of herbicides or weeding operation and execution of all controlling operations by using the sensors.

Offline method: Prior information on location specific weed data is available before performing the weed controlling measures. This data will be loaded to the system and all controlling operations are executed afterwards.

Since the weed populations are spatially varied within a field, positioning of the weed data plays an important role. To do so, a global navigation satellite system GPS (global positioning system) are widely used to mark the absolute position of the data.

Weed detection techniques

The detection of weeds through automation has the capability to increase sample point density and accuracy of the system. Mainly three approaches were used to detect the weeds under automation, they are:

1. **Biological morphology:** recognition of shape and structure of the plant species.
2. **Spectral characteristics:** recognition based on the plant reflectance. Pixel based colour or hyper spectral classifiers are used for these purposes.
3. **Visual texture:** recognition based on the grey scale calculation and colour of the images.

Machine vision based weed detection from images

Machine vision is an optical sensor based system, which discriminates and differentiates the weeds from crop. Main components of the system include image capturing device (by camera or optical sensors), microprocessors (image processing and system control) and weed control actuators.

GPS-controlled patch spraying

In GPS-controlled patch spraying, an application map has to be developed based on the interpolated maps of weed distribution and economic weed threshold value. The developed application map will be feeded into the spraying system where it sprays the herbicide according to the pre-decided dose. The herbicide application dose will be decided on the basis of weed density and area of coverage.

Variable rate herbicide applicator (VRA)

The variable rate herbicide applicators can be operated with both real time (optical sensor) and/or with interpolated maps of weed distribution and economic weed threshold (offline) method. Under optical sensor based system, the sensors detect weed, their density and area of coverage and then decides the dose of herbicide needs to be applied. In map based approach, variable dose of herbicide application will be developed on the basis of weed economic threshold value and then the herbicide will be applied accordingly.

Machine vision based online spraying system

The machine vision based online spraying system integrated with optical sensors and spraying system detects the weeds in real time, differentiates it from the crops and applies the herbicide accordingly. Herbicide application through drones (UAVs) The drones are unmanned aerial vehicles (UAVs) which are used to develop weed map and to apply the herbicide as part of the precision weed management practice. Through drones, a target oriented as well as the variable rate application of herbicides is also possible with increased field capacity and efficiency. In addition, around 96% of water can be saved in drone based spraying system.

Machine vision based/ sensor based weeding machines and robots

An automatic control of weeding operation is possible by using the RTK-GPS, optical and position sensors like camera, laser, ultra-sonic sensor and robotic arms/ weeding tools, etc. Under this system both online (real time) and offline method of weeding can be possible. In online method, a machine vision based system or an object detection based system is used to detect and differentiate the weeds from crops. Later, a decision will be made by system to perform the weeding operation. In offline method, a GPS based position detection system is used to perform the weeding operation.

Microwave weeding

The microwave weeding involves the projection of microwave energy on the weeds and on soil through suitable devices to kill the plants and their seeds. The advantage of the microwave energy is that, it is not affected by weather parameters. An ultra-high frequency (UHF) electromagnetic energy with wavelengths much greater than those of light (frequency 2450 MHz and wavelength 12.25 cm) were used in the weeded system to kill a plant and weed seed bank. A portable microwave weeded has been developed and is available commercially.

Laser thermal beam based weeders

In this system, laser thermal beams were used to kill the weeds. System involves application of laser radiations on the plants, which affect the thermal balance of a plant and partially destroys the plant tissue through thermodynamic heat transfer mechanisms. Laser beams can selectively impair the growth of plants by destroying the sensitive growth centres, the so called meristems, thus weed can be controlled effectively. However, the system requires a high spatial resolution for more accuracy. The commercially available laser beam based weeders can take out 1,00,000 plants per hour.

Conclusion

Precision weed management technologies are need of the hour to meet out the ever increasing food demand. Through sensor based site-specific weed management, we can reduce 30 to 90 % of operating cost and 50 to 80 % of herbicide usage over conventional practices. It also reduces the weed seed bank in the soil effectively. Both the online (real-time) and offline (map based approach) method of weed management can be effectively used. However, real-time based weed control systems are much effective and less complex compared to the map based approach.

3.11. Climate Smart Agriculture

Climate-smart agriculture (CSA) may be defined as an approach for transforming and reorienting agricultural development under the new realities of climate change

(Lipper et al. 2014). The most commonly used definition is provided by the Food and Agricultural Organisation of the United Nations (FAO), which defines CSA as "agriculture that sustainably increases productivity, enhances resilience (adaptation), reduces/removes GHGs (mitigation) where possible, and enhances achievement of national food security and development goals".

Nearly 690 million people around the world are hungry today (FAO et al., 2020). At the same time, more and more people around the planet, including youth, are pushing for action on climate change. Thus, the question of how to achieve food security while also adapting to climate change and reducing greenhouse gas (GHG) emissions becomes ever more relevant. Agriculture contributes a significant part of gross domestic product (GDP) in many economies; 2.5 billion people worldwide depend on agriculture for their livelihoods (FAO, 2016). Agri-food systems are believed to be responsible for one third of global anthropogenic GHG emissions (Crippa et al., 2021). At the same time, the sector offers significant potential to contribute to global climate goals (FAO, 2019). Therefore, transforming agri-food systems (including crops, livestock, fisheries, aquaculture, agroforestry and forestry) is pressing. This transformation should be achieved through integrated, multisectoral approaches that are gender-transformative, inclusive and pro-poor, and harness synergies such as those between climate change adaptation and mitigation. Agri-food systems must increase production in a sustainable manner in order to meet the Sustainable Development Goals (FAO, 2019).

Against this background, climate-smart agriculture (CSA) is an innovative approach based on the following three pillars:

- sustainably increase agricultural productivity and incomes;
- adapt and build resilience of people and agri-food systems to climate change;
- reduce or, where possible, avoid GHG emissions.

CSA interventions have been implemented successfully around the world. This success is underpinned by the five action points of CSA implementation, as formulated by FAO.

1. Expand the evidence base to identify:
    - the effects of climate change on food production and on agri-food systems as a whole;
    - vulnerabilities in the agriculture sector that affect food security;
    - institutional and financial barriers and possibilities; and
    - climate-smart adaptation and mitigation options.

2. Supporting enabling policy frameworks by, for example:
   - developing relevant policies, legislation, plans and investments to support an enabling environment for CSA;
   - modifying existing policies where needed; and
   - coordinating policymaking processes to enable effective collaboration between institutions responsible for agriculture, climate change, food security and land use.
3. Strengthen national and local institutions by, for example:
   - supporting institutions so that they can empower, enable and motivate farmers; and
   - building capacities of national policymakers so that they can effectively participate in international policy fora on climate change and agriculture, and engage with local public authorities.
4. Enhance funding and financing options by, for example:
   - accessing funding instruments such as the Green Climate Fund (GCF), the Global Environment Facility (GEF), official development assistance (ODA) and national sectoral budgets; and
   - creating innovative links between climate finance and public and private agricultural investment; and
   - integrating climate change in agricultural planning and budgeting.
5. Implement practices at field level:
   - selecting locally suitable CSA options by engaging local farmers' knowledge, requirements and priorities.

4

Climate Change and Global Warming

4.1. Climate Change

Climate change refers to long-term shifts in temperatures and weather patterns. These shifts may be natural, such as through variations in the solar cycle. But since the 1800s, human activities have been the main driver of climate change, primarily due to burning fossil fuels like coal, oil and gas. Burning fossil fuels generates greenhouse gas emissions that act like a blanket wrapped around the Earth, trapping the sun's heat and raising temperatures. Burning of crop residues also adds to climate changes

Examples of greenhouse gas emissions that are causing climate change include carbon dioxide and methane. These come from using gasoline for driving a car or coal for heating a building. Clearing land and forests can also release carbon dioxide. Landfills for garbage are a major source of methane emissions. Energy, industry, transport, buildings, agriculture and land use are among the main emitters. (Fig 4.1)

Global climate change is not a future problem. Changes to Earth's climate driven by increased human emissions of heat-trapping greenhouse gases are already having widespread effects on the environment: glaciers and ice sheets are shrinking, river and lake ice is breaking up earlier, plant and animal geographic ranges are shifting, and plants and trees are blooming sooner. The effects that scientists had long predicted would result from global climate change are now occurring, such as sea ice loss, accelerated sea level rise, and longer, more intense heat waves.

Some changes (such as droughts, wildfires, and extreme rainfall) are happening faster than scientists previously assessed. In fact, according to the Intergovernmental Panel on Climate Change (IPCC) — the United Nations body established to assess the science related to climate change — modern humans have never before seen the observed changes in our global climate, and some of these changes are irreversible over the next hundreds to thousands of years.

Scientists have high confidence that global temperatures will continue to rise for many decades, mainly due to greenhouse gases produced by human activities. The IPCC's Sixth Assessment report, published in 2021, found that human emissions of heat-trapping gases have already warmed the climate by nearly 2 degrees Fahrenheit (1.1 degrees Celsius) since pre-Industrial times (starting in 1750). The global average temperature is expected to reach or exceed 1.5 degrees C (about 3 degrees F) within the next few decades. These changes will affect all regions of Earth.

Climate change can affect our health, ability to grow food, housing, safety and work. Some of us are already more vulnerable to climate impacts, such as people living in small island nations and other developing countries. Conditions like sea-level rise and saltwater intrusion have advanced to the point where whole communities have had to relocate, and protracted droughts are putting people at risk of famine. In the future, the number of "climate refugees" is expected to rise.

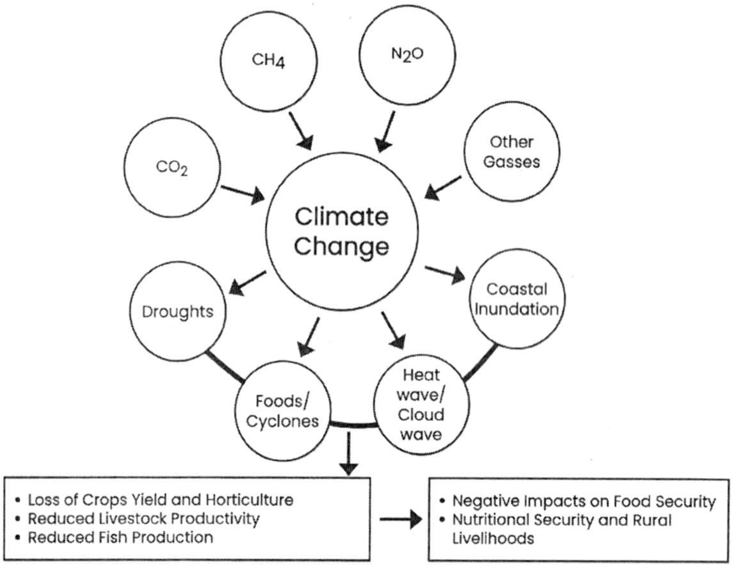

Fig. 4.1: Causes and impact of climate change on agriculture and allied sectors

Every increase in global warming matters

In a series of UN reports, thousands of scientists and government reviewers agreed that limiting global temperature rise to no more than 1.5°C would help us avoid the worst climate impacts and maintain a liveable climate. Yet based on current national climate plans, global warming is projected to reach around 3.2°C by the end of the century.

The emissions that cause climate change come from every part of the world and affect everyone, but some countries produce much more than others. The 100 least-emitting countries generate 3 per cent of total emissions. The 10 countries with the largest emissions contribute 68 per cent. Everyone must take climate action, but people and countries creating more of the problem have a greater responsibility to act first.

Solutions

Many climate change solutions can deliver economic benefits while improving our lives and protecting the environment. We also have global frameworks and agreements to guide progress, such as the Sustainable Development Goals, the UN Framework Convention on Climate Change and the Paris Agreement. Three broad categories of action are: cutting emissions, adapting to climate impacts and financing required adjustments.

Switching energy systems from fossil fuels to renewables like solar or wind will reduce the emissions driving climate change. But we have to start right now. While a growing coalition of countries is committing to net zero emissions by 2050, about half of emissions cuts must be in place by 2030 to keep warming below 1.5°C. Fossil fuel production must decline by roughly 6 per cent per year between 2020 and 2030.

Adapting to climate consequences protects people, homes, businesses, livelihoods, infrastructure and natural ecosystems. It covers current impacts and those likely in the future. Adaptation will be required everywhere, but must be prioritized now for the most vulnerable people with the fewest resources to cope with climate hazards. The rate of return can be high. Early warning systems for disasters, for instance, save lives and property, and can deliver benefits up to 10 times the initial cost.

4.2. Global Warming

Global warming is the long-term warming of the planet's overall temperature. Though this warming trend has been going on for a long time, its pace has significantly increased in the last hundred years due to the burning of fossil fuels. As the human population has increased, so has the volume of fossil fuels burned. Fossil fuels include coal, oil, and natural gas, and burning them causes what is known as the "greenhouse effect" in Earth's atmosphere. The greenhouse effect is when the Sun's rays penetrate the atmosphere, but when that heat is reflected off the surface cannot escape back into space. Gases produced by the burning of fossil fuels prevent the heat from leaving the atmosphere. These greenhouse gasses are carbon dioxide, chlorofluorocarbons, water vapor, methane, and nitrous oxide. The excess heat in the atmosphere has caused the average global temperature to rise overtime, otherwise known as global warming.

Global Warming vs Climate Change

Global warming is often mistakenly thought to be the same as climate change, but there are a few distinct characteristics of each that make them different. Global warming refers to the Earth's rising temperatures, whereas climate change includes not only warming, but the side effects of warming—events such as melting glaciers, heavier rainstorms, frigid cold snaps, or frequent droughts that lead to uncontrollable wildfires.

Another distinction between the two is when scientists or government leaders speak about global warming, they usually mean human-caused warming due to the increase in carbon dioxide and other greenhouse gases from people burning fossil fuels. Climate change, though, can mean human-caused or natural-caused, such as ice ages.

Greenhouse gas concentrations are at their highest levels in 2 million years

The Earth is now about 1.1°C warmer than it was in the late 1800s as emission continue to rise. The last decade (2011-2020) was the warmest on record. Many people think climate change mainly means warmer temperatures. But temperature rise is only the beginning of the story. Because the Earth is a system, where everything is connected, changes in one area can influence changes in all others.

The consequences of climate change now include, among others, intense droughts, water scarcity, severe fires, rising sea levels, flooding, melting polar ice, catastrophic storms and declining biodiversity.

4.3. Long Term Effects of Climate Change and Global Warming

A. Climate Change

Scientists have predicted that long-term effects of climate change will include a decrease in sea ice and an increase in permafrost thawing, an increase in heat waves and heavy precipitation, and decreased water resources in semi-arid regions.

Below are some of the regional impacts of global change forecast by the Intergovernmental Panel on Climate Change:

- **North America:** Decreasing snowpack in the western mountains; 5-20 percent increase in yields of rain-fed agriculture in some regions; increased frequency, intensity and duration of heat waves in cities that currently experience them.
- **Latin America:** Gradual replacement of tropical forest by savannah in eastern Amazonia; risk of significant biodiversity loss through species extinction in many tropical areas; significant changes in water availability for human consumption, agriculture and energy generation.
- **Europe:** Increased risk of inland flash floods; more frequent coastal flooding and increased erosion from storms and sea level rise; glacial retreat in mountainous areas; reduced snow cover and winter tourism; extensive species losses; reductions of crop productivity in southern Europe.
- **Africa:** Between 75 and 250 million people are projected to be exposed to increased water stress; yields from rain-fed agriculture could be reduced by

up to 50 percent in some regions; agricultural production, including access to food, may be severely compromised.

- **Asia:** Freshwater availability projected to decrease in Central, South, East and Southeast Asia by the 2050s; coastal areas will be at risk due to increased flooding; death rate from disease associated with floods and droughts expected to rise in some regions.

Projected Impact on Asia

- Asia-Pacific region may experience the worst effect on rice and wheat yields worldwide, and decreased yields could threaten the food security of 1.6 billion people in South Asia.
- The crop model indicates that in South Asia, average yields in 2050 for crops will decline from 2000 levels by about 50 percent for wheat, 17 percent for rice, and about 6 percent for maize because of climate change.
- In East Asia and the Pacific, yields in 2050 for crops will decline from 2000 levels by 20 percent for rice, 13 percent for soybean, 16 percent for wheat and 4 percent for maize because of climate change.
- With climate change, average calorie availability in Asia in 2050 is expected to be about 15 percent lower and cereal consumption is projected to decline by as much as 24 percent compared to a no-climate change scenario.
- In a no-climate change scenario, the number of malnourished children in South Asia would fall from 76 to 52 million between 2000 and 2050, and from 24 to 10 million in East Asia and the Pacific.
- Climate change will erase some of this progress, causing the number of malnourished children in 2050 to rise to 59 million in South Asia and to 14 million in East Asia and the Pacific, increasing the total number of malnourished children in Asia by about 11 million.
- To counteract the effects of climate change on nutrition, South Asia requires additional annual investments of 1.5 billion USD in rural development, and East Asia and the Pacific require almost 1 million USD more. Over half of these investments in both regions must be for irrigation expansion.
- The Asian countries most vulnerable to climate change are Afghanistan, Bangladesh, Cambodia, India, Lao PDR, Myanmar, and Nepal.
- Afghanistan, Bangladesh, India, and Nepal are particularly vulnerable to declining crop yields due to glacial melting, floods, droughts, and erratic rainfall, among other factors.
- Asia is the most disaster-afflicted region in the world, accounting for about 89 percent of people affected by disasters worldwide.

B. Global Warming

Regardless of whether you prefer to use the terms global warming and climate distinctly or interchangeably, you are still essentially speaking about the same phenomenon: the buildup of excess heat in the Earth's system. Both of these concepts have an impact on life on Earth that can cause long-term, irreparable damage. If we don't put forth efforts to help mitigate global warming, the effects will become more extreme and worsen over time.

- Global temperatures rose about 1.8°F (1°C) from 1901 to 2020.
- Sea level rise has accelerated from 1.7 mm/year throughout most of the twentieth century to 3.2 mm/year since 1993.
- Glaciers are shrinking: average thickness of 30 well-studied glaciers has decreased more than 60 feet since 1980.
- The area covered by sea ice in the Arctic at the end of summer has shrunk by about 40% since 1979.
- The amount of carbon dioxide in the atmosphere has risen by 25% since 1958, and by about 40% since the Industrial Revolution.
- Snow is melting earlier compared to long-term averages.

Extreme Weather

As the earth warms up, we'll experience more extreme weather and climate events such as heat waves and droughts. Over the last fifty years, the U.S. has seen prolonged periods of record-high temperatures, heavy storms, and in some places, both floods and droughts

Polar Vortexes

While the term global warming might make you assume that the only effect it has on Earth is to—well, warm it, it can also cause extremely harsh cold air to reach into more temperate climes from the fridget poles, a phenomenon known as a "polar vortex."

A polar vortex is a large area of low pressure and cold air surrounding the Earth's North and South poles, and the "vortex" refers to the counter-clockwise flow of air that helps keep the colder air close to the poles. As warming reduces the amount of Arctic sea ice, it allows more heat to escape from the ocean, disrupting and weakening the polar vortex and sending that cold, polar air southward.

Heat Waves

Heat waves are abnormally hot weather lasting from days to weeks. The number of these have been increasing in recent years—especially in 2011 and 2012, with the number of intense heat waves being almost triple the long-term average.

Heavy Downpours

Changing precipitation and heavy rain is increasing nationally, especially over the last 30 to 50 years. Areas receiving the most increases are in the Midwest and Northeast, but the frequency and intensity of extreme precipitation events are projected for all U.S. regions. Since 1991, the amount of rain falling in heavy precipitation events has been well above average. Extreme events such as these would occur twice as often if we reduce our emissions, but if they continue to increase, extreme precipitation events would occur five times as often from the years 2081-2100.

Glacial Melting

The Intergovernmental Panel on Climate Change projects that the temperatures at Earth's surface will increase by almost 12°F over the next 100 years. This means snow and permafrost will be lost entirely in many places and sea ice at the poles will keep melting. The amount of sea ice floating in the ocean in the Arctic and Antarctic is expected to decrease even more over the 21st century, which will also lead to rising sea levels.

Rising Sea Levels

Warmer climates cause sea levels to rise by two mechanisms:

- Melting glaciers and ice sheets (ice on land) add water to the oceans and
- Ocean water expands as it warms, increasing its volume and raising the sea level.

In the 20th century, the sea level rose about 4 to 8 inches. Thermal expansion and melting ice contributed to about half of that rise. By the year 2100, the sea level is projected to rise another 8 to 20 inches. Thermal expansion of seawater is predicted to account for 75% of this rise. With rising sea levels, we'll also see more acidic ocean water.

Ocean Acidification

Earth's oceans act as buffers against climate change by taking up some of the excess heat and carbon dioxide from the atmosphere. While this is a good thing in the short run, it can spell disaster in the long run. When carbon dioxide mixes with seawater, it forms a weak acid called carbonic acid. Scientists believe this process has reduced the pH of the oceans by 0.1 since pre-industrial times. By 2100, further acidification could be from 0.14 to 0.35, making life for marine organisms difficult.

Animal Migration and Extinction of Species

Of course, humans aren't the only ones in danger when it comes to global warming and climate change. The UN's Intergovernmental Panel on Climate Change (IPCC) found that many life-forms are moving north into deeper waters to survive as their habitats change. Some animals will be adaptable to climate change—including weeds and pests and some invasive species such as the Burmese python of Florida. Those facing the most danger will be species which are highly specialized when it comes to what they eat or where they live—those whose habitats may disappear completely.

Rising Costs in Coastal Cities

With rising sea levels come rising threats—and substantial costs—for large coastal cities. Low lying places such most of the state of Florida, and the cities of Houston, and Tokyo are finding ways to adapt by building seawalls, pumps, and even floating homes and buildings. Though climate change presents an interesting challenge for designers and innovators to solve, the solutions are often prohibitively expensive, costing tens of millions to billions of dollars to develop. In 2016, the U.S. Department of Housing and Urban Development spent nearly $50 million to relocate just about 100 residents of a Louisiana village, Isle de Jean Charles, that lost 98% of its land to the sea. The state of Louisiana is expecting more communities to be affected as time goes on.

In Japan, they have built tunnels that divert water away from the region's most vulnerable floodplains, costing them $2 billion for the underground flood prevention system. Houston has also put forth efforts to fight inevitable floods by asking for state and federal funds to help build a new $400 million reservoir to help keep stormwater from destroying downstream neighbourhoods.

The Increase of Health Issues and Costs

Global warming also puts human health at risk. Areas that experience wildfires and severe drought conditions suffer from reduced air quality that result in respiratory and cardiovascular hospitalizations for lung illnesses, bronchitis, and other breathing problems.

Vector-borne diseases are a growing concern—especially in areas where cold climates usually kept pests like mosquitos and ticks from gaining a foothold. Incidence of tick-borne Lyme disease is strongly influenced by climatic factors, especially temperature, precipitation, and humidity. In regions where Lyme disease already exists, milder winters result in fewer disease-carrying ticks dying during the winter. Increasing the overall tick population will increase the risk of contracting Lyme disease in those areas.

Other ways climate change threat human health

- Severe weather: injuries, fatalities, mental health impacts
- Air pollution: asthma, cardiovascular disease
- Changes in vector ecology: malaria, dengue, encephalitis, hantavirus, Rift Valley fever, Lyme disease, Chikungunya, West Nile virus
- Increasing allergens: respiratory allergies, asthma
- Water quality impacts: cholera, cryptosporidiosis, campylobacter, leptospirosis, harmful algal blooms
- Water and food supply impacts: foot shortages, malnutrition, diarrheal disease
- Environmental degradation: forced migration, civil conflict, mental health impacts
- Extreme heat: heat-related illness and death, cardiovascular failure

4.4. Impacts on Agriculture and Food Supply

Agriculture is a significant sector of the U.S. economy, accounting for more than $300 billion to the economy every year. Combine this with food-service and other agriculture-related industries, it adds up to more than $750 billion to the GDP. Agriculture is highly dependent on the climate. Increases in temperature and carbon dioxide can increase some crop yields in some places, but to see these benefits, nutrient levels, soil moisture, and water availability conditions also must be met.

As we experience more severe droughts and floods, agriculture becomes more challenging, threatening our food security. Climate change and global warming will make it more difficult to grow crops, raise animals, and catch fish in the same ways and places that we have before.

Impacts on Energy

Changes in temperature, precipitation, and sea level and how often severe extreme events take place will affect how much energy we consume and produce in the U.S. Since we rely on energy so heavily for modern life, it is important that we consider the long-lasting effects that global warming plays on our access to energy.

In a warmer climate, we will use more electricity for air conditioning. If the nation's climate warms by almost 2°F, the demand for energy used for cooling will increase by 5-20%, while the demand for energy used for heating will decrease by about 3-15%. This will affect greenhouse gas emissions, but the net effect will depend on which energy sources are used for electricity and heating. A warmer climate might also reduce the efficiency of power for many existing fossil fuel and nuclear power plants because these plants' equipment need to maintained at certain temperatures to function safely. The colder the water used to cool this equipment, the more efficient the generator will be. So, if higher air and water temperatures occur, the efficiency of these plants converting fuel into electricity will be reduced.

Water Availability and the Effect on Energy

Energy relies heavily on water systems. Hydroelectricity in itself is an important source of power in the U.S. Changes in precipitation, increased drought, reduced snowpack, and changes in the timing of snowmelt in the spring will influence the patterns of energy and water use.

In areas of the Southeast and Southwest, increased competition for water to meet the demands of population might stifle economic growth, especially when also trying to protect natural ecosystems. Due to concerns that water may be scarce in the next couple decades, some local governments have slowed or stopped plans for new power plants that require large amounts of water.

As per predicted changes, by 2050 there will be increase in CO_2 levels upto 450 to 600 ppm, rise in sea level by 10-15 cm, rise in temperature by 1 to 2°C apart from increase in precipitation and storminess (Table 4.1)

Table 4.1: Predicted effects of climate change on agriculture over the next 50 years

Climatic element	Expected changes by 2050's	Confidence in prediction	Effects on agriculture
CO_2	Increase from 360 ppm to 450 - 600 ppm (2005 levels now at 379 ppm)	Very high	Good for crops: increased photosynthesis; reduced water use
Sea level rise	Rise by 10 -15 cm Increased in south and offset in north by natural subsistence/ rebound	Very high	Loss of land, coastal erosion, flooding, salinization of groundwater
Temperature	Rise by 1-2°C. Winters warming more than summers. Increased frequency of heat waves	High	Faster, shorter, earlier growing seasons, range moving north and to higher altitudes, heat stress risk, increased evapotranspiration

Climatic element	Expected changes by 2050's	Confidence in prediction	Effects on agriculture
Precipitation	Seasonal changes by + 10%	Low	Impacts on drought risk' soil workability, water logging irrigation supply, transpiration
Storminess	Increased wind speeds, especially in north. More intense rainfall events.	Very low	Lodging, soil erosion, reduced infiltration of rainfall
Variability	Increases across most climatic variables. Predictions uncertain	Very low	Changing risk of damaging events (heat waves, frost, droughts floods) which effect crops and timing of farm operations

Source: Climate change and Agriculture, MAFF (2000)

4.5. Ways to Mitigate Global Warming

You may not think that your actions alone can affect the earth's climate, but most of us generates tons of greenhouse gas emissions into the atmosphere using the resources we need. It's always a good idea to make the better effort in making the world a cleaner place so that you and generations after you can enjoy a healthy planet. Here are some ways you can help.

- **Drive Less:** choose public transportation, carpooling, or take your bike.
- **Switch to a electric or hybrid car:** investing in an electric car can not only save the planet but save you money in the long run, too!
- **Travel Smart:** for every 1600 miles of air travel you avoid, you can save 720 pounds of carbon emissions. Try opting for a greener traveling option like a road trip or take the train.
- **Switch to renewable energy:** power plants are the single largest source of greenhouse gases. Just Energy offers green energy options in all markets that we serve, so consider sustainable energy when choosing your next energy provider.
- **Install solar panels:** the sun will continue to shine for billions of years, but today it's a carbon-neutral source of light and heat. While they might be an investment at first, federal and state governments often offer tax incentives for the carbon offsets created by renewable energy.
- **Make your home efficient:** use caulk, insulation, and weather stripping to seal air leaks in your home. Upgrade to energy-efficient LED lighting. You'll save money on your energy bill and reduce your carbon footprint!

- **Upgrade to a smart thermostat:** smart thermostats allow you to automatically adjust temperature settings in your home for peak energy efficiency. Some state and local governments even offer tax rebates when you install one.
- **Get energy-efficient appliances:** look for the EPA's ENERGY STAR label when buying new appliances. If every home in the U.S. used energy-efficient appliances, we could eliminate 175 million tons of greenhouse gas emissions and save $15 billion in energy costs every year.
- **Reduce, reuse, and recycle:** you can save 2,400 pounds of carbon dioxide every year by recycling just half of your household waste.
- **Consider carbon offsets:** these are something you pay for in order to compensate for the carbon or greenhouse gas emissions that you generate. Programs like Terrrapass fund projects that destroy greenhouse gases and produce renewable energy, countering your carbon footprint.

Climate Change – Mitigation and Adaptation in Agriculture

- Assist farmers in coping with current climatic risks by providing value-added weather services to farmers. Farmers can adapt to climate changes to some degree by shifting planting dates, choosing varieties with different growth duration, or changing crop rotations.
- An Early warning system should be put in place to monitor changes in pest and disease outbreaks. The overall pest control strategy should be based on integrated pest management because it takes care of multiple pests in a given climatic scenario.
- Participatory and formal plant breeding to develop climate-resilient crop varieties that can tolerate higher temperatures, drought and salinity.
- Developing short-duration crop varieties that can mature before the peak heat phase set in.
- Selecting genotype in crops that have a higher per day yield potential to counter yield loss from heat-induced reduction in growing periods.
- Preventive measures for drought that include on-farm reservoirs in medium lands, growing of pulses and oilseeds instead of rice in uplands, ridges and furrow system in cotton crops, growing of intercrops in place of pure crops in uplands, land grading and levelling, stabilization of field bunds by stone and grasses, graded line bunds, contour trenching for runoff collection, conservation furrows, mulching and more application of Farm yard manure (FYM).

- Efficient water use such as frequent but shallow irrigation, drip and sprinkler irrigation for high value crops, irrigation at critical stages.
- Efficient fertilizer use such as optimum fertilizer dose, split application of nitrogenous and potassium fertilizers, deep placement, use of neem, karanja products and other such nitrification inhibitors, liming of acid soils, use of micronutrients such as zinc and boron, use of sulphur in oilseed crops, integrated nutrient management.
- Seasonal weather forecasts could be used as a supportive measure to optimize planting and irrigation patterns.
- Provide greater coverage of weather linked agriculture insurance.
- Intensify the food production system by improving the technology and input delivery system.
- Adopt resource conservation technologies such as no tillage, laser land levelling, direct seeding of rice and crop diversification which will help in reducing in the global warming potential. Crop diversification can be done by growing non-paddy crops in rain fed uplands to perform better under prolonged soil moisture stress in kharif.
- Develop a long-term land use plan for ensuring food security and climatic resilience.
- National grid grain storages at the household/ community level to the district level must be established to ensure local food security and stabilize prices.
- Provide incentives to farmers for resource conservation and efficiency by providing credit to the farmers for transition to adaptation technologies.
- Provide technical, institutional and financial support for establishment of community banks of food, forage and seed.
- Provide more funds to strengthen research for enhancing adaptation and mitigation capacity of agriculture

Climate Impacts on Agriculture and Food Supply

Agriculture and fisheries are highly dependent on the climate. Increases in temperature and carbon dioxide (CO_2) can increase some crop yields in some places. But to realize these benefits, nutrient levels, soil moisture, water availability, and other conditions must also be met. Changes in the frequency and severity of droughts and floods could pose challenges for farmers and ranchers and threaten food safety. Meanwhile, warmer water temperatures are likely to cause the habitat ranges of many fish and shellfish species to shift, which could disrupt ecosystems. Overall, climate change could make it more difficult to grow crops, raise animals, and catch fish in the same ways and same places as we have done in the past. The effects of climate change also need to be considered along with other evolving factors that affect agricultural production, such as changes in farming practices and technology.

Impacts on Crops

Despite technological improvements that increase corn yields, extreme weather events have caused significant yield reductions in some years. Crops grown in the United States are critical for the food supply here and around the world. U.S. farms supply nearly 25% of all grains (such as wheat, corn, and rice) on the global market. Changes in temperature, atmospheric carbon dioxide (CO_2), and the frequency and intensity of extreme weather could have significant impacts on crop yields.

For any particular crop, the effect of increased temperature will depend on the crop's optimal temperature for growth and reproduction. In some areas, warming may benefit the types of crops that are typically planted there, or allow farmers to shift to crops that are currently grown in warmer areas. Conversely, if the higher temperature exceeds a crop's optimum temperature, yields will decline. Higher CO_2 levels can affect crop yields. Some laboratory experiments suggest that elevated CO_2 levels can increase plant growth. However, other factors, such as changing temperatures, ozone, and water and nutrient constraints, may counteract these potential increases in yield. For example, if temperature exceeds a crop›s optimal level, if sufficient water and nutrients are not available, yield increases may be reduced or reversed. Elevated CO_2 has been associated with reduced protein and nitrogen content in alfalfa and soybean plants, resulting in a loss of quality. Reduced grain and forage quality can reduce the ability of pasture and rangeland to support grazing livestock.

- More extreme temperature and precipitation can prevent crops from growing. Extreme events, especially floods and droughts, can harm crops and reduce yields. For example, in 2010 and 2012, high night-time temperatures affected corn yields across the U.S. corn belt, and premature budding due to a warm winter caused $220 million in losses of Michigan cherries in 2012.
- Dealing with drought could become a challenge in areas where rising summer temperatures cause soils to become drier. Although increased irrigation might be possible in some places, in other places water supplies may also be reduced, leaving less water available for irrigation when more is needed.
- Many weeds, pests, and fungi thrive under warmer temperatures, wetter climates, and increased CO_2 levels. Currently, U.S. farmers spend more

than $11 billion per year to fight weeds, which compete with crops for light, water, and nutrients. The ranges and distribution of weeds and pests are likely to increase with climate change. This could cause new problems for farmers› crops previously unexposed to these species.

- Though rising CO_2 can stimulate plant growth, it also reduces the nutritional value of most food crops. Rising levels of atmospheric carbon dioxide reduce the concentrations of protein and essential minerals in most plant species, including wheat, soybeans, and rice. This direct effect of rising CO_2 on the nutritional value of crops represents a potential threat to human health. Human health is also threatened by increased pesticide use due to increased pest pressures and reductions in the efficacy of pesticides.

Impact of climate change on Indian agriculture was studied under National Innovations in Climate Resilient Agriculture (NICRA). Rainfed rice yields in India are projected to reduce marginally (<2.5%) in 2050 and 2080 and irrigated rice yields by 7% in 2050 and 10% in 2080 scenarios. Further, wheat yield projected to reduce by 6-25% in 2100 and maize yields by 18-23%. Future climates are likely to benefit chickpea with increase in productivity (23-54%).

Impacts on Livestock

Americans consume more than 36 million metric tons of meat and poultry annually. Livestock and poultry account for over half of U.S. agricultural cash receipts, often over $100 billion per year. Changes in climate could affect animals both directly and indirectly.

- Heat waves, which are projected to increase under climate change, could directly threaten livestock. In 2011, exposure to high temperature events caused over $1 billion in heat-related losses to agricultural producers. Heat stress affects animals both directly and indirectly. Over time, heat stress can increase vulnerability to disease, reduce fertility, and reduce milk production.

- Drought may threaten pasture and feed supplies. Drought reduces the amount of quality forage available to grazing livestock. Some areas could experience longer, more intense droughts, resulting from higher summer temperatures and reduced precipitation. For animals that rely on grain, changes in crop production due to drought could also become a problem.

- Climate change may increase the prevalence of parasites and diseases that affect livestock. The earlier onset of spring and warmer winters could allow some parasites and pathogens to survive more easily. In areas with increased rainfall, moisture-reliant pathogens could thrive.

- Potential changes in veterinary practices, including an increase in the use of parasiticides and other animal health treatments, are likely to be adopted to maintain livestock health in response to climate-induced changes in pests, parasites, and microbes. This could increase the risk of pesticides entering the food chain or lead to evolution of pesticide resistance, with subsequent implications for the safety, distribution, and consumption of livestock and aquaculture products.
- Increases in carbon dioxide (CO_2) may increase the productivity of pastures, but may also decrease their quality. Increases in atmospheric CO_2 can increase the productivity of plants on which livestock feed. However, the quality of some of the forage found in pasturelands decreases with higher CO_2. As a result, cattle would need to eat more to get the same nutritional benefits.

Impacts on Fisheries

American fishermen catch or harvest five million metric tons of fish and shellfish each year. U.S. fisheries contribute more than $1.55 billion to the economy annually (as of 2012). Many fisheries already face multiple stresses, including overfishing and water pollution. Climate change may worsen these stresses. In particular, temperature changes could lead to significant impacts.

- Many aquatic species can find colder areas of streams and lakes or move north along the coast or in the ocean. Nevertheless, moving into new areas may put these species into competition with other species over food and other resources, as explained on the Ecosystems Impacts page.
- Some marine disease outbreaks have been linked with changing climate. Higher water temperatures and higher estuarine salinities have enabled an oyster parasite to spread farther north along the Atlantic coast. Winter warming in the Arctic is contributing to salmon diseases in the Bering Sea and a resulting reduction in the Yukon Chinook Salmon, Finally, warmer temperatures have caused disease outbreaks in coral, eelgrass, and abalone.
- Changes in temperature and seasons can affect the timing of reproduction and migration. Many steps within an aquatic animal's lifecycle are controlled by temperature and the changing of the seasons. For example, in the Northwest warmer water temperatures may affect the lifecycle of salmon and increase the likelihood of disease. Combined with other climate impacts, these effects are projected to lead to large declines in salmon populations.

In addition to warming, the world's oceans are gradually becoming more acidic due to increases in atmospheric carbon dioxide (CO_2). Increasing acidity could harm shellfish by weakening their shells, which are created by removing calcium from seawater. Acidification also threatens the structures of sensitive ecosystems upon which some fish and shellfish rely.

4.6 Future Hazards

Indicators of future hazard included in computation of risk index

S.No.	Indicator	Measurement (Unit)	Rationale	Relationship with Future Hazard
1	Change in annual rainfall	Change (%) in annual rainfall during 2020-49 relative to the baseline (1976-2005)	Increase in rainfall is favourable to agriculture productivity	Inverse
2	Change in June rainfall	Change (%) in June rainfall during 2020-49 relative to the baseline (1976-2005)	An increase in rainfall enable sowing of crops in right time	Inverse
3	Change in July rainfall	Change (%) in July rainfall during 2020-49 relative to the baseline (1976-2005)	An increase in July rainfall enables sowing of crops in right time and better establishment of crop stand	Inverse
4	Change in number of rainy days	Change (%) in number of rainy days during 2020-49 relative to the baseline (1976-2005)	An increase in number of rainy days implies a better distribution of rainfall	Direct
5	Change in maximum temperature	Change in maximum temperature (°C) 2020-49 relative to the baseline (1976-2005)	An increase in maximum temperature implies adverse effects on crop yields	Direct
6	Change in minimum temperature	Change in minimum temperature (°C) 2020-49 relative to the baseline (1976-2005)	An increase in minimum temperature implies adverse effects on crop yields, especially for *rabi* crops like wheat	Direct
7	Change in incidence of unusually hot days	Change in frequency of days during March to May when maximum temperature exceeds the normal by at least 4°C during 2020-49 relative to the baseline (1976-2005)	An increase in frequency will imply adverse yield effects	Direct
8	Change in incidence of unusually cold days	Change in frequency of days during December to February when minimum temperature fall below the normal by at least 4°C during 2020-49 relative to the baseline (1976-2005)	An increase in frequency will imply adverse yield effects	Direct

S.No.	Indicator	Measurement (Unit)	Rationale	Relationship with Future Hazard
9	Change in frequency of occurrence of frost	Change in frequency of occurrence of frost during 2020-49 relative to the baseline (1976-2005)	An increase in frequency will imply adverse yield effects	Direct
10	Change in drought proneness	Change in drought proneness during 2020-49 relative to the baseline (1976-2005)	Increase in drought proneness means higher yields risk	Direct
11	Change in incidence of dry spells of >= 14 days	Change in dry spell score3 computed for dry spells during June to October during 2020-49 relative to the baseline (1976-2005)	Higher the number of dry spells, less is productivity	Direct
12	Change in 99 percentile of daily rainfall	Change during 2020-49 relative to the baseline (1976-2005)	An increase indicates the possibility of crop productivity getting affected. Increase in the intensity of such extreme rainfall event also means higher probability of with all the attendant problem	Direct
13	Change in number of events with > mm in 3 days	Change in the number of event during 2020-49 relative to the baseline (1976-2005)	These events will adversely affect crop stand and crop productivity. Increased incidence of these events also mean higher probability of floods with all the attendant problem	
14	Change in average highest rainfall in a single day as % annual normal	Change during 2020-49 relative to the baseline (1976-2005)	An increase indicates the possibility of crop productivity getting affected. Increase in the intensity of such extreme rainfall event also means higher probability of with all the attendant problems. It is also an indicator of uneven distribution of rainfall	Direct
15	Change in average highest rainfall in 3 consecutive days as % annual normal	Change during 2020-49 relative to the baseline (1976-2005)	These events will adversely affect crop stand and crop productivity. Increased incidence of these events also means higher probability of floods with all attendant problems.	Direct

5

Latest Practices for Cultivation of Cereal Crops

A cereal is any grass cultivated for the edible components of its grain (botanically, a type of fruit called a caryopsis), composed of the endosperm, germ, and bran. The term may also refer to the resulting grain itself (specifically "cereal grain"). Cereal grain crops are grown in greater quantities and provide more food energy worldwide than any other type of crop and are therefore staple crops. Edible grains from other plant families, such as buckwheat, quinoa and chia, are referred to as pseudo cereals. In their natural, unprocessed, whole grain form, cereals are a rich source of vitamins, minerals, carbohydrates, fats, oils, and protein. When processed by the removal of the bran and germ the remaining endosperm is mostly carbohydrate. In some developing countries, grain in the form of rice, wheat, millet, maize etc. constitutes a majority of daily sustenance. In developed countries, cereal consumption is moderate and varied but still substantial, primarily in the form of refined and processed grains. The seven principal cereals grown in the world are wheat, maize, rice, barley, oats, rye and sorghum

Since 1965, 5,587 improved field crops varieties have been developed which include 2,777 of cereals, 933 of oilseeds, 1,030 of pulses, 210 of forage crops, 460 of fiber crops, 134 of sugarcane and 43 of potential crops. During 2020–21, a total of 254 varieties/hybrids including 35 special traits varieties were notified and released for commercial cultivation. (ICAR, New Delhi, 2020-21)

5.1. Rice (*Oryza sativa* L.)

Origin and History

Archaeologists and botanists have long debated the origins of rice. For many archaeologists who focus on East Asia or Southeast Asia, it has long appeared that rice agriculture began in South-Central China, somewhere along the Yangtze river, and spread from there southwards and to northeast towards Korea and Japan. However, archaeologists working in India have argued that their evidence suggests an origin of rice cultivation in the Ganges river valley, by peoples unconnected to those of the Yangtze.

For both regions there are current controversies about how early rice was cultivated, and how best to identify when rice was domesticated as opposed to being gathered wild. For example, recently in India rice grains and early pottery found at the site of Lahuradewa in Uttar Pradesh dating to ca. 6500 BC, have been suggested to indicate very early rice cultivation about 4000 years earlier than has often been assumed for this region. However, other scholars contend that these early rice finds may have been collected from wild stands and further evidence is needed to prove cultivation or domestication. Similarly, in Chinese archaeology it has been assumed that early rice finds of ca. 7000 BC were cultivated, but previous methods of sample analysis did not establish evidence either for cultivation behaviors or for the physical domestication traits in rice. For example, most studies on early rice in China have been based on recovered archaeological grains, which may not be the most informative on whether or a not a plant is domesticated.

A recent reassessment suggests that rice grains, especially from Lower Yangtze show a progressive increase in size over time between 6000 BC and 3500 BC, and that this size increase suggests a domestication process. The same period sees increase in size of rice fan-shaped phytoliths (microscope silica that forms in the rice leaves) which might indicate the same process. If so, then earlier finds at 6000 BC or 7000 BC, of small rice grains may be from wild rather than domesticated rice. Others have rebutted, however, that such small grains are also found in much later periods in more northerly regions, like the Yellow River valley (at 3000-2500 BC) and Korea (after 1000 BC), which archaeologists accept as agricultural and therefore small rice grains could be domesticated.

Area and Distribution

Rice cultivation in the world extends from 39° S latitude (Australia) to 45° North latitude (Japan) and 50° North latitude. In India it stretches from 8° North latitude to 34° North latitude.

Rice is the world's leading food crop, cultivated over an area of about 155 million hectares with a production of about 596 million tonnes (paddy). In terms of area and production it is second to wheat. It provides about 22 per cent of the world's supply of calories and 17% of the proteins. Maximum area under rice is in Asia.

Among the rice growing countries, India has the largest area (45.0 million hectares) followed by China and Bangladesh (Table 5.1) In respect of production, India ranks second with 178.3 million tonnes of paddy next to China (213.16 million tonnes of paddy). In regard to average yield per hectare, Australia ranks first with average yield of 100.31 q/ha followed by Tajikistan (88.31 q/ha)

The leading countries producing rice crop are China, India, Bangladesh, Indonesia, Vietnam, Thailand, Myanmar and Philippines. In India, rice is grown in almost all the states. Andhra Pradesh, Bihar, Uttar Pradesh, Madhya Pradesh and West Bengal lead in the area. West Bengal and Uttar Pradesh have the highest rice production.

Table 5.1: Area, Production and Average Yield of Rice (Paddy) in top 10 countries of the world

Country	Area in million ha		Country	Production in million tonnes		Country	Average Yield (q/ha)
India	45.00		China	213.61		Australia	100.31
China	30.34		India	178.31		Tajikistan	88.81
Bangladesh	11.42		Bangladesh	54.91		Egypt	88.30
Indonesia	10.66		Indonesia	54.65		Uruguay	86.17
Thailand	10.40		Viet Nam	42.76		USA	85.40
Viet Nam	7.22		Thailand	30.23		Peru	82.30
Myanmar	6.66		Myanmar	25.10		Morocco	79.95
Nigeria	5.26		Philippines	19.29		Greece	79.64
Philippines	4.72		Brazil	11.09		Turkey	78.15
Pakistan	3.34		Cambodia	10.96		El Salvador	76.52

Source: FAOSTAT 2020

Globally, India ranks first in area where it is grown on 45 mha with production of 178.31 million tonnes and average yield of 39.62 quintal/ha. The highest productivity of 213.61 m tonnes was in China whereas highest average yield was recorded in Australia (100.31q/ha).

Rice crop occupied 31.49 lakh hectares in Punjab with total paddy production of 208.83 lakh tonnes (139.92 lakh tonnes of rice) during 2020-21. The average yield of paddy was 66.31 quintals per hectare.

Climatic Requirements

In India, rice is grown under widely varying conditions of altitude and climate. Rice cultivation in India extends from 8 to 350 N latitude and from sea level to as high as 3000 meters. Rice crop needs a hot and humid climate. It is best suited to regions which have high humidity, prolonged sunshine and an assured supply of water. The average temperature required throughout the life period of the crop ranges from 21° to 37° C. At the time of tillering the crop requires a higher temperature than for growth. Temperature requirement for blooming is in the range of 26.5° to 29.50° C. At the time of ripening, the temperature should be between 20°-25° C. Photo-periodically, rice is a short-day plant. However, there are varieties which are non-sensitive to photoperiodic conditions.

Different Zones for Rice Cultivation

India has been broadly classified into the following six agro-climatic zones:

(i) **The humid western Himalaya region,** which comprises of Jammu and Kashmir, Himachal Pradesh and Kumaon and the Garhwal division of Uttarakhand. The major soil groups are sub-montane, hills and terai soils.

(ii) **The humid Bengal – Assam basin,** which includes West Bengal and Assam. The major soil groups are riverine alluvium, terai soils, lateritic soils, red- yellow loams and red sandy soils.

(iii) **The humid eastern Himalayan region and the Bay Islands,** which include Arunachal Pradesh, Nagaland, Manipur, Mizo-ram, Tripura, Meghalaya and the Andaman and Nicobar Islands. The major soil groups are red loamy soils, lateritic soils, red yellow soils and alluvial soil.

(iv) **The sub-humid Sutlej-Ganga plains** which include Punjab, Uttar Pradesh, Bihar and Delhi. The major soil groups are calcareous alluvial soils, riverine alluviums, saline and alkaline soils, red-yellow loams, mixed red and black soils and red sandy soils.

(v) **The sub humid to humid eastern and south eastern uplands** comprising of Orissa, Andhra Pradesh and eastern Madhya Pradesh. The major soil groups are lateritic soils, red-yellow loams, mixed red and black soils, deltaic alluvium, deep and medium deep black soils, red loamy soils and coastal alluvium.

(vi) **The arid western plains**, which include Haryana, Rajasthan and Dadra and Nagar Haveli. The major soil groups are alluvial soils, red-yellow soils and medium to deep black soils.

Nutrients content of Brown and White Rice

White and brown rice have similar calorie, carbohydrate, protein, and fat content. A 100-gram serving of white, short-grain, cooked rice contains the following nutrients:

Nutrient	Amount	% Daily Value (DV)
Calories	130	-
Carbohydrate	28.7 grams (g)	10%
Protein	2.36 g	5%
Fat	0.19 g	0%
	Brown rice	**White rice**
Manganese	42% DV	16% DV
Niacin	16% DV	3% DV
Thiamine	15% DV	2% DV
Selenium	11% DV	–
Magnesium	9% DV	2% DV

Benefits of brown rice

- Helps to weight loss
- Controls blood sugar level
- Aids digestion
- Controls cholesterol levels
- Neuroprotective effective
- Improves maternal health
- Maintains bone health

Processing of Rice

1. **Pre-cleaning:** When paddy comes into the mill, it contains foreign materials such as straw, weed seeds, soil, and other inert materials. If these are not removed before hulling, the efficiency of the huller and milling recovery will be reduced. The capacity of the paddy pre-cleaner is normally 1.5 times the milling capacity.
2. **Removing the husk (dehusking or dehulling):** Brown rice is produced by removing the husk from rough paddy rice. The husk is removed by friction as the paddy grains pass between two abrasive surfaces that move at different speeds. After dehusking, the husk is removed by suction and transported to a storage dump outside the mill. Husk accounts for about 30% of the paddy weight and an efficient husker should remove 90% of the husk in a single pass.
3. **Paddy separation:** The paddy separator separates unhusked paddy rice from brown rice. The amount of paddy present depends on the efficiency of the husker and should not be more than 10%. Paddy separators work by making use of the differences in specific gravity, buoyancy, and size between paddy and brown rice.

4. **Whitening or polishing:** White rice is produced by removing the bran layer and the germ from the paddy. The bran layer is removed from the kernel through either abrasive or friction polishers. The amount of bran removed is normally between 8 and 10% of the total paddy weight. To reduce the number of broken grains during the whitening process, rice is normally passed through two to four whitening machines connected in series.

5. **Separation of white rice:** After polishing, white rice is separated into head rice, large and small broken rice, and "brewers" by an oscillating screen sifter. Head rice is normally classified as kernels that are 75–80% or more of a whole kernel. To attain a higher degree of precision for grading and separation a length or indent grader is used.

6. **Rice mixing:** A good rice mill will produce 50–60% head rice (whole kernels), 5–10% large broken and 10–15% small broken kernels. Depending on country standards, rice grades in the market will contain from 5–25% broken kernels. If rice mixing is to be done properly, then a volumetric mixer is necessary.

7. **Mist polishing:** Mixing a fine mist of water with the dust retained on the whitened rice improves the luster of rice (polishes) without significantly reducing milling yield. A friction type of whitening machine, which delivers a fine mist of water during the final whitening process, is used for "final" polishing before sale.

8. **Rice weighing:** Rice is normally sold in 50 kg sacks which must be accurately weighed and labeled. While most rice mills use a manual mechanical weighing system, very accurate, and fast electronic systems are also available.

Parboiled Rice

Parboiled rice (also called converted rice and easy-cook rice) is rice that has been partially boiled in the husk. The three basic steps of parboiling are soaking, steaming and drying. These steps make the rice easier to process by hand, while also boosting its nutritional profile, changing its texture, and making it more resistant to weevils. The treatment is practiced in many parts of the world. Parboiling drives nutrients, especially thiamin, from the bran to the endosperm, hence parboiled white rice is mostly nutritionally similar to brown rice.

Rice varieties released and notified by ICAR, New Delhi (2020-2021)

Varieties	Area of adoption	Characters
JKRH 2354	Chhattisgarh, Madhya Pradesh,Maharashtra	Rice hybrid suitable for *kharif* irrigated condition, average grain yield 7.0–7.5 tonnes/ ha, mid early duration (125–130 days), resistant to stem-borer; moderately resistant to bacterial leaf blight (BLB), leaf blast, brown spot.
JKRH 2354 JKRH 2154	Chhattisgarh, Madhya Pradesh, Maharashtra Punjab, Haryana, Uttar Pradesh, Bihar, West Bengal	Rice hybrid suitable for *kharif* irrigated condition, average grain yield 7.0–7.5 tonnes/ ha, mid early duration (125–130 days), resistant to stem-borer; moderately resistant to bacterial leaf blight (BLB), leaf blast, brown spot. Rice hybrid suitable for *kharif* irrigated condition, average grain yield 6.0–6.5 tonnes/ ha, early duration (115–120 days), moderately resistant to BLB, leaf blast, brown spot.
JKRH 2154 CR Dhan 315	Punjab, Haryana, Uttar Pradesh, Bihar, West Bengal, Gujarat, Maharashtra	Rice hybrid suitable for *kharif* irrigated condition, average grain yield 6.0–6.5 tonnes/ ha, early duration (115–120 days), moderately resistant to BLB, leaf blast, brown spot. Suitable for *kharif,* average grain yield 5.0 tonnes/ ha, medium duration (125–130 days), biofortified variety having high zinc content (25.47 ppm).
TRC 2014-8 DRR Dhan 56	Karnataka, Himachal Pradesh, Punjab, Haryana	Suitable for low altitude hill ecology, average grain yield 4.2 tonnes/ha, medium duration (123–134 days), slender grain, moderately resistant to brown spot. Suitable for irrigated early transplanted ecology, average grain yield 5.3 tonnes/ha, maturity 115 days, long slender grains, resistant to leaf blast, false smut, moderately resistant to BLB and tolerant to stem-borer.
DRR Dhan 56 Telangana Vari 3	Punjab, Haryana Telangana, Kerala	Suitable for irrigated early transplanted ecology, average grain yield 5.3 tonnes/ ha, maturity 115 days, long slender grains, resistant to leaf blast, false smut, moderately resistant to BLB and tolerant to stem-borer. Suitable for irrigated mid early ecology, average grain yield 7.4 tonnes/ha, maturity 120 days, medium slender grain type, resistance to gall midge, moderately resistant to leaf blast.

Telangana Vari 3 DRR Dhan	Telangana, Kerala, Andhra Pradesh, Telangana, Chhattisgarh, Maharashtra, Jharkhand, Odisha, Bihar, Gujarat, Maharashtra	Suitable for irrigated mid early ecology, average grain yield 7.4 tonnes/ha, maturity 120 days, medium slender grain type, resistance to gall midge, moderately resistant to leaf blast. Suitable for irrigated condition, average grain yield 5.5 tonnes/ha, maturity 130–135 days, MAS derived variety with additional Xa3 gene, in the genetic background of Improved Samba Mahsuri (Xa21, xa13 and xa5), with enhanced resistance to bacterial blight.
28544 (IET 26549) (PHI-17108) 27P27 (IET 25745) (PHI-16101)	Telangana, Karnataka, Andhra, Pradesh Uttarakhand, Chhattisgarh, Maharashtra	Hybrid rice suitable for *kharif*, average grain yield 6.5–7.0 tonnes/ha, medium maturity (135–140 days), medium-slender grain type, resistant to BLB, tolerant to leaf blast and leaf folder. Rice hybrid suitable for *kharif*, average grain yield 7.0–7.5 tonnes/ha, mid-early maturity (125–130 days), moderate levels of resistance to leaf blast, neck blast, BLB and leaf folder.
Indam 100-012 (IET26999)	Uttar Pradesh, Punjab Telangana, Andhra Pradesh, Tamil Nadu, Kerala	Aromatic rice hybrid suitable for *kharif* irrigated condition, average grain yield 6.5–7.0 tonnes/ha, mid-early maturity (125–130 days), tolerant to stem- borer, resistant to false smut, moderately resistant to neck blast. Suitable for irrigated medium ecology, average grain yield 7.0 tonnes/ha, maturity 135 days, medium slender grains, multiple disease (blast, sheath rot and RTD) resistant and tolerant to plant-hopper and leaf-folder.
MTU Rice 1239 (IET 26263)	Andhra Pradesh, Tamil Nadu, Maharashtra Odisha, Bihar	Suitable for late duration areas, average grain yield 6.3 tonnes/ha, maturity 150–155 days, medium- slender grain, moderately resistant to leaf blast and bacterial blight. Suitable for rainfed, shallow, lowland ecology, average grain yield 5.0 tonnes/ha, late maturing (150 days), non-lodging, low shattering, resistant to blast and BLB.
MTU 1223 (Varsha) (IET 25856)	Odisha, Bihar Himachal Pradesh, Meghalaya, Uttarakhand	Suitable for rainfed, shallow, lowland ecology, average grain yield 5.0 tonnes/ha, late maturing (150 days), non-lodging, low shattering, resistant to blast and BLB. Suitable for irrigated condition of hill ecology, average grain yield 5.0 tonnes/ha, early maturing (115–120 days), moderate resistance to leaf blast.

VL Dhan 88 (IET 25819)	Himachal Pradesh, Meghalaya, Uttarakhand Assam	Suitable for irrigated condition of hill ecology, average grain yield 5.0 tonnes/ha, early maturing (115–120 days), moderate resistance to leaf blast. Rice hybrid suitable for *kharif*, average grain yield 5.8–6.9 tonnes/ha, mid-early duration (125–130 days), tolerant to blast and brown spot, and drought.
PNPH 24 Vikram-TCR	Assam, Chhattisgarh	Rice hybrid suitable for *kharif*, average grain yield 5.8–6.9 tonnes/ha, mid-early duration (125–130 days), tolerant to blast and brown spot, and drought. Suitable for mid-early ecology, average grain yield 5.4 tonnes/ha, maturing 118–123 days, mutant line with short stature, resistant to lodging and tolerant to leaf blast and stem-borer.
CG Jawaphool Trombay (RTR-31)	Chhattisgarh Uttar Pradesh	Improved over local Jawaphool for reduced plant height, average grain yield 4.0 tonnes/ ha, medium maturity (130–135 days), short aromatic slender grain, tolerance to BLB, leaf-folder, and stem-borer. Suitable for submerged condition, average grain yield 4.3 tonnes/ha, late maturing (140–145 days), withstand even under 12–17 days of flooding/ complete submergence and gives good yield.
Sabour Sampanna Dhan (IET 25960)	Bihar, Sikkim	Suitable for both drought and submergence condition, average grain yield 6.8 tonnes/ha (irrigated), 3.1 tonnes/ha (submergence) and 3.3 tonnes/ha (drought stress), late maturing (150 days), moderately resistant to BLB, leaf blast and brown spot diseases. Suitable for irrigated, medium hill ecology, average yield 4.9 tonnes/ha, duration 135 days, moderately resistant to leaf blast, resistant to sheath rot, brown spot, sheath blight and false smut.
Swarna Samriddhi Dhan (IET 24306)	Bihar, Andhra Pradesh	Suitable for late duration areas, average grain yield 5.7 tonnes/ha, late maturing (135–140 days), resistant to false smut and moderately resistant to BLB, leaf blast, sheath blight, brown spot, RTD and glume discoloration. Suitable for early *kharif*, average grain yield 4.8 tonnes/ha, duration 122–125 days, tolerant to high temperature.
NLR 40024	Andhra Pradesh Andhra Pradesh	Suitable for early *kharif*, average grain yield 4.8 tonnes/ha, duration 122–125 days, tolerant to high temperature. Suitable for late duration, average grain yield 5.2 tonnes/ha, maturity 135–140 days, fine medium slender grain.

CG Barani Dhan-2 Jammu Basmati 118	Chhattisgarh Jammu and Kashmir	Suitable for rainfed shallow lowland ecology, average. grain yield 3.5 tonnes/ha, early maturing. (105– 110 days). Suitable for medium-duration cultivation, average grain yield 4.8 tonnes/ha, maturity 130–135 days, aromatic/scented rice, lodging resistant due to short stature.
Jammu Basmati 138	Jammu and Kashmir Kerala	Suitable for late duration condition, average grain yield 4.5 tonnes/ha, late maturity (150–155 days), aromatic/scented rice, moderately resistant to bacterial leaf blight, leaf blast and brown spot. Suitable for irrigated mid-early condition, average grain yield 7.4 tonnes/ha, maturity 115 to 120 days, red colour kernel, moderately resistant to sheath blight, sheath rot, gall midge and brown plant-hopper (BPH), high temperature, acidity and salinity-stress tolerant.

Improved varieties for Punjab

PR 131: It is a high yielding, medium maturing and lodging tolerant variety. Its average plant height is 111 cm and matures in about 110 days after transplanting. It possesses long slender translucent grains with high total and head rice recoveries. It is resistant to all the ten pathotypes of bacterial blight pathogen prevalent in the Punjab state. Its average yield is 77.5 quintals per hectare.

PR 130: It is a high yielding, mid early and lodging tolerant variety. Its average plant height is 108 cm and it matures in about 105 days after transplanting. It possesses long slender translucent and lustrous grains with high total and head rice recoveries. It is resistant to all the ten pathotypes of bacterial blight pathogen prevalent in the Punjab state. Its average yield is 75.0 quintals per hectare.

PR 129: It is an improved version of PAU 201. It possesses long slender clear translucent grains. Its' average plant height is 105 cm and matures in about 108 days after transplanting. It is resistant to all the ten presently prevalent pathotypes of bacterial blight pathogen in the Punjab state. Its' average paddy yield is 75.0 quintals per hectare.

PR 128: It is also an improved version of PAU 201. It possesses long slender clear translucent grains. Its' average plant height is 110 cm and matures in about 111 days after transplanting. It is resistant to all the ten presently prevalent pathotypes of bacterial blight pathogen in the Punjab state. Its' average paddy yield is 76.25 quintals per hectare.

PR 127: It is a medium maturing rice variety with an average height of 104 cm. It matures in about 107 days after transplanting. It possesses long slender, clear translucent grains with high total and head rice recoveries. It is resistant to all the ten presently prevalent pathotypes of bacterial blight pathogen in the Punjab state. Its' average paddy yield is 75.0 quintals per hectare. Do not grow this variety in alkali soils and under brackish water.

PR 126: It is an early maturing rice variety. Its average plant height is 102 cm and matures in about 93 days after transplanting. It possesses long slender, clear translucent grains. It is resistant to seven of the ten presently prevalent pathotypes of bacterial blight pathogen in the Punjab state. Its' average paddy yield is 75.0 quintals per hectare. Transplant 25-30 days old nursery.

PR 122: Its' average plant height is 108 cm and matures in 117 days after transplanting. It possesses long slender translucent grains. It is resistant to all the ten 3 presently prevalent pathotypes of bacterial blight pathogen in the Punjab state. Its' average paddy yield is 78.75 quintals per hectare.

PR 121: It is a short statured, stiff strawed lodging tolerant variety with dark green and erect leaves. It attains height of about 98 cm and matures in 110 days after transplanting. Its' grains are long slender, translucent with good cooking quality. It is resistant to all the ten presently prevalent pathotypes of bacterial blight pathogen in the Punjab state. Its' average paddy yield is 76.25 quintals per hectare.

PR 114: It has narrow, dark green erect leaves. It attains an average height of about 102 cm and matures in about 115 days after transplanting. It possesses extralong, clear translucent grains. It is resistant to five of the ten pathotypes of bacterial blight pathogen presently prevalent in Punjab state. Its' average yield is 68.75 quintals of paddy per hectare.

PR 113: Its' average plant height is about 105 cm. Its' grains are bold and heavy. It matures in about 112 days after transplanting. It is resistant to eight of the ten pathotypes of bacterial leaf blight pathogen presently prevalent in Punjab state. Its' average paddy yield is 70.0 quintals per hectare

Soil Requirements

In India, rice is grown under so diverse soil conditions that it can be said that there is hardly any type of soil in which it cannot be grown, including alkaline and acidic soils. Soils having good water retention capacity with good amount of clay and organic matter are ideal for rice cultivation. Clay or clay loams are most suited for rice cultivation, such soils are capable of holding water for long time and to produce sustainable crop for a longer period. Rice being a semi-aquatic crop grows best under submerged conditions. A major part of rice crop in India is grown under 'lowland' conditions. Rice plant is able to tolerate a wide range of soil reaction, but it does have a preference for acidic soils. It grows well in soils having a pH range between 5.5 and 6.5. It can be grown on alkali soils also, after treating them with gypsum or pyrite.

Nursery Raising Systems

The choice of a particular nursery system depends on the availability of water, labor, land and agricultural implements.

There are 4 nursery systems for transplanting:

1. Wet-bed nursery
2. Dry-bed nursery
3. Dapog or mat nursery
4. Seedling boxes for mechanical transplanting

1. Wet-bed Method

The wet-bed nursery is mainly used in areas where there is enough water. Pre-germinated seeds are broadcast on a soil that is thoroughly puddled and leveled. Drainage canals for proper removal of water must be constructed. Addition of organic manure (decomposed) and small amount of inorganic fertilizer as basal dressing will increase easiness of uprooting of seedlings and seedling vigour. Total seed bed area for raising nursery for one hectare should be 400 sqm and broadcast 20kg healthy seeds in standing water. The best seedling age for transplanting is about 25-30 days after sowing. Nurseries should be free from weeds, any pest or disease incidence and nutrient deficiencies. If such conditions occur it must be treated at the nursery level.

2. Dry-bed Method

The nursery is prepared in dry soil conditions. Raised seed beds of convenient dimensions are prepared by raising the soil to a height of about 5-10 cm. A layer of half burned paddy husk could be distributed on the nursery bed to facilitate uprooting. Total seed bed area is also about 400 sqm. Sow 20kg healthy seeds in close lines (4-5cm) on the bed top and nursery is sufficient for one hectare.

The site should be free of shade and with adequate irrigation facilities. Uprooting of seedlings should be done between 25 - 30 days after sowing. Nursery should be maintained without any moisture stress. A basal fertilizer mixture can be applied and incorporated between rows if the soil nutrient supply is low.

The advantage of the dry-bed method is that seedlings are short and strong, with a longer root system compared with the wet-bed method. The seedlings can be raised even during periods of heavy rains. A disadvantage is however that seedlings of upland nurseries may also get infected with blast and are more prone to pests such as rodents etc.

3. Dapog Method (mats)

Dapog nurseries can be located anywhere on a flat firm surface but water supply/control should be very reliable. The area needed is about 100 m^2/ha or 1% of the transplantable land which is much smaller than conventional nurseries. Somewhat higher seed i.e. 30kg/ha should be preferred in order to compensate mortality due to competition.

Another advantage of the "dapog" over wet/dry-bed nurseries is that the cost of uprooting of seedling is minimal. But, because the seedlings are small, transplanting can be more difficult. Very young seedlings from dapog nurseries suffer less from the transplanting shock compared with other nurseries, thus the seedlings are more suitable for short duration varieties. Irrigation is obligatory to prevent water stress. Some dapog methods are discussed as below.

The modified mat nursery

This technology establishes seedlings in a layer of soil mix, arranged on a firm surface. It uses less land and requires fewer seeds and inputs, such as fertilizer and water, reducing nursery costs by up to 50%. After 20-25 days of sowing, the seedlings reach the four-leaf stage which favors quick establishment in the field and rapid growth and are ready for transplanting.

The bubble tray nursery

The bubble tray nursery is a good system to develop 12-15-day old seedlings with "root balls". The seedlings are raised on plastic trays of 59 cm by 34 cm with 434 embedded holes. You will need 750 trays per hectare of paddy.

4. Seedling boxes for mechanical transplanting

Mechanized transplanting requires techniques that are different from hand transplanting. Usually seedling boxes are used that are adapted to the type of transplanter. In a seedling box, seedlings are grown on a thin layer of soil in 30 cm x 60 cm trays. In some instances, seedlings are grown on larger areas and then cut into rectangular strips (mats of seedlings) that fit into the planting trays of the transplanter.

Direct Sowing of Rice (DSR)

Direct Seeded Rice (DSR) technology is sowing of rice seeds in moist field with rice planter. There is lot of saving in irrigation water in DSR method as compared to puddled transplanted rice. Also soil health (physical and chemical properties) is improved as in transplanted rice there is formation of hard layer in puddled rice which hinder downward movement of water resulting in adverse effect on succeeding wheat crop. However, the practice of DSR has many problems especially of weeds as paddy and non-paddy weeds appear in this crop resulting in higher weed losses if left uncontrolled. Also iron deficiency is more in DSR due to its non-availability in oxidized conditions. To make DSR successful, adopt following practices:

- Laser level the field meant for DSR sowing.
- Select medium to heavy soil for cultivation of DSR.
- Do not sow on light soil as iron deficiency is more in these soils.
- Ensure a proper crop stand by sowing the crop at proper moisture level with 22.5 cm row to row spacing.
- Use 20 kg seeds per hectare for sowing. Calibrate seeds drill before sowing in order to maintain optimum plant stand.
- Sowing should be done at optimum time i.e. first fortnight of June
- Prefer short duration varieties for direct sowing.
- Apply pendimethalin at 0.75 kg/ha as pre-em for controlling non-paddy weeds.
- Also spray bispyribac at 25 gm/ha as post-em (30-35 DAS) to control paddy weeds.
- Irrigate the field as and when required.

Transplanting methods

1. Manual transplanting
2. Machine transplanting
3. System of Rice Intensification

1. Manual transplanting

Manual transplanting can be done in the flat puddled fields or by transplanting two lines on top of the bed (67.5 cm) which are made in the dry field i.e. without puddling by keeping 33 hills/sq.m. There will be saving in irrigation water with bed planting techniques. Manual transplanting is most suited for labour-surplus areas and for small rice fields. Manual transplanting can be done in fields with less than optimal levelling and with varying water levels. The procedure for transplanting is as follows

- Pull out the seedlings at an average of 25 to 30 days after seeding (DAS) from nurseries and transport them to the main field.

- Transplant the seedlings soon after pulling from the nursery in puddled, leveled field (any delay will lead to slow revival and even death of some seedlings).
- Transplant 2-3 seedlings per hill in lines at 20cm x15cm spacing to get optimum stand of 33 hills/sq.m.
- Handle seedlings carefully to ensure their fast revival and rapid growth after transplanting.

Limitations

- Transplanting is tedious and time-consuming process.
- Difficult to get enough labour at peak periods to plant entire area on time.
- Low plant density with contract transplanting on area basis lowers yields.
- Sometimes in puddled field cracks may develop due to absence of rainfall or delay of irrigation.

2. Machine Transplanting

Machine transplanting involves planting young rice seedlings into puddled soil by machine with adoption of following procedures

- Raise seedlings in special mat nurseries or in seedling trays. Use required quantity of good seed per 100 m^2 of nursery for each ha.
- Seedlings will be ready for transplanting in 20-25 days after seeding (DAS). Ensure that fields are well puddled and leveled.
- Drain fields and allow mud to settle for 1-2 days after the final puddling.
- The subsurface soil layers need to be hard enough to support the transplanting machine.
- The soil is ready when a small "V" mark made in the puddled soil with a stick holds its shape. At this moisture level, the soil can hold the seedlings upright.
- Soil should not be so dry that it sticks to and interferes with planting parts or wheels of the transplanter.
- Load the seedling mats on the machine and transplant the seedlings at the selected machine setting.

Advantages

- Machine transplanting requires considerably less time i.e. (1-2 ha/person/day versus 0.07 ha/person/day) in manual transplanting.

- Fast, efficient and timely planting is ensured.
- Reduces stress, drudgery and health risks.
- Ensures uniform spacing and plant density.
- Seedlings recover fast, tiller vigorously, and mature uniformly.

Limitations

- Seedlings must be planted while still young, and so mechanical transplanting is best suited for irrigated areas only.
- Special care for nursery management is needed (mat nursery or seedling trays).
- Good land preparation, leveling and water management are required.

3. SRI (System of Rice Intensification)

The System of Rice Intensification involves cultivating rice with as much organic manure as possible, starting with young seedlings planted singly at wider spacing in a square pattern; and with intermittent irrigation that keeps the soil moist but not inundated, and frequent inter cultivation with weeder that actively aerates the soil.

SRI is not a standardized, fixed technological method. It is rather a set of ideas, a methodology for comprehensively managing and conserving resources by changing the way that land, seeds, water, nutrients, and human labour are used to increase productivity from a small but well-tended number of seeds. SRI is an amalgamation of multiple beneficial practices.

In SRI paddy cultivation, less quantity of seeds i.e. 5 kg / hectare is required. Hence fewer plants per unit area (25 x 25 cm) whereas in mainstream chemical intensive paddy cultivation requires 20 kg seed per hectare. (1 acre= approx. 0.4 ha). SRI is initially labour intensive because it

- Needs 50% more man-days for transplanting and weeding.
- Mobilises labour to work for profit.
- It offers an alternative to the resource poor, who put in their family labour.
- Once the right skills are learnt and implemented, the labour costs will be lesser.

Crop raised with SRI technique encourages rice plant to grow healthy with

- Large root volume
- Profuse and strong tillers - Maximum tillering (30 tillers/plant can be easily achieved; 50 tillers per plant are quite attainable) occurs concurrently with panicle initiation.
- Non lodging
- Big panicles
- More and well filled grain panicles and higher grain weight
- Resists insects because it allows rice to absorb soil nutrients naturally

Do-s and Don't-s in SRI

- **Early transplanting:** Transplant 10-15-day old seedlings (after germination), with only two small leaves to encourage more tillering and root growth potential.
- **Careful Transplanting:** Minimize trauma in transplanting. Remove plant from nursery with the seed, soil and roots carefully and place it in the field without plunging too deep into soil (more tillering potential)
- **Wide spacing:** Plant single seedlings, not in clumps, and in a square pattern 25cm x 25cm apart or wider.
- **Weeding and aeration:** Use simple mechanical "rotating hoe" that churns up soil; 2 weedings required, (more root growth, due to reduced weed competition, and aeration of soil, giving roots more oxygen and nitrogen due to increased microbial activity) Each additional weeding after two rounds results in increased productivity up to 2 t/ha / weeding.
- **Water management:** Regular water application to keep soil moist but not saturated, with intermittent dryings, alternating aerobic and anaerobic soil conditions (more root growth because it avoids root degeneration, enables better absorption of nutrients from the soil).
- **Compost / FYM** is applied instead of or in addition to chemical fertilizer; 10 tons/ha (more plant growth because of better soil health and structure, and more balanced nutrient supply)

Rotations

- Paddy - Wheat
- Paddy – Potato – Urd (black gram)
- Paddy – Potato – Moong (green gram)

- Paddy – Toria –Wheat
- Paddy – Pea (for pods) - Moong (green gram)
- Paddy – Lentil
- Paddy - Pea (for seed)
- Paddy - Raya / Transplanted gobi sarson

In case of intensive crop rotations where there are more than two crops in a year, a paddy variety of early duration should be selected.

Seed Rate :- For direct seeded and transplanted rice, 20 kg/ha of healthy seeds are required. However for SRI only 5 kg/ha seed is required.

Seed Treatment:- Treat healthy seeds with Sprint 75 WS (carbendazim + mancozeb) by making slurry of 3gm fungicide formulation in 10-12 ml water for one kg seed (60g fungicides in 200-240 ml water for 20 kg seed).

Time of Nursery Sowing:- Time of nursery sowing depends upon the selected rice variety as well as agro climatic region. In general nursery for coarse rice varieties should be shown in the second fortnight of May and for basmati the suitable time is second fortnight of June.

Area of Nursery Sowing:- Nursery for transplanted puddled crop can be raised under puddled or dry conditions (on raised beds). Prepare nursery bed of 160 sq.m. area for transplanting one acre of rice. Broadcast 8 kg healthy sprouted seed in puddled plot after adding the recommended dose of FYM and other inorganic fertilisers in pudded plot. In dry method, sow on top of bed in close lines.

Age and Time of Nursery Transplanting:- About 25 to 30 days old nursery of both coarse and basmati rice is suitable for transplanting. However in SRI, 10 to 15 days old nursery is suitable. Optimum time of transplanting is second fortnight June for coarse rice and whole of July for basmati.

Method of Transplanting:- Optimum population (33 hills/sq.m.) is essential for getting higher yield. This population can be achieved by keeping row to row spacing of 20 cm and plant to plant spacing of 15 cm. Under late sown condition plant population per unit area must be increased in order to compensate yield losses due to late transplanting.

Weed Management

• For Direct seeded rice

In line-planted or drilled rice, weeding can be done with a hand-hoe or with rotary weeders. The best time to weed the crop is three weeks after sowing. Later on two to three weedings at an interval of 15 days should be done as per requirements.

In case of direct seeded rice, both paddy and non-paddy weeds appear. For the control of non-paddy weeds, pre-emergence application of pendimethalin 0.75 kg/ ha can be done and for the control of typical paddy weeds. Bispyribac at 25 g/ha can be sprayed as post-emergence (25 to 30 DAT) for controlling nearly all weeds of transplanted paddy.

- **Transplanted rice**

Herbicides, such as butachlor 1.5 kg/ha, anilophos 0.375 kg/ha, thiobencarb 1.5 kg/ ha, pretilachlor 0.75 kg/ha, pendimethalin 0.75 kg/ha, oxadiargyl 90 g/ha, pyrazosulfuron-ethyl 15 g/ha etc. provides a satisfactory control of grass weeds when used at the recommended doses and at the right time (within 3 days of transplanting). Alternatively, early post emergence (10-12 DAT) application of penoxulam 24 SC at 24 gm per hectare or post emergence (20-25DAT) application of fenoxaprop 6.7 EV at 67g/ha for controlling grassy weeds of transplanted rice. Bispyribac at 25 g/ha can be sprayed as post-emergence (25 to 30 DAT) for controlling nearly all weeds of transplanted paddy.

For the control of sedges and broad leaf weeds from rice (both direct and transplanted), apply metsulfuron 20 WP at 15 g/ha or ethoxysulfuron 15 WDG at 20 g/ ha or bensulfuron methyl 60 DF 60 g/ha or azimsulfuron 50 DF 20 g/ha or metsulfuron methyl + chlorimuron ethyl @ 4 gha, 20-25 days after transplanting or sowing through spray applications. Drain out the ponded water before spray in transplanted crop as these are foliar uptake herbicides.

Note

I. Drain out excess water before spraying post-em herbicides.

II. Do not delay herbicide application.

III. Always use herbicide in rotation.

Water Management

The water requirement of rice is higher than that of other cereal crops of similar duration. Figures ranging from 37 to 75 acre-inches of water have been reported from various locations in India as the water requirement of rice, and this variation is primarily due to different soil and environmental conditions available in different parts of the country. Losses due to percolation depends upon soil type. In lighter soils, such losses amount to about 60 per cent of the total water requirement. Soil compaction and puddling help to reduce percolation losses. Losses due to transpiration account for about 40 per cent. Losses due to evaporation depend upon the climatic factors and range from 20 to 40 per cent. Considerable saving (30-40 per cent) in water is possible when shallow submergence (2-5 cm) is maintained from tillering to pre flowering and near saturation during other phases of growth.

Fertilizer and Nutrient Management

The new short-statured varieties have a high fertilizer response and give high grain yield. The major nutrients required by rice are nitrogen, phosphorus and potassium. Among the micro-nutrients, zinc and iron are becoming important. Among all the nutrient elements, nitrogen is most important for the rice crop in almost all the soils, particularly when nitrogen-responsive varieties are grown. The economic optimum dose of N for the new varieties is 100-120 kg/ha. About one third of the total fertilizer, now available in India, is used for the rice crop alone. However, the rice-growing conditions are equally conducive to rapid losses of applied nutrients. This is particularly true to nitrogen, and this element is usually deficient in most of the rice areas. Rice is a poor utilizer of applied nutrients, there is a need to improve the fertilizer-use efficiency. The added nitrogen is lost from the soil system through mineralization-nitrification-denitrification reactions. Therefore, the time and the method of nitrogen application based on the crop-growing conditions and the physiological growth stages of the crop are more important. An adequate supply of N in the early stages of growth is, however, essential for growth, tillering and maintenance of root activity and dry-matter production. In soils which are fertile, the basal dose of N may not be required, but in areas where the soil fertility is low and the average yields are less then 3 tonnes per hectare, a part of the N (about one-third) needs to be applied as a basal dose. An early application of moderate levels of N at planting and tillering are effective in increasing the formation of panicles whereas late applications just before the initiation of panicles increase the number and the size of the spikelet's.

Nitrogen can be applied either as ammonium sulphate or as urea. Apply basal dose of N as urea followed by top- dressing at tillering and about a week before the initiation of panicles is effective over a wide range of situations. The future development in fertilizer use would consist of a judicious combination of organic and inorganic components as well as biological N-fixation in an integrated nutrient-supply system. Rice also requires phosphorus for its normal growth and yield. Phosphorus and potash are added on soil test basis. However, the phosphorus requirement of lowland rice is not as pronounced as its nitrogen requirement. In the acid laterites, black clay soils and in coastal sandy soils, fertilization with phosphorus is a problem because of the soil fixation of large amounts of phosphates. The availability of P is increased under flooded conditions owing to the reduction of ferric phosphates to soluble ferrous phosphates on the one hand and the hydrolysis of iron and aluminum phosphates on the other hand.

The optimum rate of phosphorus for rice varies with the soil type, conditions of rice-culture, the source of P used and the variety. However, 30 to 40 kg of P_2O_5 per hectare can be used in soils which are deficient in phosphorus. Superphosphate is very efficient source of phosphorus in rice on especially highly acidic soils. Phosphatic fertilizers are best applied as a basal dose before planting. In acidic soils, rock phosphate can also be used. The application of rock phosphate, at least 15 days before planting, is essential for better utilization. Rice soils in India are fairly rich in potassium and its requirement by rice to a great extent is met from the irrigation water and by the incorporation into the soil of stubble and crop residues which contain a lot of potassium absorbed by the plant. However, with the high intensity cropping and the cultivation of high-yielding varieties, potash fertilization can become necessary in light-textured and coastal sandy soils.

Zinc Deficiency

Zinc deficiency is commonly associated with soils of high pH above neutrality. It is believed that one-unit increase in pH makes zinc one hundred times less soluble. However, submerged acid soils may, at times, develop Zn deficiency because of the possible rise in pH of these soils after flooding. In alkaline soils, a high content of the organic matter of the soil tends to aggravate Zn deficiency. The continuous submergence of neutral or calcareous soils leads to zinc deficiency, as the concentration of zinc in soil solution gets depleted owing to continuous submergence. An excessive use of N and P fertilizers also promotes microbial activity and the microbial fixation of zinc. The symptoms of zinc deficiency appear 2-3 weeks after transplanting. The lower leaves become rusty brown near the base and ultimately dry up. The seedling with zinc deficiency remains stunted and tiller less. Apply 62.5 kg/ha zinc sulphate heptahydrate (21%) or 40 kg/ha zinc sulphate monohydrate (33% at the time of puddling to check zinc deficiency

Iron Deficiency

Under scarcity of water, chlorosis among seedlings appears in the youngest leaf about three weeks after transplanting. Plants die and often the crop fails completely. Start giving copious irrigations as soon as chlorosis appears and give 2 to 3 sprays of 1.0% ferrous sulphate solution at weekly intervals (2.5 kg of ferrous sulphate in 250 liters of water per hectare). Iron deficiency is more severe in direct seeded rice grown on light textured soils.

Management of Paddy Straw

In Punjab, about 220 lakh tonnes paddy straw is produced annually and a large portion of the straw is being burnt by the farmers. The straw contains a significant amount of essential plant nutrients. One tonne of paddy burning causes loss of 400 kg organic carbon, 5.5 kg of N, 2.3 kg of P, 25 kg of K and 1.2 kg of S. The burning of residue causes complete loss of nitrogen and sulphur. Due to burning of paddy straw poisonous gases like carbon dioxide, carbon monoxide, methane and nitric oxide are produced which are harmful for human and animal health. To avoid burning of straw, the following techniques are recommended:

1. In-situ management of paddy straw

(a) **In standing stubbles**: Wheat sowing with Happy/Smart Seeder can be done in standing rice stubbles after harvest of rice with combines. However, combines fitted with super straw management system (SMS) should be preferred for wheat sowing with above said seeding drills.

(b) **Incorporation:** Chop the straw and stubbles with Paddy Straw Chopper cum Spreader after combine harvesting of paddy. The chopped straw can be mixed in to the soil with Mouldboard Plough. Thereafter, prepare the field using rotavators. The field comes in wattar/workable condition in 2- 3 weeks depending on type of soil when pre sowing irrigation has been applied. Sow wheat seed with zero till drill or with conventional drill. Grow short duration rice varieties for increasing window period for straw management.

2. Management after removal of paddy straw

Use straw baler for baling the paddy straw after combine harvesting. It collects the loose straw or complete straw after chopping the standing stubbles with stubble shaver. These bales can be used for different purposes as follows:

a) **Power generation:** Many bio mass power plants have been established in the country for the generation of electricity from paddy straw bales. Heat produced by burning paddy straw is used to run steam turbine which further generates electricity.

b) **Paddy straw based bio-gas plant:** Paddy straw can be used in a specially designed biogas plant for bio gas production. This biogas plant is filled one time with 16 quintal chopped paddy straw along with 4-5 quintal cattle dung and it provides 6-7 m^3 biogas daily for 3-4 months.

c) **Paddy straw geyser:** A geyser has been developed by the PAU, Ludhiana for heating water by using paddy straw bales. In this geyser about 100 litres of water can be heated to 45-50 °C in 3-4 hours. The water once filled in the geyser remains hot for up to 24 hours or even more.

d) **Phospo-compost:** Collect rice-straw from fields and bring it to the composting site near the tube well on the farm to have easy water availability. It can be made into bundles of convenient size (about 10-15 kilograms). Prepare large quantity of a "soaking solution" by thoroughly mixing one kg cow dung for every 1000 litre of water in a big tank. The volume of the tank can be calculated by measuring Length x Breadth x Height of the tank in metres. One cubic metre is equal to 1000 Litres of water. Dip the bundles one by one into the "soaking solution" for 2-3 minutes.

Drain the excess solution by placing the bundles on a slope lined with a plastic sheet. The drip should be collected and recycled into the tank again.

Make 15 cm raised beds 5-metre-long and 1.5-metre-wide on the ground. This will help in assessing the exact watering of the heap later. Draining of water out of bed is a visual indication of excess watering.

Take the wet rice-straw to the location of the compost heap. Line the bed with 2-6-centimeter diameter tree branches/sticks. This helps in aeration in the heaped rice-straw. The wet rice straw will generally have 70 per cent moisture. Place the wet rice straw on the beds uniformly until 500 kilograms has been stacked. Powdered low-grade rock phosphate should be mixed @ 6 per cent on dry weight basis of the rice straw approximately. For 500 kilogram of the rice straw, 30 kg of the rock phosphate should be sprinkled uniformly while making the heap after wetting. This will give approximately 1% phosphorus in the final decomposed product. The height of 500 kg rice-straw stack is 1.5 metre approximately. Any quantity of rice-straw can be composted in multiple heaps of 500 kg at one time leaving a passage of 1 metre between the beds.

Cover the heap with a 20-30-centimetre-thick layer of unsoaked rice-straw. This will minimize water loss while providing the necessary aeration. The major key to success is the ability to maintain about 70 per cent moisture in the heap. Any major error in this step will delay composting. Water the heaps using watering lance with the help of Tullu Pump. Watering heaps with sprinklers does not work because water generally runs down the sides, instead of going inside the heap. Ensure that the water penetrates the heap by using a lance with a sharp point to pierce the heap of rice-straw. Pierce the lance deepest possible with an aim to water uniformly. Composting can be terminated after 80-90 days by which time it is ready for processing or for field application. By this time its carbon and nitrogen ratio changes to 15:1. At this stage, strands of the rice straw are weak and twisting can readily break a hand-full of it.

e) **Prali Char:** Prali Char is a carbon rich porous product obtained after the partial combustion of rice straw at low temperatures in the presence of little or no oxygen. It can be prepared in a pyramid or dome type kiln, made up of bricks and clay. To prepare 8 quintals of Prali Char from 12 quintals of rice straw, the height of this dome should be 14 ft with 10 ft diameter. Two windows each at the top and at the bottom of the kiln are provided for loading of paddy straw. Six vents of 2-inch diameter in the upper portion and eight vents are provided at three heights on the remaining portion of the structure. Fill the kiln with rice straw up to top. Seal the lower loading window and ignite the straw from the top loading window and immediately seal it with clay. The partial combustion of rice straw will start from the top and proceeds towards the bottom. The emission of thin blue smoke from the vents indicates that the process of Prali Char formation is complete in this zone. To facilitate the combustion in the next zone, seal the vents located in the upper portion of the dome and likewise proceed to seal the vents of the middle and lower zone, respectively. The whole process usually takes about 10-12 hours. To seal the cracks developed during combustion and to cool the kiln, pour a diluted mixture of clay and water on the outer surface of the kiln. Normally it takes about two days to cool, however it can be cooled by sprinkling water for removing the Prali Char on the same day. On an average, it contains 30-36 % carbon, 0.5-0.6 % nitrogen, 0.16-0.22 % phosphorus and 1.6- 2.2 % potassium. Its application to rice and wheat @ 5.0 t/acre saves 40 kg N (87.5 kg urea), increases crop productivity and improves soil health.

f) **Straw Pellets:** The paddy straw can be used for making enriched fodder pellets and fuel pellets. The concept of turning straw into food pellets is not a new process and considered a successful animal feed. Countries like Singapore and Thailand are already using this concept. Japan is importing paddy straw from China for feeding to cattle also there is a tremendous scope of export of enriched fodder pellets to Qatar and the middle east countries from there it can be exported to other countries. Besides, paddy straw can also be used as mulch material in different crops, mushroom cultivation and as animals bedding.

Plant Protection

A. Insect Pests

Rice stem borers: The larvae of these insects bore into the stem and cause damage from July to October. The affected young plants show dead-hearts (yellowing and drying of central shoot) whereas the old ones produce empty ear heads which turn white and stand erect.

The fields showing more than 5% dead hearts (Economic Threshold Level, ETL) should be sprayed with 150 ml Coragen 18.5 SC (chlorantraniliprole) or 50 ml Fame 480 SC (flubendiamide 39.35%) or 125 g Takumi 20 WG (flubendiamide 20%) or 425 g Mortar 75 SG (cartap hydrochloride) or 2.5 litre Coroban/Dursban/Lethal/Chlorguard/Durmet/Classic/ Force 20 EC (chlorpyriphos) or 200 ml neem based bio-pesticide, Ecotin (azadirachtin 5%) in 250 litres of water per hectare. Further application of any of these insecticides may be repeated as and when damage reaches economic threshold level. Prefer Ecotin at pest initiation stage.

Leaf folder: The larvae fold the leaves, eat out the green tissue and produce whitestreaks. The damage is highest during August-October. When the leaf damage reaches 10% (ETL), adopt the following control measures:

- **Mechanical Control:** The mechanical control of leaf folder can be done only before flowering by passing the 20-30 m long coir/jute rope, forwards and then backwards, both ways while touching the crop canopy. While passing the rope, ensure that water must be standing in the crop.
- **Chemical Control:** Spray the crop with 150 ml Coragen 18.5 SC (chlorantraniliprole) or 50 ml Fame 480 SC (flubendiamide 39.35%) or 125 g Takumi 20 WG (flubendiamide 20%) or 425 g Mortar 75 SG (cartap hydrochloride) or 2.5 litre Coroban/Durmet/Force 20 EC (chlorpyriphos) or 200 ml neem based bio-pesticide, Ecotin (azadirachtin 5%) in 250 litres of water per hectare. Prefer Ecotin at pest initiation stage.

Planthoppers: These hoppers include, white backed planthopper and brown planthopper. Both nymphs and adults of these pests suck the cell sap particularly from the leaf-sheath from July to October. The crop dries up in patches. As the plants dry up, the hoppers migrate to the adjoining plants and kill them. In a few days, the area of the dry patches enlarge.

About one month after transplanting, a few plants in the field should be slightly tilted and tapped 2 or 3 times at the base at weekly interval. When minimum 5 planthoppers per hill (ETL) are seen floating in the water, spray 235 ml Pexalon 10 SC (triflumezopyrim) or 200 g Osheen/Token 20 SG (dinotefuran) or 300 g Chess 50 WG (pymetrozine) or 2.0 litre Ekalux/ Quinguard/Quinalmass 25 EC (quinalphos) or 200 ml neem based bio-pesticide, Ecotin (azadirachtin 5%) in 250 litres of water per hectare. Repeat the spray if necessary. For better effectiveness, use knapsack sprayer while directing its spray towards the base of the plants. If the damage is noticed at hopper burn stage, treat the affected spots alongwith their 3-4 metre periphery immediately as these spots harbour high population of the insect.

Grasshoppers: The adults and nymphs of the grasshoppers eat the leaves especially in nursery. Insecticides recommended for the control of planthoppers are also effective for grasshoppers.

Rice hispa. The grubs of this pest tunnel into the leaves, whereas the adults are exposed feeders. The grubs cause damage by producing bold, white streaks on the leaves. If the attack starts in nursery, clip-off and destroy the leaf tips of the affected seedlings before transplanting. On the transplanted crop, spray 2.0 litre Ekalux 25 EC (quinalphos) or 2.5 litre Dursban 20 EC (chlorpyriphos) in 250 litres of water per hectare with a manually operated sprayer. Repeat the spraying if the attack persists.

Rice root weevil: Its' white, legless grubs feed on roots in the soil from July to September. The attacked plants turn yellow, stunted and produce only a few tillers.

Rice-ear-cutting-caterpillar: The larvae of this insect are gregarious in habit and are commonly known as 'armyworm'. The young larvae feed on leaves, leaving only the midribs and stems. The old larvae cut off the panicles mostly at the base and hence the name "rice ear-cutting caterpillar". This stage of the insect causes serious loss to the paddy crop. The larvae are shy of sunlight and generally feed at night. The damage to paddy crop is caused mostly during September to November.

B. Diseases

Sheath blight (*Rhizoctonia solani*): Greyish green lesions with purple margin develop on the leaf-sheath above the water level. Later, the lesions enlarge and coalesce with other lesions. Its' severe attack results in the poor filling of the grains. Destroy the rice straw and stubbles after harvesting the affected crop. Avoid the excessive use of nitrogenous fertilizers. Keep the bunds clean by removing the grass.

At maximum tillering to boot stage of crop, as soon as the disease appears, spray 375 ml Iglare/Pulsor 24 SC (thifluzamide) or 67 g Epic 75 WG (hexaconazole) or 1.0 litre Galileo Way 18.76 SC (picoxystrobin + propiconazole) or 500 ml Amistar Top 325 SC or Tilt/Bumper/Pikapika 25 EC (propiconazole) or Folicur/Orius (tebuconazole) 25 EC or 200 g Nativo 75 WG (trifloxystrobin+tebuconazole) or 800 ml Lusture 37.5 SE (flusilazole + carbendazim) or 500 ml Monceren 250 SC (pencycuron) in 250 litres of water per hectare. Give second spray 15 days thereafter.

False smut (*Ustilaginoidea virens*): It is a fungal disease in which the individual grains transform into large yellowish/greenish velvety spore-balls. High relative humidity, rainy and cloudy days during the flowering period increase the incidence of the disease. The application of organic manures and high dose of nitrogenous fertilizers also increases the intensity of attack.

To control this disease, give spray of 1.0 litre Galileo Way 18.76 SC (picoxystrobin + 15 propiconazole) or 1.25 kg Kocide 46 DF (copper hydroxide) in 250 litres of water per hectare at boot stage of the crop in disease prone areas.

Brown leaf spot (*Drechslera oryzae*): It produces oval, eye-shaped spots with a conspicuous dark-brown dot in the centre and light brown margin. Spots are also produced on the grains. This disease occurs in poor soils, therefore, give adequate and balanced nutrition to the crop. To control the disease, give two sprays of 200 g Nativo 75 WG (trifloxystrobin + tebuconazole) in 250 litres of water/hectare. First spray at boot stage of crop and second spray after 15 days.

Blast (*Pyricularia grisea*): The fungus causes spindle shaped spots with greyish centre and brown margin on the leaves at maximum tillering. It also causes brown lesions on the neck of the panicle, showing neck rot symptoms and the panicles fall over. The disease is more severe on Basmati cultivars particularly in the submontaneous areas and under application of heavy nitrogenous fertilizers. Spray the affected crop with 500 ml Amistar Top 325 SC (azoxystrobin + difenoconazole) or 1.25 kg Indofil Z-78, 75 WP (zineb) per hectare in 250 litres of water, at the boot and ear-emergence stages.

Bunt/Kernel Smut (*Neovossia horrida*): Only a few grains in the panicle are infected. Frequently, only a part of the grain is replaced by a black powder. Sometimes, entire grain is also attacked and the black powder scatters on to other grains or leaves, and this is often the easiest way to detect the disease in the field. Also avoid heavy doses of nitrogenous fertilizers.

Sheath rot (*Fusarium moniliforme*): The rot occurs on the uppermost leaf-sheaths where oblong to irregular and grey-brown to light-brown lesions develop. The lesions often coalesce to cover the entire sheath. In severe cases, young panicles either do not emerge or emerge partially. A white-powdery growth of the fungus appears on the panicle inside the sheath. The glumes of infected florets are discoloured, dark-red or purple brown to black and often do not fill. The fungus over-winters in rice straw and grains. Destroy the rice straw after harvesting the infected crop. Use disease free seed for sowing. Give two sprays of 67 g Epic 75 WG (hexaconazole) in 250 litres of water per hectare. The first spray should be given at boot stage and second 15 days afterwards.

Stem rot (*Sclerotium oryzae*): The fungus affects the stem at earing and black lesions are produced on the sheath at water level. Later on, the stem gets infected and rots leading to withering and lodging of the plant. The incidence of this disease has declined on high yielding varieties due to improved cultural practices. Destroy the diseased debris of infected crop. Avoid excessive irrigation and use recommended dose of nitrogenous fertilizer.

Bacterial blight (*Xanthomonas oryzae pv. oryzae*): Greenish-yellow stripes appear along the leaf margins and extend both lengthwise and breadthwise. The leaf starts drying from the tip, becomes white in severe cases and dries up completely. The disease sometimes attacks the freshly transplanted seedlings which start wilting and in a few days the whole clump dries up. The bacterium perpetuates through seed, rice straw, and roots of non-host plants during the off-season. In order to mitigate the losses, adopt the following integrated measures:

- For the management of bacterial blight, grow rice varieties which are resistant to most of the pathotypes of bacterial blight pathogen.
- Use disease free seed.
- Do not grow nursery and crop under shade and near wheat straw stack (Kup)
- Do not apply excessive dose of nitrogen. Nitrogen should not be applied beyond six weeks after transplanting (except when LCC is used).
- Do not pond water in the field continuously.

Bacterial leaf streak (*Xanthomonas oryzae pv. oryzicola*): Small translucent streaks appear in the interveinal areas of the leaf. The streaks gradually enlarge and turn reddish, when the plants near maturity. Use disease free seed.

Root-knot Nematode (*Meloidogyne graminicola*): The disease first appears in uneven yellow patches. The affected seedings show poor and patchy growth with chlorotic symptoms and characteristic terminal hook or bead like galls on the roots. For the management of root knot nematode, apply mustard cake @ 40 g per square metre, 10 days before sowing of nursery with last preparatory tillage operation after rauni. Also, practice puddling for nursery sowing.

C. Rodents

Do rodent control before milky grain stage in August-September.

Harvesting

Timely harvesting prevents losses in yield due to shedding of grains. The later stage of grain ripening is a dehydration process and maturity is hastened when water is withdrawn from the field at the hardening stage of the grains. The timely draining away of water makes the field hard to facilitate mechanical harvesting. The harvesting of early and medium varieties, 25-30 days after flowering and 35-40 days in the case of late varieties is desirable. Paddy should be harvested when the moisture content of the grain is about 20-25 per cent which is to be reduced to 13 to 14 per cent before milling. The direct drying under the sun leads to an increased breakage of grains during milling. Gradual drying in shade is essential for better recovery in mills. The produce after thrashing is known as rough rice or paddy. The rough rice is milled before it is consumed.

5.2. Wheat (*Triticum aestivum* L.)

Wheat is one of the oldest and most important cereal crop belonging to genus *Triticum* (family Poaceae). Of the thousands of varieties known, the most important are common wheat (*Triticum aestivum*), used to make bread; durum wheat (*T. durum*), used in making pasta (alimentary pastes) such as spaghetti and macaroni; and club wheat (*T. compactum*), a softer type, used for cake, crackers, cookies, pastries, and flours. Additionally, some wheat is used by industry for the production of starch, paste, malt, dextrose, gluten, alcohol, and other products. It has an inflorescence (group of clusters of flowers on a branch) that is a spike, an unbranched, indeterminate inflorescence with sessile flowers arranged along an axis. Each spikelet may have several florets with grains that may be hard or soft, white, red, or purple. The grain includes the embryo and a largely starch endosperm covered by layers. It is an annual plant.

Wheat Growing Zones

The entire wheat growing areas of the country has been categorized into 6 major zones as follows

Zones	States/Regions Covered	Approx. Area (million ha)
Northern Hill Zone(NHZ)	Hilly areas of J&K(except Jammu, Kathua and Samba districts), Himachal Pradesh (except Una & Paonta valley),Uttarakhand(excluding Tarai region) & Sikkim	0.8
North Western Plains Zone(NWPZ)	Punjab, Haryana, Western UP(except Jhansi Divsion),Rajasthan (excluding Kota & Udaipur Division),Delhi, Tarai region of Uttarakhand, Una & Paonta valley of HP, Jammu, Samba & Kathua districts of J&K and Chandigarh.	11.55
North Eastern Plains Zone(NEPZ)	Eastern UP, Bihar, Jharkhand, West Bengal, Assam, Odisha and other NE states (except Sikkim)	10.5
Central Zone	MP, Gujarat, Chhattisgarh, Kota & Udaipur Division of Rajasthan & Jhansi Divsion of UP.	5.2
Peninsular Zone	Maharashtra, Tamil Nadu(except Nilgiris & Palani Hills),Karnataka & Andhra Pradesh	1.6
Southern Hill Zone(SHZ)	Nilgiris & Palani Hills of Tamil Nadu	0.1

Nutrient content in wheat

Wheat offers 327 calories per 100 grams and it is imbued with a ton of essential nutrients like protein, dietary fiber, manganese, phosphorus, iodine, potassium, calcium, copper, niacin, vitamin E and other catalytic elements. Wheat contains 13% water, 71% carbohydrates, 1.5% fat and 13% protein which is in the form of gluten. As wheat protein is deficient in certain amino acids lysine, these days wheat flour is fortified with a protein source to make up for the deficiency.

Origin and History

Wheat was already an important crop when history was first recorded and so accurate information on the exact time and place of its origin is not available. The distribution of the wild wheats and grasses, believed to be the progenitors of the cultivated wheats, supports the belief that wheat originated in Southeastern Asia. Some species were cultivated in Greece, Persia, Turkey, and Egypt in prehistoric times while the cultivation of other species may be of more recent origin. In India, evidences from Mohen-Jo-Daro excavations, indicate that wheat was cultivated there more than 5000 years ago.

Archaeological analysis of wild emmer indicates that it was first cultivated in the Southern Levant, dating back as far as 9600 BCE. Genetic analysis of wild einkorn wheat suggests that it was first grown in the Karacadağ Mountains in southeastern Turkey. Dated archeological remains of einkorn wheat in settlement sites near this region, including those at Abu Hureyra in Syria, suggest the domestication of einkorn near the Karacadag Mountain Range. With the anomalous exception of two grains from Iraq ed-Dubb, the earliest carbon-14 date for einkorn wheat remains at Abu Hureyra is 7800 to 7500 years BCE

The cultivation of emmer reached Greece, Cyprus and Indian subcontinent by 6500 BCE, Egypt shortly after 6000 BCE, and Germany and Spain by 5000 BCE. "The early Egyptians were developers of bread and the use of the oven and developed baking into one of the first large-scale food production industries." By 4000 BCE, wheat had reached the British Isles and Scandinavia. Wheat likely appeared in China's lower Yellow River around 2600 Before Common Era (BCE).

Area and Distribution

Among different countries of the world, India topped in total area of wheat (31.36 mha), China recorded highest production (134.25 million tonnes) and New Zealand stood 1st with average yield of 99.33 q/ha (Table 5.2)

Table 5.2: Area, production and average yield of wheat for top 10 countries

Country	Area in million ha		Country	Production in million tonnes		Country	Average Yield in q/ha
India	31.36		China	134.25		New Zealand	99.33
Russian Federation	28.86		India	107.59		Belgium	89.46
China	23.38		Russian Federation	85.90		Netherlands	85.56
USA	14.87		USA	49.69		Denmark	80.99
Kazakhstan	12.06		Canada	35.18		Germany	78.20
Canada	10.02		France	30.14		Ireland	77.66
Australia	9.86		Pakistan	25.25		Zambia	73.68

Country	Area in million ha	Country	Production in million tonnes	Country	Average Yield in q/ha
Pakistan	8.80	Ukraine	24.91	Sweden	71.56
Iran	7.58	Germany	22.17	UK	69.63
Turkey	6.91	Turkey	20.50	Kuwait	68.00

Source: FAOSTAT, 2020

The major wheat growing states in India are Uttar Pradesh, Madhya Pradesh Punjab with production of 35.50, 17.62 and 17.14 million tonnes and with share of 32.42, 16.08 and 15.65 percent respectively among total wheat productivity of the country.

Wheat is a major cereal crop of Punjab. It was grown on an area of 35.30 lakh hectares during 2020-21 with production of 171.85 lakh tonnes and average yield of 48.68 quintals per hectare.

Classification of Wheat

Wheat is an annual plant of Poaceae family. It belongs to genus Triticum. Although as many as 18 species of wheat have been describe and recognized by Percival (1921) only a few are of importance in agriculture. The following is the classification on the basis of chromosomal number:

Einkorn series (diploid)	Chromosome No.
Triticum aegiloploids	7
Triticum monecoccum	7
Einkorn Series (tetraploid)	
Triticum dicoccoides	14
Triticum timopheevi	14
Triticum persicum	14
Triticum polonicum	14
Triticum turgidum	14
Triticum orientale	14
Triticum durum	14
Triticum dicoccum	14
Dinkale series (hexaploid)	
Triticum spelts	21
Triticum macha	21
Triticum aestivum or vulgare	21

In India only three species of wheat namely *Triticum aestivum* (common bread wheat), *Triticum durum* (macaroni or durum wheat) and *Triticum dicoccum* (emmer wheat) are economically important. The bread wheat (*Triticum aestivum*) is the most important species accounting to a little over 87 per cent of the total wheat production of India followed by the durum wheat (about 12 per cent) and dicoccum wheat (about 1 per cent). The common wheat (*T. aestivum*) which is good for chapati making and bakery products is grown in whole of North India and also in Central and South India. The durum wheat which is good for 'suji', 'semya', 'sphagetti' and 'macaroni' is grown only in Central and South India and that too under rainfed conditions. The dicoccum wheat is grown only on limited acreage in Tamil Nadu, Andhra Pradesh, Maharashtra and Gujarat. It is good for the South India dish 'Uppumav'.

Climate Requirement

Wheat crop has wide adaptability. It can be grown not only in the tropical and sub-tropical zones, but also in the temperate zone and the cold tracts of the far north, beyond even the 60-degree north altitude. Wheat can tolerate severe cold and snow and resume growth with the setting in of warm weather in spring. It can be cultivated from sea level to as high as 3300 meters.

The best wheat are produced in areas favored with cool, moist weather during the major portion of the growing period followed by dry, warm weather to enable the grain to ripen properly. The optimum temperature range for ideal germination of wheat seed is 20-25 °C, though the seeds can germinate in the temperature range of 3.5 to 35 °C. Rains just after sowing hamper germination and encourage seedling blight. Areas with a warm and damp climate are not suited for wheat growing.

During the heading and flowering stages, excessively high or low temperatures and drought are harmful to wheat. Cloudy weather, with high humidity and low temperatures is conducive for rust attack. Wheat plant requires about 14-15 °C optimum average temperature at the time of ripening. The temperature conditions at the time of grain filling and development are very crucial for yield. Temperatures above 25 °C during this period tend to depress grain weight. When temperatures are high, too much energy is lost through the process of transpiration by the plants and the reduced residual energy results in poorer grain formation and lower yields. Wheat is mainly a *rabi* (winter) season crop in India.

Soil Requirements

Wheat is grown in a variety of soils of India. Soils with a clay loam or loam texture, good structure and moderate water holding capacity are ideal for wheat cultivation. Care should be taken to avoid very porous and excessively drained soils. Soil should be neutral in its reaction. Heavy soil with good drainage is suitable for wheat cultivation under dry conditions. These soils absorb and retain rain water well. Heavy soils with poor structure and poor drainage are not suitable as wheat is sensitive to water logging. Under these types of soils, bed planting technique is very useful for getting higher yield. Wheat can be successfully grown on lighter soils provided their water and nutrient holding capacity are improved.

Wheat varieties released by ICAR (2020-2021)

Variety	Area of adoption	Characters
GW 499	Gujarat	Suitable for irrigated late sown condition, average grain yield 4.60 tonnes/ha, maturity 95 days, resistant to brown and black rusts.
GW1339 (BANAS)	Gujarat	Suitable for irrigated timely sown condition, average grain yield 4.96 tonnes/ha, maturity 102 days, resistant to brown and black rusts, good amount of yellow pigment (5.5 ppm).
MP 3465	Madhya Pradesh	Suitable for irrigated timely sown condition, average grain yield 5.94 tonnes/ha, maturity 117 days, high protein content (>14%), resistant to brown and black rusts.
Chhattisgarh Hanse Wheat	Chhattisgarh	Suitable for restricted irrigation, timely sown condition, average grain yield 3.21 tonnes/ha, maturity 126 days, high zinc content (40.4 ppm), resistant to brown rust, good chapatti quality.
DBWH 221	Haryana	Suitable for timely sown irrigated conditions, average grain yield 6.28 tonnes/ha, maturity 135–149 days, highly tolerant to heat stress and resistant to yellow rust.
UP 2944	Uttarakhand plains	Suitable for late-sown irrigated conditions, average grain yield 5.07 tonnes/ha, maturity 119–127 days, high protein content (14.5%), resistant to brown rust.
UP 2938	Uttarakhand plains	Suitable for timely sown irrigated conditions, average grain yield 5.38 tonnes/ha, maturity 136–139 days, resistant to brown rust.
UP 2903	Uttarakhand plains	Suitable for timely sown irrigated conditions, average grain yield 5.06 tonnes/ha, maturity 129–139 days, high protein content (12.68%) and resistant to brown rust.

Varieties approved for Punjab

A. Timely sown irrigated conditions

PBW 869: This variety has been recommended for sowing with Happy Seeder/ Super Seeder in in-situ rice residue managed fields. Its average plant height is 101 cm and matures in about 158 days. It is resistant to brown rust and moderately resistant to yellow rust. Its average yield is 58.9 quintals per hectare. It possesses bold grains and longer coleoptile. Use 112.5 kg seed per hectare.

PBW 824: Its average plant height is 104 cm and matures in about 156 days. It is resistant to brown rust and moderately resistant to yellow rust. Its average yield is 58.25 quintals per hectare.

PBW 803: This variety has been recommended for cultivation in South-Western region of Punjab (Bathinda, Faridkot, Fazilka, Ferozepur, Mansa and Sri Muktsar Sahib). Its average plant height is 100 cm and matures in about 151 days. It is resistant to brown rust and moderately resistant to yellow rust. Its average yield is 56.75 quintals per hectare.

Sunehri (PBW 766): Its average plant height is 106 cm and matures in about 155 days. It is moderately resistant to yellow rust and resistant to brown rust. Its average grain yield is 57.75 quintals per hectare.

PBW 1 Chapati: This a premium quality bread wheat variety having excellent chapati making properties. Chapati made from it is whitish in color, sweet in taste and remains soft even after hours of cooking with good palatability and texture. Its average plant height is 103 cm and matures in about 154 days. It is moderately resistant to yellow rust and resistant to brown rust. Its average grain yield is 43.0 quintals per hectare.

DBW 222: This bread wheat variety has been recommended for cultivation in the Punjab state except sub-mountainous regions. Its average plant height is 103 cm and matures in about 152 days. It is moderately susceptible to yellow rust and resistant to brown rust. Its average grain yield is 55.75 quintals per hectare.

DBW 187: Its average plant height is 104 cm and matures in about 153 days. It is moderately resistant to yellow rust and resistant to brown rust. Its average grain yield is 56.5 quintals per hectare.

HD 3226: Its average plant height is 106 cm and matures in about 155 days. It is resistant to yellow and brown rust. It possesses good grain quality characteristics. Its average grain yield is 54.75 quintals per hectare.

Unnat PBW 343: It is an improved version of PBW 343 and has an average plant height of 100 cm. It is resistant to brown rust and moderately resistant to yellow rust. It matures in about 155 days. Its average grain yield is 58 quintals per hectare.

Unnat PBW 550: It is an improved version of PBW 550, has average plant height of 86 cm. It is resistant to yellow and brown rust. It matures in about 145 days. It has bold grains and its average grain yield is 57.5 quintals per hectare. Use 112.5 kg seed per hectare and sow this variety during 2nd to 4th week of November.

PBW 1 Zn: It possesses higher grain zinc concentration which is good for human health. Its average plant height is 103 cm and is resistant to brown rust and moderately resistant to yellow rust. It matures in about 151 days. Its average grain yield is 56.25 quintals per hectare.

WHD 943: It is durum wheat (Wadanak) with an average plant height of 93 cm and matures in about 154 days. It has low incidence of yellow berry. It possesses desirable quality characteristics and suitability for pasta making. It is resistant to yellow and brown rusts and less susceptible to leaf blight. It possesses field resistance to Karnal bunt disease. Its average yield is 49.5 quintals per hectare.

PDW 291: It is also durum wheat (Wadanak) with an average plant height of 83 cm and matures in about 155 days. It has low incidence of yellow berry and possesses suitable for pasta making. It is resistant to yellow and brown rusts, loose smut and flag smut and less susceptible to powdery mildew, leaf blight and head scab diseases. It possesses field resistance to Karnal bunt disease. Its average yield is 48.5 quintals per hectare.

B. Late sown irrigated conditions

PBW 771: Its average plant height is 80 cm and matures in about 133 days. It is moderately resistant to yellow rust and resistant to brown rust. Its average grain yield is 47.5 quintals per hectare.

PBW 757: This bread wheat variety has been recommended for cultivation in the Punjab state under irrigated very late sown conditions (January sowing). Its average plant height is 82 cm and matures in about 114 days. It is resistant to yellow rust and brown rust diseases. Its average grain yield is 39.5 quintals per hectare.

PBW 752: Its average plant height is 89 cm and matures in about 130 days. Its ears are medium dense with red glumes. It is resistant to yellow and brown rust. Its average grain yield is 48 quintals per hectare

Rotations

Rice-Wheat

Rice-Wheat-Summer Moong/Mash

Cotton-Wheat

Maize-Wheat

Maize/Rice-Potato-Wheat

Moong/Arhar/Mash-Wheat

Groundnut-Wheat

Green Manure-Rice-Wheat

Rice-Pea-Wheat

Soybean-Wheat

Summer Groundnut-Potato/Toria/Late *Kharif* Fodder-Wheat

Land Preparation

For good and uniform germination, the wheat crop requires a well pulverized but compact seedbed. Traditional system of land preparation for wheat includes one deep ploughing followed by two to three harrow-wings. After the harvest of *kharif* crops, the field is irrigated and at optimum workable moisture content, the field should be ploughed with disc or mould board plough followed by two to three harrowing and planking. Under late sown conditions, the pre-sowing irrigation for wheat may be applied in previous standing crop so that field may be prepared immediately after the harvest of previous crop. Alternately, a no tillage system may be followed and wheat is sown directly using zero-tillage seed drill.

In rainfed areas, collection and conservation of soil moisture and timely cultivation are key practices for the success of wheat crop. One deep chiseling once in three years should be done to break the hard pan created by the continuous use of the country plough. Two to three summer ploughings have been found useful for soaking the rain water in deeper layer. After cessation of monsoon all the possible efforts should be made to conserve moisture. The ploughing (3-4) should be done in the evening time, furrows should be kept open whole night and planking should be done after each ploughing early in the morning.

(i) After paddy

a. **Paddy straw retained:** When paddy is harvested with combine, the left over straw can be managed using following techniques:

  - The left over straw can also be chopped by using paddy straw chopper cum spreader into small pieces and the chopped straw can be mixed into the soil by tillage operations without any adverse effect on yield of wheat.
  - Wheat can also be sown in the combine harvested fields with 'Happy Seeder'/'Super Seeder'/'PAU Smart Seeder' in the standing paddy stubbles..

b. **Paddy straw removed:** The left over straw can be cut by using stubble shaver and collected after 2 to 3 days (sun drying) by using straw baler. Straw baler collects the straw and compresses it in the form of bales. In such fields and in manual harvested paddy fields, if the field has enough soil moisture, undertake tillage straightway, otherwise apply pre sowing irrigation (rauni). Use disc harrow once, followed by planking. Give another cultivation/ploughing with a cultivator. If the soil is heavy, give one more cultivation before planking the soil. Tractor drawn cultivator with pulverizing roller-cum-puddler can also be used for preparing the dry seed bed.

(ii) After other crops

After pre-sowing irrigation (rauni), give two cultivations/ ploughings followed by planking. Tractor drawn cultivator with pulverizing roller-cum puddler can also be used for preparing the dry seedbed.

Selection of seed

The seed used for sowing should be healthy with good germination capacity and free from seed borne diseases and seeds of weeds. The seed should be certified and obtained from a reliable seed agency.

Time of sowing

It depends on the climatic conditions, soil conditions and other available facilities like irrigation. Under irrigated conditions, the first fortnight of November is considered to be the optimum time for sowing of medium long duration varieties. For short duration varieties, the second fortnight of November is the optimum time. In eastern India, the sowing of wheat is often delayed owing to the late harvesting of paddy. The sowing in such case may go up to even the third week of December. In north-western India also, wheat sowings get delayed due to late harvesting of paddy or sugarcane or potato. However, wheat sowings beyond the third week of December are not very profitable. Under rainfed conditions, second half of October is the optimum time. The sowing of wheat, when the temperatures are high, results in a poor stand, poor tillering, attack by insects-pests and the early onset of flowering. All these factors depress yield. The success of rainfed wheat depends upon the amount of conserved moisture in the soil.

Seed Treatment

a. **For Termite:** In termite infested soil, treat the seed with 1 g Cruiser 70 WS (thiamethoxam) or 2 ml Neonix 20 FS (imidacloprid+hexaconazole) or 4 ml Dursban/ Ruban/Durmet 20 EC (chlorpyriphos) per kg seed and dry it in shade. Seed treated with Neonix also control smuts of wheat.

b. **For Loose Smut:** Treat 40 kg seed with 13 ml Raxil Easy/Orius 6 FS (tebuconazole) by dissolving in 400 ml water or 120 g Vitavax Power 75 WS (carboxin+tetramethyl thiurum disulphide) or 80 g Vitavax 75 WP (carboxin) or 40 g Tebuseed/Seedex/Exzole 2 DS (tebuconazole) per 40 kg seed.

c. **For Flag Smut:** Treat 40 kg seed with 13 ml Orius 6 FS (tebuconazole) by dissolving in 400 ml water or 120 g Vitavax Power 75 WS (carboxin+tetramethyl thiurum disulphide) or 80 g Vitavax 75 WP (carboxin) or 40 g Tebuseed/Seedex/Exzole 2 DS (tebuconazole) per 40 kg seed. Treatment should not be done earlier than one month of sowing as it affects seed germination. Seed treatment can be done effectively with the seed treating drum.

d. **Seed Inoculation:** Inoculate recommended quantity of seed for one acre with 500 g consortium or 250 g each of Azotobacter and Streptomyces (Azo-S) biofertilizer and one litre of water on pucca floor. Let it dry in shade and sow immediately. The use of bio-fertilizers increases grain yield as well as maintains/improves soil health.

Method of Sowing

Sowing of wheat is generally done by following methods:

Broadcasting: It is the most prevailing and simple method of wheat sowing in India. Seeds are broadcasted on soil surface, then worked in by harrowing or ploughing with cultivator and covered by planking. However, all the broadcasted seeds do not get proper contact with moist soil. Secondly some seeds left uncovered and are eaten by birds or some are not sown at proper depth. This results in poor and uneven germination. To overcome these impediments, 25% higher seed rate is recommended. Broadcasting is crude method of sowing and should only be practiced under special situations such as under very clayey if soils or soil containing morestuliles which hinder drill sowing.

Behind the plough: In this method, the seeds are dropped by hand into the furrows that have been opened with local plough. When seed is dropped in furrows by hand, it is called *kera* method and when seeds are dropped through a pore (pipe) like attachment with local plough it is called *pora* method. In this method, seeds are dropped at a desired depth and germination is satisfactory.

Drilling: Drilling of seed with the help of seed drill or seed-cum-fertilizer drill is the best method of sowing. Seed drills are calibrated at an optimum seed rate. These seed drills drop the seed at an optimum depth (4 to 6 cm) and spacing while running over the field with the help of a tractor or bullocks.

Bed sowing: Bed planter (67.5 cm with 37.5 cm bed top + 30cm furrow) cum seed drill (two rows/bed) can be used for sowing. This method is very effective on clayey soil or under high rainfall areas.

Dibbling: It is used when quantity of seed is limited. Sowing is done with the help of small implement known as dibbler. In this method dibblers are pressed into the soil to make the holes and one or two seeds are dropped by hand in each hole and then seeds are covered.

After Paddy

Wheat sowing with Happy/Smart Seeder can be done in standing rice stubbles after harvest of rice with combines. However, combines fitted with super straw management system (SMS) should be preferred for wheat sowing with above said seeding drills. Super seeder is becoming very popular in wheat growing zone of India. A heavy machinery i.e. 65 HP tractor is required for operating super seeder. Incorporation of straw of paddy harvested with SMS combine machines and sowing of wheat can be done in one operation. The paddy straw is chopped into small prices and buried in soil at the time of wheat sowing. Optimum moisture in the field is essential so that wheat seeds can germinate.

Seed Rate

Seed rate of wheat varies with varietal characters like seed size, tillering ability and germination percentage, time of sowing, moisture content in the soil at sowing time and method of sowing. Under normal conditions a seed rate of 100-125 kg/ha is sufficient. Under late sown conditions seed rate should be increased by 25 per cent. Similarly, when soil moisture is less at the sowing time, seed rate should be increased by 25 per cent. Method of sowing also affects the seed ratc of wheat. Broadcasting requires 150 kg/ha while sowing with local plough and seed drill requires a seed rate of 125 and 100 kg/ha, respectively under normal conditions. Where seed is sown with dibbler a seed rate of 25-30 kg/ha is sufficient.

Spacing and Depth of Sowing

For timely sown and irrigated wheat, a row spacing of 15 to 20 cm is followed. For bi-direction or cross sowing 20 cm is considered to be the best spacing. For cross sowing, use half of recommended seed rate in one direction and remaining half in other direction. For late but irrigated wheat, row spacing of 15-18 cm is considered to be the maximum. For dwarf varieties, planting depth should be between 4 to 6 cm. Usually deeper sowing is done in dry, rough and light soils, whereas shallow sowing is done in moist soils.

Weed Management

Weeds can be controlled by two hand hoeing's, one before first irrigation and second after first irrigation with wheel hoe, kasola or khurpa etc. Alternatively, weeds can be controlled by the following ways:

A. Control of grass weeds

i) Control of *Phalaris minor* (Gulli danda)

Cultural methods: Sow quick growing wheat varieties from last week of October to first week of November in narrow rows at 15 cm spacing. In fields having severe infestation of this weed, replace wheat with alternate crops like berseem, oats, potato, raya, gobhi sarson and winter maize for 2 to 3 years. Preparation of soil mulch at sowing helps in eliminating the first flush of gulli danda. Closer sowing with spacing of 15cm (row to row) or cross sowing helps to reduce *P. minor* infestations.

Herbicides: Gulli danda can be controlled with use of any of herbicides listed below. The herbicides can be applied within two days of sowing or before first irrigation or after first irrigation.

Brands (Herbicide)	Dose per hectare (commercial product)	Time of spray and amount of water
a. Pre-emergence		
Stomp/Dost/Penda/Markpendi/ Pendin/Bunker/ Zakiyama 30 EC (pendimethalin)	3.75 litre	In fields to be sown with Happy Seeder, mix any of three herbicides with basal dose of Urea and broadcast just before sowing. or In fields to be sown with any other method, spray any of three herbicides within two days of sowing using 500 litres of water.
Awkira/Momiji 85 WG (pyroxasulfone)	150 g	
Platform 385 SE (pendimethalin + metribuzin) • It also controls broadleaf weeds	2.5 litre	
• Good soil moisture for pre-emergence spray is desirable		
b. Post-emergence (Before first irrigation)		
Brands (Herbicide)	**Dose per hectare (commercial product)**	**Time of spray and amount of water**
Arelon/Delron/Hilproturon/ Ronak /Nocilon/ Wonder/ Milron/Agrilon/ Totalon/ Agrilon/ Shivron/Carelon/ Marklon/Jai-roturon/ Isoguard/ Dhar/ Rakshak/ Prowl/Kanak/ Isotox/ Isohit/ Ciluron /Isocin 75 WP (isoproturon) • Don't use isoproturon group of herbicides in areas where there is problem of resistance	750 g in light textured; 1.0 kg in medium textured and 1.25 kg in heavy textured soils	When crop is 20 days old and *P.minor* in 2-3 leaf stages, spray 2-3 days before first irrigation, using 300 litres of water
Leader/SF-10/Safal/ Marksulfo 75 WG (sulfosulfuron*) • Do not use sulfosulfuron group of herbicides if any broadleaf crop is sown in wheat and where jowar/ maize to be sown after wheat	33 g	

Brands (Herbicide)	Dose per hectare (commercial product)	Time of spray and amount of water
c. Post-emergence (After first irrigation)		
Isoproturon 75 WP: (All the above formulations of isoproturon)	1.25kg for all types of soil	30-35 days of sowing, using 300 litre of water
Topik/Point/Moolah/Topple/ Rakshak Plus/Jay Vijay/ Markclodina/ Columbus 15 WP (clodinafop)	400 g	
Axial 5 EC (pinoxaden)	1.0 litre	
Puma Power 10 EC (fenoxaprop-p-ethyl)	1.0 litre	
Leader/SF-10/Safal/ Marksulfo 75 WG (sulfosulfuron) • Do not use sulfosulfuron group of herbicides if any broadleaf crop is sown in wheat and where jowar/ maize to be sown after wheat	33 g	

ii. Control of wild oats (Jangli Javi): All the post-emergence herbicides recommended for the control of gulli danda also provide effective control of wild oats.

B. Control of broadleaf weeds only:

Broadleaf weeds can be controlled with use of any of below listed herbicides: -

Brands (Herbicide)	Dose per hectare (commercial product)	Time of spray and amount of water
2, 4-D sodium salt 80 WP or 2, 4-D ethyl ester 38 EC • Use in fields having bathu, billi booti, jangli hallon, pitpapra, jangli senji, maina, maini, jangli palak	625 g 625 ml	In timely sown wheat, spray at 35- 45 days after sowing and in late sown (in Dec.) at 45-55 days after sowing, using 300 litre of water
Algrip/Algrip Royal/Markgrip/ Makoto 20 WP (metsulfuron*) • Use in fields having Rumex *spinosus*	25 g	30-35 days of sowing, using 300 litre of water
Aim/Affinity 40 DF (carfentrazone-ethyl*) • Use in fields having *Malva neglecta*	50 g	25-30 days of sowing, using 300 litre of water
Lanfida 50 DF (metsulfuron + carfentrazone*) • Use in fields having *C. arvensis, R. spinosus, Lathyrus aphaca*	50 g	25-30 days of sowing, using 300 litre of water
• Do not use these herbicides, if any broadleaf crop such as raya, gobhi sarson, gram is sown in wheat.		

C. Control of both grass and broad leaf weeds

Grass and broadleaf weeds can simultaneously be controlled with use of any of below listed herbicides

Brands (Herbicide)	Dose per hectare (commercial product)	Time of spray and amount of water
Total/ Mark power 75 WG (sulfosulfuron + metsulfuron*) • Do not use where jowar/maize to be sown after wheat	40 g	30-35 days of sowing, using 300 litre of water
Atlantis 3.6 WDG (mesosulfuron + iodosulfuron)	400 g	
Accord Plus 22 EC (fenoxaprop+metribuzin) • Do not use Accord Plus on Unnat PBW 550	1.25 litre	
Shagun 21-11 (metribuzin+clodinafop) • Preferably use where there is a problem of multiple herbicide resistance in gulli danda • Do not usc on Unnat PBW 550 • Do not use on light textured soils	500 g	
ACM-9 (metribuzin+clodinafop) • Preferably use where there is a problem of multiple herbicide resistance in gulli danda • Do not use on Unnat PBW 550 • Do not use on light textured soils	600 g	

D. Integrated weed management

Prefer cultural and chemical methods or chemical and mechanical methods for managing weeds in wheat crop.

E. Weed control in durum wheat

All the above mentioned pre and post emergence herbicides at recommended dose and time can be used for weed control in durum wheat, except Total/Markpower/ Atlantis/Accord Plus/Shagun 21-11/ACM-9. Also do not use isoproturon on light textured soils.

Useful hints

- Spray should be as uniform as possible and on clear days.
- After spray, irrigation should be light as heavy irrigation reduces the efficacy of erbicides.

- After spray, flush the spray pump thoroughly with water and then with washing soda solution (0.5%) to remove traces of herbicides, as contaminated spray pump may be phytotoxic when used for spray on other crops.
- If raya, sarson or gobhi sarson is sown in wheat, use isoproturon, clodinafop, fenoxaprop or pinoxaden group of herbicides at recommended doses.
- In areas where there is problem of resistance in gulli danda, do not use isoproturon group of herbicides.
- Use flat fan or flood jet nozzle for pre-emergence herbicides and flat fan nozzle for post emergence herbicides.

Do not use cone type nozzle for herbicide spray.

- Change the herbicide group every year to prevent the development of resistance in weeds.
- Some weed plants may escape after the application of herbicides. These plants should be uprooted before they produce seeds. This practice will reduce the weed problem in next crop of wheat.

Water Management

Under irrigated conditions, 4-6 irrigations are needed for wheat. The first irrigation which is very critical, should be given at the crown root initiation stage, 20-25 days after sowing. Other rounds of irrigation should be done at the time of late tillcring, late jointing, flowering, milk and dough stages. Extra two or three irrigations are given if the soil is very light or sandy. Irrigation in case the water supply is limited should be given in the following way:

(i) If the water is available only for one irrigation, it should be given at the crown root initiation stage to the tillering stage.

(ii) If the water is available for two irrigations, the first irrigation should be applied between the crown root initiation and the late tillering stage and second irrigation 7 to 8 weeks after the first irrigation (flowering stage).

(iii) If the water is available for, three irrigations, the first should be given at the crown root initiation stage and the next two at intervals of six to seven weeks at the flowering and the milk stages, respectively.

It was considered necessary to optimize the irrigation schedule to obtain increased yield. The investigations revealed that first irrigation was most essential for high productivity, which must be applied 21 days after sowing. About 17 days after sowing, the crown node starts swelling and crown root develops from this node approximately 21 days after sowing. Tillers develop from the crown root and, therefore, this phase was directly relevant to the number of tillers and final heads and is supposed to be the most critical period of irrigation. Depending on the availability of water with the farmer a schedule for irrigation at different growth and development stages is very important. Keeping in view, the critical phenological phases, four irrigations, first at crown-root initiation, followed by at flowering, late tillering and milk stages, should be applied to normal sown crop particularly under high fertility conditions.

Mitigation of effect of high temperature at grain filling and enhancing yield: Apply two sprays of 2% Potassium nitrate (13:0:45) by dissolving 4 kg potassium nitrate in 200 litres of water at boot leaf and anthesis stages or two sprays of salicylic acid by dissolving 15 gram salicylic acid in 450 ml of ethyl alcohol using 200 litres of water per acre at boot leaf and early milk stages to mitigate the effect of high temperature at grain filling and enhance the wheat yield.

Fertilizer Management

Follow integrated use of organic, bio and chemical fertilizers for higher crop productivity and maintenance of soil health as under:

I) Organic Manures

i. **Farmyard manure (FYM):** Apply good quality farmyard manure before sowing and reduce the fertilizer quantity by 2 kg of N and 1 kg of P per tonne of farmyard manure. Similarly, where wheat follows potato which received 20 tonnes of farmyard manure per hectare, no phosphorus and only one-half of the recommended nitrogen dose i.e 62.5 kg N (137.5 kg urea per hectare) need to be applied.

ii. **Poultry manure/gobar gas plant slurry/press mud:** If 6.25 tonne per hectare of poultry manure or 6 tonne of gobar gas plant slurry after drying or pressmud at 15 tonnes per hectare was applied to rice, reduce the fertilizer N dose in wheat by 25% and fertilizer P dose by 50%.

iii. **After leguminous crops:** Reduce Nitrogen dose by 30%.

iv. **Rice husk ash/Bagasse ash:** Apply 10 tonne per hectare of rice husk ash or bagasse ash to wheat before last ploughing and reduce the dose of phosphorus to onehalf.

v. **Prali char:** Apply 5.0 tonne per hectare prali char to wheat and save 30 percent Nitrogen. It increases the yield and improves soil health.

II) Bio-fertilizers

Inoculate the seed with bio fertilizers before sowing as per recommendations.

III) Chemical Fertilizers

Apply fertilizer on soil test basis. In the absence of a soil test, add the following amounts of fertilizers to wheat grown on medium fertile soils:

a. With assured fertilizer supply

Nitrogen (N) 100-125 kg/ha

Phosphorus (P_2O_5) 40-60 kg/ha

Potash (K_2O) 20-30 kg/ha

(This quantity may be adjusted according to the soil test results).

The total quantities of phosphorus and potash and half the quantity of nitrogen should be placed 3 to 4 cm below the seed at the time of sowing. The remaining quantity of nitrogen should be applied at the time of first irrigation to be applied at crown-root initiation. Above recommendation can vary as per soil type. In alkaline soils, increase dose of nitrogen by 25%.

Zinc Deficiency

Zinc deficiency generally appears in light soils under intensive cropping system. Zinc deficiency is particularly reported from Punjab, the *tarai* areas of Uttar Pradesh, some parts of Haryana, western Uttar Pradesh and Delhi. When recommended dose of zinc sulphate has been applied to the *kharif* crop, its application may be omitted to the following wheat crop. Paddy responds more to zinc application; therefore, zinc should be applied to paddy and its residue is sufficient to meet the requirement of the following wheat crop. However, if zinc has not been applied to paddy and zinc deficiency is expected or appears showing symptoms of a stunted and bushy crop with leaves chlorotic in the middle, which later break and keep hanging, apply 62.5 kg of zinc sulphate per hectare which will be enough for 2-3 years. Zinc deficiency can also be corrected by foliar spray of 0.5% zinc sulphate (21% zinc). Prepare the solution for spray by dissolving 2.5 kg zinc sulphate and 1.25 kg unslaked lime in 500 litres of water. This solution is sufficient for spraying an hectare of wheat once. Two or three sprays at 15 days interval are needed.

Manganese Deficiency

Manganese deficiency generally appears in light soils under intensive cropping especially in rice-wheat rotation. The symptoms appear on the middle leaves as interveinal chlorosis with light greyish yellow to pinkish brown or buff colored specks of variable size confined largely to 2/3 lower portion of the leaf. Later, the specks coalesce forming a streak or band in between the veins which remain green. In acute deficiency, whole of the plant may become dry. At earing stage, the symptoms become prominent on flag leaf. In manganese deficient soils, give one spray of 0.5% manganese sulphate solution (2.5 kg manganese sulphate in 500 litres of water for one hectare) 2-4 days before first irrigation and three sprays afterwards at weekly intervals on sunny days. Do not grow durum wheat in sandy soils as it is more prone to manganese deficiency. Manganese sulphate should be sprayed only as its soil application is not profitable.

Sulphur Deficiency

Wheat crop suffers from sulphur deficiency when sown in sandy soils. The deficiency is more severe when the winter rains continue for a long time in the early growth period. The symptoms first appear on the younger leaves with fading of the normal green colour. This is followed in the veins. The topmost leaves become light yellow except for the tip, while the lower leaves retain green colour for a long time. This is distinctly different from the nitrogen deficiency where the yellowing starts with the lower leaves. The plants are short in size with fewer tillers. Where phosphorus was not applied as single superphosphate, apply 250 kg of gypsum per hectare before sowing to meet the sulphur requirement of the crop. Gypsum can also be applied in standing crop if deficiency of sulphur is observed.

Plant Protection

(a) Insect Pests

Termites damage the crop soon after sowing and near maturity. The damaged plants dry up completely and are easily pulled out. The plants damaged at later stages give rise to white ears. Treat the seed by diluting 40 g Cruiser 70 WS (thiamethoxam) or 160 ml Dursban/Ruban/Durmet 20 EC (chlorpyriphos) or 80 ml Neonix 20 FS (imidacloprid+hexaconazole) in one litre of water and spray the same on 40 kg seed spread as a thin layer on the pucca ground or tarpaulin or polythene. Seed treated with Neonix also control smuts of wheat. Seed treated with insecticide is less attacked by birds. Generally, termites cause more damage in sandy soil and irrigating the affected fields result in some control of termite damage. In case of severe infestation, broadcast 17.5 kg Mortel 0.3 G (fipronil) or 3.0 litre Dursban 20 EC (chlorpyriphos) per hectare mixed with 50 kg of moist sand before first irrigation.

Aphids damage the crop resulting in discolouration of leaves. In case aphid infestation exceeds economic threshold level of 5 aphids per earheads (recorded from 10 randomly selected ear heads in each of the 10 quarters of one hectare field), give two sprays of 5 litre neem extract at weekly interval or single spray of 50 g Actara/Taiyo 25 WG (thiamethoxam) in 250 litres of water per hectare using knap sack sprayer.

- Insecticides should be applied only if the aphid attack appears on earhead stage.

- Since the aphids appear first on borders of the crop, spray only the infested strip to check their further spread.

Method of preparation of neem extract: Boil 10.0 kg terminal parts of the shoots of neem trees including leaves, green branches and fruits in 25 liters of water for 30 minutes. Then, filter this material through muslin cloth and use the filtrate for spraying at the recommended dose.

Army worm attack usually wheat during March-April, however it is also observed in the month of December in fields having large loads of paddy straw. It damages leaves and earheads. Spray 100 ml Coragen 18.5 SC (chlorantraniliprole) or 1 litre Ekalux (quinalphos) in 250 litres of water per hectare with hand-operated knapsack sprayer or in 30 litres of water with motorized sprayer. For better effectiveness of the insecticide, spraying should be done in the evening when armyworm larvae are more active. These insecticides will also control aphids. Alternatively, broadcast 17.5 kg Mortel/Regent 0.3 G (fipronil) or 2.5 litre Dursban 20 EC (chlorpyriphos) mixed with 50 kg of moist sand in one hectare before first irrigation.

Pink stem borer (*Sesamia inferens*) generally attacks the wheat crop at seedling stage. The larva bore into the stem of young plant and kills the central shoot causing 'dead heart'. If the severe damage of pink stem borer is observed in previous rice crop, avoid sowing of wheat in the month of October. Prefer to irrigate the fields during day time to maximize predation of insects by birds. In case of severe infestation, broadcast 17.5 kg Mortel/Regent 0.3 G (fipronil) or 2.5 litre Dursban 20 EC (chlorpyriphos) mixed with 50 kg of moist sand in one hectare before first irrigation. Alternatively, spray 125 ml Coragen 18.5 SC (chlorantraniliprole) in 250 litre of water per hectare.

Gram pod borer (*Helicoverpa armigera*) attacks wheat at maturity. It feeds on the grains in the earheads. Damage is more where wheat follows cotton. Spray 2.0 litre Ekalux 25 EC (quinalphos) in 250 litres of water per hectare with hand operated knap sack sprayer.

Brown mite is a minor insect of wheat causing discolouration of leaves. It is of very small size and attack is more observed on rainfed wheat.

7. Rodents

Field rats: They cause heavy loss to wheat crop in the field and do considerable damage to the harvested crop lying in stacks in the field. For the control of field rates close all burrows in the evening and fumigate the reopened burrows of rats in the morning time with the following poison baits:

Zinc phosphide bait: Smear 1 kg of bajra or sorghum or cracked wheat or their mixture with 20 g of oil and mix it thoroughly with 25 g of zinc phosphide. Never add water in zinc phosphide bait and always use freshly prepared bait.

0.005% Bromadiolone bait: Mix 20 g of 0.25% Bromadiolone powder, 20 g of oil and 20 g of powdered sugar in 1 kg of any cereal flour.

III. 0.0375% racumin bait: Mix 50 g of 0.75% racumin powder, 20 ml of groundnut or sunflower oil and 20 g of powdered sugar in 1 kg of cracked wheat or any other cereal.

(b) Diseases

Yellowing in the seedling stage can be caused by nutrient deficiency, bad weather conditions, poor drainage, attack of Alternaria and Drechslera sp. or soil infestation with cereal cyst nematode.

A general yellowing of the leaves and necrosis occurs especially starting from the tips and along the edges. Later on, necrotic area develops within the chlorotic tissue, there is stunting and general decline. In case of zinc deficiency, the plants remain stunted and bushy; leaves become chlorotic, stocky and break in the middle. Yellowing of young leaves may be due to sulphur deficiency, whereas yellowing of older leaves could be due to nitrogen deficiency.

- In light soil, add 62.5 kg zinc sulphate per hectare. One such application will be enough for 2-3 years.
- Spray 3% urea solution (3 kg in 100 litres of water). To cover the crop thoroughly, spraying may be done cross-wise and a total volume of 750 litres of water per hectare should be used.
- Give proper irrigation to counteract the effect of frost. Avoid excessive irrigation.
- For control of cereal cyst nematode present in the soil, apply 32.5 kg per hectare Furadan 3 G at sowing time.
- Apply 250 kg gypsum per hectare for correcting sulphur deficiency.

Wheat rusts: Generally, wheat crop is infested by following rusts:

(i) Yellow or stripe rust (*Puccinia striiformis f.sp.tritici*): Yellow powdery pustules appear on leaves, forming stripes. The disease appears first in the sub-mountaineous areas of the state. Minimum temperature in the range of 7-13°C coupled with 85-100% relative humidity during night and maximum temperature in the range of 15-24°C during day are congenial for infection, development and spread of disease. Frequent rains along with wind during Feb-March may further enhance the disease severity and spread.

Intergrated management of yellow rust

- Grow rust resistant varieties.
- Don't grow unrecommended varieties especially in sub-mountainous zone.
- Do not sow the crop in the month of October particularly in the disease prone submountainous areas.
- Monitor the crop from 2nd week of December onward particularly after irrigation or rain to locate the initial foci of infection.
- Spray the crop with 500 g Caviet 25 WG (tebuconazole) or 300 g Nativo 75 WG (trifloxystrobin + tebuconazole) or 500 ml Ampect Xtra 25.5 SC (azoxystobin+cyproconazole) or 500 ml Opera 18.3 SE (pyraclostrobin + epoxiconazole) or 500 ml Custodia 320 SC (azoxystrobin + tebuconazole) or 500 ml Tilt 25 EC/Shine 25EC/Bumper 25 EC/ Stilt 25 EC/Compass 25 EC/Markzole 25 EC (propiconazole) in 250 litres of water per hectare as soon as the disease is noticed.
- Initially only spot application surrounding infected area should be done.
- Regularly monitor the crop and repeat the spray as per need.

(ii) Brown or leaf rust (*P. triticina*): Round, orange powdery pustules, irregularly arranged or in clusters on leaves, less common on the leaf sheath and stalk.

- Grow rust resistant varieties.
- Spray the crop with 500 ml Tilt 25 EC/Shine 25 EC/Bumper 25 EC/Stilt 25 EC/ Compass 25 EC/Markzole 25 EC (propiconazole) in 250 litres of water per hectare as soon as the disease is noticed.

Karnal bunt (*Neovossia indica*): In the ear only a few grains are infected. The infected grains on pressing give out black powder of spore mass. This powder gives peculiar stinking smell. Grow resistant varieties. A single spray of 500 ml Tilt 25 EC (propiconazole) per hectare using 250 litres of water at ear emergence stage for the control of Karnal bunt is recommended in wheat meant for seed production only.

Powdery mildew (*Blumeria graminis tritici*): The fungus develops numerous superficial white floury spots on all the above ground parts of the plant. The white colour of the floury spots changes to grey or reddish brown when cleistothecia develop. Infected plants become stunted due to reduction in the size and number of leaves. Spray with 300 g Nativo 75 WG (trifloxystrobin + tebuconazole) in 250 litres of water per hectare.

Loose smut (*Ustilago segetum tritici*): The fungus destroys the ears completely, turning them into a black loose powdery mass consisting of spores and leaving behind the rachis only.

- Grow resistant varieties
- Soak the wheat seed in ordinary water from 8 a.m. to 12 noon on any calm and sunny day during May/June. After 4 hours soaking, spread out the moist seed in the sun in a thin layer on cemented floor (pucca), on tarpaulin or sheets of cloth. Dry the seed completely and store in a dry place till sowing.
- Alternatively, the seed may be treated with fungicides as per recommendations.

Flag smut or leaf-smut (*Urocystis agropyri*): Long narrow lead grey or black streaks or stripes running parallel to veins are formed on the leaves. The stripes eventually rupture and expose black sooty mass of spores. Practise shallow sowing. Rougue out the affected stools and destroy them by burning. Treat the seed with recommended fungicides.

Head blight or Scab (*Fusarium spp.):* Individual spikelets or portion of earheads show premature bleaching or brown discolouration. Characteristic pinkish mycelium of the fungus may be visible on infected spikelets during humid conditions. The grain formation is either completely inhibited or the infected spikelets produce shrivelled chalky grains showing pinkish discolouration. Seed lots affected by scab should be cleaned thoroughly to eliminate shrivelled seeds. The incidence of this disease is more on durum varieties, their cultivation in highly humid areas, particularly near rivers should be avoided.

Leaf blight, glume blight and black tip of grains: Elongated brown spots and blotches appear on leaves and glumes. Ears are poorly filled, grains shrivelled, discoloured and black tipped. Use disease free seed.

Ear cockle (*Mamni*) **and Yellow ear rot** *(tundu) (Anguina tritici and Rathayi bacter tritici):* Diseased plants have spreading tendency and swollen base, leaves become crinkled and twisted. Earheads contain dark-brown, hard and roundish galls (Mamni) instead of grains. Severely diseased plants are stunted and may die at seedling stage. Occurrence of yellow, slimy mass indicates ear rot phase which prevents grain and gall formation. Put wheat seed in ordinary water and agitate vigorously for few minutes. Ear-cockle galls will float to the surface. These may be skimmed off with an ordinary sieve and burnt.

Molya (*Heterodera avenae*): Plants become stunted with yellowing of leaves, reduced tillering, absence of ears on some tillers or small ears with poorly filled 20 grains. The root system of infected plants gets reduced in size and becomes bunchy with profuse development of thin rootlets. Cysts (shining white bodies of female nematode) are seen attached to the roots at the later stage of the crop. Expose the soil to the hot sun by cultivation during May and June. Practise rotation with non-cereals in badly infested soils. Apply 32.5 kg Furadan 3G (carbofuron) per hectare at sowing time.

Harvesting

The wheat crop is harvested when the grains harden and the straw becomes dry and brittle. The harvesting time varies from zone to zone and also depends whether the wheat is under irrigated or rainfed conditions. The rainfed crop reaches harvesting stage much earlier than the irrigated crop. However, the yields under the rainfed crops are much lower compared to the irrigated crop.

Yield

The potential yield of semi-dwarf wheat varieties varies between 60-75 quintals per hectare. At farmer's field the grain yield of high yielding varieties generally ranges between 40-50 quintals per hectare. Generally, the yield of straw from rainfed crop is about the same as that of grain whereas from the irrigated crop it is about one and a half to two times the grain yield.

Management of Stored Grain Insects

Pests and Symptoms of attack: Twenty species of insects infest grains in the Punjab. Khapra beetle (*Trogoderma granarium*), lesser grain-borer (*Rhizopertha dominica*) rice weevil (*Sitophilus oryzae*) and flour beetles (*Tribolium spp.*) are serious pests of wheat jowar, rice barley and maize. Mung dhora (*Callosobruchus analis*), gram dhora (*C. chinensis*) and cowpea dhora (*C. maculatus*) attack different pulses. Grain moth (*Sitotroga cerealella*) attacks wheat, maize, jowar, oats, barley grains which lose nutritive value and germinating capacity, besides loss in weight.

Recommendations

Preventive measures

- Dry the grains properly before storage.
- Plug all cracks, crevices and holes in the godowns thoroughly.
- Store new grains in the clean godowns or receptacles.
- Use new gunny bags.
- Disinfect empty godowns or receptacles by spraying 0.05% malathion emulsion (100 ml Malathion 50 EC in 10 litres of water) on the floor, walls and ceiling or fumigate the godowns using 25 tablets of aluminium phosphide/100 cum of empty space before storing the grains. Exposure 7 days.
- Against dhora, cover the pulses stored in bulk with 7 cm layer of sand or sawdust or dung ash.

Curative measures

1. Phostoxin or Delicia or Celphos (aluminium phosphide) one tablet of 3 g/ tonne or 25 tablets/100 cum space. Exposure 7 days.

5.3. Maize (*Zea mays* L.)

Maize (*Zea mays L.)* is the world's leading crop and is widely cultivated as cereal grain that was domesticated in Central America. It is one of the most versatile emerging crops having wider adaptability. Globally, maize is known as queen of cereals because of its highest genetic yield potential. Maize is the only food cereal crop that can be grown in diverse seasons, ecologies and uses. Maize have many types like normal yellow/ white grain, sweet corn, baby corn, popcorn, waxy corn, high amylase corn, high oil corn, quality protein maize, etc. Apart from this, maize is an important industrial raw material and provides large opportunity for value addition.

Among the cereal grains, wheat, rice and maize are the major ones that are considered as stable food across the globe due to their high nutritional significance enriched with abundant amount of macronutrients like starch, fibre, protein and fat along with micronutrients like B-complex vitamins, ß-carotene and essential minerals, i.e. magnesium, zinc, phosphorus, copper, etc. Maize is also considered as low-cost-high-benefit ratio for human beings that help in the prevention of metabolic syndrome due to the presence of different antioxidants like phenols and phytosterols in it. Maize or corn can be consumed only after processing into different food items such as popcorn, flour, tortillas, cornflakes, corn germ oil, etc. Maize products are also used in supplementary nutritional programmes to feed the malnourished children and to improve their health status. However, the quality of maize products depends upon the agronomic practices and climatic conditions.

Origin and History

How corn evolved and what are its progenitor species is still a matter of investigation. Archaeological and molecular data indicates that modern maize was domesticated from annual teosinte *(Zea mays ssp. parviglumis)* in southern Mexico between 6,600 and 9,000 years ago. Selection soon followed: favourable alleles at loci controlling plant morphology and kernel nutritional quality were fixed at least 4,400 years ago, and further selection by native Americans facilitated maize adaptations to varied environments.

Area and Distribution

Maize is cultivated throughout the world from 58°N latitude to 40°S latitude, the crop spreads and cultivated over 139 million ha of area and around 600 million tonnes of maize is produced. Crop occupies the third position next to rice and wheat in area and production. USA, China, Brazil, Mexico, India, Romania, Philippines, Indonesia are some of important countries cultivate maize crop. In India, Rajasthan, UP, MP, Bihar, Karnataka, Gujarat, AP, J&K, HP and Maharashtra are important maize producing states.

Globally, highest area of maize is in China (41.29 mha), with highest production of 360.25 million tonnes in USA. However, the highest average yield of 296.50 q/ha was recorded in Jordan In India, area of maize was 9.87 mha with production of 30.16 mt during 2020. (Table 5.3)

Table 5.3: Area, production and average yield of maize for different countries of the world

Country	Area in million ha		Country	Production in million tonnes		Country	Average Yield (q/ha)
China	41.29		USA	360.25		Jordan	296.5
USA	33.37		China	260.88		Saint Vincent and the Grenadines	292.41
Brazil	18.25		Brazil	103.96		United Arab Emirates	260.23
India	9.87		Argentina	58.4		Israel	192.2
Argentina	7.73		Ukraine	30.29		Tajikistan	135.19
Nigeria	7.53		India	30.16		Kuwait	130.38
Mexico	7.16		Mexico	27.42		Qatar	124.61
Ukraine	5.39		Indonesia	22.5		Spain	122.58
Tanzania	4.2		South Africa	15.3		New Zealand	122.12
Indonesia	3.96		Russian Federation	13.88		Austria	113.45

Source: FAOSTAT, 2020

Type of Corn

Corn variation may be artificially defined according to kernel type likes: dent, flint, flour, sweet, pop and pod corn etc. Except for pod corn, these divisions are based on the quality, quantity and pattern of endosperm composition in the kernel and are not indicative of natural relationships (Brown and Darrah, 1985).

Dent Corn: Dent corn is characterized by the presence of corneous, horny endosperm at the sides and back of the kernels, while the central core is a soft, floury endosperm extending to the crown of the endosperm. It collapses to produce a distinct indentation on drying. Degree of denting varies with its genetic background. Dent corn is used primarily as animal food, but also serves as a raw material for industry and as staple food. It is still an important human food and industrial material, entering into many specialized products via the dry- or wet-milling industry. However, white dent often receives a premium price in the dry milling industry, where it utilized for certain human food products because of its whiter starch.

Flint Corn: The flint corns mostly have a thick, hard, vitreous (glassy) or corneous endosperm layer surrounded by small, soft granular center. The relative amounts of soft and corneous starch, however, vary in different varieties. Generally, the kernels are smooth and rounded, and the ears are long and slender with a comparatively small number of rows or kernels. In temperate zones, flint corn often matures earlier, germinates better has more spring vigor, more tillers and fewer prop roots than dent strains.

Flour Corn: This is one of the oldest types of corn, tracing back to the ancient Aztecs and Incas. American and Indians used to ground the soft kernels for flour. Floury maize types have soft starch throughout, with practically no hard, vitreous endosperm and thus are opaque in kernel phenotype. Kernels tend to shrink uniformly upon drying, so usually have little or no denting. When dry, they are easy to grind, but may mold on the mature ear in wet areas.

Sweet Corn: In sweet corn, the sugary gene prevents or retards the normal conversion of sugar into starch during endosperm development, and the kernel accumulates a water-soluble polysaccharide called "phytoglycogen." As a result, the dry, sugary kernels are wrinkled and glassy. The higher content of water-soluble polysaccharide adds a texture quality factor in addition to sweetness.

Popcorn: Popcorns are perhaps the most primitive of the surviving races of maize. This corn type is characterized by a very hard, corneous endosperm containing only a small portion of soft starch. Popcorns are essentially small-kernelled flint types. The kernels may be either pointed (rice-like) or round (pearl-like). Some of the more recently developed popcorns have thick pericarps (seed coats), while some primitive semi-popcorns, such as the Argentine popcorns, have thin pericarps.

Pod Corn: Pod corn (tunicate maize) is more of an ornamental type. The major gene involved (Tu) produces long glumes enclosing each kernel individually, which also occurs in many other grasses. The ear is also enclosed in husks, as with other types of corn. Homozygous pod corn usually is highly self-sterile. and the ordinary type of pod corn is heterozygous. Pod corn may be dent, sweet, waxy, pop, flint or floury in endosperm characteristics. It is merely a curiosity and is not grown commercially.

Waxy corn: Waxy corn name derives from the waxy appearance of the endosperm exposed in a cleanly cut cross-section. Common corn starch is approximately 73 percent amylopectin and 27 percent amylose, whereas waxy starch is composed entirely with amylopectin, which is the branched molecular form of starch. Ordinary corn starch stains blue with 2 percent potassium iodide solution, whereas waxy cornstarch stains a reddish brown. The waxy gene also expresses itself in the pollen with this staining reaction, which is an aid in breeding. Products made from waxy corn are used by the food industry as stabilizers and thickeners for puddings pie fillings, sauces, gravies, retorted foods, salad dressings, etc. Other waxy products are used as remoistening adhesives in the manufacture of gummed tape, in adhesives and in the paper industry. Waxy grain is also grown as a feed for dairy cattle and livestock.

High-amylose corn: Amylo-maize is the generic name for corn that has an amylose content higher than 50 percent. The endosperm mutant amylose extender (ae) was first observed by R. P. Bear in 1950, which increases the amylose content of the endosperm to about 60 percent in many dent backgrounds. Modifying factors alter the amylase contents as well as desirable agronomic characteristics of the grain. The amylose-extender gene expression is characterized by a tarnished, translucent, sometimes semifull kernel appearance. The starch from high-amylose corn is used in the textile industry, in gum candies (where its tendency to form a gel aids production), and as an adhesive in the manufacture of corrugated cardboard.

Nutrients Content of Maize

The nutrition values in a 100 g serving of corn, according to the nutrient database of the U.S.D.A (United States Department of Agriculture) are as follows:

Calories - 385

Macronutrients

Total Fat - 7%

Saturated Fat - 3%

Polyunsaturated Fat - 5%

Monounsaturated Fat - 2%

Cholesterol - 0%

Sodium - 0%

Total Carbohydrates - 25%

Dietary Fiber - 14%

Sugars - 2%

Protein - 10%

Micronutrients

Minerals:

Calcium - 26%

Iron - 11%

Potassium - 27%

Vitamins:

Thiamine - 5%

Riboflavin - 7.6%

Niacin - 3.7%

Folic Acid - 3%

Vitamin C - 7%

Vitamin E - 4.6%

Being imbued with carotenoid antioxidants, lutein and zeaxanthin, as well as flavonoids, sterols and tannins, corn significantly boosts overall health. Moreover, corn is a very rich plant source of crucial essential amino acids namely isoleucine, tryptophan, valine, methionine and threonine, thus complementing both vegetarian and vegan diets.

Uses of Maize

- **Maize flour, or meal,** is made into a thick porridge in many cultures.
- **Maize meal** is also used as a replacement for wheat flour, to make cornbread and other baked products.
- **Masa (cornmeal treated with lime water)** is the main ingredient for tortillas, atole, and many other dishes of Mexican food.
- **Cornstarch** is made from maize kernels, which are high in starch, and used as a thickening agent in soups.
- **Corn syrup** is used as a sweetener instead of sugar in thousands of products, including soda, candy, cookies and bread. Kitty litter made from maize is environmentally-friendly.
- **Aminal feed.** It can be used as feed for cows, pigs, catfish and chickens etc. Maize is also a large component of commercial chicken feed, as well as food for catfish, especially in farmed catfish.
- **As fodder or silage.** The largest market for maize is actually as food for livestock. Cows eat field corn, not the sweet corn.

Maize Zones

Maize improvement project in India demarcated maize cultivated areas in the country into four broad zones:

i. **Himalayan Zone:** It includes temperate regions above an elevation of 600 m that receive moderate rainfall. Crop duration is usually longer.

ii. **Northwest Plains Zone:** This zone is characterized by wet and arid climate below 600 m above MSL. Soil moisture stress at flowering and grain filling is common. Soils are sandy and rarely alluvial. Medium to early maturing cultivars are preferred. Ground frosts may be a problem during winter season.

iii. **Northeast Plains Zone:** Hot and humid climate with frequent waterlogging. Maize can be cultivated in three seasons: monsoon, winter and spring. Soils are Oxisols with low nutrient status.

iv. **Peninsular Zone:** This zone is characterised by tropical to subtropical climate. Coastal areas favour maize cultivation although the year, if irrigation water is available. Soils are medium deep to very deep. Deep soils need drainage provision during heavy monsoon rains.

Climatic Requirements

Maize can be cultivated in different types of climates, but thc growth, development and yield of maize is found more in the hot regions. It is a hot season crop. Night and day temperatures should be higher for its deposition. The maize crop requires sufficient moisture in the soil since its inception. Temperature of 18 to 23 degree Celsius for deposition and 28-degree Celsius temperature is considered good for growth and development.

Maize varieties released by ICAR (2020-2021)

Maize varieties	Area of adoption	Characters
HT 17169	Punjab, Haryana, Delhi, Uttar Pradesh	Suitable for both rainfed and irrigated conditions during *kharif*, average grain yield 9.64 tonnes/ha, maturity 120 days, highly responsive to high density and high fertilizer dose, resistant to maydis leaf blight (MLB) and charcoal rot (ChR), moderately resistant to Chilo partellus.
CP 858	Punjab, Haryana, Delhi, Uttarakhand (Plains), Uttar Pradesh (Eastern and Western regions), Bihar, Jharkhand, Odisha, West Bengal	Suitable for irrigated conditions during *kharif*, average grain yield 8.82 tonnes/ha, late maturity (110–115 days), stay green trait, attractive grains colour, resistant to TLB and moderately resistant to ChR, MLB, banded leaf and sheath blight (BLSB), FSR, S. inferens and Chilo partellus.

Maize varieties	Area of adoption	Characters
NUZI 260 (Sweet Corn)	Punjab, Haryana, Delhi, Uttarakhand (Plains), Uttar Pradesh (Eastern and Western Regions), Bihar, Jharkhand, Odisha, West Bengal, Maharashtra, Karnataka, Andhra Pradesh, Tamil Nadu, Telangana	Suitable for irrigated conditions during *kharif,* average green ear yield (without husk) 11.2 tonnes/ha (NWPZ), 12.1 tonnes/ha (NEPZ) and 13.4 tonnes/ ha (PZ), maturity 80–82 days, moderately resistant to MLB, ChR and P. Rust, and Chilo partellus, good sweetness of kernels with 15.4% TSS.
VL QPM Hybrid 59	Uttarakhand (Hills)	Suitable for rainfed condition of *kharif*, average grain yield 3.33 tonnes/ha under organic conditions, early maturity (90–95 days), high tryptophan (0.77%) and lysine content (3.33%), moderately resistant to TLB and MLB.
NMH 713	Assam	Suitable for irrigated conditions during *rabi*, average grain yield 9.49 tonnes/ha, maturity 155 days, resistant to common rust, moderately resistant to turcicum leaf blight (TLB), sorghum downey mildew (SDM) and Sesamia inferens.
PM 16202L	Andhra Pradesh, Telangana, Tamil Nadu, Karnataka, Maharashtra, Rajasthan, Gujarat, Madhya Pradesh	Suitable for *rabi* irrigated ecology, average grain yield 9.1 tonnes/ha, late maturing (109 days), responsive to higher plant densities and higher fertilizer level, moderately resistant to TLB, ChR and CR, Chilo partellus, S. inferens and shoot fly.
P3392	Rajasthan, Gujarat, Madhya Pradesh	Suitable for *rabi* season irrigated ecology, average grain yield 9.2 tonnes/ha, late maturing (121 days), responsive to higher plant densities and higher fertilizer level, moderately resistant to FSR, ChR, TLB, Chilo partellus and Sesamia inferens.
P3302	Punjab, Haryana, Uttar Pradesh	Suitable for *rabi* irrigated ecology, average grain yield 8.5 tonnes/ha, medium-late (100–105 days) maturity, moderately resistant to MLB, TLB, BLSB, CR, RDM, CLS, Chilo partellus; responsive to higher plant densities and higher fertilizer level.
Rasi 3499	Odisha, Jharkhand, Bihar, Uttar Pradesh (eastern part), West Bengal	Suitable for *kharif* in medium to high fertility soils under irrigation or assured rainfall conditions, average grain yield 7.5 tonnes/ha, medium-late maturity (110– 115 days), deep yellow grains, moderately resistant to MLB, TLB, BLSB and FSR, good stay green quality.
BRMH 1	Karnataka	Average grain yield 7.0–7.2 tonnes/ha, late maturity (118–120 days), good shelling percentage (80–82%), yellow kernel, moderately resistant to TLB.

Improved varieties of Maize for Punjab

Long Duration Varieties/Hybrids

PMH 13: This single cross hybrid has tall plants with medium high ear placement. Leaves are dark green and broad. Tassel is medium and open. Ears are conico-cylindrical and long with light orange flint grains. It matures in 97 days and average yield is 60.0 quintals per hectare. It is moderately resistant to maydis leaf blight, charcoal rot and maize stem borer.

ADV 9293: This single cross hybrid has tall plants with medium ear height. Stem is sturdy and leaves are broad. Tassel is medium and open. Ears are long with attractive orange flint grains. It matures in 97 days and average yield is 61.25 quintals per hectare It is moderately resistant to maydis leaf blight, charcoal rot and maize stem borer.

JC 12: The plants of this composite variety are medium tall in height with medium ear placement. It has medium thick stem which resists lodging and has heavy open tassel. Ears are medium long with good girth. This composite has semi-flint, yellow orange grains. It matures in about 99 days. It yields about 45.5 quintals per hectare. This variety is recommended for kandi areas of the state.

PMH 11: This single cross hybrid has tall plants and well-developed root system. Stem is sturdy and green in colour. Leaves light green and broad. Tassel is open and heavy. Ears are long with dark orange flint grains. It matures in 95 days and Its' average yield is 55.0 quintals per hectare.

PMH 1: This single cross hybrid has tall plants with well-developed root system. The stem is zig-zag, sturdy and purple colored. The leaves are medium broad. Tassel is open and medium in size. Ears are medium long with yellow orange flint grains. The plants remain stay green at maturity. It matures in 95 days and average yield is 52.5.0 quintals per hectare. Medium duration variety

PMH 2: It is a short duration single cross hybrid. It has medium plant height with medium ear placement. Leaves are medium sized and dark green in colour. Tassel is of medium size and semi-open. Silk is of green colour. It has medium long ears. Grains are yellow orange flint with yellow caps. The hybrid resists lodging and is tolerant to bacterial stalk rot. It matures in about 83 days and Its' average yield is 43.0 quintals per hectare.

Special Purpose Varieties

Punjab Baby Corn 1: This single cross hybrid is male sterile and having medium tall plants. This hybrid is most suitable for baby corn as it gives higher yield of uniform and good quality ears. Being male sterile this hybrid does not need detasseling. Picking of ears starts around 52 days after sowing. This hybrid gives about 3 pickings per plant. This hybrid gives an average of 53.5 quintals per hectare yield of dehusked ears. This hybrid gives 320 q/hectare fodder yield after the completion of ear picking.

Punjab Sweet Corn 1: This composite has tall plants with medium thick stem and medium ear placement. The leaves are broad and tassel is open with creamish anthers. Ears are medium long and the cob colour is white. Silk colour at the time of emergence is usually creamish. Husk cover is well developed and grains are orange in colour at maturity. This variety is highly suitable to use as sweet corn on commercial basis because Its' developing and immature grains in green ears possess high sugar content. It matures in about 95-100 days. Its' average green ear and grain yield is 125.0 and 32.5 quintals per hectare, respectively.

Pearl Popcorn: This is a composite variety of popcorn. Its' ears are long and thin and grains are small and round. The commercial value of these grains is very high. It matures in about 88 days. Its' average yield is 30.0 quintals per hectare and popping quality is very good.

Soil Requirements

Maize can be grown successfully in variety of soils ranging from loamy sand to clay loam. However, soils with good organic matter content having high water holding capacity with neutral pH are considered good for higher productivity. Being a sensitive crop to moisture stress particularly excess soil moisture and salinity stresses; it is desirable to avoid low lying fields having poor drainage and also the field having higher salinity. Therefore, the fields having provision of proper drainage should be selected for cultivation of maize.

Rotation

Since maize is a short duration crop, it conveniently fits into a wide range of crop rotations. In northern India, particularly in the states of Uttar Pradesh, Punjab, and Bihar, it is rotated with wheat, potato, barley, etc. in a one-year rotation under irrigated conditions. It is usually grown as a pure crop, but occasionally legumes, e.g. mung, arhar or beans, and quick growing vegetables, e.g. pumpkins and various types of gourds are taken as subsidiary crops. In central India, under rainfed conditions, however, cotton and arhar are grown mixed with maize. A usual *kharif* crop can also follow maize grown for green ears or fodder. In spring, very early maturing varieties of Moong and Mash (Sathi) are intercropped in maize. Given below are some of the most important crop rotations:

- Maize – Wheat
- Maize – Toria – Wheat
- Maize – Potato – Wheat
- Maize – Berseem

- Maize – Toria – Sugarcane
- Maize – Potato – Onion
- Maize – Senji – Sugarcane
- Maize – Raya /Gobi Sarson
- Maize – Gram
- Maize – Lentil
- Maize – Pea

Crops like soybean, urd, moong, cowpea etc., are also grown mixed with maize. These legume crops are grown in the space between two rows of maize. In some parts of the country maize is also grown with pigeonpea

Land Preparation

A seedbed for maize should be fine but compact. Maize kernels need a seedbed which is friable, well aerated, moist and free from weeds. It is desirable that the previous crop refuse is buried under with a mould-board plough. In due course, two or three ploughings with the wooden plough are given. In case the tractor is used, one ploughing followed by a couple of diskings is generally adequate. There is no need of preparing extremely fine seedbed for maize. The first ploughing for maize should be done with soil inverting plough so that 20-25 cm deep soil may become loose. It should be followed by two or three harrowings or three or four intercrossing ploughings with local plough. Planking should be done with each ploughing. It is important to note that while preparing the field for maize crop, levelling is not overlooked. A properly levelled and uniformly graded field is necessary for good water management.

Time of sowing

Time of sowing is indirectly determined by the soil and atmospheric temperature as well as by the supply of water. Thus, there is a general preference for sowing of the crop at the time of onset of monsoon in most parts of the country. The actual dates vary from region to region as indicated below

North Eastern Hills – Early March

North Western Hills – April to early May

Peninsular zone – May to mid-June

Under irrigated conditions in northern India, time of sowing varies from last week of May to end of June, depending on the prevailing conditions of the region. In fields which are prone to water stagnation, the crop is sown in the end of May/

early June so that the crop gets firmly established before the onset of monsoon. Maize cultivation has been extended to *rabi* season also i.e. from September end to beginning of November particularly in the Indo-Gangetic plains region. Trials conducted under All India Coordinated Maize Improvement Project revealed that yield potential of maize in *rabi* season was about twice that of the *kharif* season (Dayanand and Jain, 1994). Delay in sowing of *rabi* maize beyond first half of November in north western India results in a decline in its yield because of very cold conditions encountered during December and January. In eastern region and south India delayed sowing leads to reduction in yield due to shortening of the growing season.

Selection of Seed

It is necessary to use certified seed of improved varieties. Before planting, the seed lot must be tested for its germination percentage and the seed quantity must be adjusted accordingly. For achieving maximum yield, every year a new hybrid seed must be used. The yield may be reduced drastically by using the seed from one's own hybrid crop, by 30 per cent or even more. However, the composite maize varieties do not have this characteristic and therefore, a farmer can use the seed from his own composite crop for next year also, provided no mixture has been allowed in the field and at the threshing floor. It is advisable to change the seed of composite every three to four years. Check germination of seed before sowing.

Seed Rate and Spacing

A population of 60,000-65,000 plants at harvest per hectare would be needed to attain maximum yield. It is very important that optimum plant density is maintained in maize, because unlike tillering plants such as rice and wheat, maize cannot compensate for lost space. In *rabi* season, the recommended plant density is 85,000-90,000 plants per hectare and this could give a yield advantage of 0.3-0.5 tonnes per hectare. Maize is sown in rows, 60- 75 cm apart, whereas the plants in the row are spaced at 20 to 25 cm. Based on seed weight and required plant population, the seed rates are as follows:

Hybrids: 20-25 kg per hectare

Composites: 18-20 kg per hectare

Method of Planting

Maize should be planted across the slope and the seed be sown 3-5 cm deep. The planting depth to a large extent will depend on the moisture status of the field and the type of soil. For example, if soil is dry and sandy, it is advisable to plant deeper and under normal conditions sow at 3-5 cm deep. Normally planting is done in one of the following ways:

(a) Planting on the side of the ridge: A method adopted in high rainfall situation.

(b) Planting in narrow furrows: A method adopted in low rainfall areas.

(c) Planting in flat bed with no earthing up: In normal conditions.

(d) Planting on flat bed and earthing up after 40-50 days of plan-ting: In areas where there is heavy storm during rainy season.

(e) Zero tillage sowing can also be adopted after spraying Gramoxone 24 SL (paraquat) at 2.5 litre per hectare before sowing maize.

Weed Management

During the initial stage, the growth of the maize plants is suppressed by weeds and weeds create an acute problem in rainy season crop. The main weeds found in maize are annual grasses, sedges and broad leaf weeds. Weeding may be done between the rows with bullock or tractor drawn implements, while the weeding within the rows is done by hand. Two to three hoeings are required at 15 and 30 days after sowing, following which the crop is earthed up to provide for better stand ability. No inter-cultivation after flowering should be done, as it is likely to damage the lateral roots.

Weeds can also be controlled by growing one or two rows of fodder cowpea in between maize rows and harvest 35-45 days after sowing thereafter no weed control operation is required. Effective weed control can be obtained by spraying Atrazine 50 WP (@ 1.0 kg a.i. ha-1(2.0 kg per hectare of commercial product) before the crop germinates i.e., within two days of sowing in 500 litres of water. The dose of herbicides should be reduced by 40% in light soils. When hardy weeds like *Commelina benghalensis* (Kaun makki), *Arachne racemose* (ghas), *Bracharia reptans* (Bans patta) are a problematic weed, spray tank mix application of atrazine 0.75 kg/ha with Lasso 50 EC 1.25 kg a.i./ha or Stomp 30 EC 0.75 kg a.i./ha (1.5 kg +2.5 litre per hectare of commercial product) as pre-emergence. Alternatively, tembotrione 110 g a.i./ha (Laudis 420 SL at 260 ml per hectare) can be sprayed 20 days after sowing for the control of mixed weed flora of maize.

For the control of Dila/Motha (*Cyperus rotundus*) and broad leaf weeds from maize, apply 2,4-D amine salt 58 SL at 0.58 kg a.i./ha, 20-25 days after sowing. (400 ml per acre of commercial product)

Water Management

Maize is extremely susceptible both to excess water and moisture stress. Maize can withstand heavy rain; however, the water should not be allowed to stand in the field any time during maize growth. Water stagnation even for as small period as six hours can destroy the crop. Hence for *kharif* cultivation, it is essential that adequate drainage is provided. It may be provided in the form of very shallow surface drains at 40- 50 m apart (depending on the slope and the texture of the soil) across the slope and connected to a main outlet. Shallow drains do not obstruct the movement of cattle or tractors during cultivation. Surface drains should be provided before sowing. In regions with about than 60 cm of well distributed rainfall during the growing season, any additional irrigation is not necessary. A good crop of maize does require about 460-600 mm of water during its life cycle. It must be ensured that the maize plants never wilt because of water shortage any time during their life cycle. Tasseling and silking stage is very crucial. At this stage water shortage for even two days can reduce maize yield by 20 per cent. Also, inadequate soil moisture during flowering and post-flowering particularly during the grain-filling period markedly reduce the yield. During the grain filling period, the most susceptible stage, additional irrigation, if needed, should be applied. The total number of irrigations will depend on the rainfall distribution pattern. The spring and *rabi* crops are entirely raised under irrigation; the number of irrigations may, however, vary from 5-10, depending upon the type of the soil, the prevailing temperature and rainfall conditions.

Fertilizer Management

Fertilizers and manure both play a critical role in maize cultivation. For obtaining high yields, the maize crop should be heavily manured. More than 15 tonnes of good quality farmyard manure or compost should be ploughed into the soil before sowing. For hybrid and composite varieties of maize, 100-120 kg nitrogen, along with 60 kg P_2O_5 and 40 kg K_2O per hectare, is recommended. The precise level of application of phosphorus and potash should be modified in the background of soil analysis. One- third of nitrogen and total quantity of potash and phosphorus should be applied before sowing, while the remaining nitrogen should be applied as side dressing at the knee-high stage and at tasseling in two equal doses. Farmers planting local varieties may apply 40-60 kg per hectare nitrogen to the soil as per soil test. However, winter or *rabi* maize requires 50 % higher nutrition due to its long growing period. These nutrients can be increased or decreased depending on soil fertility.

Zinc Deficiency

Zinc deficiency is wide spread in areas where the high yielding varieties are grown. The deficiency symptoms appear within two weeks of seedling emergence. A broad band of white or very light-yellow tissue, with reddish veins appears on each side of the midrib, beginning at the base of the second or third leaf from the top of the plant. The white patch later extends in stripes towards the tip parallel to the midrib. The midrib and leaf margins remain green. The plants remain stunted and have short internodes.

Where zinc deficiency had been noticed in the preceding crop, apply 25 kg of zinc sulphate heptahydrate (21%) or 17.5 kg per hectare zinc sulphate monohydrate (33%) per hectare at sowing time or after crop emergence along the rows. Do interculture and apply irrigation. Foliar application of zinc sulphate and lime mixture (3.0 kg zinc sulphate heptahydrate with 1.5 kg unslaked lime) can be done after dissolving in 500 litres per hectare of water.

Plant-Protection

A. Insect Pests

Maize borer: The maize borer is a serious pest from June to September. Its' larvae first scrape the leaves and then bore into the stem through the whorl or leaf sheath. The central leaves of the attacked whorl get perforated. In a young plant, the growing point is killed and a dead-heart results. Adopt the following integrated control measures:

- Kill the borer larvae hibernating in plant remnants like stubbles, stalks, cobs and cores. Plough up the fields after harvesting the maize crop, collect the stubbles and destroy them. Use maize stalks, cobs and cores by the end of February. Chop the remaining stalks, if any, for subsequent use. For seed, keep only healthy cobs, free from borer attack.
- Remove and destroy the plants showing severe borer injury, while hoeing the crop.
- Use trichocards twice having 1,00,000 eggs of *Corcyra cephalonica* parasitized by *Trichogramma chilonis* per hectare. First release on 10 days old crop and second 7 days after the first release. Cut trichocards into 100 strips, each having approximately 1000 parasitized eggs. Staple these strips uniformly on the underside of the central whorl leaves in evening hours.
- Spray the crop 2-3 weeks after sowing as soon as borer injury to the leaves is noticed with 75 ml Coragen 18.5 SC (chlorantraniliprole) using 150 litres water per hectare with knap-sack sprayer.

Fall armyworm: The young larvae feed by scrapping the leaf surface making papery windows. The bigger larvae feed voraciously on the central whorl leaves causing round to oblong holes and produce a large amount of faecal matter. The larva can be identified by predominant

- Sow the crop at recommended time only.
- Avoid staggered sowing of maize in adjacent fields to minimize spread of this pest.

- Regularly monitor the field to collect and destroy egg masses of fall armyworm from leaves. Egg masses are covered with hairs and are easily visible.
- Spray the crop with Coragen 18.5 SC (chlorantraniliprole) @ 0.4 ml per litre water or Delegate 11.7 SC (spinetoram) @ 0.5 ml per litre or Missile 5 SG (emamectin benzoate) @ 0.4 g per litre using 250 litres of water per hectare, for crop up to 20 days old. Thereafter for older crop, the amount of water used per hectare needs to be increased up to 500 litres with corresponding increase in dosage of above insecticides. For effective management of this pest, direct the nozzle towards the whorl.
- If the infestation is in patches or the crop is more than 40 days old and spraying is difficult, apply soil-insecticide mixture (about half gram) in the whorls of the infested plants to manage fall armyworm. To prepare soil–insecticide mixture, add 5 ml of Coragen 18.5 SC (chlorantraniliprole) or Delegate 11.7 SC (spinetoram) or 5 g of Missile 5 SG (emamectin benzoate) in 10 ml of water and mix well in one kg of soil. Precaution: Use gloves for preparation and application of the mixture.

Jassid, thrips, pyrilla, grey weevil and leaf-feeding insects: These attack the *kharif* crop.

Armyworm and silk cutter: These insects feed on the leaves in the whorl. The pesticides used against the maize borer are effective in controlling these pests also.

Hairy caterpillars: Hairy caterpillars, if appearing in an epidemic form, cause serious damage by feeding on the leaves and the tender stems. When young, they feed gregariously. The grown up caterpillars may migrate from one field to another. Adopt the following control measures:

- Use light-traps for the destruction of moths.
- Young larvae are gregarious. Destroy them by plucking the infested leaves or by pulling out the infested plants and bury them.
- Destroy grown up caterpillars by crushing them under feet or by picking and putting them into kerosenized water.

Mite: The attack of mite is serious in June on the young crop or in September-October when the crop is nearing maturity. The affected leaves turn pale and can be recognized from the presence of dusty webs.

B. Diseases

Seed rot and seedling blight (Several fungi): Poor germination, unthrifty seedlings and seedling mortality are the symptoms. Use disease free seed.

Banded leaf and sheath blight (*Rhizoctonia solani*): Water soaked, straw coloured necrotic lesions alternating with dark brown bands develop on basal leaf sheaths. Lesions enlarge and coalesce with each other. Later, sclerotia develop on diseased sheaths, husk and cobs. In severe cases, developing ears are completely damaged and dry up prematurely with cracking of husk. To manage this disease, spray 250 ml Amistar Top 325 SC (azoxystrobin + difenoconazole) in 250 litres of water per hectare at disease appearance. If needed, repeat the spray at 15 days interval.

Maydis leaf blight (*Drechslera maydis*): This disease is characterized by the presence of spindle shaped, necrotic to brown lesions on the leaves. Such lesions may merge to form large, irregular patches. Sometimes the symptoms also appear on leaf sheaths, cob husks and ears. Late sowing, high humidity (>80%) and temperature of 25 ± 2°C favours the development of disease. Destroy the infected crop residue in the field. Grow improved varieties. Follow spray schedule as against Brown stripe downy mildew.

Bacterial stalk rot (*Dickeya zeae*): Water soaking and rotting of basal stem especially the leaf sheaths followed by rapid rotting of basal internodes. The rind loses its natural green colour and gives appearance as if boiled in water. The rotten stalks emit a characteristic fermenting odour and may break over from the second or third basal internode. The infected plants wilt. Destroy the diseased plant debris. Excessive rains and poor drainage favours the disease, so keep the fields well drained.

Brown stripe downy mildew (*Sclerophthora rayssiae var zeae*): It is characterized by the presence of long, narrow, brownish, interveinal stripes on leaves. Whitish downy fungal growth may be observed on close examination on underside of the stripes. Destroy the collateral host Takri grass (*Digitaria sanguinalis*) from the maize field. Keep the fields well drained. Spray 500 g Indofil M-45 (mancozeb) in 250 litres of water per hectare after about a fortnight of sowing. Give two more sprays at 10-day intervals. Grow recommended varieties.

Post flowering stalk rots (*Fusarium spp., Macrophomina spp., Cephalosporium spp.*): Plants wilt after flowering. The rind and basal internodes become discoloured. On splitting, the discolouration of the pith progressing upward is also seen.

Harvesting

The maize crop is harvested when the husk has turned yellow and the grains are hard enough having not more than 20 per cent moisture. The appearance of the plant may be misleading, particularly in case of high yielding hybrids and composites whose grains are dry, while the stalk and leaves may be still green. Ears are removed from the standing crop. Harvested ears are dried in the sun before shelling. In case of the late sown crop, farmers prefer to harvest the whole plant and pile them, and the ears are removed later. Maize stalks are used as a cattle-feed and fuel. In fact, no part of maize plant, even the cobs from which the grains have been removed, is left unused. Maize grown for fodder should be harvested at the milk to early dough stage; the earlier harvested crop is likely to yield less and have lower protein content. For silage, however, the late dough is preferred. Both power- and hand-operated low priced maize shellers are available indigenously. These shellers are considerably more efficient than hand-shelling or beating with sticks, the common practice in northern India. Farmers using hybrid maize should not save their own seed for the next crop, as the advanced- generation hybrid seeds are likely to lead to a yield reduction of 25-30 per cent. However, farmers can save seed from composites and the open-pollinated varieties, when grown in isolation. At least seeds from 500 to 1,000 ears of the best yielding, normally spaced plants resistant to prevalent diseases and pests should be bulked. Ears should be dried, shelled and treated with an insecticide. The treated seed is necessary, as the untreated seed on ears is at times badly attacked by stored grain pests and the germination is markedly reduced. Considerable variation in grain yield is observed.

Yield

The yield levels depend upon the variety, the amount of the fertilizer used, the rainfall pattern, etc. Under irrigated conditions and recommended cultural practices, it gives 50-60 quintals per hectare in case of hybrids and 45-50 quintals in case of composites. In case of rainfed crop yield levels are about 20-25 quintals for hybrids and 15-20 quintals for composites.

5.4. Barley (*Hordeum vulgare* L.)

Barley, (*Hordeum vulgare*), cereal plant of the grass family Poaceae and its edible grain. Grown in a variety of environments, barley is the fourth largest grain crop globally, after wheat, rice, and corn. Barley is commonly used in breads, soups, stews, and health products, though it is primarily grown as animal fodder and as a source of malt for alcoholic beverages, especially beer.

Only ten percent of barley is used as human food, while the remaining percentage is used for brewing malt beverages, including beer and whiskey. However, the majority of harvest barley is used for livestock feed.

The old English word for 'barley' was baere, which traces back to Proto-Indo European and is cognate to the Latin word farina "flour". The un-derived world 'baere' survives in the north of Scotland as bere, and refers to a specific strain of six-row barley. The word barn, which originally meant barley-house, is also rooted in these words.

Nutritional Content in Barley

100gm barley has 354 calories

Content	% Daily Value
Total Fat 2.3g	3%
Saturated Fat 0.5g	2%
Cholesterol 0 mg	0%
Sodium 12mg	0%
Potassium 452 mg	12%
Total Carbohydrate 73g	24%
Dietary fibar 17g	68%
Sugar 0.8g	
Protein 12g	24%
Calcium	3%
Iron	19%
Vitamin B6	15%
Magnesium	33%

Benefits

- Barley is rich in vitamins, minerals and other beneficial plant compounds.
- Reduces hunger and may help you lose weight.
- Insoluble and soluble fiber content improves digestion
- May prevent gallstones and reduce your risk of gallbladder surgery.
- Beta-glucans may help lower cholesterol.
- May reduce heart disease risk
- Magnesium and soluble fiber may protect against diabetes
- May help prevent colon cancer
- Versatile and easy to add to your diet

Origin and History

One of the first cultivated grains of the Fertile Crescent, barley was domesticated about 8000 BCE from its wild progenitor *Hordeum spontaneum*. Archaeological evidence dates barley cultivation to 5000 BCE in Egypt, 2350 BCE in Mesopotamia, 3000 BCE in northwestern Europe, and 1500 BCE in China. Barley was the chief bread plant of the Hebrews, Greeks, and Romans and of much of Europe through the 16th century. Genetic studies suggest that Tibet was an additional, independent centre of domestication for cultivated barley.

Area and Production

Highest area and production of barley i.e. 8.27 mha and 20.94 million tonnes was in Russia. However, Belgium topped in average yield with 74.93 q/ha (Table 5.4). Total area of barley during 2020 in India was 0.61 mha with production of 1.72 mt.

Table 5.4: Area, production and average yield of barley for top 10 countries

Country	Area in million ha		Country	Production in million tonnes		Country	Average yield in q/ha
Russia	8.27		Russia	20.94		Belgium	74.93
Australia	5.04		Spain	11.47		New Zealand	74.8
Turkey	3.09		Germany	10.77		Ireland	69.75
Canada	2.81		Canada	10.74		Switzerland	69
Spain	2.75		France	10.27		Zambia	65.26
Kazakhstan	2.73		Australia	10.13		Chile	64.79
Ukraine	2.37		Turkey	8.3		Germany	64.59
Iran	2.06		UK	8.12		Austria	64.16
France	1.97		Ukraine	7.64		Netherlands	64.15
Germany	1.67		Argentina	4.48		Denmark	63.63

Source: FAOSTAT 2020

In India, barley production is highest in Rajasthan (47% of total barley production) followed by Uttar Pradesh 23%, Haryana 9%, Madhya Pradesh 8%. Other barley growing states are Punjab, Bihar, Himachal Pradesh, West Bengal and Jammu Kashmir.

In 2020-21, barley was grown on 5.9 thousand hectares in Punjab with a production of 22.3 thousand tonnes and average yield of 37.70 quintals per hectare.

Classification

There are three main types of barley:

1. ***Hordeum vulgare:*** This six -rowed barley variety bears a spike notched on opposite sides with three spikelets on each notch. Each notch carries a floret that later matures into a kernel. This variety of barley is the mostly cultivated variety across the world.

2. ***Hordeum distichum:*** This two-rowed type of barley bears central florets which produce kernels. Its lateral florets are sterile.

3. ***Hordeum irregulare:*** This variety is not cultivated for commercial purpose. It has fertile central florets and lateral florets could be sterile or fertile or both.

Climatic Requirements

Barley may be grown in subtropical climatic condition. The crop requires around 12-15 °C during growing period and around 30 °C at maturity. It cannot tolerate frost at any stage of growth and incidence of frost at flowering is highly detrimental for yield. The crop possess vary high degree of tolerance to drought and sodic condition.

Barley variety released by ICAR (2020-2021)

Barley (Hordeum vulgare)		
DWRB 182 (Barley-1)	Punjab, Haryana, Delhi, Western Uttar Pradesh, Rajasthan (except Kota and Udaipur divisions)	Suitable for irrigated, timely sown condition, average grain yield 4.97 tonnes/ha, maturity 133 days, two- row malt barley, low beta glucan in grain (4.7%) and wort (506 ppm), higher malt diastatic power.

Improved Varieties of Punjab (irrigated conditions)

DWRB 123: It is a two rowed malt barley variety which matures in 141 days. It is medium tall variety with 101 cm height. It is a bold seeded variety with 11% protein content. It is resistant to major pathotypes of yellow rust, moderately resistant to brown rust and tolerant to leaf blight disease. Its average grain yield is 48.5 quintals per hectare. This variety is particularly suitable for brewing industry.

PL 891: It is a two rowed hulless food barley variety. It contains 4 % β-glucan content and 12% protein content. β-glucan is helpful in preventing heart diseases and type-II diabetes. It matures in 144 days. It is medium tall variety with 102 cm height. It is resistant to major pathotypes of yellow rust, brown rust and leaf blight disease. Its average grain yield is 42 quintals per hectare. It is a bold seeded variety so use 125 kg seed per hectare. Hulless barley can be consumed as whole grain cereal, barley flakes, barley flour, etc. Barley sattu can be used as energy drinks.

PL 807: It is a six-rowed and feed barley variety which matures in 137 days. It is fairly resistant to yellow rust, brown rust and leaf blight. Its average grain yield is 43 quintals per hectare.

DWRUB 52: It is a two-rowed and malt barley variety which matures in about 140 days. It is fairly resistant to yellow rust, brown rust and leaf blight diseases. Its average grain yield is 43.25 quintals per hectare. This variety is particularly suitable for brewing industry.

PL 426: It is a six-rowed and feed barley variety which matures in 124 days. It is resistant to lodging and yellow rust. It is moderately susceptible to stripe disease and aphids. Its average grain yield is 35.0 quintals per hectare.

Soil Requirements

Sandy loam to loamy stand soils of Indo-Gangetic plains having neutral to mild saline reaction and medium fertility are the most suitable types for barley cultivation, however, it may be grown on a variety of soil types, viz; saline, sodic and lighter soils. Being salt resistant, its cultivation has become possible on saline coastal areas of Sunderban in West Bengal and saline black soils of canal irrigated areas of northern Karnataka.

Rotation

The rotation for barley is very similar to that for wheat. No exact rotation is followed year after year. Maximum production is obtained when the land is not cropped in the immediately preceding *kharif* season. It is generally rotated with crops, such as bajra, maize, paddy, jowar, cotton, groundnut, smaller millets, moong and moth in different parts of the country. Double-cropping with barley is practised under assured soil moisture or irrigation facilities. On soils infested with nematodes, the rotating of barley with non-host crops e.g. sunflower, for at least one to two years would reduce the nematode population.

Barley is grown pure or mixed with other *rabi* crops, such as wheat, gram, peas and lentil. Sometimes, rape and mustard, taramira and linseed are also grown mixed with barley.

The following are common crop rotations where barley is included:

- Paddy – Barley
- Jowar – Barley
- Bajra – Barley
- Urd – Barley
- Cowpea – Barley Cultivation

- Maize – Barley
- Cotton – Barley
- Groundnut - Barley
- Moong – Potato-Barley
- Maize - Potato-Barley

Land Preparation

Barley may be grown on a rough seed bed and hence one cross cultivation with soil turning plough followed by planking, is sufficient to get a good seed bed for sowing. Conservation of late monsoon rains received during August and September is very important. In moderately saline-cum-high water table conditions, the last preparatory cultivation should be given a week to 10 days before sowing to allow the salts to accumulate near the surface to facilitate sowing of seeds in a relatively less or non- saline soil layer.

For sowing in rainfed and saline areas, soak the seeds in water overnight at room temperature for better and quicker germination. Under irrigated normal conditions soaking of seeds is not required. Before sowing, the seed should be treated with Vitavax 75 WP (Cat boxin) or Baxil 2DS (teliuconazole) @ 1.5 g/kg seed covered smut, loose smut and stripe disease of easley.

Time of Sowing

Time of sowing is an important aspect in successful cultivation of barley. Barley gives best results when sown between October 15 and November 15. There is gradual decline in yield as sowing gets delayed up to the end of December. In rainfed areas sowing should be done in third and fourth week of October but in the irrigated areas the best time of sowing is first and second week of November.

Seed Rate and Spacing

Seed rate varies according to agroclimatic conditions. In irrigated areas, for normal sowing 75 kg seed is sufficient while in late sown conditions seed rate should be increased to 100 kg per hectare. Under rainfed conditions 80-100 kg seed per hectare is required depending on soil moisture availability. For saline soils use 100 kg seed per hectare to ensure better stand of the crop.

For normal sown irrigated crop, a spacing of 20.0 cm between rows is recommended and sowing should be done at a depth of 4-5 cm. In rainfed crop a spacing of 23 to 25 cm between rows is suggested and sowing should be done at a depth of 6-8 cm depending upon soil moisture.

Method of Sowing

Method of sowing is an important aspect particularly under rainfed conditions. Seed should be dropped with the help of 'pora' or 'nal' attached to a local plough or with the help of seed drill to ensure uniform distribution of seed at the optimum depth. In irrigated areas seed may be sown by 'kera' method, where seed is dropped by hand into the furrows. There should be enough moisture in soil for proper germination.

Weed Management

Barley exhibits early vigorous growth compared to other winter crops. By the time tillering stage is reached, crop covers much of the ground area. Because of initial fast growth barley is more competitive to weeds and it smothers weed growth and thus weed problem is less in barley. However, barley crop raised with higher doses of fertilizer under irrigated conditions, weed control becomes essential. The most common weed species infesting barley are *Avena fatua, Phalaris minor, Fumaria parviflora, Chenopodium album, Anagalis arvensis, Melilotus alba, Melilotus indica, Convolvulus arvensis, Rumax spp*s. etc. Interculture is not usually practiced unless the field is very weedy. If need arises, one hand weeding is sufficient. The use of weed free seed and a thoroughly prepared seed bed are essential for controlling the weeds effectively. Weeds can also be controlled with an application of 2,4-D sodium or amine salt at the rate of 0.50 kg a.i./ha or metsulfuron 20 WG at 4 g/ha in 250 litres per hectare of water, 30-35 days after sowing the crop. Application of isoproturon 0.94 kg/ ha after dissolving in 500 litres of water as post emergence spray at 30-35 days after sowing effectively control *Phalaris minor* and wild oats. For the control of Phalaris minor (gullidanda), Puma power 10 EC (fenoxaprop-p-ethyl) @ 1.0 lit./ha at 30–35 days after sowing in 250 lit/ha of water.

Water Management

Barley is grown as a rainfed crop as it has low water requirement and can tolerate drought. However, providing irrigation is quite useful in increasing yield from fertilizer responsive short statured dwarf and semi-dwarf varieties. To give good yields, barley requires 2-3 irrigation. One or two extra irrigations will be required on sandy soils. If the supply of water is inadequate, its efficiency can be increased by a proper timing of its application at the critical stages of crop growth. If only one irrigation is available, its application near the tillering stage has proved very profitable. When two irrigations are available, one should be applied at the active tillering stage and the other at the flowering stage. On highly alkali-saline soils, frequent light irrigation gives better results than a fewer heavy irrigation.

Fertilizer Management

The rainfed barley crop is rarely manured directly. It is the preceding crop in the rotation which receives the manure. Irrigated barley is often given a basal dose of eight to ten cartloads of farmyard manure or com-post. Any subsequent application of a fertilizer, however, is not practised. Recent agronomic experiments indicate that barley responds strikingly to an application of 20-40 kg N per ha under rainfed conditions, and to 40-60 kg N per ha under irrigated conditions, depending upon the soil, climate and variety. The application of 20 kg of N per ha to the crop under dry land farming led to the realization of 2-6 q per ha of additional grain and an application of 40 kg of N per ha under irrigated conditions resulted in a yield of 7-9 q per ha more than that from unfertilized plot at different locations. Good response to phosphorus up to 20 kg P_2O_5 per ha has been obtained under irrigated and timely sown conditions, but only at some locations. Hence its application needs to be governed by a soil test value.

To maximize the efficiency of nitrogenous fertilizers, the full dose of nitrogen should be placed 8-10 cm deep in the soil at sowing time under rainfed conditions. For irrigated barley, a split application of the half dose of nitrogen plus the full dose of P_2O_5 at sowing time and the remaining half dose of nitrogen with the first irrigation as a top-dressing is advantageous. Excess of nitrogen affects the malting and brewing quality of grain adversely. Super phosphate, however, helps to neutralize its deleterious effects.

Plant Protection

(a) Insect Pests

The barley crop is attacked by a number of soil insects. The important soil insects are white ants, termites, Gujhia weevil and cut worms. The attacked plants dry up completely causing poor crop stand. Use 2% Methyl parathion or 5% Malathion dust at the rate of 20-25 kg per hectare and mix it well in the soil at the time of last ploughing before sowing.

Aphids are the major insect pest of barley. The natural predators (*Coccinella septempunctata*) become active on the appearance of the aphids.

(b) Diseases

Stripe disease (*Drechslera graminea*): Yellow to brown stripes appear on leaves. The plants become stunted and the leaves are shredded. Use disease free seed.

Loose smut (*Ustilago nuda*): The smut sori are enclosed in a fragile membrane which soon ruptures, releasing the dark dusty spore mass which is disseminated by wind, leaving the naked rachis behind. Treat seed as per recommendations given in wheat.

Covered smut (*U. hordei*): The entire ear, except the awns turn into a black compact mass of spores. Treat the seed as per recommendations. Sow the crop late and shallow to reduce seedling infection.

Yellow rust (*Puccinia striiformis*): Yellow pustules appear on leaves, forming stripes. When the attack is severe, leaf-sheath, awns and glumes are also affected. Grow recommended varieties.

Seedling blight, leaf and glume blight (*Bipolaris sorokiniana*): Infects all the parts of the plant. In earlier stages may reduce crop stand. Elongated brown spots may be chlorotic, coalescing and may extend to the whole leaves and glumes causing blight. Prefer resistant varieties.

Harvesting and Threshing

Harvesting of barley in Jammu, Punjab, Haryana, Rajasthan and western Uttar Pradesh takes place from the third week of March to the middle of April. In Madhya Pradesh, Southern and Eastern Uttar Pradesh, Bihar and West Bengal, it starts a week to ten days earlier and in Maharashtra, Gujarat and Karnataka by about the first week of February. In the hills, where it is grown during the rabi season, the harvesting time varies from the end of April to the end of May, depending upon the altitude. The spring sown crop is harvested from the end of July to the end of September.

The plants are cut with a sickle when fully ripe. In order to prevent the loss of grain through shattering, the crop should be harvested immediately after it ripens. It is also advisable to cut the crop in the early hours of morning as the head becomes too dry and fragile for handling in the afternoon. The threshing in the plains is done by treading the dried produce under the feet of cattle or by stationary threshers. The grain is separated by winnowing.

Yield

When cultivation of high yielding varieties of barley is done with improved scientific methods, they produce about 30-35 quintals of grain and 40-45 quintals of straw per hectare.

5.5. Sorghum (*Sorghum bicolor* (L.) Moench)

Sorghum is a genus of about 25 species of flowering plants in the grass family (Poaceae). Some of these species are grown as cereals for human consumption and some in pastures for animals. One species is grown for grain, while many others are used as fodder plants, either cultivated in warm climates worldwide or naturalized in pasture lands. It's small, round, and usually white or pale yellow — though some varieties are red, brown, black, or purple. Sorghum is the fifth most produced cereal crop in the world. It's rich in natural nutrients and easy to add to your diet. It's also widely used as animal feed, and as a natural and cost-effective fuel source.

It is cultivated worldwide in warmer climates and is an important food crop in semi-arid tropical areas of Africa, Asia and Central America. Sorghum grain is a small, hard caryopsis covered by glumes. In grain sorghum, panicles are compact and bear 25,000 to 60,000 seeds/kg. Forage sorghum yields 120,000-160,000 seeds/kg. The whole grain can bc boiled, roasted, popped or ground to make flour for baking (flat breads) and pastry. Sorghum grain is used for the production of alcoholic beverages, including beer and liquors. Some sorghum varieties are used for dyeing textiles or leathers. In animal nutrition, grain sorghum is mostly used as an energy source and is a good feedstuff for poultry, pigs and ruminants. The stalks remaining after harvest can be grazed as some varieties stay green for a long period of time. Sorghum may also be grown for fodder, for grazing or cut green to make silage and hay (see the Sorghum forage datasheet) (Balole et al., 2006).

Nutritional Content

100 gms of Sorghum flour has 359 calories

Content	% Daily value
Total Fat 3g	5 %
Saturated Fat 1g	3%
Trans Fat	0%
Sodium 3mg	0 %
Total Carbohydrate 77g	26 %
Dietary Fiber 6.6g	26 %
Protein 8g	17 %
Vitamin C	1%
Calcium	1%
Iron	17%

Origin and History

Sorghum originated in the north-eastern quadrant of Africa, where the greatest variability in wild and cultivated species is found to this day. It was probably domesticated in Ethiopia by selection from wild sorghum between 5,000 and 7,000 years ago. From this center of origin, it was distributed along trade and shipping routes throughout Africa, and through the Middle East to India at least 3,000 years ago. It reached China along the silk route. Sorghum was first taken to the Americas through the slave trade from West Africa. It was reintroduced in late 19th century for commercial cultivation and has subsequently been introduced into South America and Australia. Sorghum is now widely found the drier areas of Africa, Asia, the Americas and Australia.

Area, Production and Yield

Highest area under Sorghum is in Sudan (5.79 mha), with highest production in US (9.47 million tonnes). Highest average yield of 405.24 q/ha was in Oman. Area of Sorghum in India was 5.5 mha with production of 4.77 mtonnes (Table 5.5)

Table 5.5: Area, production and yield of sorghum for the top 10 countries of the world.

Country	Area in million ha		Country	Production in million tonnes		Country	Average Yield in q/ ha
Sudan	5.79		USA	9.47		Oman	405.24
India	5.5		Nigeria	6.36		Israel	210
Nigeria	5.18		Ethiopia	5.06		Jordan	181.51
Niger	3.67		India	4.77		Uzbekistan	94.09
USA	2.06		Mexico	4.7		Austria	86.9
Burkina Faso	1.86		China	3.55		Italy	70.65
Mali	1.83		Brazil	2.77		Turkey	50
Ethiopia	1.79		Sudan	2.54		Egypt	49.19
Mexico	1.45		Niger	2.13		China	48.54
Chad	1.16		Burkina Faso	1.84		Philippines	48.36

Source :FAOSTAT, 2020

In India, Maharashtra leads with 47% of total production which is followed by Karnataka (22%), Madhya Pradesh (8%), Rajasthan (4%).

Climatic Requirements

Sorghum can grow in a wide range of ecological conditions and can still yield well even under unfavorable conditions of drought stress and high temperatures. It is generally grown between 40°North and 40° South of the equator, in warm and hot countries characteristic of the semi-arid environment. Sorghum requires warm conditions but it can be grown under a wide range of conditions. It is also widely grown in temperate regions and at altitudes of up to 2300 m in the tropics. It can tolerate high temperature throughout its life cycle better than any other crop. Sorghum requires about 26-30°C temperature for good growth. The minimum temperature for the germination of the sorghum seed is 7 to 10°C. Grain sorghum does not germinate and grow well under cool soil conditions. Poor emergence and seedling growth may result if planted before soil temperatures reach 35°C.

Sorghum is best adapted to areas having an average annual rainfall between 45 to 65 cm. Although sorghum can respond to good moisture supplies, it is nevertheless one of the toughest, drought tolerant crops available and this tends to maintain its popularity in the regions where the weather is very unpredictable.

Sorghum Variety Released by ICAR (2020-2021)

Variety	Area of adoption	Characters
Sorghum		
CSH 42	Karnataka, Andhra Pradesh, Madhya Pradesh, Gujarat	Sorghum hybrid suitable for rainfed ecology during *kharif*, average grain yield 4.0 tonnes/ha, early maturity (98–105 days), moderately resistant to grain moulds.
Forage sorghum		
CSH 43 (SPH-1881)	Haryana, Punjab, Uttar Pradesh, Rajasthan, Gujarat, Uttarakhand, Maharashtra, Tamil Nadu, Telangana, Karnataka	Suitable irrigated summer (spring) and rainfed *kharif* for timely and late (up to 30 July) sowing for green fodder production through 3–4 cuttings, average green and dry fodder yields of 96.52 tonnes/ha and 21.83 tonnes/ha, respectively; tolerant to foliar diseases and stem-borer.
Mega Sweet (SPH 1890) (ADV 6681)	All India	Single cut forage sorghum hybrid suitable for irrigated condition under medium to high soil fertility for normal sowing during spring and kharif, green-fodder yield 44 tonnes/ha, tolerant to stem-borer.
Nutrifast-ADV 961 (ADV0061)	Madhya Pradesh, Maharashtra, Gujarat	Multicut sorghum variety suitable for irrigated/rainfed conditions during kharif and summer, high green (97.49 tonnes/ha) and dry fodder yield (19.06 tonnes/ha), crude protein (1.46 tonnes/ha), seed yield (2.04 tonnes/ha), high per day productivity for GFY (8.47 q/ha/day) and DFY (1.59 kg/ha/day).
Harabhara (HTBH 4902)	Irrigated summer season in medium fertility soils of Telangana, Karnataka, Andhra Pradesh, Tamil Nadu, Kerala	Suitable for multicut system in summer irrigated condition, high green (53.24 tonnes/ha) and dryfodder yield (11.38 tonnes/ha), crude protein (1.31 tonnes/ha), high per day productivity for GFY (6.63 q/ha/day) and DFY (1.39 kg/ha/day).

Improved fodder variety for Punjab

SL 45: It is a single cut late maturing variety having tall (297cm) plants with long and broad leaves. Its stem is juicy and sweet. It is resistant to red leaf spot and moderately resistant to zonate leaf spot diseases. It possesses better fodder nutritional quality especially the crude protein and in-vitro dry matter digestibility. It gives 677.5 quintals per hectare of green fodder yield

Soil Requirements

Grain sorghum can be grown on many different soils. Sorghum will yield best on deep, fertile, well-drained loamy soils. However, it is quite tolerant of shallow soil and droughty conditions.

Loamy soil

Sorghum can be grown successfully on clay, clay loam, or sandy loam soils. Fertile, well-drained soils are important to optimize yield. Soils with clay loam or loam texture, having good water retention capacity are best suited for sorghum cultivation. Grain sorghum is more tolerant of wet soils than most grain crops. Sorghum grown on deep, well drained permeable soils usually develops extensive root systems. Mature plant roots may penetrate to depths of 4 to 6 feet in an ideal soil. Root development can be severely restricted by soil conditions such as excessively high or low soil moisture levels, hard pan and compaction.

Sorghum has moderate salt tolerance - slightly less than wheat but higher than maize. It does well in pH range of 6.0-8.5 as it tolerates considerable salinity and alkalinity. Sorghum is sensitive to aluminum toxicity and soils with acid saturation higher than 20% can pose a problem.

In tropical and subtropical conditions of India, sorghum is extensively grown in light - textured red sandy, red loamy, alluvial and coastal - alluvial soils as well as on mixed black and red and medium black soils. Sorghum is also grown on medium black soils, deep alluvial loams and on sandy and gravelly soils of poor fertility with low organic matter content but the yields are low.

Sorghum Races

Cultivated sorghum has five basic races, viz.,

- bicolor
- durra
- guinea
- caudatum
- kafir

It has 10 intermediate races, viz.,

- *guinea–bicolor*

- *caudatum–bicolor*,
- *durra–bicolor*,
- *kafir–bicolor*,
- *guinea–caudatum*,
- *guinea–durra*,
- *guinea–kafir*,
- *durra–caudatum*,
- *durra–kafir*,
- *kafir–caudatum*.

These races are morphologically distinct based on panicle and spikelet traits that are maintained through spatial and ethnological isolation. The 10 intermediate races exhibit a blend of characteristics of both the parents. These cultivated types also hybridize with their wild relatives wherever they are sympatric giving rise to members of sub sp. drummondii. Cultivated sorghums may also be variously classified agronomically as grain sorghum, sweet sorghum, sudangrass, broomcorn, etc. The other four sections include 19 purely wild species; thus, the Sorghum genus as currently proscribed consists of 22 species (Dillon et al., 2007). Affinities within the genus sorghum are complex and still under investigation.

Field Preparation

Sorghum seed bed should be well prepared and free from weeds. The first ploughing should be done with soil turning plough so that 20-25 cm deep soil may become loose. It should be followed by two to three harrowing or three to four intercrossing ploughings with country plough. Thereafter planking should be done to break the clods and to level the field. In black cotton soil area, if the land is badly infested with weeds, ploughing followed by disc harrowings is usually practiced, but where land is free from weeds or with few weeds, the land is cultivated only with Cultivators or tillers.

Time of Sowing

In northern India sorghum is sown only in *kharif* season. In irrigated areas, first week of July has been found most suitable for sowing of most hybrids and improved varieties. Under un-irrigated conditions, sowing should be done preferably within one week of the onset of first monsoon showers. Planting on the receipt of first monsoon showers was found to be ideal for rainfed (*kharif*) sorghum. Under irrigated conditions planting 7-14 days in advance of monsoon rains could result in yield increase. Timely planted crop escapes the damage due to shoot fly and midge. Late planting may not fit well in multiple rotations. Normal (optimal) plating time is considered on the onset of monsoon and any delay in seeding from normal time results in progressive decrease in grain yield. The shootfly incidence increases with the delay in sowing.

Rabi sowing is done mainly in Maharashtra, Karnataka and Andhra Pradesh. *Rabi* sowing should be done from the second fortnight of September to the middle of October. Summer crop of sorghum is sown in the month of January and February in irrigated areas of Tamil Nadu, Andhra Pradesh and some areas of Karnataka.

Seed Treatment : The seed should be purchased from a reliable source. It is advisable to always use certified seed. If seed is not already treated, it should be treated with. Bavistin 50 WP or Derosal 50 WP or agrozim 50 wp @ 3g per kg seed.

Seed Rate and Spacing: In case of hybrids, new hybrid seed should be used every year. Sorghum seed should be drilled in a well-prepared seed bed. A seed rate of 12-15 kg per hectare is considered sufficient to ensure a good stand. There should be 1,50,000 plants per hectare to attain maximum yield. The seed should be sown in rows 45 cm apart. Plant to plant distance should be 12 cm. Seed should be sown at a depth of 3-4 cm. It should not be sown more than 5 cm deep in any case.

Method of Sowing: In Northern India sorghum is sown either by broadcasting or in rows behind the plough. Seeds of new hybrids and varieties should always be sown in lines to economic seed and for obtaining higher yield. Sowing in rows is also common in black cotton soil.

Thinning: In sorghum cultivation, thinning is a very important operation for maintaining the desired plant population. Twelve to fifteen cm plant to plant spacing in a row by thinning out extra plants at two stages should be ensured. First, thinning should be done 10-15 days after emergence and second, when crop is 20-25 days old. All disease and insect infested plants should be removed while thinning.

Weed Management

Sorghum is badly infested with both broad leaf and grassy weeds, because of congenial weather especially during *kharif* season because of rainfall and humidity. Some of the important weeds associated with *kharif* sorghum are: *Commelina* sp., *Trianthema portulastion* sp., *Celosia argentea*, *Amaranrthus* sp., *Elucine* spp., *Digeria arvensis* etc. During *rabi* season, *Denebra retrofhxa*, *Bracharia isachne* and *Chrozophora rottliri* are predominant. Perennial weeds *Cyperus* sp., *Cynodon dactylon* and *Convolvulus arvensis* occur in all the seasons. Sorghum seedlings are comparatively small and weak and cannot compete with weeds. Weed free environment is essential till the seedlings establish and produce enough vegetative growth that approximately takes 45-50 days.

Weed control can be accomplished either by mechanical methods, chemical methods or a combination of both. By completely controlling weeds during the first four weeks, weed problem can be tackled easily. Inter row weeds may be controlled mechanically by hand weeding with a *khurpa* or a hand hoe. Two hand weedings at 15 and 30 days after sowing effectively control the weeds. However, the cost of mechanical methods is sometimes prohibitive or situations during rainy season do not permit hoeing or hand weeding and under these circumstances chemical method can be used. Chemical method is more economical. Pre-emergence application of atrazine @ 0.25-0.75 kg a.i./ha, propazine @ 1.0-1.5 kg a.i./ha or simazine @ 0.50-0.75 kg a.i./ ha can be given. Pre-emergence spray of triazine chemicals has been very effective in controlling broad leaf weeds and annual grassy weeds without any adverse effect on sorghum. The use of atrazine in sorghum is very safe and effective. Under Punjab conditions, pre-emergence application of atrazine at 0.5 kg/ha or pendimethalin 0.75 kg/ha provided good control of weeds.

Irrigation Management

Usually, sorghum is grown as a rainfed crop. The irrigation should, however, be provided whenever, rains are not received. At the time of flowering and grain filling stages, the crop requires more water. If enough moisture is not there in the soil at the time of flowering and grain filling, it should be irrigated at once. At no stage, the plants should be allowed to wilt. Suitable drainage conditions should be provided for the removal of excess rain water from the field.

Fertilizer Management

As sorghum removes nutrients in heavy amount from the soil and hence it requires heavy doses of fertilizers. Manure and fertilizers both play important roles in the sorghum cultivation. In the rainfed areas, application of farm yard manure or compost at the rate of 10 to 15 tonnes per hectare improves the water holding capacity and microbial activities in the soil, besides providing essential nutrients to the crop. Farm yard manure or compost should be added in the field at the time of last ploughing. The quantity of fertilizers to be applied varies according to the fertility status of the soil. However, when soil test data are not available, apply 100-120 kg Nitrogen, 50 kg P_2O_5 and 40 kg K_2O per hectare for hybrids and improved varieties of sorghum under irrigated condition. Half dose of nitrogen and total amount of phosphorus and potash should be applied at the time of sowing. The basal dressing can be done with the help of fertilizer-cum-seed drill. The remaining half quantity of nitrogen should be top dressed after 30-35 days of sowing. In light soils top dressing should be done in two splits. In case of local varieties, half of the above dose should be applied for better results. For rainfed crop, quantity of fertilizer should be reduced to half of the irrigated crop and the entire quantity should be applied 10 cm deep in soil at the time of sowing.

Among several micronutrients, deficiency of zinc and iron is widespread. Response to zinc application has been noticed on vertisols at Akola, Indore, Nandyal etc. Hybrid sorghums were reported to respond to the zinc application in the soil with 0.26-0.82 ppm of zinc. Zinc supply was observed to increase the yield to the tune of about 410-540 kg/ha. Sorghum is also sensitive to iron stress and is less efficient in its absorption and translocation. Hence this element is needed to be applied. The problem of iron deficiency is more on calcareous soils because iron uptake decreases with increased calcium carbonate content of the soil. Deficiency of zinc and iron can be corrected either through soil application or through foliar sprays. For soil application 25 kg of zinc sulphate per hectare once in two to three years is applied. In case of foliar application, 2-3 sprays of 0.2% zinc sulphate or 0.5% iron sulphate is adequate.

Plant Protection

a) Insect pests

The important insect pests of sorghum are given below:

Sorghum shoot fly: It is a major pest of sorghum attacking in the early seedling stage of growth. Plants up to three-week age are more susceptible to this pest. The eggs are laid on the lower surface of the leaves, stem or soil near the base of plants. The maggots hatching from the eggs creep down under the leaf sheaths and cut the central growing point of the plant resulting in dead hearts. In areas where it is serious, use Phorate (Thimet) 10% granules at the rate of 15 kg per hectare or Carbofuran (Furadan) 3% granules at the rate of 30 kg per hectare in furrows or as broadcast before sowing. If these chemicals have not been applied at sowing time, spray the crop with oxydematon methyl (Metasystox) 25 EC at the rate of 1.5 litres in 250 litres of water per hectare a week after germination and repeat the spraying after 15 days.

Stem borer: This pest attacks sorghum crop throughout the growing season. In an early stage of the crop, larvae bore inside the tender stem and feed on the tissues resulting into the death of growing point thus forming dead hearts. Grown up plants are not killed but suffer in vigour and develop weak heads. As estimated damage up to 80% may be caused. Use higher seed rate and pull all the affected plants in an early stage of the crop. Spray the crop with 0.1% endosulfan 35 EC or 0.2% carbaryl (sevin) 50 WP after 15 days of germination followed by the application of endosulfan 4% granules or sevin 4% granules in the whorl of the plants at the rate of 15 kg per hectare after a fortnight of spraying.

Pink borer: The damage is caused by the larvae which bore into the shoot and the dead hearts are formed in the early stage of the crop. Spray 0.1% endosulfan 35 EC or 0.2% carbaryl, 15 days after germination. Repeat the spraying if necessary.

Sorghum midge: Due to its infestation grain fails to develop and in cases of severe attack the whole head may be damaged resulting in serious yield reductions. Apply 5% carbaryl or malathion dust at the rate of 25 kg per hectare or spray the crop with 0.1% carbaryl 50 WP at the rate of 250-400 litres per hectare at pre-bloom stage (50% panicle emergence) followed by another application after 7 days.

Leaf roller: The larvae fold or roll the leaves by bringing the two margins closely with the help of silken thread to hide and feed inside this roll. Due to this feeding the leaves become white papery in appearance. Dust the crop with 2% methyl parathion dust at the rate of 25 kg per hectare. Spray the crop with dichlorosphos (nuvan) 70 EC at the rate of 1 ml per 4 litres of water.

Sorghum ear head bug: Both adults and nymphs suck the milky juice of the developing grains due to which the grains remain shrivelled. When a large number of nymphs feed, the whole earhead may become blackened at first and may eventually dry up producing no grains. Remove all the plants of wild sorghum from the surroundings. Dust the crop with 5% Lindane or Carbaryl or Malathion dust at the rate of 20 kg per hectare or spray the crop with Endosulfan 35 EC at the rate of 1.0 litre in 400 litres of water per hectare.

Hairy caterpillar: The eggs are laid in clusters on the leaves of wild as well as cultivated plants. After hatching, first and second instar larvae feed gregariously and skeletonize the leaves. Later on, grown up larvae disperse and feed singly. The egg masses should be collected and destroyed. The light traps should be used as soon as the moths are visible after the rains start.

Diseases

Sorghum crop is attacked by a number of diseases resulting in huge losses in yield. The important diseases are next page:

Grain moulds: Several fungi can grow on developing ears of sorghum when the grain setting coincides with rains or excessive humidity. Discolouration of grain occurs and in severe case loss of seed weight, seed size and seed germination take place. Toxins produced by the fungi are toxic to animals. This disease can be controlled by growing early maturing and tolerant genotypes.

Sorghum downy mildew: Abundant downy white growth is produced on the under surfaces of infected portions of leaves during humid weather. Systemically infected plants show progressively more of a complete bleaching of the leaf tissue, sometimes in streaks or stripes and sometimes over the entire leaf surface. As the infected bleached leaves mature, they become necrotic and the interveinal tissues disintegrate to give the typical shredded leaf symptom. For its control eradicate, wild hosts near sorghum fields, remove infected sorghum plants in the field, destroy plant debris by deep ploughing and other methods, growing resistant varieties and hybrids like CSV5, CSV 6 etc. Treat seed with Metalaxyl at 4 g/kg of seed or with Redomyl MZ at 6 g/kg seed before sowing of sorghum. Foliar spray of Mancozeb at 2.5 g/litre or Metalaxyl MZ at 2g/litre is recommended if standing crop is infested with downy mildew.

Ergot: Ergot can be an economically devastating problem. In seed production fields, yield losses ranging between 10 to 80% due to ergot disease have been reported in India. Ergot contamination reduces grain quality and limits its use as a feedstock. Ergot or sugary disease results from the ergot fungus that infects the florets in the earhead (panicle) and prevents seed set in such florets. The first symptom is droplets of a liquid honeydew exudate from between the glumes of infected florets. If honey-dew exudation is profuse, it may drip down and smear the whole earhead, sometimes dripping down onto the soil. If the exudate is plentiful, the earhead appears with blackened grain. Soaking seeds with 5% salt solution will aid to remove ergot infested seeds, as ergot infested seeds will float in the salt solution. Seed treatment with fungicides such as Captan or Thiram. Control of ergot with fungicides such as Propiconazole or Tebuconazole has proved to be cost effective in seed production plots. Multiple sprayings at 5-7 days interval from flag leaf stage till end of ant-hesis are recommended to reduce the variation in flowering within a field.

Charcoal rot: The charcoal rot fungi infects roots of plants growing under adverse environmental conditions, particularly moisture stress and high temperature. Infected roots show water-soaked lesions that turn brown or black. Affected stalks are soft or spongy at the base and tend to lodge. Lodging is the most apparent symptom of charcoal rot. If lodged plant is split open and examined, the pith of the stalk may be found in various stages of rotting of its tissues. Maintaining soil moisture through irrigation, if possible, during the post- flowering period can minimize charcoal rot infestation. A balanced fertilizer program is beneficial because high levels of nitrogen application with low levels of potassium may lead to charcoal rot. High plant population should be avoided. Best plant density based on moisture availability need to be adopted. Growing of drought tolerant, lodging resistant, and no senescing types of sorghum can reduce yield losses from charcoal rot infestations.

Rusts: Young plants are rarely affected by rust. The typical symptoms are seen in 1.5 to 3 months old plants. Scattered purple, red or tan flecks appear on both surfaces of leaves and in resistant cultivars the symptoms don't develop further. In susceptible cultivars typical rust pustules develop, mainly on the lower leaf surfaces. In highly susceptible cultivars the pustules occur so densely that almost the entire leaf tissue is destroyed. The pustules may also occur on leaf sheaths and on inflorescence stalks. In areas where rust becomes an economically important disease, growing of less-susceptible or resistant cultivars is the only solution.

Harvesting and Threshing

Most of the high yielding sorghum hybrids and varieties take about 100-115 days to mature. The crop should be harvested immediately after it is mature. The right stage for harvest is when grains have become hard having less than 25 per cent moisture. It is advised not to wait for stalks and leaves to dry because plants of hybrid sorghum appear green even after the crop is mature. Harvesting is done by cutting the entire plant or removing the earheads first and cutting down the plants later on. In the areas where there is danger of rain at the time of harvesting, the mature earheads should be harvested first and plants are cut and heaped later on. Threshing is done with the help of threshers or by beating the earheads with sticks or trampling by bullocks. The threshed grain should be cleaned and dried in sun for about a week to bring the moisture content down to 13-15 per cent for safe storage.

Yield

With improved cultural practices it is possible to harvest nearly 50 quintals of grain and about 100-125 quintals of dry stover from one hectare of crop under irrigated conditions and about 25-30 quintals of grain and about 80-100 quintals of dry stover from a rainfed crop.

5.6. Pearl Millet (*Pennisetum glaucum* (L.) R.Br.)

Pearl millet (*Pennisetum glaucum*) is the most widely grown type of millet. It is grown in Africa and the Indian subcontinent since prehistoric times. It is generally accepted that pearl millet originated in tropical Africa and was subsequently introduced into India. The earliest archaeological records in India dates back to 2000 BC, so domestication in Africa must have taken place earlier. Pearl millet is well adapted to production systems characterized by drought, low soil fertility, and high temperature. It performs well in soils with high salinity or low pH. Because of its tolerance to difficult growing conditions, it can be grown in areas where other cereal crops, such as maize or wheat, would not survive. It is an important forage crop of the arid and semi-arid regions of the country. It is fed to the cattle either as green or dry. It hybridizes very well with elephant grass (*Pennisetum purpureum* Schum.) which is believed to be of African origin. Bajra is highly drought tolerant and can grow well in the areas with a rainfall of 25–75 cm. This is grown as a *kharif* crop in northern parts and also grown as summer crop in southern part of the country.

Origin and History

Pearl millet has its origins in West Africa, where the oldest finds in Mauritania date to 1000 BC. With a cultivation area of 26 million hectares, pearl millet is the most economically significant small-seeded millet species. It is primarily cultivated in sub-Saharan Africa.

Area and Distribution

Globally, India topped in area (9.71 mha) and production (12.49 million tonnes), whereas highest average yield of millet was recorded in Mexico (146.84 q/ha) (Table 5.6)

Table 5.6: Area, production and average yield of Millets for top 10 countries

Country	Area in million ha		Country	Production in million tonnes		Country	Average yield in q/ ha
India	9.71		India	12.49		Mexico	146.84
Niger	6.74		Niger	3.51		Uzbekistan	69.27
Sudan	2.42		China	2.3		Switzerland	33.38
Mali	2.16		Nigeria	2		Turkey	30.84
Nigeria	2		Mali	1.92		Tajikistan	26.07
Burkina Faso	1.18		Ethiopia	1.22		China	25.55
Chad	1.16		Senegal	1.14		Ethiopia	25.36
Senegal	1.02		Burkina Faso	0.96		Saudi Arabia	21.08
China	0.9		Chad	0.69		Syria	20.77
Namibia	0.49		Sudan	0.48		Morocco	20.5

Source: FAOSTAT, 2020

Pearl millet occupies 6.93 million ha with an average production of 8.61 million tonnes and productivity of 1243 kg/ha during 2018-19 (Directorate of Millets Development, 2020; Project Coordinator Review, 2020). The major pearl millet growing states are Rajasthan, Maharashtra, Gujarat, Uttar Pradesh and Haryana which account for more than 90% of pearl millet acreage in the country and commonly grown in rainy (*kharif*) season (June/July–September/October). It is also cultivated during summer season (February–May) in parts of Gujarat, Rajasthan and Uttar Pradesh; and during the post-rainy (*rabi*) season (November–February) at a small scale in Maharashtra and Gujarat.

Climatic Requirements

Being a rainfed crop, pearl millet is mostly grown in areas where annual rainfall is between 400 and 750 mm. It grows best in warm dry climates with adequate soil moisture. It can withstand drought and soil moisture stress to a great extent. During the vegetative growth of the crop, moist weather and light to medium rainfall is very useful, but at flowering and grain formation there should be no rain and the weather should be clear and dry. Continuous rains at flowering adversely affects the grain formation by causing poor seed setting. Besides, high humidity and low temperature at flowering increase the incidence of ergot disease leading to low yield. A temperature of 28-32oC is considered to be optimum for the vegetative growth of the crop. High temperature at this stage induces early flowering and low temperature increases the ergot infection. It is, therefore, recommended to follow optimum sowing time to protect this crop from incidence of disease.

Pearl millet varieties released by ICAR (2020-2021)

Pearl millet varieties	Area of Adoption	Characters
(*Pennisetum glaucum*)		
MP 7366	Rajasthan, Gujarat, Uttar Pradesh, Maharashtra, Tamil Nadu	Suitable for summer, average grain yield 5.1 tonnes/ha, maturity 89 days, highly resistant to downy mildew, smut and ergot diseases, tolerant to terminal stress condition, bio-fortified variety with high iron (69 ppm) and zinc (43 ppm).
VPMH 7	Karnataka	Suitable for *kharif*, average grain yield 2.4 tonnes/ha, maturity 85 days, moderately resistant to major diseases (downy mildew, smut and ergot), tolerant to lodging, bio-fortified variety with high iron (67 ppm) and zinc (52 ppm).
MBP 2 (Malnoor Bajra Population 2)	Karnataka	Suitable for *kharif* sowing, average grain yield 2.4 tonnes/ha, medium maturity, compact panicle, pale yellow, obovate seeds with high iron (72 ppm) and zinc (48 ppm) content, tolerant to major diseases and pests.
Finger Millet (*Eleusine coracana*)		
CFMV 2	Andhra Pradesh, Chhattisgarh, Gujarat, Maharashtra, Odisha	Suitable for rainfed condition during *kharif*, average grain yield 2.95 tonnes/ha, dry fodder yield 8.6 tonnes/ha, medium maturity (119–121 days), rich in calcium (454 mg/100 g), iron (39 ppm) and zinc (25 ppm), resistant to leaf blast, foot rot, brown spot, grain mould.
VL Mandua 378	Uttarakhand	Suitable for rainfed organic ecology, average grain yield 2.1 tonnes/ha, medium maturity (114 days), high calcium (361 ppm) and iron (45 ppm), moderately resistant to neck and finger blast.

Pearl millet varieties	Area of Adoption	Characters
VL Mandua 382	Uttarakhand	Suitable for rainfed organic ecology, average grain yield 2.28 tonnes/ha, medium maturity (112 days), white coloured grains, higher calcium (340 ppm) and protein content (8.8%), very low tannin content, suitable for processing industry.
Kodo millet *(Paspalum scrobiculatum)*		
Gujarat Anand Kodra 3 (GAK 3)	Gujarat	Suitable for rainfed organic ecology of hilly and tribal region of Dahod and Panchmahal districts of middle Gujarat, average grain yield 2.46 tonnes/ha, medium maturity (114 days), good nutritional properties, high protein content and minerals, milling recovery 53.4%, highly resistant to shoot fly, moderately resistant to head smut.

Improved varieties for Punjab

PCB 166: This is a dual purpose composite variety with more number of tillers. It is a tall variety (281 cm) with long and broad leaves. It is late maturing variety and takes 124 days to mature. The ears are very long (63.8 cm) cylindrical. It is tolerant to all the major diseases of pearl millet. The grains are medium bold and slat in color. On an average, it gives 40.75 quintals per hectare of grain yield.

PCB 165: This is a quick growing dual-purpose composite variety with more number of tillers. It is a tall variety with average plant height of 252 cm. It is a late maturing variety and takes 111 days to mature. It bears long cylindrical ears with 31.3 cm length and 11.5 cm ear girth. The grains are soft and slate in color with very good popping potential. The grains are of good quality with high iron content. It is tolerant to all the major diseases of bajra. On an average, it gives 32 quintals per hectare of grain yield.

PHB 2884: This hybrid is 230 cm tall and bears 2-3 productive tillers. It has long ears with average length of 28 cm and girth of 12 cm. Its' grains are medium bold and slate in colour. It is tolerant to downy mildew, ergot and smut diseases. It matures in 88 days and average grain yield is 33 quintals per hectare.

PCB 164: This is dual purpose composite variety having medium thick stalks and flexible stem with average plant height of 207 cm. It matures in 80 days. This variety has long cylindrical dense ears having 27-28 cm length and 8-10 cm girth. The grains are medium bold and light slate in colour. The average grain yield is 37.5 quintals per hectare. It is tolerant to downy mildew.

Soil Requirements

Pearl millet is successfully cultivated in different types of soils. Sandy loams to loamy soils with good drainage and free from alkalinity or salinity are the best suited for its cultivation. It can give fairly good yield on poor soils where other cereal crops cannot be grown successfully.

Rotations

Some of the most important cropping systems are given below:

- Pearl millet – Wheat
- Pearl millet – Barley
- Pearl millet – Chickpea
- Pearl millet – Field pea
- Pearl millet – Potato
- Pearl millet – Potato – Wheat
- Pearl millet – Toria – Wheat
- Pearl millet – Wheat – Moong

Field Preparation

In rainfed areas, the moisture conservation measures are extremely important and therefore tillage operations before and after seeding the crop assume great significance. In such areas, summer ploughing every year and deep ploughing once in 3-4 years are recommended. The deep ploughed soil is capable of releasing major nutrients and greater retention of moisture. Results of research experiments clearly indicated the beneficial effect of deep ploughing (30-40 cm) over shallow (10-12 cm) ploughing for rainfed pearl millet.

Time of Sowing

In rainfed areas the sowing of pearl millet should be done before the onset of monsoon. Sowing at the optimum time not only increases the grain yield but also reduces the chances of disease incidence. Under delayed sowing the rate of decline grain yield varies from 4 to 80 kg/day/ha under rainfed conditions. Hence, the crop should be sown at an optimum time of first fortnight of July. Sowing if delayed results in poor stand of the crop due to high rate of mortality of the seedlings, restricted vegetative growth of the crop, poor grain setting and higher incidence of diseases due to comparatively low temperature during the period.

Transplanting

Transplanting of pearl millet has been recommended if sowing gets delayed. The nursery sown in 500 m^2 is sufficient to transplant one hectare. Three weeks old seedlings are transplanted in August keeping a spacing of 50 cm between the rows and 10-12 cm between the plants. However, irrigation is necessary for transplanting. Wherever, irrigation facility is not available, seedlings can be transplanted on a rainy day.

Seed Rate and Plant Population

Recommended seed rate is 4-5 kg/ha. The seed is sown by *kera* method (behind the plough) and row to row spacing is kept at 45 cm. After three weeks of sowing at the time of thinning, the plants in the rows are spaced at 10-12 cm. The optimum plant population is 1.75-2.0 lac per hectare. In areas with uneven distribution of rainfall, plant population of 1.6 lac per hectare should be maintained. Seeds of pearl millet should be sown about 2-4 cm deep preferably by seed drill operated with tractor or with plough or *kera* method.

Weed Management

Many *kharif* weeds compete with pearl millet and may cause considerable yield reductions ranging from 25-50%. The damage is severe during 3-5 weeks after sowing. Therefore, timely control of weeds is essential to get higher yields. Interculture the crop 3-4 weeks after sowing. Avoid deep hoeing near the base of the plants so that their roots are not damaged. A wheel hoe, *triphali* or hand hoe can be used for interculture. However, sometimes or soil being too wet to permit manual weeding or due to unavailability of labour timely weeding becomes difficult. Under such circumstance's weeds can be effectively controlled with the use of herbicides. Pre-emergence application of Atrazine 50 WP at the rate of 0.25 to 0.50 kg ha^{-1} in 500 litres per hectare of water controls most of the grassy and broadleaf weeds.

Fertilizer Application

Add 25 tonnes per hectare of farmyard manure or compost before preparing the land and apply 50 kg of N (110 kg urea) per acre in two doses, first half as the basal dose and the second half, 3 weeks after sowing when the crop height is 10-15 cm.

Water Management

Being a rainfed crop, pearl millet hardly needs any irrigation. Irrigation is applied only if there are no rains. Generally, two irrigations during the growing period of the crop are enough. If moisture is limiting, the crop must be irrigated at the earhead emergence stage because it is the critical stage for moisture stress. Pearl millet does not tolerate water logging. If the field is prone to water stagnation, drainage should be provided to drain away excess water.

Plant Protection

a) Insect pests

Shoot fly: Shoot fly is a serious pest of pearl millet in north India. Three weeks old plants are more prone to be attacked by the pest. Eggs are laid on the lower surface of the leaves or near the base of the plants. Larvae after hatching creep down the leaf sheaths and cut the central growing point of the plant which results in dead hearts. Early planting with the onset of monsoon is effective in controlling shoot fly incidence. Two dustings of Malathion (5% dust) at the rate of 25 kg/ha at 10 and 20 days after germination or two sprays of neem oil (in 4 g soap/litre water) or two sprays of endosulfan (0.07%) at 10 and 20 days after germination is recommended for controlling shoot fly.

White grub: The pest damages the root system of the young seedlings and as a result seedling die. Intercropping of pearl millet with pulses like green gram, cluster bean and cowpea significantly reduces the white grub damage. Mixing of phorate or carbofuran at the rate of 12 kg/ha with pearl millet seed and applying in furrows at the time of sowing.

Midge: It is a serious pest of developing grains of pearl millet. The fly lay eggs in the flowers and the larvae develop on their ovaries, which shrivel and fail to develop into grains.

For controlling midge 250 ml of Metacid (50 EC) or 625 ml of thiodine (5 EC) in 250

litres of water per hectare is sprayed on pearl millet crop.

Grasshoppers: Dust the crop with Malathion 5% at 25 kg/hectare for its control.

b) Diseases

Green-ear or downy mildew: The green-ear disease or downy-mildew is caused by Sclerospora graminicola. The leaves of infected plants show discolouration, yellowing and whitening. Under humid conditions, the leaves are covered with a downy white growth of the fungus, which is prominent on the lower-surface. The leaves turn necrotic and are torn into shreds. The ears of the infected plants are transformed wholly or partly into green heads of small, twisted, leafy structures. Adopt following integrated measures to control the disease:

- Grow downy-mildew resistant hybrids and composite variety.
- Rogue out the diseased plants early in the season to prevent secondary infection.
- Collect the diseased ears from the crop before harvesting and destroy them by burning.
- Practice five-year rotation with other crops.

Grain smut: Smut is caused by *Tolyposporium penicillariae*. Individual grains in an ear get transformed into smut balls which may later burst open to release millions of spores which get disseminated and cause secondary infection on the portion of the ear which is enclosed by the sheath of the upper leaf. The intensity of the attack varies according to the humidity in the area. The following control measures are suggested:

- Remove the diseased ears early in the season and destroy them.
- Treat the seed before sowing with 3 g of Thiram/Captan per kg of seed to prevent the introduction of smut into new areas.

Ergot: This disease is caused by the fungus *Claviceps fusiformis*. At blossoming, a pinkish or light-colored fluid (honey dew) exudes from the spikelets in different parts of the ear. Later dark sticky patches appear on the ear and small dark-brown sclerotia appear in place of grains between the glumes. The seed set is poor or completely inhibited. The ovary is replaced by a fungal mass with many folds on its surface. The fungus perpetuates through the seed-borne and soil-borne sclerotia. The contaminated grains, if fed to cattle or used by human beings can cause poisoning. Therefore, take the following precautions:

(i) **For Cattle:** Do not feed the infected ears showing honey-dew symptoms to cattle. Even the stems and leaves of such plants are not safe as cattle feed. Cut and burn a badly affected crop to reduce the amount of inoculum.

(iv) **For Human beings:** Immerse the grains in 10 per cent salt solution. The sclerotia, being lighter than normal grains, will float. Remove them with a sieve and burn them. Repeat the process two or three times.

Prevention of Ergot: Once the disease appears, it is not possible to eliminate it.

Take the following precautions to prevent its spread.

- Immerse seed in 10 per cent salt solution and remove the sclerotia and smut- balls by skimming. Then wash the seed in ordinary water and dry it thoroughly before treating it with fungicides, as recommended under Grain Smut.
- At the boot-stage, spray the crop with Cuman L 27% (ziram) using 1.25 litre in 250 litres of water per ha. Repeat 2 or 3 times at 5-day intervals.
- Burn the ears infested with honey-dew, as soon as they are observed in the field.
- After harvesting the crop, bury the debris with a furrow turning plough so that the ergot sclerotia rot in the soil.
- After threshing the ergot affected crop, the left-over-ear-heads of bajra in the threshing floor should also be burnt.

- Avoid sowing bajra next year in a field in which the crop had suffered heavily from ergot.

Rust: The rust pustules are small orange and full of spores. In later stages, black coloured pustules can be seen. In severe infection leaves are completely dried. Spray the crop with 0.2% Zineb or Mancozeb 75 WP three to four times to provide good protection.

Harvesting

Harvesting at physiological maturity or at a little later stage when grains are hard and contain 15-20% moisture is ideal. Harvesting is done by cutting the entire plant or removing the earheads first and cutting down the plants later. The earheads after harvesting should be sun-dried at the threshing floor for 10-20 days. Threshing is done thereafter either by beating the earheads with sticks or under the feet of bullocks. Tractor operated thrashers can also be used for this purpose. The threshed grain should be cleaned and dried in sun to bring down the moisture content of grain to 12-14% for safe storage.

Yield

Under irrigated conditions, 30-35 quintals of grain and about 100 quintals of dry stover can be obtained from one hectare. The rainfed crop gives 12-15 quintals of grain and 70-75 quintals of dry stover per hectare.

5.7. Oats *(Avena sativa)*

Oats (*Avena sativa)* are an extremely versatile *rabi* cereal fodder crop of India. It can be grown in wide range of environments of diverse soil types for forages and as a companion crop. It has excellent growth habit, quick recovery after cutting and good quality herbage. It is a palatable, succulent and nutritious crop. It is a good source of protein, fibre, and minerals. On dairy farms oat fodder is a must, as it can be fed green and the surplus converted into silage or hay for use during the scarcity period. The oat crop is a high yielder with an average yield varying from 45 to 55 tonnes of green fodder per hectare. Oat remains an important grain crop for people in marginal environments throughout the developing world, while in developed economies it is meant for specialist uses. Livestock grain feed is still the primary use of oats crop.

Origin and History

Oats are a crop of Mediterranean origin, not as old as wheat and barley but their domestication dates back to ancient times. There are various school of thoughts regarding the exact timing and location for the domestication of oats. According to Thellung (1912) oats was domesticated after wheat and barley and perhaps it originated in Asia Minor. The ancient Greeks and Latins called oats 'Bromos' and 'Avena' respectively, but these names were commonly applied to wild species occurring as weeds among cereals (De Candolle, 1890). Denaiffe and Sirodot (1901) wrote "the derivation of the Latin word 'Avena' remains somewhat obscure. It seems probable that it is from the Latin word Aveo (to desire), that is to say, forage desired by all animals". The exact time of introduction of cultivated oats into India is unknown. However, green oats were mentioned as a horse fodder in Ain 49 of Ain-I-Akbari, written by Abul Fazl (1590 AD).

Classification

The genus *Avena* is large and diverse and contains both wild and cultivated species of polyploid series with a basic chromosome number of n = 7. Three naturally occurring ploidy levels known within the genus are diploids (2n = 2x = 14), tetraploids (2n = 4x = 28) and hexaploids (2n = 6x = 42). Cultivated oat is believed to have originated from *Avena byzantina* or *A. sterilis*. The genus *Avena* comprises following species which are economically important:

I. *Avena brevis* (2n=14); the short oats, is grown in southern Europe, mostly for green fodder or hay due to its high yield of stem and leaf.

II. *Avena abyssinica* (2n=28); the Abyssinian oat, grows wild and is also cultivated in Ethiopia. It is grown mainly for fodder in several parts of north Africa.

III. *Avena byzantina* (2n=42); the red oat is grown around the Mediterranean regions, Europe and north Asia and warmer parts of the subtropical world for both grain and fodder. It is also cultivated in India.

IV. *Avena chinensis* (2n=42); the Chinese naked oat is grown extensively in hilly parts of china for its grain.

V. *Avena fatua* (2n=42); the wild oat is cultivated for grain, green fodder and other purposes in India and part of south-west Asia.

VI. *Avena ludoviciana* (2n=42); the wild red oat is grown for grains in Mediterranean and south-west Asian regions.

VII. *Avena sativa* (2n=42); it is the common cultivated oats sown in several parts of the world for grain, green fodder and other purposes.

VIII. *Avena sterilis* (2n=42); the fly oat is grown mainly for fodder in parts of Mediterranean region and south-west Asia.

IX. *Avena strigosa* (2n=42); the sand oat is cultivated for fodder in the Mediterranean region.

Climate

Oats make best growth in cool moist climates and can be sensitive to hot, dry weather from head emergence through maturity. For these reasons, world oat production is generally concentrated between latitudes 35 – 65°N, including Finland and Norway, and 20 to 46°S. Oats require about 15-25°C temperature for its optimum growth. Frost and severe cold retard its growth, whereas hot and dry conditions hasten its maturity resulting in poor yield and inferior quality of fodder. They require more moisture to produce a given unit of dry matter than any other cereal except rice. They are grown in areas where the annual rainfall ranges from 38 to 114 cm, but most extensively where it does not exceed 76 cm and is well distributed.

Soil and Soil Preparation

Oats are grown on different soil types around the world except sodic and waterlogged ones. Next to rye (Secale cereal) oats are the most versatile of the cereals with regard to soil types. Almost any reasonably fertile, well-drained soil is suited to oats if temperature and moisture conditions are favourable, however, fertile and well drained loam to clay loam soils are the best. The crop suffers badly due to lodging when heavily fertilized with nitrogenous fertilizers especially on poorly drained clay soils. It can tolerate slight acid and saline conditions but not above 8.5 pH. Under non-saline conditions oats are moderately tolerant to Na, but less tolerant than wheat and barley.

For field preparation, one deep ploughing followed by three or four harrowing and planking will serve the purpose to get friable and well levelled seed bed.

The normal sowing time starts from October to November. However, middle of October to middle of November is the optimum sowing time for higher production. Multi- cut varieties if sown earlier will help to tide over the scarcity period from middle of December to middle of January. Early seeding helps to ensure efficient use of available moisture.

Oats varieties released by ICAR (2020-2021)

Varieties	Area of adoption	Characters
HFO 427	Andhra Pradesh, Odisha, Maharashtra and Bihar	Suitable for irrigated conditions during *rabi*, average yield 32 tonnes/ha (green fodder), 6.7 tonnes/ha (dry fodder) and 1.04 tonnes/ha (grain), maturity 135 days.
HFO 607	Punjab, Haryana, Rajasthan, Tarai part of Uttarakhand	Suitable for irrigated conditions during *rabi*, average green fodder yield 61.5 tonnes/ha, dry fodder yield 13.1 tonnes/ha and grain yield 2.76 tonnes/ha, maturity 155 days, moderately resistant to Helminthosporium leaf blight, 9.2% crude protein and 56% dry-matter digestibility.
Pant Forage Oat 4	Uttarakhand (Low hills and Plains)	Suitable for normal fertility and irrigated conditions during rabi, average green fodder yield 34 tonnes/ ha, dry fodder 7.5 tonnes/ha and grain yield 2.76 tonnes/ha, maturity 145 days, high leaf stem ratio (0.58), 9–11% crude protein, 58–60% dry-matter digestibility and high-protein yield (7.5–8 q/ha).
OL 1874	Punjab, Haryana, Rajasthan, Tarai part of Uttarakhand	*Rabi* fodder crop suitable for multicut system, high green fodder (73.77 tonnes/ha), high crude protein (14.7%), seed yield (1.9 tonnes/ha), moderately resistant against leaf blight.
OL 1896	Punjab, Haryana, Rajasthan, Tarai part of Uttarakhand	*Rabi* fodder crop suitable for single cut system, high green fodder (64.91 tonnes/ha), high crude protein yield (1.32 tonnes/ha), seed yield (2.70 tonnes/ha), moderately resistant to leaf blight.
OL 1876-2	Assam, Jharkhand, Odisha, Eastern Uttar Pradesh	Rabi fodder crop suitable for dual- purpose system, high green fodder (22.61 tonnes/ha), high crude protein (8.6%), seed yield (1.34 tonnes/ha), resistant to leaf blight.

Improved Varieties for Punjab

OL 15: It is a single cut variety for irrigated areas of Punjab. Its plants are tall, having long and broad leaves with more leafiness and tillering ability. Its fodder quality is better than OL 12, OL 11, Kent and at par with OL 13. On an average, it yields about 319 quintals of green fodder per acre. Its seed yield is about 24.5 quintals per hectare.

OL 14: It is a multicut variety for irrigated areas of Punjab. Its plants are tall with profuse tillering and leafy growth. The leaves are longer and broader. Its fodder quality is superior in terms of total digestible nutrients (TDN) and digestible crude protein (DCP) than OL 10. On an average, it yields about 307 quintals of green fodder per acre. Its seed yield is 27.25 quintal per hectare.

OL 13: It is a single cut variety for irrigated areas of Punjab. Its plants are tall with profuse tillering and leafy growth. The leaves are longer and broader. Its fodder quality is superior in terms of DCP than OL 12 and Kent. On an average, it yields about 305 quintals of green fodder per acre. Its seed yield is 24.25 quintal per hectare.

OL 12: It is a single cut variety for irrigated areas of Punjab. Its plants are tall, having long and broad leaves with more leafiness and tillering ability. Its fodder quality is better than OL 11, OL 9 and Kent. On an average, it yields about 258 quintals of green fodder per acre. Its seed yield is about 21.5 quintals per hectare.

OL 11: It is a single cut variety with tall plants, leafy and profused tillering ability. Its fodder quality is superior to OL 9 and Kent. On an average, it yields about 245 quintals of green fodder and 21.25 quintals of seed per hectare

OL 10: It is a multicut variety. Its plants are tall with profused tillering and leafy growth. The leaves are longer and broader. Its fodder quality is superior in terms of TDN and DCP than OL-9 and Kent. On an average, it yields about 687.5 quintals of green fodder per hectare.

Kent: It is tall, erect growing variety suitable for single/two cuttings. Seeds are bold. It gives 525 quintals of green fodder per hectare. It produces 20 quintals of seed per hectare.

Seed Rate

There is seldom one optimum seeding rate because oat plants adjust their tillering and growth depending upon plant density. Seed should be drilled at the rate of 80-90 kg/ ha for small seeded varieties, whereas it ranges 100-125 kg/ha for bold varieties e.g., Kent. A moderate reduction in seeding rate seldom results in a yield reduction. This is due to more vigorous plants and the production of more tillers. Very high seeding rate results in increased competition, weaker culms and frequently increased lodging. For maximum yields, only good quality seed of recommended cultivars should be sown.

Spacing

In case of low tillering varieties spacing between rows should be kept at 20-25 cm apart. However, high tillering varieties should be sown at 25-30 cm apart rows.

Method of Sowing

Seed should preferably be drilled in one direction with straight rows. If seed drill is not available the sowing can be done either by pora or kera method with the plough.

Weed Control

Herbicides are often used as a control measure. As a pre-emergence herbicide linuron at the rate of 0.5 kg/ha may be used to control associated weeds and this will cause 60-70 per cent reduction in weeds. Sodium salts of 2,4-D may be applied as post emergence herbicide at the rate of 0.75 kg/ha. Other herbicides selective for oats are pendimethalin and trifluralin.

Water Management

Water requirement of crop depends on soil texture, prevailing climatic conditions etc. The oat crop needs copious irrigation for succulent and quality herbage production, though, it can be grown with acceptable level of yield even under limited irrigation. Productivity of oat greatly depends on availability of water during growing season and accordingly scheduling of effective irrigation technique is thus an essential factor, keeping in view its limited availability, cost and efficient utilization. A pre sowing irrigation is required for assured moisture in seed bed to ensure good germination and early establishment. A total number of 4-5 irrigations are required for oats. Generally, the first irrigation should be given 20-25 days after sowing. Subsequent irrigations may be provided at tillering, jointing, panicle primordial initiation, flowering and at milk stage. Judicious and timely application of irrigation water could check the leaching losses of nutrients and could also save the deterioration of soil by avoiding over irrigation. Further, during boot stage, irrigation is essential due to high water requirement at flowering stage.

Fertilizer Management

To obtain maximum yields adequate nutrients must be present throughout the growing season. Oat plants will be less vigorous, have smaller leaves and panicles and will have less developed root system if nutrients are short in the first half of the growing season. In general, 60 kg of nitrogen per hectare should be applied and properly mixed in soil before sowing. Additional dose of 30 kg nitrogen per hectare should be broadcasted in two splits i.e. (10 kg and 20 kg) after first irrigation (at tillering stage) and at first cut, respectively. In light textured soils 40 kg of phosphorus per hectare should be applied at the time of sowing as it shows a favourable effect on shoot height and tillering during the earlier growth stages. Zinc sulphate may be applied at the rate of 10 kg per hectare in areas prone to zinc deficiency.

Plant Protection

(a) Insect Pests

Oat aphid (*Rhopalosiphum padi*): Oat aphid appears on oats during January to March. It sucks sap from the plants and lowers down the quality of fodder.

Coccinellid beetles, syrphid fly and heavy rainfall keep natural check on its multiplication. Do not spray any insecticide on fodder crop as it is hazardous to animals and also kills the predators of the aphid.

(b) Diseases

Loose smut (*Ustilago avenae*): The individual flowers of the oats panicle are replaced by spore masses. Use disease free seed.

Covered smut (*U. kolleri*): The smut destroys the kernels completely and replaces them with black masses of spores which remain closed within the affected spikelets till harvested. Rogue out and destroy the smutted ears. Use disease free seed.

Harvesting

For oats, number of cuttings depends upon the type of varieties and requirement of fodder for specific time and purpose. The single cut crop may be harvested at 50 per cent flowering to milk stage for better quality and quantity. In two cut management system, first cutting should be taken at flag stage (about 60 days) and last cutting at 50 per cent flowering to milk stage. For multi-cut crop the first cutting should be done at 55 days, second at 105 days and last at 50 per cent flowering stage. Delay in harvesting beyond milk stage may result in sharp decline in digestible protein. For satisfactory seed production, the crop should be left for seed after the first cutting to be taken at 50-55 days after sowing. First cut should be taken 8-10 cm above the soil surface for good regrowth.

For seed, oats are harvested when grain is in the hard dough stage and straw is slightly green (moisture content of the grain is 14% or less). If too ripe, shattering causes seed loss. Crop is usually cut with binder and left in the field until dry and then threshed. In mechanized societies, oats are combined harvested. For this type of harvesting, crop must be fully ripe, usually when the straw has lost greenness and glumes have become white. Seeds are cleaned by winnowing, and dried to below 14 per cent moisture for storage.

Yield

In case of only forage production plots, the green forage yield per hectare would be about 500-600 quintals. If crop is left for seed after the first cut then 30-35 quintals of seed and 25-30 quintals of straw per hectare may be obtained in addition to 250 quintals of green fodder of first cut.

6

Latest Practices for Cultivation of Oilseed Crops

Oilseed crops have been grown all over the world and are considered important crops due to their economic value. Oilseed crops are primarily grown for edible oil. Recently, oilseeds attracted more attention due to an increasing demand for their healthy vegetable oils, livestock feeds, pharmaceuticals, biofuels, and other oleochemical industrial uses. The increased interest resulted in an 82% expansion of oilseed crop cultivation areas and about a 240% increase in total world production over the last 30 years.

Oilseed crops are generally grown for the oil in their seeds and vary considerably in oil content, quality, and composition – factors that rely heavily on the crop species or cultivar and upon the environmental conditions in which the crop is grown. The seed meal left after the oil has been extracted can provide a high-protein product for use in either animal or human food

There are two major sources of Oilseeds i.e. Primary and Secondary. Primary sources are made combining the edible group [(Groundnut, Rapeseed (Toria, Mustard and Sarson), Soybean, Sunflower, Sesame, Safflower and Niger)] and non-edible group (Castor and Linseed). Similarly, Secondary sources have been established combining edible group (Seasonal crops : Cotton seed, Rice bran, Maize germ, Watermelon, Plantation crops : Coconut, Red-oil palm, Tree borne oilseeds: Sal seed, Mahua, Mango-kernel, Cheura / Phulwara, Kokum, Dhupa, Simarouba) and non-edible group (Seasonal crops : Mesta seed, Tobacco seed, Plantation crops : Rubber – seed, other Tree Borne Oilseeds: Neem, Karanj, Pilu or Khakan, Palash, Nahor, Undi, Pisa, Wild-apricot, Rattan-jyot, Maroti, Jojoba, etc.). Almost 72% of the total oilseeds area is confined to rainfed farming cultivated mostly by marginal and small farmers

Area

Lack of appropriate technologies, cultivation under input-starved conditions, combating the biotic and abiotic stresses are some of the major causes for poor productivity of Oilseeds. The huge drain on the import bill coupled with the above factors led to establishment of Technology Mission on Oilseeds (TMO) in 1986 for enhancing the domestic production of edible Oils. Consequent upon the setting up of Technology Mission on Oilseeds, a major breakthrough in increasing Oilseeds production was achieved through an integrated approach by introducing new crop production technologies, better supply of inputs and extension services support for marketing, post-harvest technologies and excellent coordination/cooperation between various concerned organizations/ departments and Ministries. As a result of concerted efforts by the TMOP, the production of Oilseeds increased from 108.3 lakh tonnes in 1985-86 to 361.009 lakh tonnes in 2020-21. This brought not only increase in area but also improvement in productivity from 570 kg/ha (1985-86) to 1284 kg/ha, 1224 kg/ha and 1254 kg/ha during 2017-18, 2019-20 & 2020-21, respectively. The country recorded the highest ever production of 361.009 lakh tonnes oilseeds during 2020-21 followed by 332.192 lakh tones during 2019-20 with record productivity level of 1284 kg/ha during 2017-18 followed by 1254 kg/ha during 2020-21 due to favourable weather conditions and support given by the Govt of India to the Oilseeds production/developmental programmes and policies. During last two years i.e. 2019-20 and 2020-21, country recorded highest oilseeds production due to the efforts of Government of India by implementing several programmes viz., NFSM Oilseeds, TRFA oilseeds, Seed hubs on Oilseeds, cluster demonstrations of improved technology, Special programme on rapeseed & Mustard programme during *Rabi*, 2020-21, etc. Despite impressive progress of vegetable oil sector in the last two decades, the import has been rising for last 7-8 years. India has been cultivating almost all oilseed crops. Major ones are Soybean, Groundnut, Rapeseed-Mustard, Sesame, Sunflower, Castor, Safflower, Linseed and Niger. Among the nine oilseed crops in India, the highest average contribution to total production of oilseeds is of soybean (38.41%) followed by Rapeseed-Mustard (26.24%) and groundnut (25.41%) (Average of 2016-17 to 2020-21). Similarly, highest average area contribution to total oilseed area is of soybean (44.3%) followed by Rapeseed-Mustard (23.33%) and Groundnut (19.08%). On an average, total *Kharif* oilseed crops contribution to total production is about 67% and the remaining 33% is of *Rabi* / summer oilseed crops.

Globally highest production of 363.86 metric tons was of Soybean crop which was followed by Rapeseed (70.62 mt), Sunflower seed (57.26mt) and Peanuts (50.68 mt) (Figure 6.1)

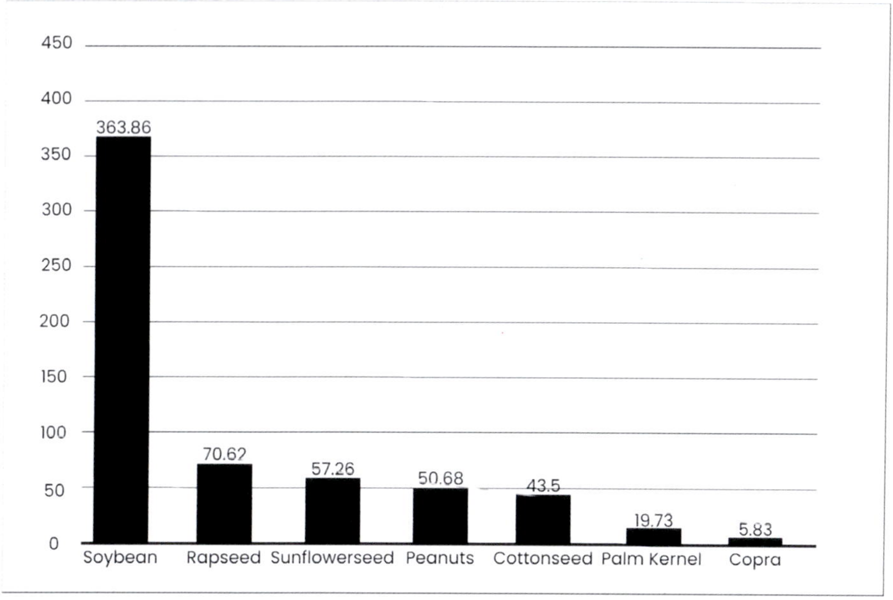

Fig 6.1: Worldwide oilseed production in 2021/2022, by type (in million metric tons)
Source: https://www.statista.com/statistics/267271/worldwide-oilseed-production-since-2008/

In India, area under all oilseed crops was 27.11 mha with production of 36.10 mt during 2020.

6.1. Rapeseed and Mustard

Rapeseed and mustard are the third most important edible oilseed crops of the world after soybean and oil palm. It belongs to the Family: Brassicaceae (Cruciferae). Two species of Rapeseed and mustard are *B. campestris* and *Brassica juncea*. The oil content varies from 37 to 49%. The seed and oil are used as condiment in the preparation of pickles, curries, vegetables, hair oils, medicines and manufacture of greases. The oil cake is used as feed and manure. In the tanning industry, mustard oil is used for softening leather.

Rapeseed (*Brassica campestris*) and Mustard (*Brassica juncea*) are the major *rabi* oilseed crops of India. India is one of the largest producers of these crops in the world. The production of rapeseed and mustard in India accounts for about 18% of the total oilseed production of the country. Sarson and toria (lahi) are generally termed as rapeseed, rai or raya or laha is termed as mustard. The seed and oil are used as condiment in the preparation of pickles and for flavouring curries and vegetables. The oil is utilized for human consumption throughout northern India in cooking and frying purposes. The oil cake is used as a cattle feed and manure. Green stems and leaves are a good source of green fodder for cattle. The leaves of young plants are used as green vegetables as they supply enough sulphur and minerals in the diet.

The Brassicaceae, contains about 3500 species and 350 genera, is one of the 10 most economically important plant families (Warwick et al 2000). It is distinguished on the basis of the presence of conduplicate cotyledons (i.e. the cotyledons are longitudinally folded around the radical) and/ or two-segmented fruits (siliquae), which contain seeds in one or both segments, and only simple hairs, if present. Crop Brassicas encompass many diverse types of plants, which are grown as vegetables, fodder or sources of oils and condiments. The oleiferous Brassica species, commonly known as rapeseed-mustard, are one of the economically important agricultural commodities. Rapeseed-mustard comprising eight different species viz. Indian mustard, toria, yellow sarson, brown sarson, gobhi sarson, karan rai, black mustard and taramira, are being cultivated in 53 countries spreading all over the globe

Although they look similar, canola and rapeseed plants and oils are very different. Canadian scientists used traditional plant breeding in the 1960s to practically eliminate two undesirable components of rapeseed — erucic acid from oil and glucosinolates from meal — to create "canola," a contraction of "Canadian" and "ola." Canola oil is prized for its heart-healthy properties with the least saturated fat of all common culinary oils. To earn the name canola, products must meet this internationally regulated standard: "Seeds of the genus Brassica (*Brassica napus, Brassica rapa or Brassica juncea*) from which the oil shall contain less than 2% erucic acid in its fatty acid profile and the solid component shall contain less than 30 micromoles of any one or any mixture of 3-butenyl glucosinolate, 4-pentenyl glucosinolate, 2-hydroxy-3 butenyl glucosinolate, and 2-hydroxy- 4-pentenyl glucosinolate per gram of air-dry, oil-free solid."

Origin and History

Historically the brassicas are one of the earliest domesticated crop plants by man. It is mentioned in several ancient scripture and literature and might have been cultivated as early as 5000 BC. There is evidence of its cultivation in Neolithic age. Seeds of mustard were found from the Channhudaro of Harrapan civilization ca.

2300-1750 BC. Aryans used Brassica species as condiments and for oil. However, the use of oil was still commoner with the non-Aryans than the Aryans. Thus, it is evident that in a period of over 3500 years, mustard came to occupy an important place in the diets of Indian people as a source of oil and vegetable.

The origin of *B. juncea* is conflicting. Middle East seems to be the place of origin since the putative parent species; *B. nigra* and *B. campestris* would have crossed. From there it soon spread to Europe, Africa, Asia, India and Far-East. It had arisen more than once with different progenitors and in different localities i.e., China, Eastern India and the Caucasus the present day two centers of diversity. Biochemical and molecular studies support this finding and further provide evidence for the existence of two geographical races, Chinese and Indian.

Area and Distribution

The estimated area, production and yield of rapeseed-mustard in the world was 36.59 million hectares (mha), 72.37 million tonnes (mt) and 1980 kg / ha, respectively, during 2018-19. Globally, India account for 19.8 % and 9.8% of the total acreage and production (USDA). During the last eight years, there has been a considerable increase in productivity from 1840 kg/ha in 2010-11 to 1980 kg/ha in 2018-19 and production has also increased from 61.64 mt in 2010-11 to 72.42 m t in 2018-19.

Rapeseed–mustard crops in India are grown in diverse agro climatic conditions ranging from north-eastern / north western hills to down south under irrigated/ rainfed, timely/late sown, saline soils and mixed cropping. Indian mustard accounts for about 75-80 % of the 6.23 m ha under these crops in the country during 2018-19 crop season.

Globally Canada, China and European union contributed highest in area with 24.2, 21.23 and 19.09% respectively and production share was highest in European union (30.27%) followed by Canada 26.3% and China 20.40% (Table 6.1). India ranks fourth in area (17.19%) and production (8.54%).

Table 6.1: Percent Contribution of different countries in area and production for rapeseed and mustard crops (AV 2013 to 2018)

Country	% contribution in area	Country	% contribution in production
Canada	24.27	European Union	30.87
China	21.23	Canada	26.36
European Union	19.09	China	20.41
India	17.19	India	8.54
Australia	7.47	Australia	4.93
Others	10.75	Others	8.89

As indicated from Table 6.2, highest area wise contribution of rapeseed and mustard was from Rajasthan (40.74%), Madhya Pradesh (11.97%) and Uttar Pradesh (11.20%) and highest production percentage was for Rajasthan (44.97%), Haryana (12.44%) and Madhya Pradesh (11.32%).

Table 6.2: Percent contribution of different states of India in area and production of rapeseed and mustard crop (Av. 2013 to 2018)

States	% contribution in area	States	% contribution in production
Rajasthan	40.74	Rajasthan	44.97
Madhya Pradesh	11.97	Haryana	12.44
Uttar Pradesh	11.20	Madhya Pradesh	11.32
Haryana	8.94	Uttar Pradesh	10.60
West Bengal	8.70	West Bengal	7.53
Jharkhand	4.02	Gujarat	4.27
Assam	4.82	Assam	2.46
Gujarat	3.33	Jharkhand	2.17
Bihar	1.41	Bihar	1.27
J&K	0.86	Punjab	.57
Chhattisgarh	0.76	J&K	.43
Punjab	0.52	Chhattisgarh	.27
Others	2.73	Others	1.70

Classification of mustard-rapeseed crop

English name	Vernacular name	Botanical name	Identification characteristics of seeds
Indian mustard / Brown mustard	Rai, ryada, raya, laha, lahta, sasve, herbo	*Brassica juncea (L.)* Czern. & Coss.	Seeds are medium sized round and dark brown or black in colour.
Indian rape / Rapeseed/ Toria	Toria, tori, lahi	*Brassica rapa* L. var. toria *(syn. B. campestris* L. var. toria.)	Seeds are dark brown, bold and large sized.
Brown sarson, / Rapeseed	Brown sarson, Bhoori- sarson	*Brassica rapa L* . var. brown sarson *(syn. B. campestris L.* var . brown sarson	Seeds are light reddish in colour, bold, large sized.
Yellow sarson / Colza/ Rapeseed	Yellow sarson, Pilli sarson	*Brassica rapa L* .var. yellow sarson *(syn. B. campestris L.* var. yellow sarson)	Seeds are slightly smaller than sarson, size is ovoid in shape, yellow in colour.
Rapeseed	Gobhi sarson	*Brassica napus L.*	Seeds are brownish black and large sized.

English name	Vernacular name	Botanical name	Identification characteristics of seeds
Abyssinian mustard / Ethiopian mustard	Karan rai	*Brassica carinata* A. Br.	Seeds are small, round and reddish brown in colour.
Rocket Salad	Duan, tera, tara, saundh, taramira	*Eruca sativa* Mill	Seeds are light reddish brown coloured and distinctly ovoid shape.

Source: 1) National Research Centre on Rapeseed-Mustard, Sewar, Bharatpur (Rajasthan). 2) Department of Botany, Nagpur University.

Climatic Requirements

Require cool and dry weather and a fair supply of soil moisture during the growing period and a dry clear weather at the time of maturity. In India they are grown in Rabi season from September-October to February-March.

Rapeseed and Mustard varieties released by ICAR (2020-2021)

Indian mustard varieties	Area of Adoption	Characters
Pusa Mustard 32	Rajasthan (Northern and Western Parts), Punjab, Haryana, Delhi, Western Uttar Pradesh, Plains of Jammu & Kashmir, Himachal Pradesh	Low erucic acid (<2%) mustard variety, suitable for timely-sown irrigated condition, average seed yield 2.71 tonnes/ha, oil content 38%, maturity 145 days.
Radhika	Jammu, Punjab, Haryana, Delhi, Rajasthan	Suitable for irrigated late sown conditions (November) under irrigated condition, average seed yield 1.79 tonnes/ha, oil content 40.7%, maturity 131 days, tolerant to high temperature at terminal stage.
Brijraj	Jammu, Punjab, Haryana, Delhi, northern Rajasthan	Suitable for late sown (November) under irrigated condition, average seed yield 1.73 tonnes/ha, oil content 39.9%, maturity 132 days, tolerant to high temperature at terminal stage.
SVJH 108	Haryana	Suitable for irrigated conditions under high and low fertility, average grain yield 2.4 tonnes/ha, oil content 41.3%, black and bold seed (6.1 g/1000-seed).
Azad Mahak	Uttar Pradesh	Suitable for timely sown irrigated conditions, average seed yield 2.05 tonnes/ha, oil content 41.6%, maturity 120–125 days, tolerant to high temperature.

Indian mustard varieties	Area of Adoption	Characters
Toria (*Brassica rapa*)		
Azad Chetna	Uttar Pradesh	Suitable for early sowing (Mid-September) under irrigated conditions, average seed yield 1.44 tonnes/ha, oil content 42.2%, escapes infestation/ exposure to Alternaria blight, aphid and fog due to early maturity.
Raj Vijay Toria 2	Madhya Pradesh	Suitable for double cropping system under irrigated and rainfed conditions, average seed yield in irrigated condition 1.04–1.2 tonnes/ ha (irrigated), 0.94 tonnes/ ha (rainfed), oil content 41.9%, early maturing (108 days).

In trade, toria, gobhi sarson and taramira are categorized as rapeseed while raya and African sarson are categorized as mustard. Rapeseed and mustard were grown on 31.0 thousand hectares with a production of 46.5 thousand tonnes during 2019-20 in Punjab. The average yield was 14.82 quintals per hectare (6.00 quintals per acre). Toria, gobhi sarson and African sarson are sown exclusively under irrigated conditions, whereas raya can be grown under both irrigated and rainfed conditions. Taramira is grown as rainfed crop only.

Improved varieties for Punjab

Toria

TL 17: This variety is suitable for multiple cropping systems due to its early maturity. It matures in 90 days. It has profuse branching and siliqua bearing capacity. Its average yield is 13.0 quintals per hectare. Its seeds contain 42.0 per cent oil.

TL 15: This variety takes about 88 days to mature. Because of its early maturity, it fits well in multiple cropping systems. It yields 11.3 quintals per hectare. Its oil content is 41.0 per cent.

Raya

RCH 1: It is the first canola quality ('00') hybrid of raya in the country. It is recommended for general cultivation in south-western region (Bathinda, Faridkot, Ferozepur, Muktsar and Mansa) of Punjab under timely sown irrigated conditions. This hybrid is tall with perfuse siliquae bearing. Its average seed yield is 23.0 quintals per hectare with 39.4 per cent oil content. It matures in 152 days.

PHR 126: It is medium tall hybrid with perfuse branching and siliquae bearing. It is recommended for general cultivation in south-western region (Bathinda, Faridkot, Ferozepur, Muktsar and Mansa) of Punjab under timely sown irrigated conditions. Its average seed yield is 22.75 quintals per hectare with 40.2 per cent oil content. It matures in 145 days.

Giriraj: This variety is recommended for general cultivation in the state under timely sown irrigated conditions. This is a bold seeded variety and has tolerance to terminal heat stress. It is moderately resistant to white rust. Its average yield is 19.25 quintals per hectare and it matures in 144 days. It contains 40.3 per cent oil.

RLC 3: This is the first canola quality ('00') variety of raya in the country. It is a yellow seeded, medium tall variety which is recommended for general cultivation in the state under timely sown irrigated conditions. It is resistant to white rust. Its average seed yield is 18.25 quintals per hectare with 41.5 per cent oil content. It matures in 145 days.

PBR 357: This variety is recommended for general cultivation in the state under timely sown irrigated conditions. It is a medium tall variety with more number of branches per plant. It is bold seeded variety, moderately tolerant to Alternaria blight and white rust diseases. Its average yield is 21.25 quintals per hectare. It contains 39.0 per cent oil content. It matures in 145 days.

PBR 97: This variety is recommended for sowing under rainfed conditions in the state. It has profuse branching and long main shoot with dense siliquae. The grains are medium bold with oil content of 39.8 per cent. On an average, it yields 13.0 quintals per hectare. It matures in 136 days.

PBR 91: This variety is recommended for cultivation in south-western region (Bathinda, Faridkot, Ferozepur, Muktsar and Mansa) of Punjab under timely sown irrigated conditions. This variety has profuse branching habit. It has broad, thick, dark green leaves. Its seeds are bold and dark brown in colour having 37.6 per cent oil. Its average yield is 20.25 quintals per hectare. It matures in 145 days.

RLM 619: This variety is recommended for general cultivation in the state for timely and late sown conditions under irrigated as well as rainfed conditions. It is very suitable for sowing after cotton. It possesses bold seeds and has oil content of 43 percent. It has comparatively greater resistance to white rust, Alternaria blight and downy mildew. Its average yield is 20 quintals per hectare. It matures in 143 days.

Gobhi Sarson

PGSH 1707: It is the canola quality ('00') hybrid of gobhi sarson. It is recommended for general cultivation in Punjab under timely sown irrigated conditions. It is resistant to white rust. This hybrid is tall with profuse siliquae bearing. Its average seed yield is 22.0 quintals per hectare with 41.0 per cent oil content. It matures in 162 days.

GSC 7: This canola quality ('00') variety is recommended for general cultivation in the state under timely sown irrigated conditions. It is a medium tall variety which is free from white rust and is tolerant to Alternaria blight. It has lusterous brownish-black seeds. Its average yield is 22.25 quintals per hectare with 40.5 per cent oil content. It matures in 154 days.

GSC 6: This canola quality ('00') variety is recommended for cultivation in the state for timely sowing under irrigated conditions. It has lustrous bold seeds. Its average yield is 15.25 quintals per hectare with oil content of 39.1 per cent. It matures in 145 days.

Hyola PAC 401: This canola quality ('00') hybrid is recommended for general cultivation in Punjab under timely sown irrigated conditions. Its seeds are brownish black and lustrous. It contains 42 per cent oil. The average seed yield is 16.75 quintals per hectare. It is medium statured and matures in 150 days.

GSL 2: This is the first ever atrazine herbicide resistant variety of gobhi sarson in India suitable for atrazine based weed management. It is recommended for cultivation under timely sown irrigated conditions in Punjab. Its leaves are dark green and plants have comparatively close canopy. It has tall plants with stiff stem and as such does not lodge. On an average, it yields 17 quintals per hectare. Its oil content is 44.5 per cent. It takes about 160 days to mature.

GSL 1: This variety is recommended for general cultivation in the state under timely sown irrigated conditions. Its leaves are thick, smooth and sweet in taste and give quality saag. Its initial growth is slow and plants remain in vegetative phase till early February. As such, the crop escapes frost injury. Plants are stout and do not lodge easily. Its yield is 16.75 quintals per hectare. The seed possesses 44.5 per cent oil. It takes about 160 days for maturity.

African sarson

PC 6: This variety is the world's first determinate oilseed Brassica variety. It is recommended for general cultivation in the state under timely sown irrigated conditions. This medium tall variety is resistant to seed shattering and is suitable for combine harvesting. It is free from white rust and has better tolerance to Alternaria blight and mustard aphid. Its average yield is 19.25 quintals per hectare with 40.1 per cent oil. It matures in 157 days.

Taramira

TMLC 2: This variety is recommended for cultivation in Bathinda, Sangrur, Ferozepur and Kandi areas of Hoshiarpur, Gurdaspur, Rupnagar and Shaheed Bhagat Singh Nagar districts. It has longer main shoot length, more number of siliquae on main shoot and more number of seeds per siliqua. It gives an average seed yield of 7.25 quintals per hectare with 36.6 per cent oil content. It matures in 150 days.

Soil Requirements

Rape seed and mustard are capable of growing under a wide range of soil conditions varying from sandy loam to clay loam soils but they thrive best on light loam soils. They neither tolerate water logging conditions nor do well on heavy soils. Plants can tolerate moderate salinity reasonably well but a soil having neutral pH is ideal for their proper growth and development.

Land Preparation

Brassica has small seeds and therefore, for good germination and stand, a fine tilth is required. Cloddy fields lose the stored soil moisture rapidly and since most oilseed Brassica in India are grown on stored soil moisture, the formation of clods must be prevented. Generally, 3-4 ploughings or harrowing followed by planking gives a good fine tilth.

Crop rotation

Rice – Gobi Sarson / Raya / African Sarson

Maize - Gobi Sarson / Raya / African Sarson

Cotton - Raya / Transplanted Gobi Sarson / Transplanted African Sarson

Maize – Toria / Wheat

Maize – Taramira

Date of Sowing

The sowing time of rapeseed mustard begins from September and goes on to end of November. Taramira is sown from mid-September to end of October. Sowing of Gobi Sarson can be done from 10th to 30th October whereas Raya and African Sarson are sown from mid-October to mid-November. About 1-month old seedlings of Gobi or African Sarson can be transplanted during November to mid-December for getting higher yields in the fields which are vacated late in the *kharif* season.

Seed Rate

For sole crop, optimum seed rate 4.0 kg/ha for all the Brassicas. Profuse branching compensates for missing plants or low plant population. Seed rate as mixed or intercrop depends on the desired proportion of Brassica crop relative to the component crop. It usually ranges from 1 to 2 kg/ha. For transplanted crop 1.5 kg/ha seed is sufficient which is sown on 500 sq.m area.

Method of Sowing

Brassicas are usually broadcasted and the field planked after sowing to give a good contact between seed and moist soil for optimum germination. Low seeding also permits inter-cultivation for minimizing weed infestation. Toria, Raya, African Sarson and Taramira are sown 30 cm apart to get higher yield. Sow these crops 4 to 5 cm deep. Gobhi sarson should be sown in rows, 45 cm apart with plant to plant distance of 10 cm. Also maintain same spacing for transplanted crop.

Nutrient Management

Nitrogen requirement of raya or mustard ranges from 60 to 100 kg/ha on medium soil fertility under irrigated conditions. Under dry land conditions apply 30-60 kg N/ha, depending on soil moisture availability during crop season. Also add 30 kg P_2O_5 and 20 kg K_2O/ ha on soil test basis. All the fertilizer should be applied as basal dose at seeding by drilling. Among the field crops brassicas have highest requirement for Sulphur. In general, 40 kg/ha of Sulphur is optimum under different situations.

Water Management

Flowering and pod formation stages are most sensitive to soil moisture deficits. If irrigation water is not limiting factor, two irrigations at 60 and 90 DAS are ideal for optimum yield. If there is scope for one irrigation only, it should be given 60 DAS.

Among Brassica crops, taramira can withstand water stress.

Weed Management

Brassica crops are fast growing crops and suppress the weed growth. Critical period for crop- weed competition is from 30-45 DAS. Dinitroaniline herbicides such as trifluralin and fluchloralin are very safe for these crops and isoproturon can be applied as post-em on raya and gobhi sarson only.

Plant Protection

(a) Insect Pests

Painted bug (*Bagrada hilaris*): It is serious on the germinating crop in October and again on mature crop in March-April. The nymphs and adults suck the sap from the foliage and siliquae which subsequenty dry up. Application of first irrigation 3-4 weeks after sowing reduces the painted bug population significantly.

Mustard sawfly (*Athalia lugens proxima*): The larvae attack the young crop, bite holes into the leaves and may eat all the leaves in case of heavy attack. It can be managed by spraying 625 ml Ekalux 25 EC (quinalphos) in 150-200 litres of water per hectare.

Mustard aphid (*Lipaphis erysimi*): Cold and cloudy weather is very favourable for the development of mustard aphid. The green plant lice become innumerable, covering the inflorescence and siliquae. They suck the plant sap in huge quantities and as a result, the plants remain stunted, siliquae shrivel up and seeds do not develop. The following integrated pest management programme is recommended for its effective and economic control:

i. Sow the crop at recommended time preferably up to 3rd week of October.

ii. Apply recommended dose of fertilizers.

iii. Apply insecticides on the basis of need by following any of the economic threshold levels given below. For determining aphid infestation, observe 12-16 widely scattered plants from an acre twice a week starting from first week of January.

   - Spray, when pest population reaches, 50- 60 aphids per 10 cm terminal portion of the central shoot or
   - When an average of 0.5 to 1.0 cm terminal portion of central shoot is covered by aphids or
   - When plants infestation by aphids reach 40-50% (observe 100 plants per acre).

Spray any of the following insecticides using 200-300 litres of water per hectare depending upon the stage of the crop, and repeat application whenever the above mentioned economic threshold values are noticed:

Actara 25 WG (thiamethoxam) 100 g /ha

Metasystox 25 EC (oxydemeton methyl) 1.0 litre/ha

Rogor 30 EC (dimethoate) 1.0 litre/ha

Dursban/Coroban 20 EC (chlorpyriphos) 1.5 litre/ha

Green peach aphid (*Myzus persicae*): The green peach aphid attacks taramira from December to March with peak activity during February. After the appearance of inflorescence, the aphid congregates on terminal buds and feeds there. As a result there is flower shedding, poor siliqua formation and shrivelling of grains. For its control, sow the crop in second week of October and spray 500 ml Rogor 30 EC (dimethoate) or 625 ml Metasystox 25 EC (oxydemeton methyl) in 250 litres of water per hectare in third week of February when this aphid starts congregating on top flower buds.

Leaf miner (*Chromatomyia horticola*): Larvae feed by making mines into the leaves and cause heavy damage. Systemic insecticide Rogor 30 EC (dimethoate) recommended at 500 ml per hectare for control of mustard aphid should be used for controlling the leaf-miner or apply 32.5 kg Furadan 3 G (carbofuran) per acre followed by light irrigation.

Hairy caterpillar (*Spilosoma obliqua*) **and cabbage caterpillar** (*Pieris brassicae*): The caterpillars feed on leaves, young shoots and green siliquae. When young, they feed gregariously but the grown-up caterpillars migrate from one field to the other. When in the gregarious stage, they can be easily controlled by picking and destroying the infested leaves.

Note

- Generally the mustard aphid becomes serious around mid-January, hence greater vigilance is needed from first week of January for monitoring its population at regular intervals for deciding the need and time of control measures to be followed.
- Spray in the afternoon when the pollinators are less active.
- Assure a waiting period of 7 days for thiamethoxam 25 WG, 20 days for chlorpyriphos 20 EC and dimethoate 30 EC and 30 days for quinalphos 25 EC after spray on Saag crop.

(b) Diseases

Alternaria blight (*Alternaria brassicae*): Brown to blackish spots with concentric rings appear on leaves, siliquae and stems. In case of severe attack, the upper parts of the stem and siliquae wither and contain shrivelled seeds. Temperature in the range of 15-25 °C and relative humidity of more than 75% favour the disease development. To avoid incidence of this disease, destroy diseased debris from the previous crop.

Downy mildew (*Peronospora brassicae*): Small light green lesions appear on the lower side of the leaves, which later enlarge, become more greyish, dry up, shrivel and tear easily. All aerial parts of the plants are attacked. The disease is severe on foliage and seed bearing parts. Greater deformity occurs in the stem. Relative humidity of more than 90% with temperature in the range of 10-20° C are conducive for disease development and spread. To avoid incidence of this disease, destroy diseased debris from the previous crop.

White rust (*Albugo candida*): Prominent white creamy-yellow, scattered pustules appear on the under surface of the leaves. The swelling of affected parts often occurs. Flowers get malformed, become sterile and club shaped. All parts are attacked, except roots. Weather conditions of 13-25° C temperature and more than 75% relative humidity favour the disease development. To avoid incidence of this disease, destroy diseased debris from the previous crop. To manage white rust, spray 625 g of Ridomil Gold (metalaxyl 4%+ mancozeb 64%) by dissolving in

250 litres of water per hectare at 60 and 80 days after sowing. If needed, repeat the spray after 20 days interval.

Sclerotinia Stem rot (*Sclerotinia sclerotiorum*): This soil borne disease appears as water soaked lesions on the stem which later on are covered with white cottony growth of mycelium. Black coloured sclerotia develop on these lesions as well as in the pith of the stem. Upon harvesting, these sclerotia are mixed in soil where they can survive for longer period. It appears generally in the month of January when the temperature is quite low and soil has abundance of moisture. The sclerotia lying under the soil germinate and the mycelium enter the plants. The affected plants look wilted, then dry up and stand erect with whitish appearance. If plant canopy is heavy then there is breaking of stem also. To manage this disease, avoid irrigation to the rapeseed-mustard crop during 25th December to 15th January.

Phyllody (*Mycoplasma Like Organisms*): Plant parts are malformed. Numerous shoots arise and the plants become bushy or broom shaped. Floral parts are transformed into leafy structures. Avoid early sowing. Rogue out the affected plants to minimise further spread of this disease

Harvesting

Brassica crops are harvested when the pods turn yellow, the seeds become dark and they rattle inside the pod when shaken. Toria takes about 90-100 days to mature. It is the earliest crop to be harvested from mid-December. Raya takes 130 to 150 days for maturity and the harvesting is from end February to March. Yellow sarson take 140 to 160 days and brown sarson 120 to 140 days for maturity. In India, average yield is about 670 kg/ha. These crops should be harvested at proper time as these are highly prone to shattering.

6.2. Soybean (*Glycine max* (L.) Merrill)

Soybean, (*Glycine max*), also called soja bean or soya bean, annual legume of the pea family (Fabaceae) and its edible seed. The soybean is economically the most important bean in the world, providing vegetable protein for millions of people and ingredients for hundreds of chemical products.

Although the U.S.A. and Brazil account today for most of the soybean production of the world, the introduction of this crop to Western agriculture is quite recent. Soybeans are, primarily, an industrial crop, cultivated for oil and protein. Despite the relatively low oil content of the seed (about 20% on moisture-free basis), soybeans are the largest single source of edible oil and account for roughly 50% of the total oilseed production of the world.

With each ton of crude soybean oil, approximately 4.5 tons of soybean oil meal with a protein content of about 44% are produced. For each ton of soybeans processed, the commercial value of the meal obtained usually exceeds that of the oil. Thus, soybean oil meal cannot be considered a by-product of the oil manufacture. The soybean is, in this respect, an exception among oilseeds.

It can be calculated that, the quantity of protein in the yearly world production of soybeans, if it could be totally and directly utilized for human consumption, would be sufficient for providing roughly one third of the global need for food protein. This makes the soybean one of the largest potential sources of dietary protein. However, the bulk of soybean oil meal is used in animal feeds for the production of meat and eggs. Despite considerable public and commercial interest in soybean products as food, the proportion of soybean protein consumed directly in human nutrition is still relatively small.

Dry soybean contains 36% protein, 19% oil, 35% carbohydrate (17% of which dietary fiber), 5% minerals and several other components including vitamins.

Origin and History

The soybean originated in Eastern Asia, probably in north and central China. It is believed that cultivated varieties were introduced into Korea and later into Japan about 2000 years ago. Soybeans have been grown as a food crop for thousands of years in China and other countries of East and South East Asia and constitute to this day, an important component of the traditional popular diet in these regions.

Area and Distribution

The total world production for soybeans in 2020 was 353.46 million tonnes, up by 5.1% from 336.33 million tonnes in 2019. Brazil was the largest producer, accounting for 34% of world production, followed by the United States at 32%. Area of soybean in India was 12.2 mha with production of 11.22 mtonnes during 2020.

Soybean Growing States

The state wise details of area, production and yield during 2022 for major Soybean growing states (Table 6.3). Highest area and production of soybean was in Madhya Pradesh and Maharashtra whereas lowest was recorded in Chhattisgarh.

Table 6.3: Area, production and yield of soybean in different states of India

S.No.	States	Sowing Area (Lakh hectare)	Production (lakh tonne)	Yield (kg/ha)
1	Madhya Pradesh	50.645	53.248	1051
2	Maharashtra	43.446	46.911	1080
3	Rajasthan	10.344	9.855	953
4	Karnataka	4.376	4.398	1005
5	Gujarat	2.216	2.409	1087
6	Telangana	1.754	1.763	1005
7	Chhattisgarh	0.42	0.441	1050
8	Others	1.286	1.358	1056
	Grand Total	114.487	120.383	1051

Source : SOPA Databank

Products of Soybean

Edamame (Green Soybeans): These are soybeans harvested when the beans are green and sweet tasting. They are usually located in the frozen section of the supermarket, and are available either in the pod or shelled. They cook quickly and can be eaten cold or warm, but not with the pod. Shelled beans can easily be added to salads, soups, pasta. They could be snacked on the go. One can eat them right out of the pod. Edamame are also a healthy substitute for popcorn and chips.

Soy Beans (Dry Soybeans): They are a species of legume native to East Asia. They can be found as canned or dry in black and white varieties. Their characteristic beany flavor and the presence of anti-nutritional factors are the main reasons for the non-utilization of soybean at domestic level in rural and urban areas. A study conducted on making soybeans completely free from the inhibitor with simple and inexpensive methods of heat treatment which can be adopted at home level revealed that pressure cooking or twelve minutes of roasting at 85 to 90 degrees or boiling the beans for 20 minutes was found sufficient for complete inactivation of the inhibitors.

Soy Flour : In recent times soy foods such as soy milk, soy 'paneer' or soy cottage cheese, soy cheese, nuggets find a good place in the market. But in India most of them are beyond the reach of common man due to higher prices. In such a case soy flour becomes a common inexpensive variant that can be easily used to prepare traditional recipes, roti blends, biscuits/snacks, germinated soy bread, fermented products, supplementary foods and therapeutic foods.

Soy Nuggets and Granules: Soy is processed to make soy nuggets and granules. These are creamish in color and easy to digest as compared to beans and flour. These just have to be soaked in warm water before use and therefore have a varied application. They could be added to curries, cutlets, parathas, stir fries, pulavs and various other recipes.

Soy Nuts: Soy nuts are produced when whole soybeans are soaked and then baked until browned. Roasted soy nuts are available in most grocery stores and are great for snacking.

Soy Milk: Soybeans, soaked, ground, strained and fermented produce a fluid called soybean milk, which is a good substitute for dairy milk. It is rich in protein and calcium. It can be used in place of cow's milk for cooking and drinking. One can easily make smoothies, soups, or sauces using it. Soymilk comes in non-refrigerated recyclable boxes found on grocery shelves. Soy milk comes in chocolate, vanilla or plain flavors too. It also can be found in single serving sizes which makes it a great on-the-go snack and needs no refrigeration before opening. Soy milk stands out as it is easy to digest, high on protein and enriched with calcium, vitamin D and B12 which are some of the most essential nutrients for health.

Soy Yogurt: Soy yogurt is made from soymilk. Fermentation improves the nutritive value of the product. Yogurt's creamy texture makes it an easy substitute for sour cream, cream cheese or dairy yogurt. Soy yogurt is rich in protein, calcium, and is a source of probiotics.

Tofu: Very similar to cottage cheese, known as soybean curd, tofu is a soft cheese-like food made by curdling fresh hot soymilk with a coagulant. Tofu is a neutral-flavoured product and easily absorbs the flavours of other ingredients with which it is cooked. Tofu can be used instead of cottage cheese in any recipe. Tofu is found in water packed tubs in the refrigerated section or on grocery shelves of supermarkets. Silken tofu, which is smooth and creamy, can be used in place of cream in soups, as a substitute for mayonnaise or sour cream in salad dressings and dips. Soft tofu is moist and firmer than silken tofu. It can be substituted for soft cheese. Firm and extra firm tofu will hold its texture and shape and can be used in place of meat in salads, for stir fries, Chinese curries, fajitas etc. Unlike cottage cheese it has a greater amount of protein and nil cholesterol.

Tempeh: Whole soybeans, sometimes mixed with another grain such as rice or millet, are fermented into a rich cake with a smoky or nutty flavor called as Tempeh. Tempeh is sold in vacuum packed rectangular shaped cakes/patties. Tempeh can be used in stir-fries or pasta dishes. It can also be grilled or baked and used for sandwiches and salads. It is a good source of protein, calcium, and probiotics.

Miso: It is a smooth paste, made from soybeans and/or grains such as rice or barley, with salt and a mold culture, aged in cedar vats for one to three years. Miso is the main ingredient in the traditional Japanese soup. Miso can be found in many grocery stores and it can be used as a soup base, in salad dressings, and Sandwich spreads.

Climatic Requirements

Soya bean is best grown in warm and moist climates and does not require intense care for hardy growth, much like other members of the oil seed and pulses family. This is because it can grow even with minimum agricultural inputs and requires very little management to result in a profitable crop for its grower. Soyabean is considered one of the most stable *kharif* crops of India. Unlike other *kharif* crops, soyabean yields are a cost-effective production in varied agro-climatic regions. Scientifically, soyabean belongs to legume family, hence is capable of meeting its nitrogen requirements from the surrounding atmosphere and also helps in building soil fertility. The temperature ranging from 26.5 to 30°C is considered favorable for most of the soyabean varieties. It requires a minimum 15.5°C soil temperature for rapid germination and growth. The minimum temperature for effective growth of soyabean crop is 10°C. Below optimum temperature for soybean might result in delayed flowering. Its growth also depends on day length as they are short-day plants.

Soybean varieties released by ICAR (2020-2021)

Varieties	Area of adoption	Characters
MACS 1520	Madhya Pradesh, Bundelkhand region of Uttar Pradesh, Rajasthan, Gujarat, Marathwada and Vidarbha region of Maharashtra	Suitable for irrigated and rainfed conditions during kharif, average grain yield 2.21 tonnes/ha, oil content 19%, maturity 98–102 days, resistant to charcoal-rot.
NRC 132	West Bengal, Bihar, Jharkhand, Chhattisgarh, Odisha, Southern Maharashtra, Karnataka, Telengana, Andhra Pradesh, Tamil Nadu	Suitable for irrigated and rainfed conditions during kharif, first variety of India with less beany flavor (null lipoxygenase 2), developed through marker-assisted selection, average seed yield 2.29 tonnes/ha (SZ), 1.65 tonnes/ha (EZ); 18–19% oil content, maturity 98 days (SZ),104 days (EZ).
NRC 128	Punjab, Uttar Pradesh (except Bundelkhand region of Uttar Pradesh), Delhi), West Bengal, Bihar, Jharkhand, Chhattisgarh, Odisha	Suitable for irrigated and rainfed conditions during *kharif*, average seed yield 1.87 tonnes/ha (EZ), 2.27 tonnes/ha (NPZ); oil content 18–19%, maturity 106 days (EZ), 118 days (NPZ); resistant to pod blight and Indian bud blight, moderate antixenosis against Spodoptera litura.

Varieties	Area of adoption	Characters
RSC 11-07	Eastern Zone (West Bengal. Bihar, Jharkhand, Chhattisgarh and Odisha) Southern Zone (Southern Maharashtra, Karnataka, Telangana, Andhra Pradesh and Tamil Nadu excluding rust prone areas on banks of river Krishna)	Suitable for irrigated and rainfed conditions during *kharif*, average seed yield 1.92 tonnes/ha (EZ), 2.52 tonnes/ha (SZ); 18–19% oil content, maturity 102 days (EZ), 97 days (SZ), resistant to Indian bud blight and pod blight.
RSC 10-46	Eastern Zone (West Bengal, Bihar, Jharkhand, Chhattisgarh and Odisha, Central Zone (Madhya Pradesh, Bundelkhand region of Uttar Pradesh, Rajasthan, Gujarat and Marathwada and Vidarbha region of Maharashtra)	Suitable for irrigated and rainfed conditions during *kharif*, average seed yield 2.04 tonnes/ha (CZ), 1.95 tonnes/ha (EZ), Oil content 19.8% (CZ), 17.8% (EZ); maturity 102 days (CZ), 104 days (EZ); resistant to charcoal rot and Indian bud blight.
RSC 10-52	Madhya Pradesh, Bundelkhand region of Uttar Pradesh, Rajasthan, Gujarat and Marathwada and Vidarbha region of Maharashtra	Suitable for irrigated and rainfed conditions during *kharif*, average seed yield 2.05 tonnes/ha, oil content 18%, maturity 101 days, resistant to charcoal-rot, bacterial pustule, TLS, MLS, IBB, RAB.
AMS 2014-1	West Bengal, Bihar, Jharkhand, Chhattisgarh, Odisha	Suitable for irrigated and rainfed conditions during *kharif*, average seed yield 1.8 tonnes/ha, 18.5% oil content, maturity 106 days, resistant to charcoal rot, RAB, BLB, ALS.
AMS-MB-5-I8	Madhya Pradesh, Bundelkhand region of Uttar Pradesh, Rajasthan, Gujarat, Maharashtra	Suitable for irrigated and rainfed conditions during *kharif*, average seed yield 2.04 tonnes/ ha, oil content 19%, maturity 100 days.
Raj Vijay Soybean	Madhya Pradesh, Bundelkhand Region of Uttar Pradesh, Marathwada and Vidarbh Region of Maharashtra, Rajasthan, Gujarat	Suitable for irrigated and rainfed conations, average yield 2.07 tonnes/ha, maturity 101 days, resistant to charcoal rot.
SL 1028	Punjab, Haryana, Uttar Pradesh (except Bundelkhand region of Uttar Pradesh), Delhi	Suitable for irrigated and rainfed conditions, average yield 2.2 tonnes/ha, 19.8% oil content and 39.2% protein, maturity 124 days, moderately resistant to YMV.

Improved varieties for Punjab

SL 958: It has shining, light yellow coloured grains with black hilum. Its' grains contain 41.7% protein and 20.2% oil. It is highly resistant to yellow mosaic virus and soybean mosaic virus. It takes about 142 days to mature. Its' average seed yield is about 18.3 quintals per hectare.

SL 744: It has shining, light yellow coloured grains with grey hilum. Its' grains contain 42.3% protein and 21.0% oil. It is resistant to yellow mosaic virus and soybean mosaic. It takes about 139 days to mature. Its' average seed yield is about 18.3 quintals per hectare.

SL 525: It has uniformly bold, shining, cream coloured grains with light black (grey) hilum. Its' grains contain 37.2% protein and 21.9% oil. It is resistant to yellow mosaic virus and tolerates stem blight and root-knot nematode. It matures in about 144 days. Its' average seed yield is about 15.3 quintals per hectare.

Soil Requirements

Well drained and fertile loam soils with a pH between 6.0 and 7.5 are most suitable for the cultivation of soybean. Sodic and saline soils inhibit germination of seeds. Water logging is also injurious to the crop. An increase in protein concentration by 20% and 16% decrease in oil concentration was observed as soil pH rouse from 4.5 to 7.0. Soybeans can tolerate slightly acid soil with the optimal soil pH range of 5.5 to 6.5. But they are not tolerant of strongly acid soils (below pH 4.5) as aluminium (Al) and manganese (Mn) toxicity are likely to be a problem on such soils. At the other end of the scale, it is not recommended to grow soybean in soils with pH greater than 8 as micronutrient deficiencies such as zinc (Zn) and iron (Fe) can occur.

Soybean lacks drought tolerance as its roots are relatively shallow and the root structure limits water absorption during dry periods. Hence soybean performs poorly on sandy soils and soils with low water storage, such as gravelly or shallow soils mainly due to drought stress. Low rainfall on clay soils will reduce the chances of seed germination and plant establishment. The soil having good stored soil moisture capacity is recommended to grow soybeans. Soybeans are susceptible to water logging between emergence and the four-leaf stage. However, after this stage soybean has good tolerance to water logging compared to other non-rice crops, which is another reason to grow it. Soybean may also tolerate flood irrigation techniques more effectively than other crops

Cropping systems

Soybean is ideally suited for fitting in various sequence

- Soybean-Wheat
- Soybean-Barley
- Soybean-Mustard
- Soybean-Pea\
- Soybean-Lentil

Land Preparation

Being a leguminous crop, soybean requires fine seed-bed with least clods. Deep ploughing by soil-turning plough during summer is desirable. It should be followed by 2-3 cross harrowings or cultivator operations to ensure ideal seed-bed in June. To avoid water stagnation, seed-bed should be provided with little gradient.

Planting Time: Photoperiodic response dictates the time of planting for maximum yield of soybean. Early planting gives too great canopy to produce maximum yield, whereas late planting results in canopy development too restricted for maximum yield. Soybean planting dates depend on the variety and the agroecological region. Obviously, photo-periodic effects, occurrence of drought and flood decide the date of sowing of the crop. Sowing time should be adjusted in such a way that one can escape from these natural calamities. In India, soybean is mostly a rainy season crop, sown from last week of May depending upon the climatic conditions

Seed Rate and Spacing: Seed rate depends on seed size (test weight), variety, germination percentage and date of planting. Generally, 75-80 kg seeds/ha is recommended. Row spacing of 45 cm and plant-to-plant spacing should be 5 cm. The ideal plant population varies from 0.4 to 0.6 million/ha.

Method of Sowing: Line sowing is the best method of sowing as it facilitates hoeing/interculture operations easily. Moreover, stand of crop will be uniform under medium to heavy textured soils, bed planting is preferred as there will be minimum or no loss due to water stagnation as compared to flat sowing. This crop is very shy in germination when there is rainfall after sowing and with bed planting technique, its germination will be normal.

Raised Bed Sowing: Sowing of soybean in medium to heavy textured soils should be done on beds spaced 67.5 cm apart (37.5 bed top, 30 cm furrow) by using wheat bed planter. Sow two rows per bed using same quantity of seed, fertilizer and following other cultivation practices as in flat sown soybean. Irrigation should be applied in furrows by taking care that beds are not inundated. This practice not only saves the crop from damage by rains especially at emergence, but also saves about 20-30% irrigation water along with increased yield over conventional flat sowing method. Ensure good moisture conditions while sowing and in case it is not so, apply irrigation in furrows within 2-3 days after sowing for optimum germination and emergence.

Depth of Planting: A seed depth of 2.5 to 4 cm is optimum. Shallower depth may be justified to crust-prone areas and deeper in sandy soils. Some cultivars have inhibitors of hypocotyl elongation and for them shallow depth is recommended. Under conditions of optimum moisture and heavy soils, the crop should be sown at 2-3 cm depth, whereas for optimum moisture, medium to light textured soils, it should be planted at 3-4 cm depth. Planting can be done with the help of seed drills or with tractor-operated planters.

Seed Treatment and Seed Inoculation: In order to get rid of seed-borne diseases, seeds must be treated with Thiram @ 3 g/kg seed. Moisten the seed of one acre with water and mix thoroughly one packet of culture with it and let it dry in shade. Sow the seeds immediately. Grain yield will increase up to 8%. For higher yield, seed inoculation with *Bradyrhizobium japonicum* culture (LSBR3) is recommended.

Manures and Fertilizers: Manure and fertilizer application must be decided depending on soil-test value, productivity potential, expected yield, cropping system, management and method of fertilizer application. In addition to 10 tonnes FYM/ha, application of 30 kg N, 60-80 kg P_2O_5 per hectare, and K_2O and S are applied as per soil test basis. The FYM must be applied at the time of land preparation. All the fertilizers must be applied at the time of sowing.

Weed Management: Normally, weed control in soybean is accomplished using mechanical intercultivation (hoeing) or by the use of herbicides. Hand weeding is preferred under small holding situations. Proper rotation may also reduce weed intensity and flora. To avoid the competition during early growth stage, soybean field should be kept free of weeds for the first 30-40 days after sowing. Two manual weedings, 20-25 and 40-45 days after sowing are sufficient for control of weeds. During rainy season, due to continuous rains, the hand weeding may not be possible at proper time. Under such circumstances, use of herbicides has been found very effective. Fluchloralin or Trifluralin 1 kg/ha as pre-plant incorporation or Alachlor 2 kg/ha or Pendimethalin 1 kg/ha or Metolachlor 1 kg/ha or Clomazone 1 kg / ha as pre- emergence. Alternatively, Imazethapyr 100 g /ha or Quizalofop ethyl 50 g /ha can be used as post-emergence (15-20 DAS) in 200-250 litres water/ ha.

Water Management: One of the most frequent and severe environmental limitations on soybean seed yield is water supply and drainage of excess water from the field. Soybean is often cultivated during rainy season, but supplemental irrigation becomes necessary if larger dry spells occur. Soil moisture stress at seed emergence and pod filling stages are most sensitive. Soil moisture deficit shows relatively less loss of grain yield as compared to other crops, because soybean has an extended period of flowering. The early flower drops which is caused due to moisture stress could be compensated by the late flowers maturing into pods.

Avoid flooding of soybean fields at any stage of the crop. Flooding at floral initiation is detrimental because it retards nodule and overall root activity, which is most crucial at this stage. In case of *kharif* crop, irrigation is generally not needed and it is grown rain fed provided the rains are well distributed, otherwise the crop needs water at critical growth stages. However, during summer, the crop can be grown only under assured irrigation and it needs around 5-6 irrigations. Water stress at the end of flowering and pod initiation is most critical because it leads to flower and pod abortion. Stress during the grain filling reduces the seed size.

Plant Protection

Insect Pests

Jassid, Aphid, Whitefly and Flea beetle: These insects can be controlled by spraying the crop with 940 ml Malathion 50 EC (malathion) or 625 ml Rogor 30 EC (dimethoate) using 200-250 litres of water per ha with a manually operated sprayer. Whitefly can also be controlled by spraying 100 g Actara 25 WG (thiamethoxam) using 200-250 litres of water per ha. Repeat the spray at 10 days interval, if necessary.

Hairy caterpillar: The body of the caterpillar is covered with hair. The caterpillar eats the green matter of leaves, leaving behind only the midribs. The crop may be totally denuded due to severe attack. When young, they feed gregariously but on few plants in scattered spots. Adopt the following control measures:

- Young larvae are gregarious. They can be destroyed by pulling out the infested plants along with larvae and burying them underground.
- The grown-up caterpillars can be destroyed by crushing them under feet or by picking and putting them into kerosenized water. If the population is high, spray 1.25 litre of Ekalux 25 EC (quinalphos) in 200-250 litres of water per ha with a manually operated knapsack sprayer.

Semi looper: The larvae are green in colour measuring 2-4 cm in length. When touched, they form a loop. The larvae feed extensively on the leaves of mash and moong. In case of severe damage, the plants are totally defoliated within a few days.

Tobacco caterpillar (*Spodoptera litura*) : It is a polyphagous pest. The small larvae are black whereas grown up larvae are dark green with black triangular spots on body. Its moth lays eggs in masses covered with brown hairs on the lower side of leaves. After hatching, first and second instar larvae feed gregariously and skeletonize the foliage. Later on, the grown-up larvae disperse and feed singly. Besides leaves, they also damage buds, flowers and pods.

- **Mechanical control:** Egg masses and young larvae of tobacco caterpillar feeding gregariously should be collected along with leaves and destroyed.
- **Chemical control:** It can be controlled by spraying 375 ml Rimon 10 EC (novaluron) or 2 kg Asataf 75 SP (acephate) or 3.75 litres Dursban 20 EC (chlorpyriphos) using 250 litres of water per ha with a manually operated knapsack sprayer. Spray the crop as soon as the pest appears in the field and repeat after 10 days, if necessary.

Blister beetle: Blistle beetle is diurnal and general feeder. Adult beetles are robust with bright black and red stripes on the forewings. When disturbed, the beetles emit a fluid containing cantharidin that causes blisters on human skin. The adult beetles attack arhar, moong, mash and other pulse crops. The major damage is caused at the flowering stage. They feed on tender buds and flowers of the plant, thus preventing grain formation.

Blister beetle can be controlled by spraying Decis 2.8 EC (deltamethrin) @ 500 ml or Kingdoxa 14.5 SC (indoxacarb) @ 500 ml per ha using 200-250 litres of water with manually operated knapsack sprayer. Spray the crop on pest appearance in the evening hours and repeat the spray after 10 days, if necessary.

Pod borer (*Helicoverpa armigera*): The larvae damage the crop by feeding on leaves, flower buds, flowers, pods and seeds in the pods, thus causing heavy loss in yield. The larvae may be pale green, yellow, brown or black in colour measuring about 3-5 cm in length when fully grown. Larval presence can be observed from damage to plant and from dark green faeces below the plants on the soil. The larvae fall on the ground when plants are shaken vigorously. Spray the crop at the appearance of larvae with Kingdoxa14.5 SC (indoxacarb) @ 500 ml or Asataf 75 SP (acephate) @ 2 kg or Tracer 45 SC (spinosad) @ 150 ml in 250 litres of water per ha using a knapsack sprayer at the start of pod formation and repeat after two weeks, if necessary.

Mite: The mite causes webbing on the underside of the leaves which turn pale. Such infested leaves turn light-brown to dark reddish-brown. It can be controlled by spraying the crop with 375 ml of Rogor 30 EC (dimethoate) in 200-250 litres of water per ha.

Diseases

Yellow mosaic virus: It is a viral disease and is transmitted by whitefly. Disease appears as a blend of yellow and green patches on the infected leaves. Infected plants bear a few pale pods. Grow resistant varieties SL 958, SL 525 and SL 744. Follow the control measures against whitefly as given under moong.

Harvesting and threshing

When pods become golden yellow, black or gray, the crop is ready for harvest. At harvest, moisture content of the seed should be 15-17%. The maturity period ranges from 90 to 140 days depending on the varieties. When the holdings are small, manual harvesting close to the ground is followed.

Use of reapers and combine harvesters is common on large area for soybean harvesting. Threshing can be done either with the mechanical soybean thresher or some conventional methods used in other legumes.

Yield potential

With good crop management, seed yield of about 1,600-2,000 kg/ha under rainfed conditions and 2,000-2,500 kg/ha under irrigated conditions can be expected. Under high input management, yield levels of 3,000 kg/ha can also be realized.

6.3. Groundnut (*Arachis hypogaea* L.)

Groundnuts are popular source of food throughout the world, consumed either as peanut butter or crushed and used for groundnut oil or simply consumed as a confectionary snack. Groundnuts are cultivated in the tropical and subtropical regions of the world on sandy soils. Yields of groundnut vary from about 400 kg to several tonnes per hectare, depending on the production system, but on the average, the global yield is 2500–2700 kg per hectare.

The groundnut is particularly valued for its protein content (26%). On equal weight basis (Kg for Kg), groundnuts contain more protein than meat and about two and a half times more than eggs. Being an oil seed crop, it contains 40 to 49% oil. It also contains Starch (11.5%), Soluble sugar (4.5%) Crude Fiber (2.1 %) and Moisture (6%). In addition to protein, groundnuts are a good source of calcium, phosphorus, iron, zinc and boron. The groundnuts also contain vitamin E and small amounts of vitamin B complex.

History and Origin

Groundnut is a native New World crop. Early explorers found it cultivated extensively in both Mesoamerica and South America. Remnant pericarp (fruit hull) tissue recovered from archaeological sites in Peru dates its purposeful agricultural use there at approximately 3900–3750 years before the present (YBP).

Area and Distribution

The major groundnut-producing countries of the world are India, China, Nigeria, Senegal, Sudan, Burma and the USA. Out of the total area of 18.9 million hectares and the total production of 17.8 million tonnes in the world, these countries account for 69% of the area and 70% of the production. India occupies the position, both in regard to the area and the production, in the world. Groundnut is cultivated on 6.1 mha with production of 9.95 mtonnes (excluding shelled). About 70% of the area and 75% of the production are concentrated in the four states of Gujarat, Andhra Pradesh, Tamil Nadu and Karnataka.

Groundnut was grown on 1.5 thousand hectares during 2020-21 in Punjab. Its' production was 3.0 thousand tonnes with an average yield of 19.80 quintals per hectare.

Climatic Requirements

Climatic conditions such as temperature and rainfall significantly influence the groundnut production. Warm and moist conditions are very favorable than cool and wet climate, which results in slow germination and seedling emergence, increasing the risk of seed rot and seedling diseases. Temperature is a major environmental factor that determines the rate of crop development. Temperatures above 35°C inhibit the growth of groundnut. Optimum mean daily temperature to grow is 30°C and growth ceases at 15°C. For rapid emergence, soil temperature above 21°C is needed. The optimum temperature for the most rapid germination and seedling development is about 30°C. A minimum 100 - day optimum temperature growing season is necessary for successful groundnut crop production. Adequate and well distributed rainfall during the growing season, especially during flowering, pegging and pod formation stages, is essential for maximum yield and quality of groundnut.

Groundnut varieties released by ICAR (2020-2021)

Varieties	Area of adoption	Characters
K1719	Andhra Pradesh, Telangana, Tamil Nadu	Suitable for *rabi*-summer condition under both high and low fertility conditions, average pod yield 2.17 tonnes/ha, maturity 110–113 days, tolerance to collar rot, peanut bud necrosis disease (PBND) and moderately resistant to thrips.
DH 257	Karnataka, Maharashtra	Suitable for rabi-summer condition, average pod yield 3.28 tonnes/ha, maturity 120–125 days.
K 1812	Andhra Pradesh, Telangana, Tamil Nadu, Karnataka	Suitable for rainfed conditions, average pod yield 3.89 tonnes/ha, maturity 107–113 days, resistant to collar rot, stem rot, dry root rot, ELS, LLS, Alternaria leaf blight, Spodoptera, leaf miner and thrips.
J 87	Punjab, Uttar Pradesh	Suitable for rabi-summer condition, average pod yield 4.17 tonnes/ha, maturity 107–109 days, large seeded with 100-kernel weight of 63 g.

Improved Varieties for Punjab

J 87: It is recommended for sowing in both spring and kharif seasons. It is bunch type variety. The kernels of this variety are bold, oblong in shape and light pink in colour. Each pod bears 2-3 kernels. Its' 100-kernel weight is 79 g. It has 69 per cent shelling outturn. Its' average pod yield is 38.3 q per hectare in spring season and 32.0 q per hectare in *kharif* season. It has oil content of 49 per cent and sucrose content of 5.2 per cent. It has 65.7 per cent oleic acid and is suitable for confectionery. It matures in 112 days.

TG 37A: It is an early maturing bunch type variety which is suitable for cultivation during spring season. It has 65 per cent shelling outturn with 100 kernel weight of 42.5 g. Each pod bears 2-3 kernels. The kernels are spherical in shape and pink in colour. Its' average pod yield is 30.8 quintals per hectare. It has 48.6 per cent oil content. It has soluble sugar content of 5.8 per cent. It matures in 101 days.

SG 99: It is a bunch type variety. Its' average pod yield is 25 quintals per hectare and is tolerant to bud necrosis disease. The pods are medium in size with slight beak and moderate constriction. The pods are borne around the main root, which allows easy harvesting of crop with minimum pod losses. It has 66 per cent shelling outturn, with 100 kernel weight of 54 g and 52 per cent oil content. The kernels have light brown colour. It matures in about 123 days.

M 522: It is a semi spreading type variety. It yields about 22.5 quintals per hectare of pods. The pods are medium bold in size with mostly two kernels per pod. The pods are borne towards the main root. It has 68 per cent shelling outturn, 65 g 100 kernel weight and 51 per cent oil content. The kernels have light brown colour. It matures in about 120 days.

Soil Requirements

Groundnut grows best in deep, well-drained soils with a sandy or very loose surface layer. If the moist soil is rubbed between the index finger and the thumb should not form a ribbon but should fall apart easily. Such soil is best for groundnut crop. In addition to soil fertility, soil - texture is vitally important aspect of groundnut production. Well - drained, light - textured, loose, friable sandy - loam or sandy clay loam soils, well - supplied with calcium and a moderate amount of organic matter are ideal for groundnut cultivation.

Seed treatment and Sowing

Seed treatment with Thiram (5g/kg seed) or Indofil M45 (3.0 g/kg seed) is essential against soil and seed borne diseases for adequate stand establishment. In general, for irrigated crop May to mid-June is the optimum seeding time for *kharif* Groundnut. For *rabi* groundnut, first fortnight of November is ideal time. Spacing of 30cm x 10cm for bunch types during kharif and 22.5 cm x 10 cm in *rabi* is ideal. A spacing of 30cm x 15cm and 22.5cm x 15cm is optimum for semi spreading type during *kharif* and *rabi* respectively. Recommended seed rates for bunch types are 150 and 180 kg/ha for *kharif* and *rabi* respectively. Semi spreading types require 120 and 150 kg/ha in *kharif* and *rabi* respectively. Bold seeded requires 20 kg/ha additional seed rate than small seeded varieties. Optimum depth of seeding is 5-7 cm in light soils and 4-5 cm in heavy soils.

Nutrient Management

Recommended fertilizer schedule for rainfed groundnut is 15 kg N, 20 kg P_2O_5 and 25 kg K_2O per ha in addition to about 10 ton/ha FYM. For an irrigated crop, an additional dose of 10 kg N/ha is recommended 30 DAS. Recommended fertilizer schedule must be applied by placement at the time of seeding for rainfed groundnut. Gypsum 200 kg/ha is recommended at full bloom stage (30 DAS) both for rainfed and irrigated groundnut. Most common source of phosphorus for rainfed groundnut is single super phosphate. For an irrigated crop, DAP is the most commonly used fertilizer as it can meet the needs of both N and P.

Water Management

Peak flowering and early pod development stage are most sensitive to soil moisture stress. There is no necessity for irrigation up to 25 DAS the groundnut. Two to three irrigations may be necessary depending upon the seasonal rainfall. Irrigation at flowering is very beneficial. If the rainfall during the preceding period is not adequate, apply one to two more irrigations, depending upon the time of recession of monsoon, during the pod formation period for proper development of pods. Another irrigation a few days before the harvest may be given for full recovery of pods from soil. Residue moisture can be used for sowing succeeding crops.

Weed Management

In groundnut, critical period for crop-weed competition is up to 35 DAS. There should not be any disturbance to pegs 45 DAS by mechanical or field operations. Give two hoeing, three and six weeks after sowing to check the weeds. Weeds can also be controlled with pre-plant application of Basalin 45 EC at 0.75 kg a.i./ha or pre- em application of Lasso 50 EC at 2.5 kg a.i./ha after dissolving in 500 litre of water. Integration of one hoeing (45 DAS) with either pre-emergence application of Lasso 1.25 kg a.i./ha or Stomp 30 EC 0.75 kg a.i./ha or Treflan 48 EC at 0.75 kg a.i./ha as pre-plant application provides very effective control of hardy weeds like *Acrachne racemose Commelina benaalensis* etc. in groundnut.

Plant Protection

A. Insect-Pests

Aphid: The pest becomes serious when the rainfall is low. It weakens the plant by sucking the cell sap, particularly from the growing points.

White-grub: This insect is serious in some areas only. The adult beetles emerge from the soil during June-July with the first showers of rain. They congregate on the nearby trees such as ber, guava, rukmanjani, grapevines, almonds etc. and feed on their leaves during night. The eggs are laid in the soil and the larvae (grubs) hatching from them eat away the rootlets or root hairs of the groundnut plants. The damaged plants look pale, wilted and ultimately die. Adopt the following integrated approach for its effective management:

- Plough the field twice during May-June to expose the beetles resting in the soil.
- Sow the crop early wherever possible.
- Treat the seed with 2 ml Neonix 20 FS (imidacloprid 18.5%+hexaconazole 1.5%) per kg seed before sowing. It will also control termite.

Hairy Caterpillar: Hairy caterpillar, if appearing in epidemic form cause serious damage by feeding on the leaves and tender stems. When young, they feed gregariously. The grown-up caterpillars migrate from one field to another. For management, monitor the fields regularly. Collect and destroy the egg masses and young gregariously feeding larvae. The grown up caterpillars can be destroyed by crushing them under feet or by picking and putting them into kerosenized water.

Termite: This insect damage the crop after sowing and feed on roots and shoots by making tunnels. Damaged plants wilt, dry and can be easily pulled out. Damaged plants can be differentiated from white grub damage in that the tunnels made by termites and damaged pods are filled with soil. Treat the seeds with 2 ml Neonix 20 FS (imidacloprid18.5%+hexaconazole 1.5%) per kg seed before sowing.

B. Diseases

Collar-rot and seed-rot: This disease is caused by *Aspergillus niger* and other seed and soil borne fungi which may cause seed rot before germination or the germinating seedlings may develop rotting lesion at the collar region.

Tikka or Cercospora leaf-spot: This disease is caused by *Cercospora personata* and *C. arachidicola*. Disease develops rapidly at a temperature range of 25-30°C coupled with high relative humidity. Control the above diseases by adopting the following integrated measures:

- Select healthy and unblemished kernels for seed.
- Treat the kernels before sowing with 2 ml Neonix 20 FS (imidacloprid 18.5%+hexaconazole1.5%) or 5 g Seedex 2 DS (tebuconazole) or 5 g of Thiram (tetramethyl thiurum disulphide) or 3 g of Indofil M-45 (mancozeb) per kg of kernels. Seed treatment with Neonix 20FS takes care of collar rot and white grub with single seed treatment and it also controls termites.

- Spray the crop with 1.25 to 1.88 kg per hectare Sultaf 50 WP* (wettable sulphur) in 500 to 700 litres of water. Give 3 or 4 sprays at fortnightly intervals, starting from the first week of August.
- Alternatively, spray the irrigated crop with Bavistin/Derosal/Agrozim 50 WP @ 125-150 g in 250 litres of water per hectare. Give three sprays at fortnightly intervals, starting when the crop is 40 days old.
- Do not grow groundnut repeatedly in same field.

Root knot: This disease occurs in patches and is caused by the nematode *Meloidogyne arenaria* and *M. javanica*. The affected plants show poor growth with chlorotic leaves. The root knots become stubby. The plants affected early in the season become stunted and bushy. Disease can be checked by exposing the soil to the sun during May and June to reduce nematode population. Practise green-manuring wherever possible or add organic manures to the soil.

Harvesting

The optimum harvesting time is when good percentage of pods is well developed and is fairly intact with vines, the plant turns yellows and leaf shedding starts. Maturing of pods is determined by inside of the shell turning dark, kernel reaching its normal size and the taste having normal color of the variety. When around 75% of pods mature, harvest the crop. At the time of harvesting, pods usually have moisture content around 40-50%. The pods are to be dried to about 10% moisture content for safe storage. Drying should be done rapidly to prevent fungal molding. Sun drying is usual method of drying. Summer groundnut should be dried in shade to prevent loss of seed viability, if it is for seed purpose.

6.4. Linseed (*Linum usitatissimum*)

Linseed is the oilseed crop with the highest content of the omega-3 fatty acid called alpha-linolenic acid. It is also a rich source of protein, dietary fibers, and phytoestrogens. Cultivated linseed varieties contain 45–50% oil on a dry-weight basis. The versatility of linseed is exemplified by its multiple end uses as human food, animal feed, and industrial applications. Omega-3-enriched eggs and meat are obtained by feeding linseed to poultry and livestock, and enriched beverages are produced by linseed oil supplementation. Many health benefits have been associated with the consumption of linseed.

Linseed contains anti nutritional compounds, therefore, not fit for edible purpose. Linseed oil is used for paint related industries. It contains 20-34% protein, 37-42% oil and 15-29% carbohydrates.

Origin and History

Linseed (*L. usitatissimum)* is a slender plant with narrow leaves and blue flowers that grows anywhere from 8– 45 in (20–130 cm) tall. The plant originated in India but has been farmed across the world for thousands of years. Archeologists discovered evidence that flax was cultivated in ancient Babylon as early as 3,000 B.C. Today, in Europe and Asia, a tall variety of flax is grown primarily for its fibers, which are used to make linen.

Area and Distribution

In India it is grown on an area of 0.2 mha with production of 0.12 mt during 2020. Linseed cultivation is mainly confined to Gurdaspur, Hoshiarpur and Rupnagar districts adjoining the main linseed growing area of Himachal Pradesh.

Climatic Requirements

Linseed is purely a cool season *rabi* crop. Temperate and cool climate conditions are best suited for growth. The minimum temperature regime is 10°C while the maximum is 38°C. Thus, the main season for sowing linseed is October to November depending upon the availability of soil moisture. Early sowing helps the crop to escape attack of powdery mildew, rust and linseed bud fly in different regions. The crop takes 120 to 140 days to mature depending on the cultivar. The oilseed crop needs about 25 to 30 °C during germination and vegetative phase and 15 to 20 °C during seed formation. An annual rainfall of 700 to 750 mm is optimum. Under dry climates irrigation is essential. A well-distributed rainfall of 450 to 500 mm can also suffice to grow good crop of linseed. The elevation of 400 to 500 mm from sea level is quite suitable. Drought and high temperature at early and seed filling stages are detrimental causing yield and quality reduction.

Linseed varieties released by ICAR (2020-2021)

Varieties	Area of adoption	Characters
LMS 2015-31	Rajasthan, Madhya Pradesh, Maharashtra, Odisha, Chhattisgarh, Karnataka.	Suitable for irrigated condition, average seed yield 1.26 tonnes/ha, oil content 32%, maturity 120–122 days, moderately resistant to bud fly, powdery mildew and Alternaria leaf blight.
LCK 1611	Jammu & Kashmir, Himachal Pradesh, Punjab, Haryana	Suitable for irrigated condition, average seed yield 1.34 tonnes/ha, oil content 37%, maturity 168 days, moderately resistant to rust and bud fly.
RLC 164	Jammu & Kashmir, Himachal Pradesh, Punjab, Chhattisgarh	Suitable for rainfed condition, average seed yield 1.16 tonnes/ha, oil content 32.6%, maturity 161 days, resistant to rust, moderately resistant to Alternaria blight and bud fly.
RLC 167	Jammu & Kashmir, Himachal Pradesh, Punjab, Chhattisgarh	Suitable for irrigated condition, average seed yield 1.30 tonnes/ha, oil content 34.3%, maturity 160 days (Jammu, Himachal Pradesh, Punjab), 120 days (Chhattisgarh); resistant to rust, moderately resistant to bud fly and Alternaria leaf blight.
Kota Alsi 6	Uttar Pradesh, Jharkhand, Bihar, West Bengal, Assam	Suitable for irrigated conditions, average seed yield 1.18 tonnes/ha, oil content 35%, maturity 130–135 days, moderately resistant to wilt, powdery mildew, rust and Alternaria blight.
RL 15584	Himachal Pradesh, Jammu, Punjab	Suitable for rainfed conditions, average seed yield 0.92 tonnes/ha, oil content 32%, maturity 150–155 days, moderately resistant to wilt, powdery mildew, Alternaria blight and bud fly.

Improved Varieties for Punjab

LC 2063: It is a tall variety with profuse branching and blue flowers. It is tolerant to wilt, rust, Alternaria blight and moderately resistant to powdery mildew. It has lustrous brown, bold seeds with 38.4 per cent oil content. Its average yield is 12.3 q per hectare. It matures in about 158 days.

LC 2023: It is recommended for both rainfed and irrigated conditions. It is a tall variety having blue flowers and profuse capsule bearing. It has brown medium sized seeds. It is tolerant to wilt, rust, Alternaria blight and moderately resistant to powdery mildew. Its seed contains 37.4 per cent oil. Its yield is about 11.3 q per hectare. It matures in 158 and 163 days under rainfed and irrigated conditions, respectively

Soil Requirements

Linseed can be raised in almost all types of soils, where sufficient moisture is available, but it also does better on heavier soils having greater water-retention capacity. It is also tolerant to wide range of soil pH (5.0-7.0). However, it grows best on well drained loam to clay loam soils rich in humus. In Madhya Pradesh and Maharashtra, linseed is largely raised on black cotton soils having high clay and lime content. It is also raised on light alluvial soils of Uttar Pradesh, Bihar, Jharkhand and West Bengal.

Cropping System

Usually intercropped with Chickpea, Wheat, Lentil and Potato. As pure crop it can be grown after harvesting maize, rice, moong, mash or *kharif* fodder.

Preparatory Tillage

The land should be prepared by giving two or three ploughings, depending on the intensity of weeds.

Sowing Time and Method of Sowing

Linseed is sown in the first fortnight of October with drill or pora at a depth of 4 to 5 cm in rows 23 cm apart. Plant spacing in the row is 7 to 10 cm. Linseed can also be raised without any tillage operation with zero till drill after rice.

Seed Rate

Use 37.5 kg seed per hectare. Fertilizer Application: Apply 62.5 kg N (137.5 kg of urea) and 40 kg P_2O_5 (200 kg superphosphate) per hectare at sowing. Prefer phosphorus from superphosphate.

Nutrient Management

Rainfed: 40 kg N: 20 kg P_2O_5 : 20 kg K_2O per ha

Irrigated: 80 kg N: 40 kg P_2O_5: 30 kg K_2O per ha

Under irrigated conditions, 1/2 dose of N should be applied along with Phosphorus and Potassium as basal dose at seeding by placement method. Top dressing of the remaining half of N should be applied at 35 DAS, preferably at first irrigation. These fertilizer doses depend upon soil fertility and climatic conditions.

Water Management

Flowering and grain development stages are most critical for irrigation.

Weed Management

Critical period for crop-weed competition in linseed is between 20-30 DAS. Two manual weedings, 3 and 6 weeks after seeding can keep linseed crop free from weeds. Weeds can also be controlled with the pre or post- emergence (30-35 DAS) application of isoproturon at 0.94 kg/ha.

Plant Protection

(a) Insect Pest

Lucerne caterpillar: This insect may cause damage by feeding on the leaves.

(b) Diseases

Rust (*Melampsora lini*): Pink fungal sori appear on the surface of leaves, stems and pods. Grow resistant varieties.

Wilt (*Fusarium oxysporum*): Young seedlings die when attacked. Full grown plants show yellowing of leaves and later wilt and die. Grow tolerant varieties.

Powdery Mildew (*Erysiphe cichoracearum*): A greyish white powdery growth on the youngest growing tips is the first visible symptom of this disease. In case of severe infection, the fungus attacks branches, leaves and flowers. It causes heavy loss owing to the defoliation of infected plants which leads to the shrivelling of grains.

6.5 Seasme (*Sesamum indicum)*

Sesame (*Sesamum indicum*) is crop that is grown for the oil in its seed. Compared to similar crops, such as peanuts, soybean, and rapeseed, the seeds of sesame are believed to have the most oil. Sesame seeds are also rich sources of protein, vitamins, and antioxidants.

Sesame also called benne, erect annual plant of the family Pedaliaceae, grown since antiquity for its seeds, which are used as food and flavouring and from which a prized oil is extracted. Widely cultivated, the sesame plant is found in most of the tropical, subtropical, and southern temperate areas of the world. The aroma and taste of sesame seed are mild and nutlike. The chief constituent of the seed is its fixed oil, which usually amounts to about 44 to 60 percent. Noted for its stability, the oil resists oxidative rancidity. The seeds are also high in protein and are rich in thiamin and vitamin B6.

Origin and History

The sesame plant likely originated in Asia or East Africa, and ancient Egyptians are known to have used the ground seed as grain flour. The seeds were used by the Chinese at least 5,000 years ago, and for centuries they have burned the oil to make soot for the finest Chinese ink blocks.

Area and Distribution

In India, it is grown on an area of 1.52 mha with production of 0.66 mt during 2020. Sesame (Til) was grown on 2.5 thousand hectares with a production of 1.0 thousand tonnes during 2020-21 in the Punjab. The average yield was 3.84 quintals per hectare (1.55 quintals per acre).

Climatic Requirements

Sesame requires fairly hot conditions during growth for optimum yield. A temperature of 25°C to 27°C is ideal for rapid germination, initial growth and flower formation. Low temperature at flowering results in pollen sterility and premature flower drop

Sesame varieties released by ICAR (2020-2021)

Variety	Area of adoption	Characters
RT 372	Rajasthan, Haryana, Punjab, Gujarat, Himachal Pradesh, Uttar Pradesh, Maharashtra, Nagaland	Suitable for rainfed condition, white seeded with average seed yield 0.65–0.7 tonnes/ha, maturity 85–87 days, moderately resistant to Macrophomina stem and root-rot, phyllody, Alternaria leaf spot, bacterial leaf spot and powdery mildew.
SVT 222	Haryana	Suitable for high and low fertility rainfed areas, shinning white seed colour with average seed yield 0.6–0.7 tonne/ha, oil content 46–47%, maturing 84–90 days, suitable to early and timely sown conditions, moderately resistant to Macrophomina stem and root-rot, phyllody and powdery mildew.

Improved Varieties for Punjab

Til No. 2: This variety yields 7.0 quintals per hectare. It has profuse branching and dense capsule bearing. Capsules arranged opposite to each other are long and non-hairy. Its' seeds are white, bold and have 49% oil content. Seeds have less crude fibre, are soft and better in palatability. This variety is tolerant to phyllody and cercospora leaf blight. It matures in 90 days.

RT 346: This variety yields 6.5 quintals per hectare. It has profuse branching, long non-hairy capsules arranged alternately. It has white, bold seeds which contain 49% oil. It is moderately resistant to Antigastra capsule borer. It matures in 87 days.

Soil Requirements

Sesame grows on a variety of soils but thrives best on those, which are moderately fertile with good internal drainage. Soils with neutral reaction are preferred. It will grow well on soils with a pH ranging from 5.5 to 8.0

Land Preparation

The crop requires a well-prepared seedbed. Give two or three ploughings followed by planking for getting ideal seed bed.

Time of Sowing

The crop should be sown in the first fortnight of July after receipt of adequate rain or with the application of pre-sowing irrigation. The crop sown early in June suffers from phyllody disease.

Seed Rate and Method of Sowing

The seed rate is 2.5 kg per hectare. Sow the seeds at row to row spacing of 30 cm. The seeds should be sown 4 to 5 cm deep with a pora or tube attached to the desi plough. After complete germination extra plants should be thinned to maintain plant to plant spacing of 15 cm.

Fertilizer Application

Drill 60 kg N (130 kg urea) per hectare before sowing. Avoid excessive manuring as it induces heavy vegetative growth.

Weed Management

Sesame is sensitive to weed competition during the first 15-25 DAS. Two weedings 15 and 35 DAS are optimum to minimize weed growth. Pre-emergence application of alachlor 50 EC at 1.5 kg/ha provides effective control of weeds in this crop.

Water Management

It is a low water requiring crop. In case of failure of monsoon, apply one to two irrigations to get good yield.

Plant-Protection

Insect-Pests

Sesame leaf webber and capsule borer: The larvae of these insects roll the leaves and feed inside these rolls; or bore into the capsule and feed on developing grains. If the infestation takes place at early stage, the plants die. It can be controlled by spraying the crop twice (first at pest appearance and then at flowering stage) with 250 ml Sumicidin/ Fenval/Agrofen 20 EC (fenvalerate) or 375 ml Decis 2.8 EC (deltamethrin) or thrice (at pest appearance, flowering and pod formation stages) with 500 ml Ripcord 10 EC (cypermethrin) in 250 litres of water per ha. Sow the crop at recommended time.

Jassid: It causes considerable damage to the crop by sucking the sap and transmitting mycoplasma, which induces the malformation of the inflorescence. It can be controlled by spraying the crop with 1000 ml of Malathion 50 EC (malathion) in 250 litres of water per hectare with manually operated sprayer at least two times at 2-3 weeks intervals. Avoid early sowing of crops.

Diseases

Phyllody: It is caused by a Mycoplasma like organism (MLO). The flowers are modified into leaf like structures and do not bear pods. It is transmitted by Jassid (*Orosius albicinctus*). Therefore, control of jassid is required. Also rogue out the diseased plants to prevent further spread.

Blight (*Cercospora sesami*): It appears at flowering stage as dark brown, angular lesions with grey centre on leaves. It also appears on petiole, stem and pods. The diseased plants give blighted appearance followed by defoliation. To control the disease, spray the crop thrice with Bavistin 50 WP* @ 250 g per hectare in 250 litres of water at 10 days interval, just at the appearance of the disease.

Harvesting

Depending on the variety, sesame crop takes 100-120 days for maturity. The crop is harvested when the leaves, stems and capsules begin to run yellow and the lower leaves start shedding. Average grain yield under rainfed conditions as sole crop, range between 5-8 q/ha.

6.6. Safflower (*Carthamus tinctorius* L.)

Safflower, flowering annual plant, *Carthamus tinctoris,* of the Asteraceae family; native to parts of Asia and Africa, from central India through the Middle East to the upper reaches of the Nile River and into Ethiopia. The safflower plant grows from 0.3 to 1.2 metres (1 to 4 feet) high and has flowers that may be red, orange, yellow, or white. Safflower has been used as an adulterant of the condiment saffron. Oil obtained from the seed is the chief modern use of the plant. Safflower oil does not yellow with age, making it useful in preparing varnish and paint. Most of the oil, however, is consumed in the form of soft margarines, salad oil, and cooking oil. It is highly valued for dietary reasons because of its high proportion of polyunsaturated fats. The meal, or cake residue, is used as a protein supplement for livestock.

Origin and History

Based on the existence of its close related species, safflower was believed to be originated in an area bounded by the eastern Mediterranean and Persian Gulf, encompassing southern parts of former USSR, Western Iran, Iraq, Syria, Southern Turkey, Jordan and Israel. Presently, safflower is cultivated on a commercial scale mostly in India, USA, Mexico, Ethiopia, Australia, China, Argentina, Russia and to a limited extent in Pakistan, Italy, Spain, Portugal and Iran.

Area and Distribution

In India, it is grown on an area of 0.085 mha with production of 0.044 mt during 2020. India has the larger coverage but lower yield among the major safflower producing countries. Maharashtra and Karnataka are major safflower growing States, which contribute more than 90% of India's production.

Quality Consideration

Its seeds contain 28-32% oil (traditional safflower seeds) while thin hull types contain 35-45%. Protein content ranges from 11 to 24%. Safflower oil is valued for its high linoleic acid.

Climatic Requirements

Safflower is a cool *(rabi)* season crop. The optimum temperature for germination is about 15.5°C. The day temperature in the range 24-32°C at flowering are congenial for higher yields. It is cultivated from sea level to an elevation of 1000 m above sea level. The seed yield and oil content reduce with increasing elevation. When flowering has just commenced, temperature below 0°C may cause considerable damage in the form of sterile heads. At seedling stage, however, the crop can tolerate much lower temperatures (-12 to –10 °C). High temperatures at the time of flowering are harmful to the crop. At all the stages of growth, excessive rainfall or humidity increases the infestation of fungal diseases. It is a day neutral plant. However, temperatures are more important than photoperiod in safflower. Short day lengths prolong rosette stage.

The crop is not fit for tracts of heavy rainfall. The crop grows well in areas having rainfall between 60-90 cm. Waterlogging due to poor drainage or prolonged rains can cause substantial reduction in yield. Frost is also harmful to the crop, especially at seed formation.

Safflower varieties released by ICAR (2020-2021)

Variety	Area of adoption	Characters
IGKV Kusum (RSS 2016-03)	Chhattisgarh, Madhya Pradesh	Suitable for late-sown irrigated conditions, average seed yield 2.71 tonnes/ha, oil content 34.26%, maturity 138–140 days.

Some of the other improved varieties are Annigiri, Bhima, Malavya, Manjira, Nira, Co 1, Sagarmutyalu, Sharda, Swetha, A 300, JSF 7, K 1, S 144, Girna etc.

Soil

Safflower requires fairly deep, moisture retentive and well drained soils. The crop is fairly tolerant to saline condition and grows well under residual moisture in paddy fallows.

Tillage

Tillage is very important to keep the weeds under check and to pulverize the soil for better crop emerged. Generally, 2 to 3 ploughings followed by planking are sufficient.

Sowing time

Safflower is a *rabi* oilseed crop both under rainfed and irrigated conditions. Optimum time of sowing for Safflower is Sept/October.

Seed Rate and Spacing

Seed rate vary from 7.5 to 12 kg/ha in South India. Optimum spacing is 45 × 20 cm and at depth of sowing is 5 cm. Overnight soaking of seed in water will ensure better germination.

Nutrient Management

Nutrient requirement by safflower is N = 40 kg/ha P = 25 kg/ha (P_2O_5) Drill all the fertilizer at sowing.

Water Management

Safflower is a drought resistant crop. Flowering and seed filling stages are most critical for soil moisture stress. In deep soils, two irrigations (pre-sowing and flowering are adequate for optimum yield. Flowering is very sensitive stage and stress at this stage adversely affects seed yield.

Weed management

First 60 days after planting are critical for competition from weeds. Two inter cultivations at 25-30 and 45-50 days after sowing in between the rows are optimum for keeping the crop free from weeds. Pre-emergence herbicides such as alachlor 2.0 kg/ha, fluchloralin 0.70 kg/ha or Stomp 30 EC 0.75 kg/ha may be used for weed management.

Plant Protection

Insect pests

Two species of aphid attack safflower crop from January to March.

Green peach aphid (*Myzus persicae*) It feeds on the central whorl and the plants show stunted growth and burnt appearance.

Safflower aphid (*Uroleucon compositae*): Feeds on leaves, tender shoots and stems of the plant. The infested plants become weak, remain stunted and sometimes dried up.

For the control of aphid, spray 250 ml of Rogor 30 EC (dimethoate) or Monocil 36 SL (monocrotophos) or 625 ml Coroban 20 EC (chlorpyriphos) in 250 litres of water per hectare and repeat the spray after 15 days, if necessary.

Diseases

Wilt and head rot (*Sclerotinia sclerotiorum*): Plants become yellowish, turn brown and ultimately die. Large black sclerotia of the fungus are formed on the crown, inside the stem, floral heads and adjoining roots. Shredding of the stem takes place. For its effective control, destroy the crop debris. Take the seeds from healthy crop. Do not pile up soil around stem during rains. In heavily infested soils, grow non-host crops.

Harvesting and Threshing

Safflower crop matures in 4-5 months and it should be cut with sickles at the base and stocked in the field in the form of small and well pressed heaps until they are fully dried. Threshing can be done by power operated threshers. On an average it yields 12-15 q/ha of seed.

6.7. Sunflower (*Helianthus annuus* L)

The sunflower (*Helianthus annuus*) is an annual plant with a large daisy-like flower face. Its scientific name comes from the Greek words helios ("sun") and anthos ("flower"). The flowers come in many colors (yellow, red, orange, maroon, brown), but they are commonly bright yellow with brown centers that ripen into heavy heads filled with seeds. Sunflowers are heliotropic, which means that they turn their flowers to follow the movement of the Sun across the sky east to west, and then returns at night to face the east, ready again for the morning sun. Heliotropism happens during the earlier stages before the flower grows heavy with seeds. Sunflower is grown as a crop for its edible oil and seeds. This sunflower species is also used as wild bird food, as livestock forage (as a meal or a silage plant), in some industrial applications, and as an ornamental in domestic gardens.

Origin and History

Common sunflower was one of several plants cultivated by Native Americans in prehistoric North America as part of the Eastern Agricultural Complex. Although it was commonly accepted that the sunflower was first domesticated in what is now the southeastern US, roughly 5,000 years ago, there is evidence that it was first domesticated in Mexico around 2600 BCE. These crops were found in Tabasco, Mexico, at the San Andres dig site. The earliest known examples in the

US of a fully domesticated sunflower have been found in Tennessee, and date to around 2300 BCE. Other very early examples come from rockshelter sites in Eastern Kentucky. Many indigenous American peoples used the sunflower as the symbol of their solar deity, including the Aztecs and the Otomi of Mexico and the Incas in South America. In 1510, early Spanish explorers encountered the sunflower in the Americas and carried its seeds back to Europe. Of the four plants known to have been domesticated in eastern North America and to have become important agricultural commodities, the sunflower is currently the most economically important.

Area and Distribution

In India, Sunflower was grown on an area of 0.32 mha with production of 0.21 m tonnes during 2020. In Punjab, during 2020-21, sunflower was grown on 2.5 thousand hectares with a production of 4.5 thousand tonnes and average yield of 17.95 quintals per hectare.

Climatic Requirements

Sunflower is basically a crop of temperate regions but adaptable to both cold and high temperature conditions. The crop requires cool temperature during germination and seeding growth and warm weather during flowering and seed development. Optimum ambient temperature regime is 20° to 35°C but can tolerate up to 30°C. Sunflower crop is sensitive to water-logging even for short period.

Sunflower varieties released by ICAR

Varieties	Area of adoption	Characters
Kaveri Champ	Punjab, Haryana, Bihar, West Bengal and Odisha	Tolerant to downy mildew and leaf hoppe
PDKVSH-952 (Hybrid)	Maharashtra	Rainfed and irrigated area in kharif and rabi; moderately resistant to powdery
Prabhat (Hybrid)	Andhra Pradesh	Tolerant to downy mildew

Improved varieties for Punjab

PSH 2080: It is a short duration medium tall hybrid with average plant height of 151 cm. The average seed yield of this hybrid is 24.5 quintals per hectare. The seeds of this hybrid are black and elongated with 100 seed weight of 5.8 g. The oil content of this hybrid is 43.7 per cent. It matures in 97 days.

PSH 1962: It is a short duration, medium tall hybrid with average plant height of 165 cm. The average seed yield of this hybrid is 20.5 quintals per hectare. The seeds of this hybrid are black and bold with 100 seed weight of 6.4 g. The oil content of this hybrid is 41.9 per cent. It matures in 99 days.

DK 3849: It is a tall hybrid with an average plant height of 172 cm. The average seed yield of this hybrid is 21 quintals per hectare and 100 seed weight is 4.5 g. The seeds of this hybrid contain about 34.5 per cent oil. It matures in 102 days.

PSH 996: It is a short duration medium tall hybrid with an average plant height of 141 cm. The average seed yield of this hybrid is 19.5 quintals per hectare. The seeds are black and bold with 100 seed weight of 6.8 g. The oil content of this hybrid is 35.8 per cent. It matures in 96 days. This hybrid is also suitable for growing under late sown conditions.

PSH 569: It is a short duration medium tall hybrid with an average plant height of 162 cm. The average seed yield is 18.5 quintals per hectare. The seeds are bold with 100 seed weight of 6.8 g. The seeds contain 36.3 per cent oil content. It matures in about 98 days. This hybrid is also suitable for growing under late sown conditions.

PSFH 118: It is a medium tall, short duration hybrid with an average plant height of 155 cm. The average yield of this hybrid is 19 quintals per hectare. It contains 40.5 per cent oil. It matures in about 98 days. This hybrid fits well in the late sown conditions.

SH 3322: It is a long duration medium tall hybrid with an average plant height of 160 cm. Its average yield is 20.8 quintals per hectare. It contains 43.0 per cent oil. It takes 120 days to mature

Soil Requirements

Sunflower can be grown on a variety of soils ranging from sandy loams to heavy clay soils. Deep soils with good internal drainage are ideal for rainfed crop. It grows well on soils with moderately alkaline reaction. Sunflower requires well prepared seedbed for optimum seed germination, stand establishment and growth.

Sowing Time

Sunflower can be grown although the year, provided soil moisture/ irrigation is not limiting factor. Sowing time should be decided in such a way that flowering and seed development stages do not coincide with continuous rains or temperature above 38°C. Its sowing is done both in *Rabi* as well as in *Kharif* season in India according to conditions.

Kharif: In light soils - Mid June to mid-July

Rabi **(rainfed)** – First fortnight of September to first fortnight of October.

Rabi **(irrigated)** – November to December

Spring season - Second fortnight of January to end of February.

Seed Treatment

Treat the seed before sowing with Tagron 35 WS (metalaxyl) @ 6g per kg seed.

Seeds Rate and Method of Sowing

For quick germination and optimum stand establishment, especially under rainfed conditions, the seed should be soaked in fresh water for about 14 hrs allowed by shade drying. The seed should be treated with Thiram or Captan at 2-3 g/kg to protect from seed borne diseases.

Line sowing, using seed drills are the best method of sowing. The seed is sown is furrows behind the country plough. Hand dibbling in plough furrows is also common. Optimum sowing depth is 5 cm. Recommended seed rates are given below:

Cultivars	Rainfed	Irrigated
Varieties	8-10 kg/ha	6-7 kg/ha
Hybrids	5-6 kg/ha	4-5 kg/ha

Spacing

In India, optimum row spacing vary from 45 to 60 cm. Depending upon soil types indicated below:

Soil type	Spacing (cm)
Light	45 × 20
Medium	45 × 30
Heavy	60 × 30

Wider row spacing of 60 cm is preferred for hybrids and long duration varieties. For short duration and dwarf varieties a row spacing of 45 cm is adequate.

Fertilizer Requirement

Sunflower crop responds to applied nutrients effectively. Application of FYM at 10 t/ha is very beneficial. On an average fertility soils add 60 kg N, 30 kg P_2O_5 per hectare for its successful cultivation. Apply 1/2 N and whole of P_2O_5 and K_2O at the time of sowing and remaining nitrogen 30 days after sowing. Single superphosphate is the best source of phosphorus supply. Sulphur is essential for increasing yield and oil content of seed. In sulphur deficient soils, addition of 25 kg /ha of Sulphur can significantly increase the yield.

Water Management

Sunflower crop has deep root system, extending beyond 2 m depth especially on deep soils. Sunflower crop is sensitive to water logging even for a short period. Flower opening stage is the most sensitive for soil moisture stress followed by seed filling and bud initiation stages. Sunflower has the highest water requirement during anthesis stage.

Weed Management

Critical period of crop-weed competition in sunflower is 30 to 45 DAS (days after sowing). Two inter-cultivations at 15 days interval commencing from 25 DAS hoeings effectively check the weed growth. Earthing up also help in smothering weeds Pre-emergence herbicides like Lasso at 2.0 kg a.i./ha or Stomp 30 EC at 0.75 kg a.i./ha are very effective.

Plant Protection

(a) Insect Pests

Cutworms (*Agrotis spp.*): This insect may be serious during March-April in fields where sunflower follows potato. Caterpillars cut the seedlings at the ground level. Sow the crop on ridges to avoid cutworm damage in the germinating seedlings.

Tobacco caterpillar (*Spodoptera litura*) **and Bihar hairy caterpillar** (*Spilosoma obliqua*): These are serious during April-May and defoliate the plants. Young larvae feed gregariously but the grown-up caterpillars spread throughout the field and migrate from one field to the other. To control these pests, egg masses and young larvae feeding gregariously should be collected along with infested leaves and destroyed. Grown up larvae of hairy caterpillar can also be managed by spray of 750 ml Superkiller 10 EC (cypermethrin) per hectare in 250 litres of water.

Caution: Spray should be done in the evening when the non-target organisms including honeybees and predatory arthropods are less active.

Head borer or American bollworm (*Helicoverpa armigera*): It is a serious pest of sunflower and causes heavy damage by feeding on the tissues and developing grains in the head.

(b) Diseases

Stem rot (*Sclerotium rolfsii, Sclerotinia sclerotiorum*): The pathogens attack basal part of the stem. However, when *S. sclerotiorum* is predominant, it can cover the entire plant parts, including the head as evident by the white cottony growth. The sclerotia of this pathogen, almost of the size of sunflower seed, are produced in abundance which are visible in the pith of rotten stem and infected head. The affected tissue turns black and shredded. Warm (12-20°C) and moist weather conditions favour the disease development. Use disease free seed.

Root rot or Charcoal rot (*Macrophomina phaseolina*): Infection at early stages kills the seedlings rapidly. At later stages, it causes premature ripening, small sized heads, poor filling and reduction in yield. Avoid moisture stress to the crop.

Head rot (*Rhizopus oryzae, R. stolonifer, Sclerotinia sclerotiorum*): The symptoms are visible on any part of the receptacle, usually starting from the stalk. Severely affected heads show shredding of the tissue and incomplete filling. The rot may be partial or complete. Seeds formed in affected heads are highly shrivelled causing considerable reduction in yield and oil content.

Downy mildew (*Plasmopara halstedii*): Plants remain severely stunted, leaves become thick with green and yellow patches visible from upper side. White cottony shining downy growth of fungus appears on lower surface of the leaves. Leaves become necrotic and plants develop small and empty heads. Treat sunflower seeds with Tagron 35 WS (metalaxyl) @ 6 g per kg seed. Fungicide should be applied by making slurry in water and seeds should be dried before sowing.

Harvesting

Sunflower crop matures in 90-100 days. The crop has to be harvested when the lower side of the head turns yellow, some of the bracts dry up and bottom leaves start drying and withering. Mature heads are cut with a sickle and dried in sun for 2-3 days.

6.8. Castor (*Ricinus communis* L)

Basically, Castor (*Ricinus communis L.*) is grown for its seeds. The oil extracted from Castor seed is being used widely for various purposes. It is used as a lubricant in high-speed engines and aero planes, in the manufacture of soaps, transparent paper, printing-inks, varnishes, linoleum and plasticizers. Castor seeds are also being used for medicinal and lighting purposes. The Castor may be used to make manure and plant stalks as fuel or as thatching material or for preparing paper-pulp. Castor leaves are being fed to silkworms. Castor belongs to the family of "Euphorbiaceae and genus of *"Ricinus L."*. Castor is indigenous to the Eastern Africa, south-eastern Mediterranean Basin and India, but is grown throughout tropical regions and sub-tropical regions of the world. Castor seeds contain between 45 % and 60 % of oil that is rich in triglycerides, mainly ricinolein.

Commercial growing of Castor seeds is very much successful due to its demand in local and international markets. In India, this crop is grown where annual rainfall is expected below the national average and the soil is not that fertile. Castor is a hardy plant, requires low input, tolerates marginal soils, is easy to establish in the field, is resistant to drought, and gives yield 350–900 kg oil per hectare. Castor oil shows great functional value in energy sector, industry, and pharmaceutical.

Origin and History

Varied opinions are held regarding the exact land of origin of the castor plant. Some workers view that though castor is cultivated throughout India, yet it is indigenous to Africa. Castor is native to India by some workers. They hold this view mainly on the basis of knowledge of the medicinal uses of this plant as found in Sanskrit literature. Some workers toured to India and found some evidence only at the foot of Himalayas to show that castor is a native of India. They therefore believe that it may have originated in India as well as in Africa. Hindus have known castor oil from very remote periods. This oil has been mentioned in Susruta Ayurveda, one of the oldest works on Ayurveda. It is therefore possible that castor has originated both in India and Africa.

Area and Production

Castor was grown on an area of 0.85 mha with production of 1.84 m tonnes during 2020.

Quality Considerations

Oil content of castor seed ranges from 40-45%. It contains 12-16% protein, 2-7% carbohydrates and 23-27%, fibre. Dehulled seed contains 60-70% oil and 18-26% protein. It contains 85-90% linolenic acid. Castor cake contains 6% N 2.5% Phosphorus and 1.5% Potassium. It is unfit for edible purpose because of toxic constituents i.e. an alkaloid recinine which is mildly toxic and a protein, which is extremely poisonous.

Climatic Requirements

Castor being warm season crop, grown in temperate and tropical regions of the world. Castor crop require moderately high temperature of 20°C-26°C with low humidity throughout the growing season for maximum yield. Castor crop is very sensitive to high temperature and excessive soil moisture.

Castor variety released by ICAR (2020-2021)

Variety	Area of adoption	Characters
Jawahar Castor 26 (JC 26)	Madhya Pradesh	Suitable for both rainfed and irrigated conditions, average seed yield 1.88 tonnes/ha, oil content 46– 49%, medium maturity (100–130 days), moderately resistant to Fusarium wilt and root-rot.

Seeds Treatment and Seed Rate

Seeds may be treated with Thiram or Bavistin (3 g/kg seed) to protect from seed borne alternaria blight and wilt. Ideal time for seeding *Kharif* castor is immediately after the receipt of monsoon rains. Optimum seed rate is 10-12 kg/ha. For improved castor varieties/hybrids recommended spacing is 90×90 cm for irrigated crop and 75×75 cm spacing for rainfed crop. Under delayed sowing plant population should be increased.

Nutrient management

For rainfed castor varieties: 40-60 kg N/ha: 40 kg P_2O_5/ha

Irrigated castor: 80-100 kg N/ha: 40 kg P_2O_5/ha

For irrigated *rabi* castor 30 kg N must be applied at sowing and the remaining recommended 60 kg in three splits of 20 kg each around 30-35, 60-65 and 90-95 DAS. Phosphorus and Potassium must be applied at sowing by placement. For rainfed crop all P_2O_5 and 1/2 N at sowing and remaining nitrogen with subsequent rainfall.

Water Management

Castor is usually a rainfed crop. However, it responds well to irrigation. If one or two irrigations are given during drought spells, castor yield can be doubled. Critical period is flowering and seed development stages and irrigation at these stages should never be missed.

Weed Management

Critical period of crop weed competition is 20-60 DAS. Two to three hoeing / cultivations starting from 20 DAS are required for optimum yield. Pre-emergence application of alachlor 2.0 kg/ha or pendimethalin 0.75 kg/ha are effective against weeds in castor. Also, fluchoralin at 0.75 kg/ha as pre-plant incorporation is equally effective for weed control in castor.

Plant Protection

Spraying Thiodicarb 1.0 g/l, Spinosad 45 EC at 0.3 ml/l or Phosalone 35 EC at 2 ml/l can control castor semilooper and capsule borer. Thiram at 3 g/kg seed can minimize seedling blight infection. Alternaria leaf blight can be managed by spraying Mancozeb (0.25%) at an interval of 15 days starting from 90 DAS,

Harvesting and Threshing

Harvesting of the mature spikes over a span of 180-240 days, each at an interval of 30 days. Harvested spikes are placed in heaps and sun dried for a couple of days. Threshing is done by beating with sticks or under the feet of cattle and then stored in gunny bags.

7

Latest Practices for Cultivation of Pulses

Pulses are high in protein, fibre, and various vitamins, provide amino acids, and are hearty crops. They are most popular in developing countries, but are increasingly becoming recognized as an excellent part of a healthy diet throughout the world. Pulse crops are one of the most sustainable crops a farmer can grow. It takes just 163 litres of water to produce 0.45 kg (1 pound) of pulses, compared with 818 litres for soybeans. They also contribute to soil fertility by fixing nitrogen in the soil.

Though pulses are a very popular crop in the developing world, there is a massive gap in productivity between pulse crops inside and outside the developing world. With the introduction of improved varieties and promotion of better management techniques, pulse crops can continue to be an excellent choice for farmers in the developing world. Up to 25% of pulses are used as feedstuff, particularly for pigs and poultry. As a steady source of nutrition, feed for animals, and soil sustainability, pulse crops play a major role in food security, a role which will only grow in the future.

In 2018, the world produced an estimated 92.4 million MT of pulses. Between 1998 and 2018, pulse production increased by nearly 36 million MT (or 63 percent). That compares to an increase of only 12.5 million MT between 1978 and 1998. The expansion in pulse production over the past two decades was particularly impressive. Substantial increases have been achieved in the production of common beans (+14 million MT), chickpeas (+8.3 million MT), lentils and pigeon peas (+3.6 million MT each), cowpeas (+3.5 million MT) and dry peas (+1.2 million MT). In terms of regions, production increased by 17.3 million MT in Asia (with South Asia contributing 12 million MT of that total), 10.5 million MT in sub-Saharan Africa, 4.8 million MT in North America and 2.2 million MT in Latin America and the Caribbean.

Total area under all pulse crop during 2020 was 33.18 mha with production of 23.37 mt in India.

7.1 Chickpea (*Cicer arietinum* L. / *Cicer kabulium*)

The name chickpea (*Cicer arietinum)* comes from the Latin word cicer, referring to the plant family of legumes, Fabaceae. It is also known by its popular Spanish-derived name, the garbanzo bean. Kidney beans, black beans, lima beans, and peanuts are other familiar foods found in this legume family. These plants produce edible seeds, called pulses, that have high nutritional value. Two main varieties of chickpeas are the larger round light-coloured Kabuli-type, common in the United States, and the smaller dark irregularly shaped Desi-type often used in India and the Middle East. Chickpeas appear in early recordings in Turkey about 3500 BCE and in France 6790 BCE. India produces the most chickpeas worldwide but they are grown in more than 50 countries. An excellent source of carbohydrate, protein, fiber, B vitamins, and some minerals, they are a nutritious staple of many diets.

History and Origin

Chickpeas are one of the earliest cultivated legumes. Remains of chickpeas from the Middle East have been found that are roughly 7,500 years old. These remains were found in the aceramic levels of Jericho and Çayönü, Turkey, meaning that humans had been cultivating chickpeas since before they could produce pottery. Other samples have been found in Neolithic pottery in Hacilar, Turkey, and appear throughout history in Greece, France, and other areas of Europe.

Area and Distribution

Chickpea was grown on an area of 10.95 mha with production of 11.08 mt in India during 2020. In Punjab it was grown on 2.0 thousand hectares with production of 2.6 thousand tonnes during 2020-21. The average yield was 13.22 quintals per hectare.

Health Benefits

Chickpeas and all pulses contain several components that, when eaten as part of a balanced plant-rich diet, may help prevent the development of various chronic diseases.

- **Diabetes:** Both dried and canned chickpeas have a low glycaemic index and low glycaemic load, and contain amylose, a resistant starch that digests slowly. These factors help to prevent sudden surges in blood sugar and insulin levels, which can improve overall blood sugar control in people with type 2 diabetes.
- **Gut Flora:** Chickpeas contain a soluble fiber called raffinose, a type of oligosaccharide that is fermented in the colon by beneficial bacteria called Bifidobacterium. As bacteria break down this fiber, a short chain fatty acid called butyrate is produced. Butyrate plays a role in reducing inflammation in the cell wall of the colon, promoting regularity in the intestines, and possibly preventing colorectal cancer by promoting cell apoptosis (death).
- **Heart Disease:** Chickpeas contain a plant sterol called sitosterol that is structurally similar to cholesterol in the body. It interferes with the body's absorption of cholesterol and thereby can help to lower blood cholesterol levels. The fiber and unsaturated fats in chickpeas may also favourably affect blood lipid levels.
- **Obesity:** High fiber foods can help to promote a feeling of fullness and satiety by delaying digestion and adding bulk to meals. The satiating effect of the high fiber and protein content of chickpeas may help with weight management

Climatic Requirements

Chickpea is a cool season food legume and grown as a winter crop in the tropics and as a summer or spring crop in the temperate environments. It likes cool, dry and bright weather. Temperature, day length and availability of moisture are the three major abiotic factors affecting flowering. In general, flowering is delayed under low temperatures and also under short-days. Genetic variability exists in chickpea germplasm for response to variation in day length (photoperiod sensitivity) and also for response to variation in temperature (thermal sensitivity) and has been exploited in development of short-duration cultivars. Chickpea is sensitive to high (maximum daily temperature >35°C) as well as low (mean of maximum and minimum daily temperatures <15°C) temperatures at the reproductive stage. Both extremes of temperatures lead to flower drop and reduced pod set.

Chickpea varieties released by ICAR (2020-2021)

Variety	Area of adoption	Characters
Jawahar Gram Kabuli 6 (JGK 6)	Punjab, Haryana, Uttar Pradesh, Rajasthan, Jammu and Kashmir	Large seeded (40.3 g/100 seeds) Kabuli chickpea variety suitable for timely-sown irrigated condition, average seed yield 1.70 tonnes/ha, maturity 143 days.
Raj Vijay Gram 204 (RVG 204)	Madhya Pradesh, Maharashtra, Chhattisgarh, Parts of Rajasthan, Gujarat	Suitable for timely-sown irrigated condition, average seed yield 2.25 tonnes/ha, maturity 110 days, 100-seed weight 23.8 g, amenable to mechanical harvesting.
RG 2015-08	Chhattisgarh, Madhya Pradesh	Suitable for cultivation under timely-sown irrigated condition, average seed yield 2.23 tonnes/ha, maturity 106 days, 100-seed weight 25.46 g.
PBG 9	Uttar Pradesh, Bihar, Jharkhand, Assam, West Bengal	Suitable for cultivation under timely sown irrigated condition, average seed yield 1.75 tonnes/ha, maturity 132 days, 100 seed weight 25 g, moderately resistant to wilt.
PDKV-Kanak	Maharashtra, Gujarat, Madhya Pradesh	Suitable for cultivation under timely-sown irrigated condition, average seed yield 2.48 tonnes/ha, maturity 132 days, 100-seed weight 21.74 g, moderately resistant to wilt.
Pusa Chickpea 20211	Madhya Pradesh, Maharashtra, Gujarat	Suitable for timely-sown irrigated condition, average seed yield 2.29 tonnes/ha, maturity 108 days, 100-seed weight 19.5 g, Fusarium wilt resistant introgression line (MAS) of desi chickpea variety Pusa 391
Raj Vijay Gram 210 (RVG 210)	Madhya Pradesh	Suitable for semi-irrigated/irrigated condition, average yield 1.81 tonnes/ ha, maturity 105 days, 100-seed weight 26.7 g, resistant against Fusarium wilt.
Raj Vijay Kabuli Gram 121 (RVKG 121)	Madhya Pradesh	Kabuli variety suitable for semi- irrigated/ irrigated areas, average seed yield 1.97 tonnes/ ha, maturity 114 days,100-seed weight 27.8 g, resistant against Fusarium wilt.
RLB Chana Kabuli 1 (RLBGK 1)	Andhra Pradesh, Telangana, Karnataka, Tamil Nadu, parts of Odisha	Suitable for irrigated timely-sown condition, average seed yield 1.5 tonnes/ha, maturity 94–104 days, 100-seed weight 36 g, moderately resistant to wilt.

Improved Varieties for Punjab

Desi gram (Irrigated)

PBG 8: The plants are tall with semi-erect growth habit. It has medium bold seeds with brownish colour and 100-seed weight is 16.5 g. This variety is moderately resistant to Botrytis grey mould and wilt. It matures in about 158 days. Its average yield is 21.0 quintals per hectare.

PBG 7: The plants are tall, semi-erect. It has medium bold seeds with brownish colour and 100-seed weight of 16 g. This variety is moderately resistant to Ascochyta blight (Chanani) and fairly resistant to wilt and dry root rot. It matures in about 159 days. Its average yield is 20 quintals per hectare.

PBG 5: It has thick stem with dark green foliage. It matures in about 165 days. Its seeds are medium bold (18 g/100 seeds weight) with dark brown appearance. This variety is fairly resistant to Ascochyta blight (Chanani) and wilt complex (wilt, root rot and foot rot). Its average yield is 17 quintals per hectare.

GPF 2: The plants are semi-erect with lush green leaves and long fruiting branches with two seeds per pod. It has bold seeds and matures in about 170 days. Its average yield is 19 quintals per hectare.

Desi gram (Rainfed)

PDG 4: Its plants are erect with dark green foliage. It bears long fruiting branches arising from the base. It has bold seeds and matures in about 160 days. It is fairly resistant to wilt, foot rot, root rot and Ascochyta blight. Its average yield is 19.5 quintals per hectare.

Kabuli gram

L 552: It is early in flowering and matures in 157 days. It is tall and erect variety. It has large pods and bold seeds (33.6 g/100 seeds weight). The seeds are creamy white in appearance with good culinary properties. Its average yield is 18.3 quintals per hectare.

Classification

The Indian gram has been classified into two broad groups:

1. **Desi or Brown gram (*Cicer arietinum L.*):** In this group the colour of the seed ranges from yellow to dark brown. Seed size is small. Plants are small with good branching ability.
2. **Kabuli or white gram (*Cicer kabulium*):** In this group the colour of the seed is white. Grains are bold and attractive. Yield potential of this group is poor as compared to desi or brown gram. Plants are generally taller than the desi gram and stand more or less erect.

Soil Requirements

Chick pea is grown on a wide range of soils in India. In the north, gram is generally grown on moderately heavy soils. In Maharashtra and on the Deccan plateau, black cotton soils are used. Light soils, mostly sandy loams are preferred in Punjab, Uttar Pradesh, Haryana and Rajasthan. Though gram is grown on all kinds of soils, sandy loam to clay loam is considered to be most suitable. The best type of soil for chick pea is one that is well drained and not too heavy. On dry and light soils, the plants remain short while on heavy soils having high water retention capacity, the vegetative growth is abundant, light becomes limiting and fruiting is retarded. The soil chosen for its cultivation should be free from excessive soluble salts and near neutral in reaction. However, it is not suited to soils having a pH higher than 8.5.

Cropping Systems

Chickpea is sown after the harvest of *kharif* crops. The most common cropping systems are as below:

- *Kharif* fallow-Chickpea
- Maize-Chickpea
- Pearl millet-Chickpea
- Sorghum-Chickpea
- Rice-Chickpea
- Rice-Chickpea-Summer moong beans

Chick pea can be grown mixed with wheat, barley, linseed, rapeseed and mustard crops. It is grown mixed with toria in tarai region.

Field Preparation

A rough seedbed is required for chick pea. In case the chick pea crop is taken after a *kharif* fallow, it would be desirable to go for a deep ploughing during the monsoon as it helps in conservation of rain water in the soil profile for use by this crop. Very fine and compact seedbed is not good for chick pea. It requires a loose and well aerated seedbed.

Time of Sowing

Date of sowing is a very important factor affecting the yield of chick pea in all chick pea growing States. Second fortnight of October is the optimum time for sowing chick pea in northern India. For peninsular India, first fortnight of October is the best time for chick pea sowing. Delay in planting beyond optimum period can cause considerable yield reductions as well as reduction in nodule dry weight. Under humid subtropical conditions of tarai, which are characterized by shallow water table and relatively more winter rainfall, first fortnight of November is most suitable. Early sowing of chick pea results in excessive vegetative growth and poor setting of pods. The early sown crop suffers more from wilt owing to high temperature at that time. While delayed sowing restricts vegetative growth and pod-bearing branches, decreases biological nitrogen fixation and may also lead to forced maturity as well as increased incidence of pests.

Seed Treatment

I. Fungicide: Treat the seed with a fungicide Bavistin or Captan at 3.0g or Rovral 2.5g per kg of seed.

II. Insecticide: In termite infested soil, treat the seed before sowing with Chlorpyrifos 20 EC at 10 ml per kg seed.

Inoculation: Inoculate the seed with the recommended Rhizobium (RB-1) and Mesorhizobium (LGR-33) biofertilizer at the time of sowing to supplement the nitrogen supply for getting higher productivity of gram. Moisten the seed recommended for one acre with minimum amount of water. Mix thoroughly one packet each of biofertilizer cultures with seed on clean pucca floor and let it dry in shade. Sow the seed immediately. Rhizobium and fungicide can be applied simultaneously. If the seed is to be treated with biofertilizer, insecticide and fungicide, then first apply insecticide followed by fungicide and biofertilizer.

Seed Rate and Sowing

A seed rate of 60-100 kg depending upon the seed size and sowing time is sufficient for one hectare. The crop may be sown with seed drill or local plough at a row spacing of 30-45 centimetre. In rainfed areas, deep sowing (10 cm) of chick pea proved better than shallow depth (5 cm). Further, deep planting promotes better root growth and can reduce incidence of Fusarium wilt as compared to shallow planting. Increase seed rate under late sown conditions to obtain good yield. In case chickpea is to be sown after paddy, prefer to sow on beds for higher yield.

Nutrient Management

Chick pea being a leguminous crop fulfils the major part of its nitrogen requirement (about 75 per cent) through the process of symbiotic nitrogen fixation which starts effectively from three to four weeks after sowing. However, it requires 20-25 kg per hectare of nitrogen as starter dose which can meet plant requirement before the formation of nodules. The chick pea responds to phosphorus application and its requirement is more under irrigated conditions than that of rainfed conditions. Generally, chick pea has been found to respond up to 30-60 kg P_2O_5/ha. The response of chick pea to potassium application is meagre as most of the soils in chick pea growing areas are rich in available potassium. All the fertilizers should be drilled in furrows at a depth of 7-10 centimetre at the time of sowing. On S-deficient soils, the response to sulphur has been found in the range of 20-60 kg S/ha. Amongst micronutrients, the significant response to application of Zn, B, Mo and Fe can be found at many locations depending upon soil type. Apply 25 kg/ha zinc sulphate or 10 kg/ha Borax or 10 kg/ha Ammonium molybdate as per need.

Water Management

Chick pea is capable of drawing water from depths greater than 150 cm (Singh and Bhushan 1979) but a major proportion of total extractable water comes from the top 60 cm of the soil profile which has the main concentration of active roots. Chick pea is mostly sown as a rainfed crop. However, where irrigation facilities are available, give a pre-sowing and also irrigation which will ensure proper germination. If winter rains fail, give one irrigation at pre-flowering and one at pod development stage. In no case first irrigation should be given earlier than four weeks after sowing. Irrigation should not be given at the flowering stage. Excess of irrigation enhances vegetative growth and depresses chick pea yield.

Weed Management

Chick pea being short statured crop suffers severely with the infestation of weeds. One hand weeding or interculture with hand hoe or wheel hoe after 25-30 days and second hoeing if needed can be done after 60 days of sowing. Trifluralin (Treflan 48 EC) 1.2 kg per hectare in 500-600 liters of water as pre-planting spray may be used as an effective herbicide. It should be well incorporated in soil before sowing. In case Treflan is not available use pendimethalin (Stomp 30 EC) at 0.75 kg/ha as pre- emergence by dissolving in 500 to 600 liters per hectare of water.

Plant Protection

(a) Insect Pests

Termite: Termites attack the crop especially at seedling stage and also near maturity. The pest can generally be observed feeding on roots or near the root zone of the damaged plants. The affected plants dry up and can be pulled out easily. The incidence of pest is more in light soils.

Gram caterpillar (*Helicoverpa armigera*): The larvae damage the gram crop by feeding on leaves, flower buds, flowers, pods and grains in the pods. Monitor the gram crop at pod initiation stage to record larval population of gram caterpillar on per meter row length basis. Record the observations by gently shaking the plants from a minimum of 25 randomly selected spots per hectare. If a total of 40 or more larvae are observed from 25 spots, spray 800 g *Bacillus thuringiensis* var. kurstaki 0.5 WP (DOR Bt-1) or 500 ml Helicop 2% AS (HaNPV) or 125 ml Coragen 18.5 SC (chlorantraniliprole) or 200 g Proclaim 5 SG (emamectin benzoate) or 400 ml Rimon (novaluron) in 200-250 litres of water per hectare and repeat after two weeks, if necessary. Prefer to use biopesticides as first spray for younger larvae and repeat the spray after a week if necessary.

Precaution: Before consuming leaves and green grains, ensure a waiting period of 3 days after the spray of Coragen 18.5 SC (chlorantraniliprole).

(b) Diseases

Ascochyta blight/Chanani (*Ascochyta rabiei*): Dark-brown spots studded with black dot-like bodies are produced on the stem, branches, leaflets and pods. The lesions are elongated on stems, branches and twigs. On pods and leaves, these black dot-like bodies are concentrically arranged in the spots. Shoot terminals are especially liable to attack. Even the seeds in the pods are infected. In the event of excessive rains, whole crop may be blighted and killed rapidly. A temperature of 20° C, relative humidity above 85% and cloudy weather with intermittent rains is highly conducive for development of disease. Grow comparatively resistant varieties. After harvest, the diseased plants should not be allowed to stand in the field and should be destroyed. Use disease free seed.

Grey mould (*Botrytis cinerea*): Small water-soaked spots are produced on leaflets. Spots on infected leaves become dark brown. Under high humid conditions, white mycelial strands and erect sporophores of the fungus are produced on flowers, leaves, growing tips, branches and pods. At the point of infection, soft rotting of the tissue occurs. Flowers and growing tips are more vulnerable. The fungus forms dark grey to black sporodochia on infected tissue. The most suitable weather conditions for development of disease are temperature around 25°C, high relative humidity and rain during flowering. To manage this disease, sow disease free seed.

Wilt (*Fusarium oxysporum f. sp. ciceri*): Affected plants show drooping of petioles and dull green colour in the initial stage. Slowly, all the leaves turn yellow and later become straw coloured. The most characteristic symptom of this disease is vascular dis-colouration of roots which is dark brown to black. Sometimes, only one-sided branches are affected resulting in partial wilting. The soil temperature of 24-27°C favours disease development. Grow wilt tolerant varieties of desi and kabuli gram. Conserve soil moisture in rainfed areas. Plough the field with a furrow turning plough to eradicate weeds and promote water infilteration. Early sowing of the crop should be avoided.

Stem rot (*Sclerotinia sclerotium*): The disease attacks all the above ground parts of the plant. The main stem is usually affected at the soil level. The affected parts are shredded and covered with white mycelium embedded with black sclerotia. The whitish, fluffy fungal growth can also be seen on the debris at soil surface. The disease is predominant in cool and moist conditions. The seed should be free from sclerotia. Sow the gram crop from 25 October to 10 November to escape the disease attack. Collect and destroy the diseased debris soon after harvesting the crop. In the month of May or June, flood the field after deep ploughing and do not leave it fallow. Rotate gram with non-susceptible crops such as wheat and barley.

Foot rot: (*Opercullela padwickii*): Light brown to dark-brown lesions appear on the collar region of the plants. Later, these lesions become black and affect the basal tap root. Complete girdling in the collar region takes place. Sow the crop from 25 October to 10 November. Follow crop rotation with non-susceptible crops such as wheat and barley.

7.2. Pigeonpea (*Cajanus indicus*)

Pigeonpea (*Cajanus cajan*) is the sixth most important edible legume crop. Pigeonpea is a member of the Fabaceae family and has many wild relatives. These include *C. scarabaeoides, C. sericeus, C. acutifolius, C. albicans, Rhynchosia aurea, R. bracteata* and *Flemingia bracteata.* Pigeonpea is found in both annual and perennial forms and can reach a height of 1–3 meters. Because it has deep roots, it is widely adapted and grows well in semi-arid areas. It has slender, pointed trifoliate leaves and yellow or yellow and red flowers. Fresh pigeon peas can be used similarly to fresh peas, in both raw and cooked preparations. Use them in salads, slaws or eaten raw out of hand. The dried beans can be substituted for any other dried pea, bean or lentil. They should be soaked first, and then can be simmered, sauteed, steamed and fried.

India, Myanmar, Malawi, Tanzania and Kenya are among the world's top producers for pigeonpea. It is used for both human nutrition and animal fodder. As with other legumes, it is sometimes used as green manure. Due to its tolerance to dry and hot conditions, it is an important crop in semi-arid conditions.

History and Origin

Pigeonpea was domesticated in India about 3500 years ago. It spread rapidly to East and West Africa and thence via the slave trade to the Americas.

Area and Distribution

In India, it is grown on an area of 5.01 mha with production of 3.89 mtonnes during 2020. Arhar was grown on 1.8 thousand hectares and the total production was 2.1 thousand tonnes during 2020-21 in Punjab. It gave an average yield of 11.63 quintals per hectare.

Nutritional content of Pigeon Pea (per 100 gm)

Principle	Nutrient Value	Percent of RDA
Energy	343 Kcal	17%
Carbohydrates	62.78 g	48%
Protein	21.70 g	39%
Total Fat	1.49 g	7.50%
Cholesterol	0 mg	0%
Dietary Fiber	15 g	39%
Vitamins		
Folates	456 µg	114%
Niacin	2.965 mg	18.50%
Pyridoxine	0.283 mg	22%
Riboflavin	0.187 mg	14%
Thiamine	0.643 mg	54%
Vitamin-C	0 mg	0%
Electrolytes		
Sodium	17 mg	1%
Potassium	1392 mg	30%
Minerals		
Calcium	130 mg	13%
Copper	1.057 µg	117%
Iron	5.23 mg	65%
Magnesium	183 mg	46%
Manganese	1.791 mg	78%
Phosphorus	367 mg	52%
Selenium	8.2 µg	15%
Zinc	2.76 mg	25%

Climatic Requirements

Pigeonpea is predominantly a crop of tropical areas mainly cultivated in semi-arid regions of India. Pigeonpea can be grown with a temperature ranging from 26°C to 30°C in the rainy season (June to October) and 17°C to 22°C in the post rainy (November to March) season. Pigeonpea is very sensitive to low radiation at pod development, therefore flowering during the monsoon and cloudy weather, leads to poor pod formation.

PigeonPea varieties released by ICAR (2020-2021)

Variety	Area of adoption	Characters
LRG 133-33	Tamil Nadu, Karnataka, Andhra Pradesh and Telangana	Suitable for rainfed and irrigated condition in kharif, average seed yield: 1.54 tonnes/ha, maturity 164–174 days, resistant to fusarium wilt and sterility-mosaic disease, tolerant to pod-fly and pod-borer.
Hybrid: IPH 09-5	Punjab, Haryana, Western Uttar Pradesh, Rajasthan, Delhi	Suitable for rainfed condition in kharif, average seed yield 1.82 tonnes/ha, maturity 150 days, resistant to fusarium wilt and tolerant to sterility mosaic disease, Phytophthora stem blight, pod-fly and pod- borer complex.
IPA 15-2	Eastern Uttar Pradesh, Bihar, Odisha, Jharkhand, West Bengal, parts of Assam	Suitable for rainfed condition in kharif, average seed yield 2.27 tonnes/ha, maturity 240–255 days, resistant to fusarium wilt and sterility mosaic disease and tolerant to pod-borer.
WRGE 121	Karnataka, Telangana, Andhra Pradesh, Odisha, Tamil Nadu	Suitable for rainfed condition in kharif, average seed yield 1.7 tonnes/ha, maturity 155–165 days, resistant to fusarium wilt and sterility mosaic disease and tolerant to pod-borer complex, resistant to H. cajani nematode
KRG 33	Madhya Pradesh, Maharashtra, Gujarat, Rajasthan, Chhattisgarh	Suitable for irrigated/rainfed condition in kharif, average seed yield 2.0–2.1 tonnes/ ha, matures in165 days, resistant to wilt and sterility mosaic disease and tolerant to pod-fly and pod- borer.

Improved Varieties for Punjab

AL 882: It is short statured and early maturing variety with semi-determinate growth habit. It matures in about 132 days and vacates the field well in time to sow the succeeding wheat crop. Its' plants are compact and about 1.6 to 1.8 metre tall. Pods are borne at the top of the plant in multiple and sparse clusters. Podding is profuse and each pod contains 3-5 yellowish brown and medium sized seeds. The average grain yield is about 13.5 quintals per hectare.

PAU 881: It is an early maturing variety with indeterminate growth habit. It matures in about 132 days and vacates the field well in time to sow the succeeding wheat crop. Its' plants are about 2 metre tall. Pod formation is profuse and each pod contains about 3-5 yellow brown and medium sized seeds. Its' average grain yield is about 12.8 quintals per hectare.

Classification

Numerous types of *Cajanus spps.* are known, differing in height, habit, time of maturity, colour, size and shape of pods and seeds. All these cultivated types fall into two groups:

1. ***Cajanus indicus* var. *bicolor*:** This group includes generally late maturing varieties, having tall bushy plants, flowers yellow, purple streaked pods dark coloured each with four or five seeds.
2. ***Cajanus indicus* var. *flavus*:** This group includes early maturing varieties having smaller plants, yellow flowers, plain pods with two to three seeds.

Soil Requirements

Pigeon pea may be grown well, on a wide range of soils varying from sandy loams to clay loams. It does best on fertile and well drained loamy soils. The saline-alkaline and waterlogged soils are unfit for its cultivation, as they adversely affect nodulation. Well drained, alluvial and loamy soils of Madhya Pradesh and Maharashtra are highly suitable. It grows well in well drainage red clay loam soil too.

Crop Rotation

Pigeon pea is grown in summer and *kharif* in northern India and *kharif* and *rabi* in southern Indian conditions. It may, therefore, be included in several crop rotations. The important rotations are given as under:

- Pigeon pea - Wheat
- Pigeon pea – Barley
- Pigeon pea – Raya
- Sorghum/Bajra/Maize – *Rabi* Pigeon pea

Field Preparation

Pigeon pea is a very deep tap rooted crop capable of penetrating root system in hard soils. It does not require special land preparation. Deep ploughing to a depth of 15 cm is adequate to obtain a good germination. When the crop is grown in rainy season, care should be taken to level the land to prevent waterlogging. The land should be levelled and soil should be pulverized for placement of seeds at uniform depth. In general, a deep cultivation either by ploughing or disking, followed by 2-3 harrowings and levelling's will be sufficient.

Time of Sowing

Pigeon pea is a traditionally *kharif* crop sown in June-July with onset of monsoon in various agro-climatic zones of India. Short duration varieties of pigeon pea are now becoming popular in the irrigated area. These varieties are harvested much earlier than the occurrence of frost/cyclones and can also be fitted well in multiple cropping systems. Planting of early pigeon pea before the onset of monsoon in the month of June is recommended for higher yields. Planting with onset of monsoon was the optimum time because with earlier plantings, the reproductive phase coincides with the period of heavy rainfall which ultimately caused drop of flowers and pods. Delay in planting reduces the yield of pigeon pea. Delay in planting, caused reduction in the duration of crop, plant height, number of branches and pods.

Seed Treatment

Treat the seed with Captan or Thiram @ 3.0 g/kg seed against seed-rot and seedling-blight. Also, inoculate the seed with the recommended Rhizobium culture before sowing. Mix thoroughly recommended dose of Rhizobium culture with seed on a clean pucca floor after wetting the seeds with water. Dry seeds in shade and sow in the field immediately. The inoculation of seed increased the seed yield by 5 to 7%. Rhizobium and fungicide can be mixed simultaneously.

Seed Rate and Spacing

In general, a seed rate of 8-10 kg/ha for long duration, 10-12 kg/ha for medium duration and 12-15 kg/ha for short duration varieties is sufficient in pigeon pea. Long duration varieties of pigeon pea are tall, spreading and occupy the field for about 250-270 days. These varieties are planted at wider row spacing of 90-120 cm and about 30 cm between the plants particularly under rainfed conditions. Under irrigated conditions, early maturing varieties are more popular as they fit well in double cropping systems with other crops. These varieties are planted at a row spacing of 50-75 cm and plant to plant spacing of 15-20 cm.

Method of Sowing

Line sowing is superior over broadcasting. Sow the crop 50 cm apart with plant to plant spacing of 25 cm. Broadcasting results in uneven plant population which ultimately results in low yield. In such areas where temporary water logging take place, planting on ridges have been found superior.

Mixed/Intercropping

Pigeon pea is mainly grown as mixed with other traditional crops like urdbean, greengram, groundnut, small millets etc. In this system, seeds of different crops are mixed together and broadcasted. Pigeon pea can profitably be grown as intercrop with several widely spaced crops like black gram, maize, sorghum, etc. One or two rows of pigeon pea could be sown in between two rows of maize. Paired planting can also be adopted in pigeon pea where two rows are made closer and thus this extra space is provided in between two pairs of rows. The overall plant population on area basis is not affected but by the adjustment of rows more space is provided in between two pairs which is utilised by any intercrop.

Weed Management

The growth of pigeon pea is quite slow during first 45-60 days after sowing. On the other hand, weed growth is very fast due to frequent rains. During this period, if weeds are not controlled, they cover the pigeon pea plants. Weeding by mechanical means sometimes becomes very difficult because of frequent rains due to which soil does not come in physical condition to permit use of the implements. Under such conditions, use of herbicides to control the weeds becomes more important. Two weedings, one at 25 days and another at 45 days after sowing, were found to be sufficient. Among herbicides, pre-emergence application of Stomp 30 EC at 0.75 kg a.i. ha^{-1} or Lasso 50 EC at 2.0 kg a.i./ha were found to be quite effective. Intercropping with short duration legumes like urdbean, green gram and cowpea is another biological way of controlling weeds in pigeon pea. However, this practice can only reduce the infestation of weeds to some extent by occupying the space in between the rows of pigeon pea. However, weeds can cause losses to a great extent under intercropping system also.

Fertilizer Management

Being a leguminous crop, pigeon pea possesses atmospheric nitrogen fixing ability. Therefore, the response to added nitrogen is limited. However, in soils poor in available nitrogen status, a starter dose of nitrogen to support initial growth till nodule formation and biological nitrogen fixation is essential. Pigeon pea when grown on soils having higher available nitrogen reserves generally do not respond to nitrogen. In general, application of about 15-20 kg N/ha applied at sowing has been found adequate. On an average, application of 50-60 kg P_2O_5/ha is optimum. The dose may vary with the soil type i.e. light soils need higher dose and heavy soils need lower dose. Phosphorus applied to preceding crop may also affect the rate of phosphorus application in pigeon pea.

Water Management

Pigeon pea is mainly grown in the monsoon season under rainfed conditions. Being a deep-rooted crop, it is capable of extracting moisture from deeper layers in the soil. However, under long period of moisture stress, it may respond to irrigation as well. Two irrigations, one at pre-flowering and other at flowering stage are beneficial in increasing the yield of pigeon pea under rainfed conditions.

Plant Protection

A. Insect Pests

Blister beetle: Blistle beetle is diurnal and general feeder. Adult beetles are robust with bright black and red stripes on the forewings. When disturbed, the beetles emit a fluid containing cantharidin that causes blisters on human skin. The adult beetles attack arhar, moong, mash and other pulse crops. The major damage is caused at the flowering stage. They feed on tender buds and flowers of the plant, thus preventing pod formation.

Pod borer complex: Pod borers, especially spotted pod borer (*Maruca vitrata*) and gram pod borer (*Helicoverpa armigera*) are the most important insect pests of arhar. *Maruca vitrata* appears at the initiation of flowering. Full grown larva has a pale body lined by rows of conspicuous black or brown spots on its dorsal surface. Its' larvae damage the flower buds, flowers and green pods. The larva feeds from inside a webbed mass of leaves, flowers, flower buds and pods webbed together with silken threads. For control of spotted pod borer, spray the crop at flower initiation stage using 150 ml Coragen 18.5 SC (chlorantraniliprole) or 100 ml Fame 480 SC (flubendiamide) or 500 ml Kingdoxa 14.5 SC (indoxacarb) or 150 ml Tracer 45 SC (spinosad) using 250-300 litres of water per hectare with manually operated knapsack sprayer. Repeat the spray at pod initiation/podding stage of the crop. The insecticides used for the management of spotted pod borer also control other pod borers infesting arhar.

Precautions: Because honey bees and other pollinators may be killed by the use of above insecticides it is, therefore, advised to spray the crop during evening as the population of these pollinators is minimum at that time.

Pod sucking bug (*Clavigralla gibbosa*): The adults are brown-grey in colour. Both nymphs and adults use their mouthparts to pierce the pod wall and suck the sap from the developing seeds. The attack is noticeable from the white patches that appear on the surface of pods and leaves. As a result of continuous sap feeding, premature drying of pods is noticed. When such pods are opened, shrivelled and malformed seeds are observed. Further, such seeds easily succumb to secondary infection by fungal pathogens and pose problems in post harvesting process. Seeds spoiled by the pod bug neither germinate nor are acceptable as human food. To control this, spray the crop at seed filling stage of the crop upon pest appearance using 3.1 litre per hectare homemade neem extract using 200-300 litres of water per hectare with manually operated knapsack sprayer. To prepare neem extract, boil 5.0 kg mixture of neem leaves and fruits in 10 litres of water for 30 minutes. Then, filter this material through muslin cloth and use the filtrate for spraying at the recommended dose.

B. Diseases

Phytophthora stem blight: It is caused by *Phytophthora drechslera* f.sp. cajani. The disease affects young seedlings as soon as they emerge and get killed. On stem, brown to black necrotic lesions are produced which have definite margin and are slightly depressed. In some cases, stem swells into a cankerous structure at the edge of the lesion and may break at the lesions site. On leaflets, lesions are circular to irregular in shape and whole foliage can become blighted. Avoid sowing arhar in soil with poor drainage and follow rotation in badly infested fields.

Sterility mosaic: It is a viral disease transmitted by an eriophyid mite (*Aceria cajani)*. Typical symptoms are mild mosaic and either no or little flowering and pale green colour of the leaves. The leaves are crowded and auxiliary buds give rise to bushy growth. Do not allow any arhar plant standing around sugarcane, cotton and other fields during winter season in order to check the primary source of infection.

Cercospora leaf spot: Leaf spots are caused by *Cercospora cajani*. Greyish brown to dark spots are produced on the under surface of the leaf. Often several spots coalesce to form irregular blotches. Sometimes lesions occur on petioles and stems. Infection causes premature defoliation of leaves. Use disease free seed to reduce the infection. Bacterial leaf spot: This disease is caused by *Xanthomonas campestris pv. cajani*. Angular dark-brown spots appear on the leaf surface and usually concentrated on one side of the mid-rib. Spots may develop on veins, petioles, main stems and branches. Use disease free seed to reduce infection.

7.3. Green Gram/ Moong (*Vigna radiata* (L.) Wilczek)

Moong bean (*Vigna radiate*) is native to India. Also known as Mung, Moong beans, Green Gram, Whole Green Gram. The beans are small, ovoid in shape, and green in colour. They are generally eaten either whole (with or without skins) or as bean sprouts, either in raw, soaked or boiled form. Moong beans are used to make salads, soups, dal, and other delicious dishes. The starch of moong beans is also extracted from them to make jellies and transparent/cellophane noodles. Hundred grams (g) of boiled mung beans contain: 7.02 g of protein 19.15 g carbohydrate, including 2 g sugar 7.60 g dietary fiber.

Area and Distribution

India is the major producer of green gram in the world and grown in almost all the States. It is grown in about 4.5 million hectares with the total production of 2.5 million tonnes with a productivity of 548 kg/ha and contributing 10 % to the total pulse production. According to Government of India 3rd advance estimates, greengram production in 2020-21 is at 2.64 million tonnes.

Moong occupied 2.6 thousand hectares and the total production was 2.5 thousand tonnes during 2020-21 in Punjab. Its' average yield was 9.60 quintals per hectare.

Classification

A classification of green gram in India was made by Bose in 1932, who grouped them into 40 different types, mainly on the basis of leaf size, flower colour and colour of the seed. All the types have 2n = 24 chromosomes. It has the following characteristic variations.

Seed colour – Seeds are either green, black brown or yellow

Seed surface – Either dull or shining

Flower colour – Either light yellowish-olive or olive yellow

Pod colour – Ripe pods are either iron grey, olive grey or snuff brown

Climatic Requirements

Moong grows best at an altitude of 0 to 1600 m above sea level and under warm climatic conditions (28 to 30°C). They are drought tolerant and will give reasonable yields with a minimum of 650 mm of yearly rainfall. Heavy rainfall results in improved vegetative growth with reduced development and pod setting. Green gram requires dry climate during maturity & excess rainfall is harmful to the crop. Areas where annual rainfall is up to 60-75 cm, suitable for green gram cultivation.

Moong bean varieties released by ICAR (2020-2021)

Variety Mung bean (Vigna radiata)	Area of adoption	Characters
Pusa 1641	National Capital Region of Delhi	Suitable for summer, average seed yield 1.3 tonnes/ ha, maturity 62–64 days, resistant to MYMV and low incidence of major insect pests.

Improved varieties for Punjab

A. *Kharif* Moong

ML 1808: Its' plants are erect and medium statured (71 cm). Pod formation is profuse and each pod contains 11-12 seeds. It is resistant to mungbean yellow mosaic virus, cercospora leaf spot and bacterial leaf spot diseases. It matures in about 71 days. Grains are shining green and medium bold with good cooking quality. The average grain yield is 12.0 quintals per hectare.

ML 2056: Its' plants are erect and medium statured (78 cm). Pod formation is profuse and each pod contains 11-12 seeds. It is tolerant to mungbean yellow mosaic virus, cercospora leaf spot and bacterial leaf spot diseases. It matures in about 71 days. Grains are shining green and medium bold with good cooking quality. The average grain yield is about 11.5 quintals per hectare.

ML 818: Its' plants are erect and medium statured (75 cm). Each pod contains 10-11 seeds. It is tolerant to mungbean yellow mosaic virus, cercospora leaf spot and bacterial leaf spot diseases. It matures in about 72 days. Its' average grain yield is about 10.5 quintals per hectare. Its' grains are medium bold, shining green with good cooking quality

B. Spring Moong

SML 1827: It has erect plant type with medium stature. It bears pods in clusters and has synchronous maturity (about 62 days). Each pod contains about 10 seeds. It is resistant to yellow mosaic disease. Grains are shining green and medium sized with good culinary properties. Its average yield is 12.5 quintals per hectare.

SML 832: It has erect plant type with medium stature. It bears pods in clusters and possesses early and synchronous maturity (about 61 days). Pods are of blackish brown colour at maturity. Each pod contains about 10 grains. Grains are shining green and medium sized with good culinary properties. Its average yield is 11.5 quintals per hectare.

SML 668: It has erect plant type with short stature. It bears pods in clusters and possesses early and synchronous maturity (about 60 days). Pods are long with thick coat and each pod contains 10-11 seeds. Grains are bold with good cooking quality. Its average yield is 11.25 quintals per hectare

Soil Requirements

Green gram can be grown in the *rabi*/summer season on a variety of soils but the most important thing to keep in mind is the selection of fertile, well drained & healthy land as it will certainly help in better production. Green grams are well suitable to red sandy loam soils but also do reasonably well on not too exhausted sandy soils. Green grams are not tolerant of poorly drained, wet soils. Waterlogged and saline soils are not suitable for cultivation. Loam to sandy loam soils is considered the best soil for mung bean cultivation. The soil must be well-drained and even temporary waterlogging may damage the crop. Acidic and saline-alkali soils are not suitable for mung bean cultivation. The optimum soil pH level is from 6.5 to 7.5 and fairly tolerant of soil salinity.

Intercropping and Crop Rotations

Green gram can be grown mixed with pigeon pea, sorghum, pearl millet, maize and cotton during *kharif* season. Intercropping of green gram can also be done with spring planted sugarcane. With this technology, additional grain yield of 5-6 quintals per hectare can be obtained without any adverse effect on the performance of main crop of sugarcane. Intercropping of green gram in pigeon pea is more profitable than sole pigeon pea. It is also intercropped with cowpea and sorghum. The following are the important rotations:

- Green gram - Wheat/Barley
- Green gram - Mustard
- Green gram -Toria - Wheat
- Green gram - Potato - Onion
- Rice - Wheat - Green gram
- Maize - Potato - Green gram
- Maize - Potato - Wheat - Green gram

Field Preparation

Green gram does not require a fine seedbed. One or two ploughings followed by harrowing is adequate for a *kharif* crop. During *rabi*, green gram is usually sown on deep soils on *kharif* fallows that are kept weed free by repeated harrowings. For a summer crop, tillage depends on the time available between the harvest of the standing crop and seeding of green gram. In highly intensive cropping system, minimum tillage is practiced.

Time of Sowing

In *kharif* season, sowing is done with the onset of monsoon during June-July. The optimum date of sowing varies from place to place depending on the monsoon rains actually received. Both early and late plantings have adverse effects on the performance of green gram. In early sowings, crop suffers due to rains at maturity and late sown crop suffered due to poor growth and diseases infestation. Sowing in the first fortnight of July has been found to be optimum in the arid zone of

Madhya Pradesh while early July sowing recorded the highest seed yield in south-western Punjab. In central and southern India, Madhya Pradesh, Andhra Pradesh, Tamil Nadu and Orissa green gram is sown in *rabi* season too. *Rabi* season green gram is sown during the post monsoon period up to December. Optimum time of sowing for green gram in *rabi* season is from first and second week of October, reduction in yield is possible with advance or delay in sowing. In Uttar Pradesh, Punjab, Haryana, Delhi, Bihar and West Bengal, green gram is grown during summer/spring season also under irrigated conditions. Sowing is done in the middle of March to middle of April. Only short duration varieties maturing in 60-70 days can be grown with adequate irrigation facilities in this season. Timely sowing of summer green gram is more important. The delay in sowing affects the yield adversely. The late-sown crop may be caught by the early monsoon rains at harvesting time which may create a problem for harvesting and threshing and ultimately the yield is reduced. If fields are already lying vacant this crop can be sown at any time after mid-February.

Seed Treatment

Treat the seeds with Captan or Thiran @ 3.0 g/kg of seed against seed borne diseases. Also inoculate the seed with Rhizobium culture. Mix thoroughly one packet of culture with seed of one acre on a clean pucca floor and let it dry in shade and sow immediately. A yield increase from 12 – 16% is common the culture and fungicide can be applied simultaneously.

Seed Rate and Spacing

A seed rate of 15-20 kg ha^{-1} for *kharif* and 25 kg ha^{-1} for summer or spring season has been recommended for green gram. However, as the seed rate is increased, the grain yield of green gram has also been found to increase. Rainy season green gram is generally broadcasted. But, sowing of green gram in rows has been found better as it ensures proper spacing for canopy. An inter-row spacing of 25-30 cm and intra-row spacing of 5 cm is optimum for summer and *rabi* season green gram. The *kharif* crop needs a spacing of 30-45 cm between rows and 5-10 cm within plants. Sowing can be done behind the local plough or with seed drill. Seed should be treated with some fungicides (Thiram or Carbendazim) before sowing. For increasing yield, treatment of seed with rhizobium culture should be done after the seed is treated with fungicide. After treatments, seed should not be kept in diffused sunlight. More uptake of nitrogen and increased protein content was possible in wider spacing. High temperatures, bright sunshine hours and low relative humidity where light was probably not a limiting factor, rectangular planting proved better than square planting pattern. For rainy season crop, prefer bed planting to ensure germination and to avoid crop mortality due to excessive rainfall.

Nutrient Management

Green gram is generally grown without application of fertilizers. However, experiments conducted at various places have revealed that yield of green gram can be increased to great extent with the application of fertilizers. Response of green gram to excessive N is generally absent or negative. In general, a starter dose of 10-20 kg N/ha has been recommended. Rhizobium culture responses are quite variable. Responses are comparatively higher in summer season in comparison to *kharif.* Responses of green gram to phosphorus are quite conspicuous. Significant increase in the yield of green gram due to application of 80 kg P_2O_5/ha and higher response of green gram on red soils followed by that on laterite soils. Responses of green gram to potassium are negligible or absent as potassium content in most of the Indian soils is high. Overall, application of 100 kg diammonium phosphate (18-46-0) per hectare is generally recommended for green gram. This fertilizer is comparatively cheaper on nutrient basis and can supply required quantity of N and P to the crop. Fertilizers placement (total quantity) should be done at the time of sowing.

Water Management

Rainy season green gram does not require irrigation unless there is a prolonged drought. Draining out of extra water from the field is essential for successful cultivation of green gram. *Rabi* and summer green grams may be grown only in the areas where adequate irrigation facilities are available. In general, 3-5 irrigations are required for green gram in summer season. Increase in moisture supply due to irrigation significantly increases the yield. Giving more than three irrigations does not have any significant impact on yield.

Weed Management

Green gram being a dwarf statured crop suffers to great extent if the weeds are not controlled in time. The problem of weeds is more severe in *kharif* as compared to *rabi* and summer seasons. Traditionally weeds are removed by mechanical means. One hand weeding after 20-25 days of sowing followed by another after about 20 days may be sufficient. Weeds can also be controlled with the use of herbicides. Fluchloralin (Basalin) at 0.75-1.0 kg/ha or trifluralin (Treflan 48 EC) 0.75-1.0 kg ha^{-1} were found effective. Alternatively, pre-emergence application of pendimethalin at 0.75 kg ha^{-1} or post-emergence application of haloxyfop-methyl at 0.24 kg ha^{-1} (20 days after sowing) can also be used. Fluchloralin, pendimethalin both at 0.75 kg/ha applied alone and in combination with one hand weeding will result in effective weed control. The specific recommendations of herbicides may vary in different situations owing to weed population, environmental conditions, soil type and crop rotations etc.

Plant Protection

A. Insect Pests

Whitefly, Jassid and Aphid: Nymphs and adults of whitefly and aphid suck sap from the leaves, thus lowering the vitality of the plants. Whitefly excretes honeydew on which sooty mould develops, resulting in blackening of leaves and decrease in the normal photosynthesis. In case of severe attack, there is total blackening of the crop, resulting in drying of leaves and ultimately total crop failure. Whitefly is a vector of mungbean yellow mosaic virus. Jassid nymphs and adults suck sap from the leaves resulting in yellowing and drying of leaves.

Hairy caterpillar: The body of the caterpillar is covered with hair. The caterpillar eats the green matter of leaves, leaving behind only the midribs. The crop may be totally denuded due to severe attack. When young, they feed gregariously but on few plants in scattered spots. Since, the young larvae are gregarious, they can be destroyed by pulling out the infested plants along with larvae and burying them underground. The grown up caterpillars can be destroyed by crushing them under feet or by picking and putting them into kerosenized water.

Semilooper: The larvae are green in colour measuring 2-4 cm in length. When touched, they form a loop. The larvae feed extensively on the leaves of mash and moong. In case of severe damage, the plants are totally defoliated within a few days.

Tobacco caterpillar (*Spodoptera litura*): It is a polyphagous pest. The small larvae are black whereas grown up larvae are dark green with black triangular spots on body. Its' moth lays eggs in masses covered with brown hairs on the lower side of leaves. After hatching, first and second instar larvae feed gregariously and skeletonize the foliage. Later on, the grown up larvae disperse and feed singly. Besides leaves, they also damage buds, flowers and pods. For management, egg masses and young larvae of tobacco caterpillar feeding gregariously should be collected along with leaves and destroyed.

Blister beetle: Blistle beetle is diurnal and general feeder. Adult beetles are robust with bright black and red stripes on the forewings. When disturbed, the beetles emit a fluid containing cantharidin that causes blisters on human skin. The adult beetles attack arhar, moong, mash and other pulse crops. The major damage is caused at the flowering stage. They feed on tender buds and flowers of the plant, thus preventing pod formation.

Pod borer (*Helicoverpa armigera*): The larvae damage the crop by feeding on leaves, flower buds, flowers, pods and seeds in the pods, thus causing heavy loss in yield. The larvae may be pale green, yellow, brown or black in colour measuring about 3-5 cm in length when fully grown. Larval presence can be observed from damage to plant and from dark green faeces below the plants on the soil. The larvae fall on the ground when plants are shaken vigorously.

Mite: The mite causes webbing on the underside of the leaves which turn pale. Such infested leaves turn light-brown to dark reddish-brown.

B. Diseases

Yellow mosaic virus: It is a viral disease transmitted by whitefly and is more severe on *kharif* moong. The leaves of the diseased plants develop irregular yellow and green patches. Infected plants bear no or only a few pale pods. Rogue out the diseased plants early in the season. Grow yellow mosaic virus tolerant varieties of moong.

Cercospora leaf spot: It is caused *by Cercospora cruenta* and *C. canescens*. The disease spots are circular, brown and necrotic which coalesce to cover bigger area and cause defoliation. Intermittent rains are favourable for disease development. Grow disease resistant varieties of moong.

Root rot: Root rot caused by *Macrophomina phaseolina* produces dark lesions on leaves, branches, stems and roots. The tissue of the affected portion become weak and shred easily. Pycnidia can be seen on the affected portion.

Anthracnose: It is caused by *Colletotrichum lindemuthianum*. Dark brown spots (horse shoe shaped) appear on the leaves and other parts of the plant. Under severe conditions lesions rapidly coalesce to girdle stems, branches, peduncles and petioles.

Bacterial leaf spot: Bacterial leaf spots caused by *Xanthomonas campestris pv. phaseoli* are circular to irregular and brown. Use disease free seed and fairly resistant/ tolerant varieties to this disease.

Web blight: It is caused by *Rhizoctonia solani*. It starts from leaf margins or petioles or young branches. Eventually, the tops of plants become blighted and patches of such plants are conspicuously seen in the field. Whitish web-like growth develops on leaves in humid weather. Dark brown sclerotia develop on infected tissue. Infection on crop comes from the weeds in the field. Keeping the field weed-free helps to check the disease.

7.4. Black Gram /Mash (*Vigna mungo* (L) Hepper)

Black gram *(Vigna mungo)* is a South Asian legume belonging to the *Fabaceae (Leguminosae)* family. This popular pulse goes by many names: urad, urid, black matpe, black gram, mash bean, mash daal, among others.

Black gram is essentially a black seed that is cylindrical and narrow in shape. A black coating or hull surrounds the white inside. This legume is sold in whole, split, or dehulled forms. The split form is the seed divided in half with a black and a white side, which is often called 'white lentil'.

Black gram is one of the most nutritious beans and is commonly used in India for its wide health benefits. It is both, consumed by cooking and used in Ayurvedic medicine, and is equally beneficial when used either way. It offers a ton of health benefits. Black gram can improve your digestion as it is filled with fibers that help with the bulking up and movement of your stool. It can therefore be used to combat both, constipation and diarrhoea. It boosts your energy by helping provide more oxygenated blood to your organs. It increases your bone mineral density, making your bones stronger and healthier as you grow older. It is very good for patients with diabetes as it regulates the glucose levels in the blood. It is good for your skin and can help fight dark spots, acne, and marks. It helps with joint pain, is extremely good for the heart, and is also a diuretic that helps keep your body clean. Adding black gram to your diet can boost your overall health by a significant margin.

Area and Distribution

India is the world's largest producer as well as consumer of blackgram. It produces about 23.4 lakh tonnes of blackgram annually from 46.7 lakh hectares of area, with an average productivity of 501 kg per hectare in 2020-21 (agricoop.nic.in). Blackgram area accounts for about 15.7 per cent of India's total pulse acreage and contributes 9.09 per cent of total pulse production. Madhya Pradesh, Uttar Pradesh and Andhra Pradesh are major blackgram growing states area-wise.

Climate

During *kharif*, it is cultivated throughout the country. It is best suited to rice fallows during *rabi* in southern and south-eastern parts of India. Blackgram needs relatively heavier soils than greengram. It is also successful crop of spring season in north India.

Black gram varieties released by ICAR (2020-21)

Varieties	Area of adoption	Characters
Pant Urd 12	Uttar Pradesh, Punjab, Haryana, rajasthan, Delhi, Uttarakhand and Jammu and Kashmir	Suitable for rainfed/irrigated, conditions of kharif season, yield 14.3 q/ha, maturity 79 days, resistant against MYMv, web blight, leaf crinkle, leaf curl virus and root-rot diseases, white- fly, pod-borer, pod bug, maruca, and pod-fly.
Kalinga Urd 41 (OBG 41)	Odisha, Andhra Pradesh, Tamil Nadu and Karnataka	Suitable for rainfed condition, average yield 11.0 q/ha maturity 72 days, resistant against MYMv, leaf crinkle virus, CLS and moderately resistant to web blight, powdery mildew, Macrophomina blight, anthracnose, and root-rot diseases.
Kota Urd 5 (KPU 52-87)	Andhra Pradesh, Telangana, Tamil Nadu, Karnataka and Odisha	Suitable for normal sown kharif season, yield 16.50 q/ha, maturity 74 days, moderately resistant to MYMv, Cercospora leaf spot (CLS), powdery mildew, web blight, anthracnose, exhibited less incidence of aphids, defoliators, bod- borer and pod bug.

Improved Varieties for Punjab

A. *Kharif*

Mash 883: This variety is recommended for the whole State. Plants are dwarf, erect and compact. It matures in about 77 days. Podding is profuse and each pod contains 6-7 seeds which are medium bold, black and possess good culinary properties. Average grain yield is about 10.5 quintals per hectare. It is resistant to yellow mosaic disease and tolerant to other foliar diseases.

Mash 114: This variety is recommended for the whole State. Plants are dwarf, erect and compact. It matures in about 83 days. Podding is profuse and each pod contains about 6-7 seeds, which are bold, black and possess very good culinary properties. Average grain yield is about 9.0 quintals per hectare. It is fairly resistant to yellow mosaic virus, bacterial leaf spot and cercospora leaf spot diseases.

Mash 338: This variety is recommended for the whole State. Plants are dwarf, erect and compact. It matures in about 90 days. Podding is profuse and each pod contains about 6 seeds, which are bold, black and possess very good culinary properties. Average grain yield is about 8.75 quintals per hectare. It is tolerant to yellow mosaic virus, bacterial leaf spot and cercospora leaf spot diseases.

B. Spring

Mash 1137: This variety is recommended for sub-montaneous zone of Punjab state. It has erect and compact plant type with short stature (30 cm). It matures in about 74 days. Pod bearing is profuse and each pod contains 6-7 seeds. It is resistant to yellow mosaic disease. Average grain yield is about 11.25 quintals per hectare. Grains are medium bold, blackish in colour and possess good culinary properties.

Mash 1008: It has erect and compact plant type. It is a short statured (25 cm) variety. It matures in about 72 days. Pod bearing is profuse and each pod contains 6-7 seeds. It is fairly tolerant to yellow mosaic virus and leaf crinkle virus. Average grain yield is about 10.5 quintals per hectare. Grains are medium bold, blackish in colour, contain about 24 per cent protein and possess good culinary properties.

Soil Requirements

Black gram can be grown on variety of soils ranging from sandy soils to heavy cotton soils. The most ideal soil is a well-drained loam with pH of 6.5 to 7.8. Black gram cannot be grown on alkaline and saline soils. Land is prepared like any other kharif season pulse crop. However during summer it requires a thorough preparation to give a pulverized free from stubbles and weeds completely.

Time of Sowing

Kharif black gram is sown with the onset of monsoon in the last week of June to early July. This crop is mainly grown under rainfed conditions. Early planting in first week of July results in higher yield and any delay in sowing beyond this date causes reduction in yield. In case of spring/summer black gram, sowing may be started in the last week of February till the end of March. Plantings beyond March may be caught by monsoon rains in the second fortnight of June and result in drastic yield reduction. During *rabi* season, sowing during mid-August and mid-September results in higher yield. Late sowings are unfavourable due to low temperatures in the winter.

Seed Rate, Spacing and Seed Treatment

During rainy season, the crop is generally sown by broadcasting the seed followed by planking. Sowing could be done in rows with the help of local plough or tractor drill Sowing is done in wider rows because vegetative growth of the plants is more due to favourable climate during *kharif* season. An inter-row spacing of 30 cm and intra-row spacing of 5-10 cm is optimum for *kharif* crop and seed rate of 15 to 20 kg/ha is sufficient depending upon the variety. For spring season crop, use 50 kg seed per hectare with row to row spacing of 22.5 cm. During summer season, vegetative growth of black gram is comparatively less, hence higher plant population by reducing the spacing (22.5 cm) and increasing the seed rate is desirable. A seed rate of 40 to 50 kg/ha is recommended for sowing during summer. Sowing can be done behind local plough or with the help of seed-drill. Before sowing seed should be treated with Agrosan GN or Thiram @ 2.5g/kg of seed. Seed should also be inoculated with suitable Rhizobium culture, if black-gram is being taken for the first time in the field or after a long duration.

Seed Treatment

It is same as for moong.

Fertiliser Management

The crop requires 15-20 kg/ha nitrogen as basal dose to meet the growth requirements during early growth stages particularly on soils which are poor in organic matter. The combined application of nitrogen and phosphorus further increased seed yield compared with individual applications. The seed inoculation with appropriate *Rhizobium* culture also increases seed yield and uptake of nutrients. Application of phosphorus has been found to be quite effective in increasing the seed yield of black gram. A dose of 40-50 kg P_2O_5 per hectare has been found optimum in most of the soils. Black gram does not respond to applied potassium and application of phosphorus and potassium should always be made on soil test basis. Application of sulphur at the rate of 30 kg ha^{-1} either as gypsum, pyrites or elemental sulphur increases yield.

Water Management

The crop cannot tolerate long dry spells and therefore, irrigations are very useful for black gram. The crop should be irrigated at flowering and pod filling stages but the water should not accumulate in the field. One pre-sowing irrigation at the time of seedbed preparation is required to ensure good germination, if the land is dry. Blackgram is grown purely as a rainfed crop during *kharif* and on residual moisture in rice-fallows during *rabi*. In case there is a long dry spell, the crop is benefited if one light irrigation at podding/seed filling stage is applied. On the other hand, number and frequency of irrigation depends upon the soil type and weather conditions during the growth period. The summer season crop should get irrigation at an interval of 10-15 days. From flowering to pod development stage, there is a need of sufficient moisture in the field. Excess water at any stage affects the crop adversely.

Weed Management

The first 4-5 weeks after sowing are critical for crop-weed competition. The extent of damage done by weeds can be as high as 50-60%. Good seedbed preparation, one or two inter-cultivations or manual weeding 3-5 weeks after sowing may effectively control the weeds. Fluchloralin at the rate of 1 kg ha^{-1} as pre-plant incorporation or pendimethalin at the rate of 0.75 kg ha^{-1} as pre-emergence application are effective treatments. Integration of one hand weeding with these herbicides results is improved yield. Post-emergence application of Sethoxydim at the rate of 1 kg ha^{-1}, 10 days after sowing is also effective in blackgram. When blackgram is grown as an intercrop with sorghum, pigeon pea, pearlmillet, etc., the interculture given to the main crop is adequate. When sown as a sole crop 1 or 2 weedings are required in the initial stages to keep the field free from weeds.

Plant Protection

A. Insect Pests

Hairy caterpillar: The body of the caterpillar is covered with hair. The caterpillar eats the green matter of leaves, leaving behind only the midribs. The crop may be totally denuded due to severe attack. When young, they feed gregariously but on few plants in scattered spots. Adopt the following control measures:

- Young larvae are gregarious. They can be destroyed by pulling out the infested plants along with larvae and burying them underground.
- The grown-up caterpillars can be destroyed by crushing them under feet or by picking and putting them into kerosenized water.

If the population is high, spray 1.25 litre of Ekalux 25 EC (quinalphos) in 200-250 litres of water per hectare with a manually operated knapsack sprayer.

Whitefly, Jassid, Aphid, Semilooper, Tobacco caterpillar, Blister beetle, Pod borer and Mite: Their control measures are same as in moong.

B. Diseases

Yellow mosaic virus: It is a viral disease transmitted by whitefly and is more severe on kharif moong. The leaves of the diseased plants develop irregular yellow and green patches. Infected plants bear no or only a few pale pods. Rogue out the diseased plants early in the season. Grow yellow mosaic virus tolerant varieties.

Cercospora leaf spot: It is caused by *Cercospora cruenta* and *C. canescens*. The disease spots are circular, brown and necrotic which coalesce to cover bigger area and cause defoliation. Intermittent rains are favourable for disease development. Adopt the following measures to check the disease. Grow disease resistant varieties.

Bacterial leaf spot: Bacterial leaf spots caused by *Xanthomonas campestris pv. phaseoli* are circular to irregular and brown. Use varieties which are fairly resistant/ tolerant to this disease. Root rot, Anthracnose and Web blight: Regarding disease details, see under moong

Harvesting and Threshing

Blackgram should be harvested when most of the pods turn black. Over maturity may result in shattering. Harvested crop should be dried properly on threshing floor for few days and then threshed. Threshing can be done either manually or by trampling under the feet of bullocks, or with tractor operated thrashers.

Yield

A well-managed crop may produce about 15-20 quintals of grain per hectare.

7.5. Lentil (*Lens culinaris*)

The lentil (*Lens culinaris* or *Lens esculenta*) is an edible legume. It is an annual plant known for its lens-shaped seeds. It is about 40 cm tall, and the seeds developed in pods, usually with two seeds in each. As a food crop, the majority of world production comes from Canada (33%) and India (25%), producing 58% combined of the world total. In cuisines of the Indian subcontinent, where lentils are a staple, split lentils (often with their hulls removed) known as daal are often cooked into a thick curry/gravy that is usually eaten with rice or rotis.

They resemble a tiny bean, grow in pods, and come in red, brown, black, and green varieties. They also contain high levels of protein and fiber. Lentils are relatively quick and easy to prepare, and their low cost makes them an accessible form of high quality protein for many people around the world.

Origin and History

Evidence of domesticated lentils dating to around 8000 B.C. has been found on the banks of the Euphrates River in what is now northern Syria. By 6000 B.C., lentils had reached Greece, where the legumes were regarded as poor man's food. The opposite was true in Egypt, where remains of lentils were found in the royal tombs at Thebes dating to 2400 B.C. A second-century fresco illustrates the preparation of lentil soup.

Area and Distribution

In India, it is grown on an area of 1.35 m ha with production of 1.18 m tonnes during 2020. In Punjab, lentil occupied an area of 0.6 thousand hectares with production of 0.4 thousand tonnes during the year 2020-21. The average yield was 6.7 quintals per hectare.

Lentil varieties released by ICAR (2020-2021)

Variety	Area of adoption	Characters
RKL 58 F 3715 (Kota Masoor 4)	Chhattisgarh, Gujarat, Madhya Pradesh, Maharashtra, Uttar Pradesh (Bundelkhand region), South-East Parts of Rajasthan	Suitable for normal sowing conditions, average seed yield 1.87 tonnes/ha, maturity 112 days, resistant to rust and Stemphylium blight, moderately resistant to wilt.
IPL 329	Uttar Pradesh	Suitable for timely-sown rainfed condition, average seed yield 1.01 tonnes/ha, maturity period 115–134 days. resistance to rust and Fusarium wilt diseases, a large seeded (2.9 g/100 seeds) variety with brown dotted seed coat and red cotyledons.
IPL 225	Uttar Pradesh	Suitable for timely-sown condition, average seed yield 1.03 tonnes/ha, maturity 115–120 days, highly resistance to rust and Fusarium wilt diseases, a small seeded (2.1 g/100 seeds) variety with brown dotted seed coat and red cotyledons.
Pant Lentil 11	Uttarakhand	Suitable for timely-sown rainfed/irrigated condition, average seed yield 1.2 kg/ha, maturity 101–132 days, resistant to rust disease and moderately resistant to pod- borer.

Improved Varieties

LL 1373: Its plants are short, erect with profuse branching and bear more number of pods. It has light green leaves, pink flowers, non-pigmented light green pods and rudimentary tendrils. It matures in 140 days. It is resistant to rust and possesses good tolerance to pod borer. Its seeds are bold and 100-seed weight is 3.5 g. It has good culinary properties. Its average yield is 12.75 quintals per hectare.

LL 931: Its plants are short, erect with profuse branching and bear more number of pods. It has dark green leaves, pink flowers, non-pigmented green pods and rudimentary tendrils. It matures in 146 days. It is fairly resistant to rust and possesses good tolerance to pod borer. Its seeds are medium bold with greyish brown colour and light flecks. It has good culinary properties. Its average yield is 12 quintals per hectare.

Classification

Lentil belongs to genus *Lens* of Leguminoseae family. It can be classified into two main groups according to the size of the seed:

1. ***Bold seeded group***: This group includes sub-species var. *macrosperma.*
2. ***Small seeded group***: This group includes sub-species *microsperma.*

Climate

Lentil is adapted to cool growing conditions. It is sown as winter season crop. It is very hardy and can tolerate frost and severe winter to a great extent. It can be grown successfully up to a height of 3000 metres. It remains unaffected by rain at any stage of its growth including flowering and fruiting. It can be grown with the moisture conserved in the soil during the rainy season. It requires cold temperature during its vegetative growth and warm temperature at the time of maturity. The optimum temperature for growth is 18-30°C.

Soil

Lentil is adapted to all soil types, from sand to clay loam, if there is good internal drainage. Lentil does not tolerate flooded or waterlogged soils, and does best on deep, sandy loam soils high in phosphorus and potassium. Good drainage is required, because even short periods of exposure to waterlogged or flooded field conditions kill plants. A soil pH near 7.0 is best for lentil production

Cropping Systems

Lentil is grown generally after the harvest of *kharif* crops or as sole crop of the year. The most common rotations are given below:

Rice-Lentil

Maize-Lentil

Cotton-Lentil

Pearl millet-Lentil

Sorghum-Lentil

Groundnut-Lentil

Lentil can also be grown as an intercrop in autumn sugarcane. Two lines of lentil may be sown 30 centimetre apart in the centre of two sugarcane rows. The distance between sugarcane and lentil row will also be 30 centimetre.

Field Preparation

The soil should be made friable and weed-free so that seeding could be done at a uniform depth. On heavy soils, one deep ploughing followed by two to three cross harrowings should be given. In light soils less tillage is needed to prepare an ideal seedbed. There should be proper moisture in the soil at the time of sowing for proper germination of seeds.

Time of Sowing

Timely planting is a very important factor in the full realization of the yield potential of improved varieties of lentil. Middle of October is the most suitable time for sowing lentil. Delay in planting causes reduction in yield. Under delayed sowing the reduction in yield could be minimized up to certain extent by closer spacing and use of higher seed rate.

Seed Rate and Sowing

Sowing should be done in rows 30 centimetre apart. The optimum seed rate for normal sown crop is 30-40 kg per hectare. Under late sowing the seed rate should be increased to 50-60 kg per hectare and row spacing reduced to 20-25 centimetre. Seed should be placed at a shallow depth of 3-4 centimetre.

Nutrient Management

Generally lentil is grown without fertilizers and manures. However, good yields may not be harvested if lentil crop is made to feed on symbiotically fixed nitrogen only. Some quantity of fertilizer nitrogen as starter dose is required during few weeks of the initial growth. On light and poor fertility soils, it is even more important to give the starter dose of nitrogen fertilizer. On sandy soils with low organic matter content, the response to applied nitrogen to the tune of 10-25 kg per hectare and of phosphorus to 50-60 kg per hectare have been obtained. Lentil when grown in rotation with paddy in late sown conditions may give symptoms of zinc deficiency. The deficiency can be rectified by spraying a solution of 0.5 per cent zinc sulphate and 0.25 per cent lime at the appearance of early symptoms.

Water Management

Lentil is mostly grown under unirrigated conditions. It can tolerate drought to some extent. If irrigation is available, then first irrigation should be given at 45 days of planting and second at pod filling stage.

Weed Control

Lentil is a slow growing crop and thus suffers considerably with competition from weeds particularly in the early stages of growth. The period from 30 to 60 days after sowing is most critical for competition with weeds. Under heavy infestation of weeds, there is drastic reduction in yields. A weed-free period of early 30 to 60 days is important. Thereafter, the crop covers the ground and weeds are effectively smothered. Weeds in lentil can be controlled with mechanical as well as chemical methods. Two manual weedings can be done at 30 and 50 days after sowing. Pre-plant incorporation of Basalin (fluchloralin) at the rate of 0.75 kg a.i. per hectare in 500-600 litres of water is very effective in controlling weeds. Alternatively pre-plant application of Treflan 48 EC (trifluralin) at 0.5 kg a.i./ha integrated with one hand weeding provides effective control of weeds in this crop. The pre-emergence application of Stomp 30 EC (pendimethalin) at (0.75 kg a.i. per hectare) + one hand weeding at 45 days after sowing is also very effective.

Plant Protection

Insect Pests

The lentil pod-borer causes damage to the crop by feeding on leaves, flower buds, flowers, pods and grains in the pods.

Diseases

Blight (*Ascochyta fabae f. sp. lentis*): Dark brown spots are produced on stem, branches, leaves and pods. The lesions on the stems and branches are elongated. The characteristic symptoms of the disease are black pinhead-like pycnidial bodies of the fungus arranged on the spots in the form of concentric rings. On tender branches, spots rapidly girdle the branch and parts of the plant above the girdle, wilt and dry. The cool and humid conditions favour disease development and spread. Use disease free seed. Destroy the diseased plant debris after harvest to reduce the inoculum of the fungus.

Rust (*Uromyces fabae*): Yellowish white pycnial and aecial pustules develop on stems, branches, leaves and pods. These may appear singly or in small groups. Later on brown uredia are formed. Small pustules may coalesce to form large pustules. In later stage of the crop, dark brown to black telia are formed. In severe cases, affected plants dry up and give burnt appearance. Cloudy weather with intermittent rains and temperature around 20-22°C favours the disease development. Use clean seed free from diseased plant debris.

7.6. Field Pea *(Pisum sativum)*

Field pea (*Pisum sativum*) is a cool-season grain legume from the diverse Pisum genus, grown in over 100 countries for dry or fresh seed and fodder. The growing need for affordable, high-quality protein and the role of field pea in rotations for sustainable agriculture is driving renewed interest in this crop. Field pea is grown primarily in rotation with cereals to capture the benefits of symbiotic nitrogen fixation and reduced crop water use. There is increasing interest in field pea as a component of intercrops.

Field pea seed is a rich source of protein, carbohydrate, and some minerals, although the nutritional content of the seed varies with the environment and genetic factors. By far the highest proportion of the nutrient value of field pea is contained in the cotyledons, with the embryo and seedcoat contributing less than 10% to the nutritional value.

Origin and History

While the origins of domesticated peas have not been definitely determined, the pea is one of the oldest cultivated crops. The wild plant is native to the Mediterranean region, and ancient remains dating to the late Neolithic Period have been found in the Middle East. European colonization introduced the crop to the New World and other regions throughout the globe.

Area and Distribution

In India, during 2020, the area dry pea was 0.72 mha with production of 0.80 m tonnes.

Climatic Requirements

Pea is a cool weather loving plant. It can germinate even at 4-5 °C and can tolerate frost. However, under severe continuous frost, its flowers and young pods are likely to be damaged. Optimum germination takes place at 20-25°C. At temperature of 30 °C and above, the germinating or emerging seedlings get killed. The high temperature conditions prevailing at the time of planting results in the build-up of wilt and stem fly complex which results in considerable losses in yield. Hence pea attain an ideal growth and development in areas where there is a slow transition from cool to warm weather.

Field pea varieties released by ICAR (2020-2021)

Variety	Area of adoption	Characters
HFP 1428	Punjab, Haryana, Delhi, Rajasthan, Uttarakhand, parts of Jammu and Kashmir and Western Uttar Pradesh	Suitable for rabi, average seed yield 2.6 tonnes/ha, maturity 123 days, resistant to powdery mildew, Ascochyta blight and root-rot and moderately resistant to rust and less incidence of pod-borer and aphids.
Pant Pea 347	Punjab, Haryana, Delhi, North West and Central Rajasthan, Western Uttar Pradesh and plains of Uttarakhand and Jammu and Kashmir	Suitable for rainfed/ irrigated condition, average seed yield 2.54 tonnes/ha, maturity 124 days, resistant to powdery mildew and Ascochyta blight and moderately resistant to rust and root-rot diseases.
Pant Pea 195	Uttarakhand plains	Suitable for timely-sown rainfed/irrigated condition, average seed yield 1.48 tonnes/ ha, maturity 114–136 days, resistant to powdery mildew and rust diseases and moderately resistant to pod-borer.

Improved Varieties for Punjab

Early Maturing

AP-3: This variety derived from the cross of Azad P-1 × Arkel through pedigree method. Plants are dwarf and straight. The pods are of medium length (8.85 cm), deeply curved near distal end and are borne singly or in doubles. Each pod contains 7-8 grains and shelling out turn is 50 per cent. Seeds are bold, wrinkled with green seed coat. It is an early maturing variety and gives first picking in 60-65 days after sowing, if sown in the second week of October. Its green pod yield is 78.75 q/hectare.

Matar Ageta-7 : The plants are vigorous with 15-18 well filled pods having 7-9 grains per pod. The pods are of medium length (9.57 cm), slightly curved from tip and are borne singly or in doubles. Its shelling out turn is 48 per cent. It is an early maturing variety which gives first picking in 65-70 days after sowing. Average green pod yield is 80.0q/hectare.

Matar Ageta-6: Its plants are dwarf (40 cm), erect, vigorous and green. It can tolerate high temperature and can be sown early in the season. First picking is possible after 7 weeks of sowing. Each plant bears 12-15 pods. One or two pods are borne per bunch. Each pod contains up to 6 grains and shelling out turn is 44.6 per cent. Grains have high protein and dry matter content. On drying, per cent seeds are light green, smooth with slight dimples. This variety gives 50 per cent of the total production in the first picking. Its green pod yield is 60 q/hectare.

Arkel: Arkel is an early maturing variety and takes about 60-65 days from sowing to marketable green pod stage. It is suitable for sowing in the first week of October under the irrigated conditions. Plants are dwarf (30-45 cm). Pods are borne both in double and single and are attractive, dark green, well filled, 8-10 cm long and each contains 7 to 8 dark green grains with a shelling percentage of 40. Dry seeds are light green and wrinkled. It is good for dehydration. Its green pod yield is 45 to 50 q/hectare and seed yield is 8.75 q/hectare.

Main Season

Punjab 89: The plants of this variety are medium dwarf, vigorous, having more number of well filled pods (28-30 per plant). The pods borne in doubles and are dark green, long, very attractive having 9-10 grains per pod. It takes about 85-90 days for first picking. Shelled peas are very sweet and the shelling out turn is more than 55 per cent. Average green pod yield is 150 q/hectare.

Mithi Phali: It is an edible podded variety and its shelling is not required. Plants are tall (140 cm) with green foliage. The first picking takes place 90 days after sowing and subsequent pickings at 15 days interval. Pods are light green and 12-13 cm long. Its consumable yield is double than that of standard shelled pea varieties. It is rich in protein and total sugar. Average green pod yield is 117.5 q/hectare.

Classification

Two types of peas are generally cultivated all over the globe. One is the garden pea and other is field pea.

1. **Garden Pea (*Pisum sativum* var. hortense):** It is also called table pea or vegetables peas. In this type young, green seeds are used mostly in vegetables and also for canning purposes. Seeds are bold and wrinkled. These are generally, white flowered. Leaf axils are generally green. Seeds are yellowish, whitish or bluish-green.

2. **Field Pea (*Pisum sativum* var. arvense):** In this type the ripe, mature seeds are used as the pulse (dal). They are also grown as forage or green manure crop. Field peas are hardy plants and grown on a large scale without irrigation. These plants are also able to withstand frost. Seeds are rounded and little angular. They have generally coloured flowers. Leaf axils are often pigmented. Seeds are greyish-green, greyish-brown or greyish-yellow.

Soil Requirements

It can be grown on all types of soils but well drained fertile loamy soils are best for the crop. Peas do best in soils having pH 6.0 to 7.5.

Sowing and Seed Rate

Since the incidence of wilt in September is very high, somewhat late sowing from mid-October to mid-November give the best crop in the plains. Seed rate is 112.5 kg for early maturing varieties and 75 kg for main season varieties per hectare. Line×Plant spacing should be 30×7.5 cm for early and 30×10 cm for main season varieties. The sowing of pea can also be done with Seed-cum- Fertilizer pea drill on ridges which are 60 cm wide. This drill sows two rows of pea which are 25 cm apart on each ridge. This drill can sow one acre per hour.

Inoculation

In areas where pea crop has not been sown earlier, it is advisable to treat the seed with bacterial culture (*Rhizobium leguminosarum*) to ensure nodule formation and quick growth. It increases the yield and quality of pods. One acre culture packet should be mixed with half litre of water. Rub the mixture thoroughly on seed to give a fine covering of the culture to every seed. Thereafter, spread the seed in shade for drying and sow it immediately afterwards.

Manures and Fertilizers

Apply 8 tonnes of farmyard manure, 50 kg of N (112.5 kg Urea) and 62.5 kg P_2O_5 (387.5 kg Superphosphate) per hectare before sowing.

Weed Control

The field should be kept free from weeds by giving two hoeings after four and eight weeks of germination respectively. For chemical weed control use Stomp 30 EC (pendimethalin) @2.5 litre per hectare within 2 days of sowing. Dissolve the herbicide in 400-500 litres of water and spray uniformly over the entire field. Application of this herbicide control many of the annual broad leaved and grass weeds, including guli danda (*Phalaris minor*).

Irrigation

Furrow Irrigation : Seed should be sown in proper soil moisture condition. First irrigation should be given after 15 days of sowing. Next irrigation should be given at flowering and then at fruit set if necessary. Pea can be grown as rainfed crop with limited irrigations. The total number of irrigations required are 3-4 depending upon the soil type and weather conditions.

Drip irrigation : Drip irrigation in pea not only increases the yield but also saves 30% water as compared to conventional method of irrigation. For drip irrigation, sow two rows of pea at 25 cm row to row distance on 60 cm wide bed. Under this system, irrigation should be applied at an interval of 3 days. Do not apply irrigation before 15 days of sowing and before 10 days of harvesting of crop. The crop should be irrigated with one lateral pipe per bed and keep distance lateral and dripper to dripper 30 cm having discharge of 2.2 litres per hour as per the following schedule:

Month	Time of irrigation (minutes)
November	15
December	15
January	20
February	25

If discharge rate is less than 2.2 litres per hour, time of irrigation may be adjusted proportionally by the formula

$$= \frac{2.2 \times \text{Time of irrigation (minute)}}{\text{Discharge of dripper (litre/hour)}}$$

Fertigation

Apply 80 kg to 100 kg Mono Ammonium Phosphate per hectare. The 20% of the fertilizer should be applied during the first month after germination in 5 split doses. The remaining amount of fertilizer should be applied in nine equal doses during the second month. Fertigation saves 20% fertilizer.

Harvesting

Harvest the crop at the proper edible maturity and do not allow the pods to over mature which impairs its quality. Harvesting is to be completed in number of pickings depending upon the maturity group of peas.

Seed Production

For seed production, seed rate and spacing are same as for table crop. Sow seed crop in second fortnight of November. An isolation distance of 5 metres is kept between two varieties grown for seed production.

Plant Protection

A. Insect Pests

Pests	Symptoms	Control Measures
1. Pea thrips *(Thrips indicus)*	They cause severe damage to the young crop by sucking the cell Sap.	• Remove weeds growing in and around the field • Avoid excessive use of nitrogenous fertilizers
2. Pea-leaf miner *(Chrotomyia horticola)*	The larvae feed by making tunnels in the leaves. These cause serious damage during December-March.	• Maintain good plant health with judicious use of fertilizers
3. Pea-aphid *(Acrythosiphon pisum)*	Aphid sucks the cell sap, Owing to which the leaves turn pale and dry.	• Eradicate weeds growing on field bunds, waste lands, road side and irrigation channels/ canals as they serve as reservoir for aphids.
		• Use fertilizers judiciously, as aphid populations tend to be higher in plants that are fertilized liberally with nitrogenous fertilizers
4. Pea-stem fly *(Ophiomyia phaseoli)*	Pea-stem fly sometimes causes serious damage at the seedling stage.	i) Apply 25 kg Furadan 3G (carbofuran) granules per hectare in furrows at sowing. ii) Sow the crop in the second fortnight of October to escape the damage by the pest. iii) Avoid direct contact with chemicals while treating the seed. iv) Use rubber gloves for application of granules.

B. Diseases

Diseases	Symptoms	Control Measures
1. Powdery mildew *(Erysiphe polygoni)*	White floury patches covering large areas, appear on stem, branches, leaves and pods.	Spray the crop with 1.5 kg Sulfex in 500 litres of water per hectare. Three sprays may be given at 10 days interval.

Diseases	Symptoms	Control Measures
2. Wilt, root rot and collar rot *(Fusarium oxysporum* and *Rhizoctonia solani)*	Root rotting and yellowing of lower leaves, followed by wilting. Reddish brown cankers may appear in the collar region.	i) Avoid early sowing in badly infested areas. ii) Treat the seed with 15 g Talc based formulation of Pseudomonas fluorescens per kg seed before sowing.
3. Rust *(Uromyces viciae fabae)*	Yellowish, reddish brown, spherical, raised pustules appear mainly on the lower side of leaves during December-January. This disease is more serious in the late sown crop.	i) Keep the field free from Rewari weed which serves as a source of inoculum. ii) For controlling of powdery mildew and rust together 500g Sulfex in combination with 1.0 kg Indofil M-45 per hectare.
4. White rot *(Sclerotinia sclerotiorum)*	Disease appears as irregular shaped water soaked spots on leaves, stem and pods; However, the disease is most frequently observed on _ pods which turn papery brown in colour in the later stage. White mycelial growth appears on the infected portion under wet and cool weather. Black hard sclerotial bodies embedded in mycelium are formed within the pods. The disease is high at flowering and pod formation stage.	i) Avoid growing of pea after the susceptible crops like brinjal, cauliflower, carrot and pea, instead tomato and chilli can be grown. This will help in reducing the disease and spread of inoculum in the field. ii) The plant debris of the crop should be collected and burnt.

C. Rodents

For managing rodents in pea crop, regularly remove weeds from the field along with manual killing of rats coming out of flooded burrows during irrigation. If the rodent burrows are more than five/acre, do single burrow baiting with 0.005% bromadiolone bait.

7.7. Cluster Bean /Guara (*Cyamopsis tetragonoloba*)

Cluster bean (*Cyamopsis tetragonoloba*) also known as Guara is one of the important legume crops which is grown for different purposes such as green fodder, grain, green manure and vegetable. It is known for its drought resistance and soil renovation qualities. Guara seed is used as concentrate for animals and for extraction of gum which is an important foreign exchange earner. The gum has several uses in industry and in various food products. The guara meal which remains after extraction of the gum from the seed is a high protein cattle feed.

Rotation

Adopt guara - wheat rotation in light textured soils for saving of water and maintenance of soil health in south-western region of Punjab.

Cluster bean varieties released by ICAR (2020-2021)

Variety	Area of adoption	Characters
X 10	Haryana	Suitable for rainfed under both high and low fertility condition, average seed yield 1.8–2.0 tonnes/ha, maturity 90–95 days, resistant to root-rot and bacterial blight.

Improved Varieties for Punjab

HG-365: It is an early maturing branched variety. It matures in 105 days. Its' average seed yield is 13.3 quintals per hectare.

Ageta Guara 112: Its' plants are erect, hairy, unbranched and medium in height (1-1.5 m). It bears clusters of pods at each node and has bold grains. This variety matures from the last week of October to mid-November. Its' average yield is 20 quintals per hectare.

Guara 80: It is recommended for rainfed cultivation throughout the state. It is tall, quick growing, hairy and profusely branched type. It is resistant to guara leaf blight and stem breakage. This variety does not possess bunches on each node. Its' pods are medium sized and seeds are roundish flat in shape and light grey in colour. It is late maturing variety and yields 17.5 quintals per hectare.

Soil Type

Guara grows well on all types of soil. Well drained, medium to light soils are very suitable.

Land Preparation

One or two ploughings and one planking are sufficient.

Seed Rate and Sowing

Use 20-25 kg seed per hectare. The crop under rainfed and low fertility soil conditions should be sown with the onset of rains, but under irrigated and high fertility conditions, the crop should be sown in the first fortnight of July. Irrigated crop may be sown by using kera method but the rainfed crop should be sown by pora method in rows 45 cm apart. The seed-cum-fertilizer drill can be used for sowing.

Fertilizer Application

Apply 20 kg of N (43.5 kg of Urea) and 47.5 kg of P_2O_5 (3.0 quintal of single superphosphate) per hectare before sowing in irrigated areas.

Irrigation

The crop matures without any irrigation if the rains are normal and timely, otherwise one or two irrigations may be needed. However, the crop should not be irrigated after the third week of September as it delays the maturity of the crop and adversely affects the seed quality

Plant Protection

Jassid suck the sap and damage the crop.

7.8. Cowpea (*Vigna unguiculata*)

The cultivation of cowpea is recommended in the irrigated areas. It can also be grown mixed with maize, bajra and sorghum to get not only higher yield but also increased nutrients from the green fodder. It enables the dairy cattle to maintain good milk yield during the hot and dry summer.

Soil Type

Well-drained loamy soil is conducive to a proper growth of the crop.

Land Preparation

Two ploughings followed by plankings are sufficient.

Seed Rate and Sowing Time:

Use 30 kg/ha seed rate. Sowing time varies from March to July. When sown mixed with maize, seed rate is 15 kg of cowpea and 37.5 kg of maize per hectare.

Cowpea Varieties released by ICAR New Delhi 2020-21

Varieties	Area of adoption	Characters
Pant Lobia 7 (PGCP 24)	Northern part of the country	Suitable for rainfed and irrigated, conditions, yield 11.0–12.0 q/ha, protein content 27%, maturity 70–75 days, and resistant to YMv.
Jammu Lobia Super 60	Jammu Province of Union Territory of Jammu and Kashmir	Suitable for rainfed, kharif season, yield 12.0–12.5 q/ha, maturity 60–65 days, resistant to wilt, YMv and pod-borer, and lodging resistant due to dwarf in stature.
PGCP 6	Karnataka	Suitable for early and late kharif sowing condition, yield 9.0–10.0 q/ha and early maturity (70–75 days).

Improved Varieties for Punjab

CL 367 : It is dual purpose variety suitable for fodder as well as pulse purpose. Its' plants are erect with dark green leaves. It is tolerant to yellow-mosaic virus and anthracnose diseases. Its' fodder quality is superior in terms of total digestible nutrients (TDN) and digestible crude protein (DCP). It bears large number of pods. Its' seeds are small in size and creamish white in colour. The variety is also suitable for human consumption because it has very good cooking quality. On an average, it yields about 108 quintals of green fodder and 12.25 quintals of grains per hectare.

Seed Inoculation

Dilute the 250 ml of liquid biofertilizer (*Burkholderia seminalis*) in 2.5 litres of water and thoroughly mix with the seed for one hectare on a cemented floor. Air dry the treated seeds under shade and sow seed within two hours of inoculation.

Method of Sowing

Sow with a pora or seed-cum-fertilizer-drill or zero till drill in rows, 30 cm apart.

Fertilizer Application

Drill 18.75 kg N (41.25 kg urea) as a starter dose and 55 kg of P_2O_5 (350 kg single superphosphate) per hectare at sowing. In case, cowpea fodder succeeds wheat which had received recommended level of P, omit the application of P to cowpea.

Irrigation and Drainage

The crop sown in May needs fortnightly irrigation till the advent of the monsoon. In all, 4 or 5 irrigations will be sufficient. Adequate drainage results in a good yield.

Harvesting

Harvesting stage of cowpea is between 55-65 days after sowing, which corresponds to pre-flowering stage and would produce good quality fodder.

Plant Protection

A. Insect-Pests

Jassid and Black Aphid: These insects attack this crop.

Bihar hairy caterpillar (Bhaboo Kuta): It attacks cowpea from August to November. To keep cowpea free from attack of this pest, plant a row of til (Sesamum) around cowpea field at the time of its sowing. Females prefer to oviposit on til rows from where larvae in gregarious phase should be collected and destroyed mechanically.

B. Diseases

Seed rot and Seedling mortality: It is caused by various seed-borne microflora. The infected seeds are shrivelled and discoloured. Infected seedlings are killed before these emerge out of soil and cause poor stand of the crop.

8

Latest Practices for Cultivation of Commercial Crops

8.1. Sugarcane (*Saccharum officinarum)*

Sugarcane, (*Saccharum officinarum*), perennial grass crop of the family Poaceae, primarily cultivated for its juice from which sugar is processed. Most of the world's sugarcane is grown in subtropical and tropical areas. The plant is also grown for biofuel production, especially in Brazil, as the canes can be used directly to produce ethyl alcohol (ethanol). The by-products from cane sugar processing, namely the straw and bagasse (cane fibres), can be used to produce cellulosic ethanol, a second-generation biofuel. Other sugarcane products include molasses, rum, and cachaça (a Brazilian alcohol), and the plant itself can be used as thatch and as livestock fodder.

The sugarcane plant produces a number of stalks that reach 3 to 7 metres (10 to 24 feet) high and bear long sword-shaped leaves. The stalks are composed of many segments, and at each joint there is a bud. When the cane becomes mature, a growing point at the upper end of the stalk develops into a slender arrow bearing a tassel of tiny flowers.

Origin and History

Sugarcane was an ancient crop of the Austronesian and Papuan people. It was introduced to Polynesia, Island Melanesia, and Madagascar in prehistoric times via Austronesian sailors. It was also introduced to southern China and India by Austronesian traders around 1200 to 1000 BC. The Persians and Greeks encountered the famous "reeds that produce honey without bees" in India between the sixth and fourth centuries BC. They adopted and then spread sugarcane agriculture. Merchants began to trade in sugar, which was considered a luxurious and expensive spice, from India. In the 18th century, sugarcane plantations began in the Caribbean, South American, Indian Ocean, and Pacific island nations.

Area and Distribution

The largest sugar-producing countries are Brazil, India, Thailand, China, and the United States. In India, sugarcane was grown on an area of 4.79 mha with production of 37050 m tonnes during 2020. In India, Maharashtra, Uttar Pradesh, Karnataka, Tamil Nadu, Bihar and Gujarat are important sugarcane producing states. The three States namely Maharashtra, Uttar Pradesh and Karnataka together contribute to 80% of the total sugarcane production in India.

Sugarcane occupied 89.3 thousand hectares in Punjab during 2020-21. The average cane yield was 838 quintals per hectare. The average sugar recovery was 9.03 per cent.

Climatic Requirements

Sugarcane is best suited to regions having tropical climate, but it can be grown successfully in sub-tropical areas also. In the Punjab, about 80 per cent of the total growth of the crop takes place during July, August and September owing to favourable temperature and humidity.

Uses of Sugarcane

Sugar

Extracting sucrose, the sugar found in the stems, consists in isolating it from the rest of the plant. On entering the mill, each sugarcane batch is weighed and its sugar content is analysed. The stems are then crushed into rough fibre, using a hammer grinder. To extract the juice, the fibres are simultaneously watered with hot water and pressed in a roller mill. The fibrous residue left after juice extraction is called bagasse, and can be used to fuel boilers to generate electricity.

The juice is heated, decanted and filtered after adding slaked lime, then concentrated by heating. This produces a "syrup" free of its "non-sugar" impurities, or scum, which can be used as a fertilizer. The syrup is heated in a pan, until it turns into "massecuite", containing a syrupy liquid, the liquor, and sugar crystals. That massecuite is then heated another twice, alternated with stirring and centrifugation operations, so as to obtain as great a volume of sucrose crystals as possible. The crystals are then sent for drying. The first sugars obtained are various grades of brown sugar. White sugar is produced by refining brown sugar, which is re-melted, decolourized and filtered, before being crystallized and dried. The sugars are then stored in watertight bins. White sugar is the favourite type for both consumers and industrialists.

Making Rum

For thousands of years, sugarcane growing has been associated with fermented drinks. Rum as we know it started out in the Caribbean and Brazil in the 17th century. Traditional rums are made by fermenting molasses, which produces "industrial rum", or by fermenting cane juice, which gives "agricultural rum". The sugary wort is produced by diluting molasses or filtering cane juice.

Fermentation yeasts are multiplied in a mother tank and then transferred to other large tanks in which they convert the sugar in the wort into alcohol (ethanol). It is at this stage that rum aromas are formed. This produces a cane wine with 8 to 10° of alcohol. Fermentation takes between 12 and 36 days for light rums, and up to 10 days for "grand arôme" rums. In Guadeloupe, one tonne of sugarcane stems produces 85 to 120 litres of 55° proof agricultural rum.

Once fermentation is completed, the cane wine is transferred to stills, where it is heated to separate the alcohol from the water. At the top of the still, the alcohol vapours are cooled and collected in condensers. The distiller's art consists in keeping the most volatile aroma molecules, also known as the "spirit" of the rum.

The distillation residue, known as vinasse, is a pollutant. It can be recycled as a fertilizer or as an energy source. The 30 kilos of molasses obtained per tonne of stems give some 110 litres of vinasse.

The rum aromas continue to develop as it matures or ages. Spring water is added to obtain the desired degree of alcohol, and the rum is kept for several weeks in stainless steel vats, resulting in white rum. Rums less than three years old are matured in wooden barrels. Old rums are matured for 3 to 10 years in oak barrels. The degree of alcohol for most rums varies between 37.5 and 62°.

Bagasse, Molasses, Ethanol

Bagasse, which is primarily made up of cellulose, hemicellulose and lignin, is a source of energy (it can be burnt), and of fodder for livestock. It is also a raw material for making paper, cardboard, thermal insulation, films and textiles. Molasses is used to feed animals, grow yeasts are produce acids. The sucrose undergoes a range of chemical processes to make ethers and esters, which in turn serve to make plasticizers, adhesives, cosmetics, varnishes, etc.

The ethanol obtained after fermenting the juice or molasses can be used pure as a biofuel. In Brazil, more than half of the sugarcane crop is used to make fuel ethanol, and most cars run on ethanol. Cane varieties have been specially bred for ethanol production. The ideal solution for larger-scale production would be to use the whole plant: research is under way to produce ethanol directly from cellulose.

Press Mud

Sugarcane press mud is the residue of the filtration of sugarcane juice. The clarification process separates the juice into a clear juice that rises to the top and goes for manufacture, and a mud that collects at the bottom. The mud is then filtered to separate the suspended matter, which includes insoluble salts and fine bagasse.

Sugarcane press mud contains many nutrients, so it can be applied as organic manure to the crop and composted for value addition for easy uptake by the plants. Generally, the compost sample contains C/N ratio of 725.95, total potassium of 1.43%, and total organic carbon of 12.53%.

Sugarcane Fodder

Whole sugarcane may be used as green forage and can provide a valuable dry season fodder. Sugarcane forage can be used in a wide range of ways. Sugarcane stalks may be grazed, cut for cut-and-carry systems or directly chopped in the field for further use in the green form or for silage.

Sugarcane, along with the green fronds at the top, is being cut into small billets and fed to the cattle as fodder.

Sugarcane Vinegar

Sugarcane Vinegar is obtained from raw materials containing sugarcane juice by a two-step fermentation process: one involves ethanol formation by yeasts (usually *Saccharomyces* spp.) by the conversion of fermentable sugars (alcoholic fermentation) and the second involves the oxidation of ethanol to acetic acid (acetic acid fermentation or acetification)

Sugar Cane Vinegar is rich in phosphorus, calcium, iron, magnesium and potassium. Sugarcane is an alkalizing substance, and therefore is beneficial at fighting cancer. Studies have shown that it has proven an effective at fighting prostate and breast cancer cells. It is exceptionally beneficial for rehydration as it not only cools the body but also energizes with a high quantity of useful carbohydrates and protein. It is helpful to inner organ functioning, including the kidneys, heart and brain.

Gur (Jaggery) Making

Gur is a natural sweetener rich in minerals like calcium, iron, phosphorus etc. Making of gur and shakkar from sugarcane involves juice extraction, clarification, boiling and concentration, cooling and moulding. Use an efficient cane-crusher for extracting juice. For clarifying the juice, add Sukhlai emulsion. Sukhlai is a shrub which grows in the Shivalik Hills. For preparing emulsion, soak the dry bark of Sukhlai in a bucket of water for 24 hours. Then rub the bark to obtain a thick mucilaginous fluid. Add about one litre of this fluid to 100 litres of cane juice, when the scum begins to rise in the pan. Towards the end of boiling process, heating is regulated to avoid charring. The concentrated juice is transferred to the cooling pan when its temperature reaches the striking point (114-116°C) for gur making and (120-122°C) for shakkar making. Gur is then moulded into desired shapes and sizes using moulding frames. For shakkar, at the time of solidification the material is made into powder manually using wooden scrappers. Powder is dried to 10.5- 11.5 % moisture content, sieved and packed in polythene bags.

Sugarcane varieties released by ICAR (2020-2021)

Variety	Area of adoption	Characters
Co 15023	Uttar Pradesh (central and western), Uttarakhand, Haryana, Punjab, Rajasthan, Delhi	Suitable for early group under irrigated condition, average cane yield 89.17 tonnes/ha, CCS 12.16 tonnes/ha and sucrose 19.41% in juice, resistant to moderately resistant to red-rot.
Co 13013	Tamil Nadu, Kerala, Interior Andhra Pradesh, Telangana, Karnataka, Gujarat, Maharashtra, Madhya Pradesh	Suitable for mid-late group under irrigated condition, average cane yield 121.96 tonnes/ha, CCS 14.75 tonnes/ha and sucrose 19.01% in juice, tolerant to drought and salinity, moderately resistant to red-rot.
GNS-11 (Navya)	Gujarat	Suitable for early maturing group under irrigated condition, average cane yield 133.28 tonnes/ha, moderately resistant to wilt and red-rot, resistant to whip smut, a good ratooner.
VSI 12121	Gujarat, Madhya Pradesh, Chhattisgarh, Maharashtra, Karnataka, Telangana, Andhra Pradesh, Tamil Nadu, Kerala	Suitable for growing under medium black to well drained with good fertility soil in Peninsular Zone under irrigated condition, fit for planting during autumn and spring seasons, average cane yield 124.7 tonnes/ ha, CCS 18.22 tonnes/ ha, sucrose 20.07% in juice, maturity 360–365 days, tolerant to drought and salinity stresses, moderately resistant to red-rot.
Haryana Ganna 160	Haryana	Suitable for autumn and spring planting, average cane yields 85 tonnes/ha, early maturity group (300 days), resistant to red-rot, non-lodging and multi-ratooner variety.

Improved Varieties for Punjab

1. Early-Maturing Varieties

CoPb 95: The canes of this variety are tall, thick with zigzag internode alignment and purplish green in colour with broad leaf canopy. Its' juice contains 17% sucrose in the month of December. It has field resistance to the prevalent pathotypes of red rot disease, moderately resistant to smut and also less susceptible to top borer. It is a good ratooner, non-lodging and frost tolerant. The average cane yield is about 1063 quintals per hectare.

CoPb 96: The canes of this variety are medium thick, cylindrical and yellowish green in colour. Its' juice contains 16-17% and 18% sucrose in the months of November and December, respectively. It is a good ratooner. It is tolerant to the prevalent pathotypes of red rot disease. Quality of gur is very good. The average cane yield is about 955 quintals per hectare.

Co 15023: The canes of this variety are medium thick, cylindrical and yellowish green in colour. Its' juice contains 16-17 and 18% sucrose in the month of November and December, respectively. It is tolerant to the prevalent pathotypes of red rot disease. The average cane yield is about 775 quintals per hectare.

CoPb 92: The canes of this variety are tall, medium thick and purple green in colour. Its juice contains 16-17% sucrose in the month of November and 18% in December. It is a good ratooner. It is tolerant to most of the prevalent pathotypes of red rot disease and frost. Quality of gur is also good. The average cane yield is about 838 quintals per hectare.

Co 118: It has thick, purple green coloured canes. Its juice contains 16% sucrose in the month of November and 17% in December. It has field resistance to the prevalent pathotypes of red rot disease. It is a shy tillering variety and an average ratooner. Its gur quality is excellent. The average cane yield is about 800 quintals per hectare.

CoJ 85: It is a shy tillering variety with thick green coloured canes. Its juice contains 16-17% sucrose in the month of November and 18-18.5% in December. It is tolerant to most of the prevalent pathotypes of red rot disease and is tolerant to frost. It is an average ratooner and susceptible to red stripe disease. Due to its heavy canes it is prone to lodging, hence requires proper earthing up. The average cane yield of plant crop is about 763 quintals per hectare.

CoJ 64: This variety is a good germinator, with profuse tillering and medium compact growth. Its canes are medium thick, greenish yellow and solid. The sucrose content in juice in November is 16-17%. It has become highly susceptible to red-rot disease and also to top borer. Its quality of gur is excellent. The average cane yield is about 750 quintals per hectare.

2. Mid-Late Maturing Varieties

CoPb 98: The canes of this variety are tall, thick, cylindrical and yellowish green in colour. Its' juice contains 17% and 19% sucrose in the months of January and March, respectively. It is a good ratooner. It is tolerant to the prevalent pathotypes of red rot disease. The average cane yield is about 1000 quintals per hectare.

CoPb 93: The canes of this variety are tall, thick and yellowish white in colour. Its juice contains 17% sucrose in the month of January and 19% in March. It is a good ratooner. It is tolerant to the prevalent pathotypes of red rot disease. Quality of gur is also good. The average cane yield is about 875 quintals per hectare.

CoPb 94: The canes of this variety are tall, thick, cylindrical and yellowish green in colour. Its juice contains 16% and 19% sucrose in the months of January and March, respectively. It is a good ratooner. It is tolerant to the prevalent pathotypes of red rot and smut diseases. The average cane yield is about 1000 quintals per hectare.

Co 238: The canes of this variety are tall, medium thick and yellow green in colour. Its juice contains 17% sucrose in the month of January. It is susceptible to top borer and tolerant to the prevalent pathotypes of red rot disease. It is a good ratooner. Quality of gur is also good. The average cane yield is about 913 quintals per hectare. The crushing of this variety by sugar mills can be done along with early maturing varieties.

CoPb 91: The canes of this variety are tall, thick and yellowish green in colour. Its juice contains 17% sucrose in the month of January. It is tolerant to the prevalent pathotypes of red rot disease. It is a good ratooner. Its average cane yield is about 1025 quintals per hectare.

CoJ 88 : The canes of this variety are tall, medium thick and greyish green in colour. Its juice contains 17-18% sucrose in the month of December. It tends to behave as early-mid in maturity and is also suitable for saline water conditions. It is tolerant to the prevalent pathotypes of red rot disease. It is an excellent ratooner. It is good for cogeneration. Quality of gur is excellent. The average cane yield of the plant crop is about 838 quintals per hectare.

1. Spring Sugarcane

Rotations

Rice (short duration) – Rabi fodder/Potato - Sugarcane - ratoon I - ratoon II – Wheat

Rice / kharif fodder - Toria / Raya - Sugarcane - ratoon I - ratoon II – Wheat

Maize/Cotton – Senji (fodder) - Sugarcane - ratoon I - ratoon II – Wheat

Arhar - Oat/Senji (fodder) - Sugarcane - ratoon I - ratoon - II – Wheat

Land Preparation

Give three to four ploughings, each followed by planking. Use a furrow turning plough for the first ploughing.

Time of Planting

Mid-February to the end of March is the optimum time for planting sugarcane in north India. Do not plant early maturing varieties after March. Avoid late planting. If at all late planting has to be done, adopt the following practices:

- In case of late planting, any mid late maturing variety should be preferred.
- Use higher seed-rate, viz. 75 thousand three-budded sets per hectare.
- Control the early shoot-borer effectively as it is particularly serious in the late-planted crop.

Seed Selection

The seed should be free from red-rot, wilt, smut, ratoon-stunting and grassy shoot diseases. Use only the top two-third portion of the selected canes for planting.

Seed Rate

Use 50 thousand three-budded setts or 37.5 thousand four-budded sets or 30 thousand five-budded setts per hectare. Longer sets are particularly good for rain-fed conditions. In other words, 75-87.5 quintal of seed, depending upon the variety, is required for sowing one hectare and increase seed rate by 10% higher (on weight basis) of thick cane varieties.

Seed Treatment

To improve the germination, soak the setts in ethrel solution overnight. To prepare the solution, dissolve 25 ml of Ethrel 39 SL in 100 liters of water. Alternatively, soak the setts in water for 24 hours before planting.

Spacing and Planting Techniques

1. **Trench Planting:** Plant crop in rows 75 cm apart and 20-25 cm deep trenches. After placing the setts in trenches, cover the setts with 5 cm soil. Apply irrigation immediately after planting, if the planting is not done under proper moisture (watter)conditions. Repeat the irrigation at 4-5 days interval.
2. **Paired Row Trench Planting:** Adopt paired row trench planting for saving irrigation water. Plant two rows of sugarcane 30 cm apart in 20-25 cm deep trenches. Place the cane setts at the bottom of the trenches and cover with

the soil left in between two rows. Distance between two trenches should be 90 cm. In addition to water saving, this method gives comparatively higher cane yield, easy propping up operation and reduces lodging. Use tractor operated paired row trench planter to sow paired rows of sugarcane in trenches. The complete sugarcane, is fed by two labourers sitting on the machine, is cut automatically into pieces before dropping in the furrows, fertilizer is applied simultaneously. Length of setts varies from 36 to 38 cm. The speed of operation is 2-3 km/h. Capacity of machine varies from 2-3 acres per day. For ease of mechanical harvesting with sugarcane harvester, the crop should be planted at either 120 cm row spacing or by 120:30 cm paired row trench method.

3. **Furrow Irrigated Raised Bed Planting (FIRB):** Sugarcane can also be planted in standing wheat crop sown by furrow irrigated raised bed (FIRB) planter. The furrow should be reshaped in January to loosen the soil. Apply irrigation in reshaped furrows preferably in the evening before planting. Plant treated sugarcane setts next day by pressing under feet. Sugarcane is planted in pre-opened furrows between the beds, using treated 3 budded setts, during the second fortnight of February to March.

4. **Sugarcane Cutter Planter:** Use two-row tractor operated whole cane cutter planter. The complete sugarcane which is fed by the labour sitting on the machine is cut automatically into pieces before dropping into the furrows. Fertilizers and chemicals are also applied simultaneously. The seed rate varies from 80 to 87.5 q per hectare. The labour requirement is 33 man-h/hectare. Length of setts varies from 23 to 42 cm. The speed of operation is 1.20 to 1.90 km/h. The capacity of machine varies from 2-3 acres/day. The machine can save about 25% cost of operation in comparison to traditional method. Use this machine on custom hiring basis.

Intercropping

Intercrop one row of the recommended varieties of summer moong or summer mash in between two rows of sugarcane to get an additional grain yield of 9 to 12.5 q/hectare of summer moong/summer mash. This does not affect the cane yield and improves the soil fertility. The details are as under:

Practice	Summer moong	Summer mash
Seed rate per hectare	10 kg	12.5 kg
Time of sowing	March 20 to April 10	March 15 to April 7

Mentha can also be grown as an intercrop. Plant one row of mentha between two rows of sugarcane. Mentha and sugarcane can be planted simultaneously in the first fortnight of February. Use 2.5 quintal of mentha suckers per hectare. In addition to fertilizers recommended for sugarcane, apply 45 kg N (98 kg Urea) and

25 kg P_2O_5 (155 kg Super Phosphate) per hectare. Half N and full phosphorus may be applied at planting and remaining half N about 40 days after planting.

Okra can also be grown as an intercrop. Intercrop two rows of okra 45 cm apart in between the two rows of sugarcane planted either at 90:30 or 120:30 cm paired row planting systems in the second fortnight of February. Use 27.5 kg seed of okra per hectare for 90:30 cm planting system and 25 kg for 120:30 cm. Fertilizers to the intercrop should be applied as recommended for sole crop of okra. Terminate the okra crop in the second fortnight of June to avoid the intercrop competition.

Fertilizer Application

Organic and Bio-fertilizers: Apply 20 tonnes of FYM/press-mud per hectare 15 days before planting and mix into the soil with a plough or tiller. In case of trench planting, apply press-mud at the base of the trench and mix it into soil with a kasola. In case FYM/pressmud is applied, use 100 kg N/hectare (225 kg urea). However, on coarse textured soils if FYM is applied along with recommended dose of nitrogen, approximately 10% higher yields can be obtained. On sandy soils, nitrogenous fertilizer may be applied after irrigation in moist soil and the number of splits may be increased. Application of Azotobacter/Consortium biofertilizer @ 10 kg/hectare in the furrows at the time of planting would be helpful in increasing the cane yield. If FYM is in limited quantity, apply 10 kg consortium bio-fertilizer with 10 tonnes FYM per hectare to plant as well as ratoon sugarcane crop.

Chemical Fertilizers: Apply fertilizers on soil test basis. In the absence of a soil test, apply the following quantity of fertilizers on medium fertility soils:

Crop	Nutrients (kg/hectare)		Fertilizers (kg/hectare)	
	N	P2 O5	Urea	Single Superphosphate
Plant crop	150	#	325	#
Ratoon crop	225		487.5	

#If the soil is low in available phosphorus, apply 30 kg P_2O_5 (187.5 kg single super phosphate) per hectare at sowing.

• Apply only 250 kg urea per hectare if, sugarcane is grown after potato crop.

Method of Fertilizer Application

(i) **Plant Crop:** Apply one half dose of nitrogen as top dressed/drilled alongside the cane rows with first irrigation after germination. Top dress or drill the remaining half dose of nitrogen along-side the cane rows in May or June. Apply full dose of phosphorus (based on soil test) in furrows below the cane setts at the time of planting.

(ii) **Ratoon Crop**: Top-dress one third of nitrogen in February with first hoeing, one third in April and the remaining one-third in May. Phosphorus (on soil-test basis) should be drilled along the cane rows at the time of first cultivation in February.

(iii) **Rainfed Crop:** If the moisture in the soil is optimum, apply one half of the dose of nitrogen at planting. In case the moisture is deficient, the whole dose should be applied with the onset of rains.

Iron deficiency: Iron deficiency has been observed both in the ratoon and plant crops on light-textured and calcareous soils. Deficiency symptoms first appear in young leaves as yellow stripes between the green veins. Later, the veins also turn yellow. In severe cases, leaves become white and the plants remain stunted. Spray the crop 2 or 3 times with 1% solution of ferrous sulphate (1 kg ferrous sulphate in 100 litres of water) at weekly intervals soon after the symptoms appear.

Weed Control

Cultural Control: Two or three hoeings can be done with a triphali or tractor mounted tiller or tractor operated rotary weeder. The spreading of trash-blanket between the cane rows after the emergence of the shoots helps to suppress weeds. This practice has the added advantage of conserving soil moisture, particularly in rainfed areas.

Chemical Control: Pre-emergence application of 2.0 kg per hectare Atrataf/ Solaro/ Masstaf/Markazine 50 WP (atrazine) or Sencor 70 WP (metribuzin) or Karmex/Klass 80 WP (diuron) or 2.5 kg per hectare Authority NXT 58 WP (sulfentrazone+clomazone) in 500 litres of water within 2-3 days of planting effectively controls the broadleaf weeds and annual grasses. For control of hardy weed like Bans Patta, use only Sencor 70 WP or Karmex/ Klass 80 WP. For fields infested with dila, post-emergence application of 2.0 kg per hectare 2, 4-D sodium salt 80 WP in 250 litres of water is recommended. In fields infested with *Ipomoea spp.* (lapeta vel) and other broad leaf weeds, apply 2.0 kg of 2,4-D sodium salt 80 WP or 1.0 litre per hectare 2,4-D amine salt 58 SL by dissolving in 250 litres of water when these weeds are at 3 to 5 leaf stage.

Straw Mulching: After complete germination by mid-April, uniformly spread paddy straw or rice husk or sugarcane trash or tree leaves at the rate of 50-62.5 quintals per hectare between the rows. Mulching reduces soil temperature and conserves soil moisture. It also suppresses weeds and reduces the incidence of shoot-borer. Straw-mulching increases the yield of sugarcane both under rainfed and irrigated conditions.

Irrigation and Drainage

Hot and dry period during April to June is the most critical period for the growth of sugarcane. During this period, irrigate the crop at 7 to 12 days interval. During the rainy-season, adjust the frequency of irrigation according to rainfall. Drain away excess water from the sugarcane fields, if these get flooded during the rainy season. During winters (November to January), irrigate the crop at monthly intervals. To save the crop from frost, apply one irrigation around mid-December and another in the first week of January.

Prevention of Lodging: To prevent the crop from lodging, adopt the following measures:

- Earth up heavily the flat-planted as well as the trench-planted crop at the end of June, before the onset of the monsoon. Trench-planting is particularly effective in preventing lodging.
- Prop up the crop in the end of August or in the beginning of September by using the trash-twist method which consists of the tying of a single cane row instead of tying two rows together. Twist the leaves and the trash to make a rope and pass it alternately along the cane clumps in the row. This method does not hinder the growth and photosynthesis of the plants as in the case when two cane rows are tied together.

Protection from frost: Protect the crop from frost as under:

- Grow frost-resistant varieties of sugarcane.
- Raise a bumper crop with adequate fertilization, irrigation and plant protection measures, because a poor and stunted crop suffers more from frost.
- Prevent lodging. A lodged crop is more damaged by frost.
- Irrigate the crop, as adequate soil-moisture during the frosty period keeps the soil comparatively warm and saves it from frost. Irrigate the harvested fields if meant to be ratooned. Plough the patch of land between the cane rows.
- In frosty areas, plant setts only from the top portion of the cane, as these buds are less damaged by frost. The top portion can be buried in the soil during the frosty spell and taken out in the spring for planting.

2. Autumn Sugarcane

Sugarcane can also be grown successfully as an autumn crop with various intercrops. Growing intercrops in autumn cane will enhance the total productivity/ net profit per unit area per unit time.

Rotations

Kharif fodder/Green manure/Maize/Rice (short duration)/Moong-Sugarcane with intercrop.

Time of Planting

The best time of planting autumn sugarcane is from 20 September to 20 October. The planting should not be delayed, otherwise yield will decrease.

Seed Rate

Plant 50 thousands of three budded or 37.5 thousands of four budded or 30 thousands of five budded setts per hectare. The seed for autumn planting should be obtained from a well grown spring or autumn crop.

Spacing and Method of Planting

Flat planting in rows 90 cm apart. Adopt paired row trench planting method as described under spring planting.

Intercropping

For higher returns per unit area, adopt intercropping of wheat, raya, gobi sarson or toria in autumn sugarcane.

Fertilizer Application

Apply fertilizers on soil test basis. In the absence of a soil test, apply the following quantity of fertilizers on medium fertility soils:

Nutrients (kg/hectare)		Fertilizers (kg/hectare)	
N	P_2O_5	Urea	Single Superphosphate
225	#	487.5	#

If the soil test is low in phosphorus, apply 30 kg P_2O_5 (187.5 kg single super phosphate) per hectare at sowing. Sugarcane does not respond to potash in Punjab.

Apply urea in three equal doses, 1/3 dose of N should be applied at planting, 1/3 at the end of March and the remaining 1/3 by the end of April.

Irrigation

Apply first irrigation one month after planting, followed by three irrigations upto February and subsequent irrigations as per prevailing weather conditions, soil type etc.

Ratoon Management

Improve the yield of the ratoon crop as under:

- Do not harvest the crop to be ratooned before the end of January. If the crop is harvested earlier, there will be poor sprouting of the stubbles due to low temperature during December and January.

- Soon after the harvesting of early varieties in November or December, remove the trash and irrigate the field. When the soil attains the optimum moisture conditions, loosen it by hoeing, ploughing or interculture. Do not cover the stubble with cane trash.
- Harvest the canes as close to the ground as possible to promote better sprouting. If still some big stubbles are left, shave or lop them off close to the ground. Also remove late tillers or water-shoots, as they inhibit full sprouting of the stubbles.
- Plough the harvested field twice with a tractor-drawn tiller to check weeds or adopt chemical weed-control measures.
- The stand of the ratoon crop can be improved by planting the gaps with three budded setts in the beginning of March.
- Nitrogen requirement of the ratoon crop is one and a half-times than that of the plant crop. Apply 225 kg N (487.5 kg urea) per hectare to the ratoon crop in three split doses i.e. one third in February-March, one-third in April and the remaining one-third in the beginning of June. Drill phosphorus along the cane rows in March on the basis of a soil test.
- Shoot-borer, top-borer and black bug appear in the ratoon crop. Control them as soon as noticed.

Plant Protection

A. Insect-Pests

Termite: The termite appears during April to June and again in October. It destroys the germinating buds and causes the drying up of shoots after germination. To avoid its attack, apply only well-rotten farmyard manure and remove the stubbles and debris of previous crop from the field. For the control of termites, apply 500 ml Coragen 18.5 SC (chlorantraniliprole) per hectare using 1000 litres of water over seed setts in furrows before covering them with soil or spray 112 ml Imidagold 17.8 SL (imidacloprid) in 1000 litres of water per hectare with sprinkler along the rows at post germination stage (about 45 days after planting).

Early shoot-borer: This pest appears from April to June and causes dry dead-hearts which can be easily pulled out. To control it:

- Plant the crop early, i.e. before the middle of March.
- Apply 25 kg granules of Regent/Mortel/Rippen 0.3 G (fipronil) per hectare before the setts are covered with soil by planking. or
- Apply 25 kg Regent/ Mortel/Rippen 0.3 G (fipronil) mixed in 50 kg moist sand/soil or 375 g Takumi 20 WG (flubendiamide) or 375 ml Coragen

18.5 SC (chlorantraniliprole) or 5.0 litres of Durmet/Classic/Dursban/Markpyriphos 20 EC (chlorpyriphos) in 1000 litres of water/hectare along the rows at post-germination stage (about 45 days after planting). Earth up slightly followed by light irrigation.

OR

Use Tricho-cards having 50,000 eggs of *Corcyra cephalonica* parasitized (seven days old) by *Trichogramma chilonis* per hectare at 10 days interval from mid-April to end-June. These eggs are fixed on cards of 25x15 cm size. Cut the cards into 100 pieces/strips, each having approximately 500 parasitized eggs. Staple these pieces/strips on the lower surface of the leaves uniformly at 100 spots per hectare during evening hours. Normally 8 releases are required. The tricho-cards should not be stapled during rainy days. Use of 25 pheromone traps per hectare from April to June along with Tricho-card application. Change the pheromone lure at one month interval.

Black bug: This pest is active during April to June. The attacked crop looks pale. The black adults and pink young nymphs suck the sap from the leaf-sheaths. Spray the crop with 875 ml of Dursban/Lethal/Massban/Goldban 20 EC (chlorpyriphos) in 500 litres of water per hectare with manually operated sprayer. Direct the spray into the leaf-whorl.

Pyrilla: Pyrilla reduces cane yield and sugar recovery heavily. This pest appears in April-May and again in August-September. The leaves of the damaged crop turn yellow. Later due to the development of a fungus, the crop turns black and the tops become unfit for feeding to cattle. The incidence of this pest is particularly high in a luxuriant crop and in the interior of field. When the attack is severe it becomes difficult to make gur. Spray the crop with 1.5 litre Dursban 20 EC (chlorpyriphos) in 500 litres of water per hectare with manually operated sprayer.

Top-borer: This pest appears from March to October and causes severe damage during July-August. The typical symptoms are the shot-holes in the leaf, white or red streaks on the upper side of the leaf midrib and bunchy tops from July onwards. The central leaf of the cane top dries up and turns dark. To control it:

- Collect and destroy the moths and egg-clusters.
- Cut the attacked shoots at the ground level from April to June.
- Use Tricho-cards having 50,000 eggs of *Corcyra cephalonica* parasitized (seven days old) by *Tricogramma japonicum* per hectare at 10 days old interval from mid-April to end June. The method to use these cards is given under early shoot borer.
- Use of 25 pheromone traps per hectare from May to August along with recommended Tricho-card application. Change the pheromone lure at one month interval.

- Apply 25 kg granules of Ferterra 0.4 GR or 30 kg Furadan/Diafuran/ Furacarb/ Carbocil/ Fury encapsulated 3G (carbofuran) per hectare at the base of the shoots in the last week of June or in the first week of July only if the top borer damage exceeds 5% level. Earth up slightly to prevent the granules from flowing with the irrigation water and irrigate the crop immediately. This operation will control the third brood of the top-borer which does the maximum damage.

Take the following precautions in using carbofuran:

- Use rubber gloves while applying carbofuran granules. Never handle these granules with bare hands. A mask should be used to cover the face.
- Mix it with the moist soil to reduce the chances of its falling into the eyes of the person applying it.
- The person applying these granules should not eat or drink anything without washing his hands thoroughly with soap.
- Carbofuran treated sugarcane leaves and weeds should not be fed to cattle for about one month after the treatment.

Stalk borer or Tarai borer: This pest is active throughout the year. The larvae overwinter in the stubble and water-shoots. The attack remains low during April-June and increases in July. Its incidence is highest during October-November. There are no outward symptoms of the attack of this pest. Entrance or exit holes on the attacked canes can be seen only after stripping. A larva sometimes damages upto 3 nodes and the cane may be attacked at several places. The cane yield and sugar recovery are adversely affected in the case of serious attack. The control measures against the pest are as under:

- Do not use the cane-seed from the infested field.
- Use tricho-cards having 50,000 eggs per hectare on of *Corcyra cephalonica* parasitized by *Trichogramma chilonis*, 10-12 times at 10 days interval from July to October.
- Use of 25 pheromone traps per hectare from July to October along with recommended Tricho-card application. Change the pheromone lure at one month interval.
- At harvest, do not leave the water-shoots in the field.
- Do not ratoon a heavily infested crop. Plough the affected fields, collect and destroy the stumps.

Whitefly: The damaged crop looks pale during August-October. The underside of the leaves is full of nymphs and pupae which suck the sap from the leaves. The leaves turn black due to the development of a fungus.

Sugarcane thrips: The thrips damage the crop from April to June. This pest suck the sap from the partly opened leaf and tips of the younger leaves, resulting in withering and drying of leaftip, which get rolled inwardly. The thrips prefer plant crop than ratoon crop.

Sugarcane mite: The mite appears from April to June and feeds on the lower side of the leaves under a fine web. The leaves turn red and later appear to be burnt. The growth of the affected crop is retarded during the pre-monsoon period. Baru (*Sorghum halepense*) is the alternative host plant from which this mite spreads to the sugarcane crop. So destroy the weed, if growing near the sugarcane fields.

Gurdaspur borer: This borer appears from June to October and causes the withering of the central leaves (notably the 5th leaf) followed by the total drying up of the tops. The affected canes break at the point of attack with a slight jerk. Rogue out canes showing withered tops in the afternoon every week from June to September. The tops should be cut off well below the point of attack. The timely rogueing of affected plants is very important for controlling the pest. Bury the rogued-out plants. Do not ratoon a heavily affected crop. Plough up the fields not meant for ratooning and destroy the stubbles before June.

B. Diseases

Red rot: Red rot is caused by the fungus *Colletotrichum falcatum*. The disease appears from July till the crop is harvested. The third or fourth leaf of cane from top shows yellowing at first, while rest of the leaves also loose colour afterwards and dry up. Later, the whole clump dries up. On splitting open the cane, the tissue is found to be reddened but the discoloration is not uniform and is interspersed with white patches running across the width of the split cane. The pith of affected cane emits alcoholic smell. The control measures against this disease are as under:

- Use seed from absolutely disease free seed plot.
- Do not plant sugarcane in the disease affected fields for one year.
- Grow varieties fairly resistant to red rot.
- Crush the affected crop early and plough up the fields soon after harvesting the crop. Collect and burn the stubbles.
- Rogue out and bury the diseased canes. Uproot the entire clumps and not merely the affected stalks.
- Do not ratoon the diseased crop.

Wilt: This disease is caused by *Cephalosporium sacchari* or *Fusarium moniliforme*. It appears from July till the crop is harvested. The leaves of the affected cane at first turn yellow and finally the top dries up. On splitting open a diseased cane, the pith shows a dirty red discoloration near the nodes. The discoloration is invariably darker than that in the remaining portion of the internodes. The affected stalks become light and hollow. The control measures against this disease are the same as those for red rot. As the causal fungus persists in the soil over long period, the affected field should not be put under sugarcane for 3 years.

Smut: Smut is caused by *Ustilago scitaminea*. This disease is prevalent throughout the year but is severe from May to July and again in October-November. Its incidence increases in the ratoon crop. It is easily recognised by the appearance of long whip-like shoots covered with dusty black mass of spores. These whips may arise from the top of the canes as well as from the lateral sprouted buds. Adopt the following control measures:

- Use only smut free canes for seed. Reject even the healthy looking canes in the diseased stools or those growing in the vicinity of the smut infected clumps.
- Remove the smut whips gently (without shaking) after putting them inside a closely woven drill bag. Then uproot the entire diseased clumps and bury them deep. Immerse the bag used for collecting the smut whips in boiling water for 5 minutes after every roguing of the crop.
- Do not ratoon the smut infected crop.

Ratoon Stunting: A coryniform bacterium (*Clavibacter xyli*) has been found to be associated with the disease. The affected crop remains stunted with thin canes. The leaves are comparatively pale and the roots are poorly developed. The disease can be identified by slicing mature canes longitudinally a little below the rind with a sharp knife. In the lower part of the node, parallel to the zone of the whitish waxy band, the pith shows discolored dots, commas and straight or bent streaks upto 2 to 3 mm in length. They may be yellow, orange, pink, red or reddish brown. Do not use the diseased crop for planting. Select the cane-seed from a vigorously growing and healthy crop. The moist hot air treatment of seed canes at 54°C for 4 hours is effective in destroying the causal organism. Do not ratoon the diseased crop.

Grassy Shoot Disease: The disease is caused by mycoplasma like bodies. The affected plants give rise to numerous thin tillers. The leaves become reduced in size, thin, narrow and usually turn chlorotic. If the attack is light, one or two weak canes may be formed. Uproot and destroy the affected clumps immediately after appearance. The moist hot air treatment of the seed-canes at 54ºC for 4 hours inactivates the causal organisms of this disease. Its incidence increases in the ratoon crop, therefore, do not ratoon the diseased crop.

Red Stripe: Red stripe is a bacterial disease caused by *Pseudomonas rubrilineans*. It appears during June-August. The affected leaves show bright red streaks which are long, narrow and run longitudinally on the leaf-blade, causing the rotting of tops in severe cases. Rogue out the affected canes and burn or bury them.

Top Rot (Pokkah boeng): This disease is caused by *Fusarium moniliforme*. It appears during the rainy season from July–September. The young leaves in the top portion of the plant become chlorotic at the base and get distorted and shortened. They turn dark red and fall off gradually. In severe cases, the rotting of the top portion of the cane occurs. Remove the affected clumps and bury them.

Leaf Scald: The disease is caused by the bacterium *Xanthomonas albilineans*. Whitish or cream coloured one or two narrow stripes are observed on the leaf extending sometimes down to leaf sheath. The affected plants produce side-shoots starting first from lower nodes with similar stripes on young leaves. The stripes become reddish and later the leaves start withering from top downwards giving scalded appearance. On splitting open the affected canes, reddish brown vascular streaks are observed in the internodes. Sometimes affected plants suddenly wilt and die without any obvious internal symptoms. As the disease is settborne, healthy and disease free seed should be planted. Treatment of seed-cane with moist hot air at 54°C for 4 hours inactivates the bacterium. Sterilization of cutting knives by flaming or by dipping in 2% Lysol solution during seed preparation should be practised to minimise spread of the disease. Rogue out the diseased clumps.

C. Rodents

Being a long duration crop, sugarcane provides shelter to rodents and suffers heavy damage. The rat, *Bendicota bengalensis*, which digs extensive burrows with characteristic soil heaps, is often abundant in sugarcane. A lodged crop gets highly damaged.

Management of Seed Crop

To obtain disease-free seed, a separate seed nursery should be maintained. Do not use the commercial crop for seed, as many pests and diseases go unnoticed in such a crop. Alternatively, tissue culture raised plants can also be used to raise a healthy seed crop. The plants should be spaced 60 cm apart with a row to row spacing of 90 cm, followed by immediate irrigation. The crop thus raised, should be used for raising a subsequent seed crop by planting three budded setts, following conventional practices. For seed production, the following package of practices are recommended:

- Plant the crop in the last week of March.
- Obtain the seed stalks propagated from moist hot-air-treated seed. The treated seed is planted at the Research Stations and is further propagated at the sugar factory farms and the farms of selected cane growers. This seed is supplied to the growers to raise the seed-crop.
- Give a fertilizer dose of 225 kg of N (487.5 kg urea) per hectare. Apply nitrogen in 3 equal doses i.e. at planting, in May and in mid-July. High dose of nitrogen will result in good quality of immature seed cane.

- Follow plant protection schedule strictly, to keep the crop free from insect pests and diseases.
- Frost injury results in low germination of sugarcane. Therefore, protect the seed crop against frost by irrigating it frequently during December and January.

8.2. Sugarbeet (*Beta vulgaris* L.)

Sugarbeet (*Beta vulgaris*) is mainly used for production of sucrose, a high energy pure food. Sugarbeet pulp and molasses, the by-products of sugarbeet are widely used as feed supplements for livestock. Beet tops (leaves and petioles) can also be used as silage. It accounts for nearly 21.8% of the world's total sugar production. It is also a potential source of ethanol which is now blended in automobile fuel. Sugarbeet is essentially a crop of temperate origin. It is grown for commercial sugar production mainly in European countries; U.S.A and Canada in North America, Chile in South America, Egypt and Morocco in Africa and China, Iran, Japan, Syria and Pakistan in Asia. In India, Maharashtra, Karnataka, Andhra Pradesh, Tamil Nadu, Gujarat, Punjab, Haryana, Rajasthan and Western U.P. are suitable States for its cultivation.

Climate and Season

Sugarbeet can be produced in hotter and more humid environments. Optimum temperature are 60-80°F. The most favourable environment for producing a sugarbeet crop from 90 days after emergence to harvest, is bright, sunny days with 60-80°F temperatures followed by night time temperatures of 40 to 50°F.

Varieties

Diploids- 11SR Comp. 1. L S6, Ramonskaya 06

Diploid Hybrids- Shubra, Cauvery, Indus 84- 32

Soil Requirements

Sugarbeets is well adapted to wide range of soil types. Soil should be free of clods because clods cause problems for sugarbeet planting, thinning, harvesting and processing equipment. Dryland sugarbeet production is generally limited to soils with high water holding capacities. It can tolerate pH of about 6.5 – 8.0.

Land Preparation

Main objective of tillage is to keep the field free from weeds and to improve the soil structure to meet the needs of the crop. Other objectives are to preserve seedbed moisture, maintain enough crop residues on the soil to stop erosion and that is why there is a less chance of wind to damage the seedlings.

Time for Sowing

Time of sowing is first fortnight of October.

Seed Rate and Spacing

Seed rate is 3.6 kg/ha. Sow the crop at 50 × 20 cm spacing. Population of the plants is 1-1.2 lakhs plants/ha. It should not be planted greater than 3 cm deep

Weed Management

Weeds can be controlled by giving 2-3 hand weedings. Alternatively pretilachlor 1.0 Kg/ha as pre- emergence can be used for controlling weeds.

Fertilizer Management

Crop requires fertile field rich in organic matter. Add well rotten FYM @15 tonnes per hectare and 100 kg N/ha which can be applied in 2-3 splits.

Water Management

Sugarbeet requires many irrigations. Three to four irrigations are required during winter season i.e from October to February. However, during spring season it requires 6-7 irrigations depending upon soil type and climatic conditions.

Insects Pests

Sugarbeet crop is damaged by cutworm, army worm, Bihar hairy caterpillar, flea beetle and grass hoppers. In early stage of crop growth, cutworm may infest the crop. So apply chloropyriphos at 1.0 kg a.i /ha in soil at the time of sowing. Army worm appears about 100 days after sowing can be controlled with the spray of quinalphos (0.05%).

Diseases

Powdery mildew, leaf spots and root rot damage this crop. Seed treatment with Thiram 0.25% or Carbendazim (0.1%) gives good protection against seed borne fungi. Powdery mildew can be effectively controlled with the application of wettable sulphur at 2.0 Kg/ha. Cercospora leaf spot can be controlled by fortnightly spray of Carbendazim (0.1%).

Harvesting

The October sown crop of sugarbeet starts showing symptoms of maturity during mid-April. Two conspicuous changes occur in the crop.

I. Drying of lower leaves

II. Sugar accumulation (15-16%) in roots, which continues till end of May. Roots are dug manually or mechanically. Tops are removed and roots are immediately transported to factory for processing, as delay at harvest site coupled with high summer temperatures favour microbial decomposition of roots.

8.3. Cotton (American cotton – *Gossypium hirsutum* L., Desi cotton - *Gossypium arboreum* L.)

Cotton is a soft, fluffy staple fiber that grows in a boll, or protective case, around the seeds of the cotton plants of the genus *Gossypium* in the mallow family Malvaceae. The fiber is almost pure cellulose, and can contain minor percentages of waxes, fats, pectins, and water. Under natural conditions, the cotton bolls will increase the dispersal of the seeds. The plant is a shrub native to tropical and subtropical regions around the world, including the Americas, Africa, Egypt and India. The greatest diversity of wild cotton species is found in Mexico, followed by Australia and Africa

Cotton is one of the most important fiber and cash crop of India and plays a dominant role in the industrial and agricultural economy of the country. It provides the basic raw material (cotton fibre) to cotton textile industry. Cotton in India provides direct livelihood to 6 million farmers and about 40 -50 million people are employed in cotton trade and its processing.

Cultivated Species

There are four cultivated species of cotton viz. *Gossypium arboreum, Gossypium herbaceum, Gossypium hirsutum* and *Gossypium barbadense.* The first two species are diploid (2n=26) and are native to old world. They are also known as Asiatic cottons because they are grown in Asia. The last two species are tetraploid (2n=52) and are also referred to as New World Cottons. *G.hirsutum* is also known as American cotton or upland cotton and *G.barbadense* as Egyptian cotton or Sea Island cotton or Peruvian Cotton or Tanguish Cotton or quality cotton. *G.hirsutum* is the predominant species which alone contributes about 90% to the global production. Perhaps, India is the only country in the world where all the four cultivated species are grown on commercial scale.

Origin and History

The use of cotton for fabric is known to date to prehistoric times; fragments of cotton fabric dated to the fifth millennium BC have been found in the Indus Valley civilization, as well as fabric remnants dated back to 6000 BC in Peru. Although cultivated since antiquity, it was the invention of the cotton gin that lowered the cost of production that led to its widespread use, and it is the most widely used natural fiber cloth in clothing today.

Climate Requirements

Cotton, a semi-xerophyte, is grown in tropical & sub-tropical conditions. A minimum temperature of 15°C is required for better germination at field conditions. The optimum temperature for vegetative growth is 21-27°C & it can tolerate temperature to the extent of 43°C but temperature below 21°C is detrimental to the crop. Warm days of cool nights with large diurnal variations during the period of fruiting are conducive to good boll & fibre development.

Area and Distribution

In India during 2020, cotton was grown on an area of 12.86 mha with production of unginned cotton was 17.83 m tonnes. American cotton was grown on 2.48 lakh hectares in Punjab during 2020-21. The total production was 10.14 lakh bales with an average yield of 6.94 quintals lint per hectare. In Punjab, desi cotton was grown on 3.3 thousand hectares during 2020-21. The total production was 9.6 thousand bales with an average lint yield of 4.92 quintals per hectare

In India, there are ten major cotton growing states which are divided into three zones, viz. north zone, central zone and south zone. North zone consists of Punjab, Haryana, and Rajasthan. Central zone includes Madhya Pradesh, Maharashtra and

Gujarat. South zone comprises Andhra Pradesh, Telangana, Karnataka and Tamil Nadu. Besides these ten States, cotton cultivation has gained momentum in the Eastern State of Orissa. Cotton is also cultivated in small areas of non-traditional States such as Uttar Pradesh, West Bengal & Tripura.

Cotton varieties released by ICAR (2020-2021)

Variety	Area of adoption	Characters
LAHB Cotton 1	Karnataka, Tamil Nadu, Andhra Pradesh	Extra Long Staple (ELS) HxB hybrid suitable for irrigated conditions during *kharif*, average seed cotton yield 1.79 tonnes/ha, maturity 180–195 days, tolerant to drought, Alternaria leaf spot (ALS), bacterial leaf blight (BLB), grey-mildew (GM).
LHDP Cotton 1	Karnataka, Tamil Nadu, Andhra Pradesh	Compact variety for irrigated ecosystem under high- density planting system during *kharif,* seed cotton yield 2.41 tonnes/ha, maturity: 145–155 days, resistant to BLB and grey-mildew and moderately resistant to rust.
BS 279	Odisha	Suitable for irrigated upland ecosystem during *kharif*, seed cotton yield 2.56 tonnes/ha, maturity 150– 160 days, moderately resistant to BLB, ALS and GM.
BS 30 (Irrigated/ Rainfed conditions)	Maharashtra, Gujarat, South Rajasthan, Madhya Pradesh, Odisha	Suitable for irrigated/ rainfed ecosystem, seed cotton yield 1.89 tonnes/ha, maturity 150–160 days, moderately resistant to bacterial leaf blight and grey-mildew and to sucking pests.
KR 111	Haryana	Suitable for irrigated conditions under high and low fertility, seed-cotton yield 2.47 tonnes/ha, maturity 160–165 days, non-spinnable desi cotton hybrid.
Bt-cotton/ Hybrids		
NBC 1821 BG II	Telangana, Andhra Pradesh, Karnataka, Tamil Nadu	Suitable for irrigated conditions, contains cry1Ac and cry2Ab genes, seed-cotton yield 2.80 tonnes/ha, maturity 150–160 days.
NBC 1111 BG II	Maharashtra, Madhya Pradesh, Gujarat	Suitable for irrigated conditions, contains cry1Ac and cry2Ab genes, seed-cotton yield 2.30 tonnes/ha, maturity 160–165 days.
NEO 1635 BG II	Maharashtra, Madhya Pradesh, Gujarat	Suitable for rainfed conditions, contains cry1Ac and cry2Ab genes, seed-cotton yield 2.11 tonnes/ha, maturity 130–140 days.
NEO 1655 BG II	Andhra Pradesh, Telangana, Karnataka, Tamil Nadu	Suitable for rainfed conditions, contains cry1Ac and cry2Ab genes, seed-cotton yield 2.11 tonnes/ha, maturity 135–140 days.

PRCH 2799 Bt2	Gujarat, Maharashtra, Madhya Pradesh	Suitable for irrigated conditions, contains cry1Ac and cry2Ab genes, seed-cotton yield 2.21 tonnes/ha, maturity 140–150 days.
ARCH 888 BGII	Telangana, Andhra Pradesh, Karnataka, Tamil Nadu	Suitable for irrigated conditions, contains cry1Ac and cry2Ab genes, yield 2.51 tonnes/ha, mid-late maturity.
KCH 307 BG II	Haryana, Punjab, Rajasthan	Suitable for irrigated conditions, contains cry1Ac and cry2Ab genes, seed-cotton yield 3.73 tonnes/ha, maturity 165–170 days.
ARCH 501 BG II	Madhya Pradesh, Maharashtra, Gujarat	Suitable for rainfed conditions, contains cry1Ac and cry2Ab genes, seed-cotton yield 1.93 tonnes/ha, mid-late maturity.
SP 7670 BG II	Telangana, Andhra Pradesh, Karnataka, Tamil Nadu	Suitable for rainfed conditions, contains cry1Ac and cry2Ab genes, seed-cotton yield 2.05 tonnes/ha, maturity 160–170 days.
ACH-955-2 BG II	Punjab, Haryana, Rajasthan	Suitable for irrigated conditions, contains cry1Ac and cry2Ab genes, seed-cotton yield 3.63 tonnes/ha, maturity 150–160 days.
RCH 956 BG II	Maharashtra, Madhya Pradesh, Gujarat	Suitable for rainfed conditions, contains cry1Ac and cry2Ab genes, seed-cotton yield 1.87 tonnes/ha, maturity 160–170 days.
RCH 846 BG II	Punjab, Haryana, Rajasthan	Suitable for irrigated conditions, contains cry1Ac and cry2Ab genes, yield 3.69 tonnes/ha, maturity 160–170 days.
MC 5408 BGII	Punjab, Haryana, Rajasthan	Suitable for irrigated conditions, contains cry1Ac and cry2Ab genes, seed-cotton yield 3.8 tonnes/ha, maturity 150–155 days.
MC 5516 BGII	Andhra Pradesh, Telangana, Karnataka, Tamil Nadu	Suitable for irrigated conditions, contains cry1Ac and cry2Ab genes, seed-cotton yield 2.47 tonnes/ha, maturity 160–170 days.

Improved Varieties for Punjab

Bt cotton variety

PAU Bt 3: It is a Bt cotton variety with inbuilt resistance against spotted and American bollworms. Its average seed cotton yield is 25.5 quintals per hectare. Its average fibre length is 26.2 mm and ginning out turn is 36.5 %. It is tolerant to jassid and cotton leaf curl disease.

PAU Bt 2: It is a Bt cotton variety with inbuilt resistance against spotted and American bollworms. Its average seed cotton yield is 25.0 quintals per hectare. It possesses average fibre length of 27.6 mm and ginning outturn of 34.4 %. It matures in 160-165 days. It is tolerant to jassid and cotton leaf curl disease.

PAU Bt 1: It is the first Bt cotton variety developed by public sector with inbuilt resistance against American and spotted bollworms in the country. It has bigger boll size of 4.3 g and ginning out turn is 41.4%. It possesses superior fibre properties with 28.2 mm fiber length, 28.6 g/tex bundle strength and 4.5 μg/inch micronaire value. Its' average seed cotton yield is 28.0 quintals per hectare. It is moderately resistant to cotton leaf curl disease.

Non-Bt cotton

F 2228: Its' maturity period is 180 days with an average seed cotton yield of 18.5 quintals per hectare. It is moderately resistant to jassid and bacterial blight.

LH 2108: It matures in 165-170 days with an average seed cotton yield of 21 quintals per hectare.

Improved Varieties for Desi Cotton

LD 1019: This is a shattering tolerant variety requiring 2 or 3 pickings as compared to atleast 5 pickings needed by other desi cotton varieties. Average seed cotton yield of LD 1019 is 21.5 quintals per hectare. It possesses ginning out-turn of 35.7% and its fibre length is 22.6 mm. It has green, broad leaves and cream flowers. LD 1019 is tolerant to jassid, whitefly, Fusarium wilt and bacterial blight.

LD 949: Its' plants are reddish-brown with narrow- lobed deep cut leaves and pink flowers. It possesses lint percentage of 40.1. Its' fibres are short, coarse and suitable as absorbent cotton. This variety is moderately resistant to Fusarium wilt and bacterial blight. It is tolerant to whitefly and jassid. Its' average seed cotton yield is 24.75 quintals/hectare.

FDK 124: It has green plant body and narrow lobed leaves. It is synchronous in maturity and takes about 160 days to mature. It is short staple, coarse fibre variety with 2.5% span length of 21.0 mm and ginning outturn of 36.4%. It produces an average seed cotton yield of 23.25 quintals per hectare. It is resistant to jassid and whitefly

Soil Requirements

Cotton is grown on a variety of soils ranging from well drained deep alluvial soils in the north to black clayey soils of varying depth in central region and in black and mixed black and red soils in south zone. Cotton is semi-tolerant to salinity and sensitive to water logging and thus prefers well drained soils.

Rotations

Cotton–Wheat/Barley

Cotton–Sunflower

Cotton–Senji/Berseem/Oats

Cotton-Raya/Transplanted Gobi Sarson

Cotton–Sunflower-Paddy-Wheat

Land Preparation

A fine seed-bed is essential for securing a good plant stand.

Time of Sowing

Sow the crop during 1 April to 15 May in Northern India and from July to August in Southern India. Sowing during this period ensures better yield and escapes the attack of insect pests and diseases. Sowing should be done in morning and evening hours.

Seed Rate

Use 7.5 kg/ha seed rate for desi cotton and 4.0 kg/ha for hybrids/varieties of American Cotton. However seed rate can vary with variety/hybrid/bt or non-bt cotton.

Acid Delinting of Seed

Mix 100 g commercial grade concentrated sulphuric acid with 1 kg cotton seed in earthen/ plastic container by stirring it vigorously for two to three minutes with a thick wooden stick. As soon as the fuzz gets dissolved, add 10 litres of water, stir well and drain out water through the perforated plastic basket. Repeat these washings three times to make the seed free from sulphuric acid residue. Dip the washed seed for about one minute in sodium bicarbonate solution (12.5 g sodium bicarbonate in 2.5 litres of water) to neutralize the acid residue on the cotton seed. Give one more washing with water and remove light, damaged and rotten inviable seeds floating on the surface. Dry the healthy fuzz-free seed in the shade by spreading in a thin layer. Adopt following precautions:

- Metal or wood container should not be used.
- The operator should wear the plastic gloves.
- The water containing acid and alkali residue should be properly disposed off in the waste land.
- Inadequate washing and delayed washing of the seed after acid treatment and residual acid on the seed if not neutralized may impair the germination of seed.

Or rub the non-delinted seed with fine earth, cow-dung or ash to remove its fuzz and ensure uniform sowing.

Seed Priming

Soak the seed in a solution of 0.5 g succinic acid and 5 litres of water for 2-4 hours in case of acid delinted seed or 6-8 hours in case of non-delinted seed to promote good establishment of plant stand, better early growth and more yield. In soils irrigated with sodic water (RSC > 2.5 meq per L), treat the seed with the liquid bioformulations (Azo+PSB+ZnSB) along with gypsum (25% of Gypsum Requirement) application. This will reduce the adverse effects of sodic water irrigation while sustaining soil health and cotton productivity in cotton-wheat system. Seed should be smeared with 5 g Gaucho 70 WS (imidacloprid) or 7 g Cruiser 30FS (thiomethoxam) per kg seed for preventing damage by cotton jassid.

Sowing and Spacing

Sow in lines 67.5 cm apart with a cotton sowing drill or cotton planter in order to facilitate interculture with tractor. Plant to plant spacing may vary from 45 cm to 75 cm depending upon growing habits of hybrids/varieties. Sowing can be done with cotton planter or cotton sowing drill.

Intercropping

Intercrop one row of maize/cowpea for fodder in cotton sown at row to row spacing of 67.5 cm for getting higher income as compared to sole cotton. Apply recommended fertilizers to cotton and intercrops on area basis. Harvest maize/ cowpea fodder at 45-55 days after sowing.

Ridge Sowing

Sowing of cotton on ridges prepared with cotton planter and irrigating the crop in furrows saves considerable amount of irrigation water without reduction in seed cotton yield.

Transplanting of Cotton Seedlings

For gap filling, 3 week old nursery grown in 4”x6” polythene bags, filled with 1:1 mixture of soil and FYM, can be transplanted.

Weed Control

Hoe the crop two or three times. The first hoeing should be done before first irrigation. Use tractor mounted cultivator/ tractor operated rotary weeder/triphali or wheel hand hoe for weeding. Their use after fruiting should be avoided.

For control of weeds particularly itsit, madhana/makra, apply 2.5 litre per hectare Stomp 30 EC (pendimethalin) as pre-emergence within 24 hours of sowing. In situations where weeds emerge after first irrigation or with the rain shower, Stomp 30 EC can also be applied as post-emergence after first irrigation in 500 litre of water. If some weeds emerge before the application of the herbicide, a light hoeing/interculture may be done. The herbicide can also be sprayed with tractor mounted sprayer fitted with flat fan nozzle either in morning or evening hours. Ensure a fine seed bed free from plant residues and clods, adequate moisture in the field at the time of spray of herbicides.

Alternatively, spray 1.25 litre per hectare Hitweed Maxx 10 MEC (pyrithiobac sodium 6%+quizalofop ethyl 4%) by dissolving in 250 litres of water after first irrigation, in moist soil, to control annual grass and broadleaf weeds. This herbicide also provides effective control of lapeta (guara) vel (*Ipomoea sp.*) when weed plants are at 2 to 5 leaf stage. Alternatively, at 6-8 weeks after sowing when the crop is about 40-45 cm in height, spray 1.25 litre per hectare Gramoxone 24 SL (paraquat) or 2.25 litre per hectare Sweep Power 13.5 SL (glufosinate ammonium) in 250 litres of water as a directed spray to control weeds in between the crop rows. The directed spray can be done by using a protective hood. Paraquat and glufosinate are non-selective herbicides and can cause injury to the crop if these fall on the crop leaves.

Fertilizer Application

Apply fertilizer on soil test basis. The fertilizer recommendations for medium fertility soils are as under:

	*Nutrients (kg per hectare)		Fertilizers (kg per hectare)		
	N	P_2O_5	Urea	DAP	or Single Superphosphate
Non-Bt varieties	75	30	112.5	67.5	187.5
Bt varieties	92.5	30	200	67.5	187.5
Bt hybrids	105	30	225	67.5	187.5

* These nutrients can also be supplied from other fertilizers available in the market

Note:

- Omit application of phosphorus to cotton when it follows wheat which had received recommended dose of phosphorus. Where 27 kg DAP is used, reduce the urea dose by 10 kg.
- Apply 50 kg muriate of potash and 25 kg zinc sulphate heptahydrate (21%) or 16.25 kg zinc sulphate monohydrate (33%) per hectare to cotton in light soils.

Drill all phosphorus at sowing. Apply 62.5 kg per hectare magnesium sulphate as basal dose at the time of sowing. Apply half nitrogen at thinning and remaining half at the appearance of flowers. If the soil is low in fertility, the first half dose of nitrogen may be applied at sowing instead of at thinning.

Apply 1.0 kg boron (10 kg borax) per hectare at sowing to boron deficient (<1.25 kg available boron per hectare) calcareous soils having 2% or more calcium carbonate. However, boron should not be applied indiscriminately, as excessive boron application may cause toxicity.

To get higher yields, give 4 sprays of 2% potassium nitrate (13:0:45) at weekly interval starting at flower initiation. For high yield and management of leaf reddening in Bt cotton, give 2 sprays of 1% magnesium sulphate (2.5 kg magnesium sulphate in 250 litres of water per hectare) at 15 days interval during full bloom and boll development stages.

Use of growth retardant

In heavy soils, cotton attains excessive vegetative growth during rainy season. Thick crop canopy prevents the penetration of sunlight which results in shedding of flower buds, flowers or bolls and ultimately causes yield reduction. To check excessive vegetative growth in heavy soils, give 2 sprays of 750 ml per hectare Chamatkar (mepiquat chloride 5% w/w) at 60 and 75 days after sowing using 200-250 litres of water

Irrigation and Drainage

Cotton requires 4-6 irrigations depending upon the seasonal rainfall. The first irrigation should be given 4 to 6 weeks after sowing and the subsequent ones at interval of two or three weeks. However on light soils or in crop sown on ridges, the first irrigation may be advanced, if necessary. Sowing cotton on ridges and irrigation in furrows save considerable amount of water. Under poor quality irrigation water conditions, give presowing irrigation with canal water and subsequent irrigations can be applied with poor quality tube well water in alternate furrows. In soils irrigated with saline water (EC upto 10 dS/m), application of 40 quintal per hectare of rice-residue biochar reduces adverse effect of salinity and increases seed cotton yield. The crop must not be allowed to suffer from water stress during the flowering and fruiting stages, otherwise a lot of shedding of flowers and bolls will take place resulting in low yield. Cotton during its early growth is very sensitive to water stagnation. Therefore, drain out the stagnant water if such a situation arises. To hasten boll opening, give the last irrigation by the end of September.

Water stress management through Salicylic acid: To minimize loss of cotton yield owing to water stress (due to no rainfall or sudden canal closures), dissolve 31.25 g Salicylic acid in 938 ml of Ethyl alcohol and then add it to 300 litres of water for spraying crop per hectare on stress appearance.

Caution: Application under well-watered conditions may not increase yield.

Plant Protection

A. Insect-Pests

i. Cultural and Mechanical Management

- Grow only recommended Bt cotton cultivars.
- Prefer to grow desi cotton in area of high infestation of whitefly and leaf curl.
- Even the apparently healthy seed-cotton (kapas) may be harbouring larvae of pink bollworm. It should be acid delinted or thoroughly dried in the sun in a thin layer for 3-4 consecutive days in April.
- Complete the sowing up to 15th May.
- Avoid excessive use of nitrogenous fertilizer.
- Eradicate weeds like kanghi buti, peeli buti, puth kanda, congress grass, itsit growing on field bunds, waste lands, road side and irrigation channels/ canals to avoid further spread of whitefly, mealybug, tobacco caterpillar and spotted bollworm to cotton fields
- Regular surveillance on alternate host crops like brinjal, cucurbits (cucumber, long melon, chappan kadu), tomato, chilli, okra from February onwards and on cotton, moong from April onwards should be carried out for timely management of whitefly on these crops.
- Use low cost yellow sticky traps @ 100 per hectare during initial phase of cotton crop to check early infestation of whitefly.
- Grow bajra, maize and jawar as barrier crops, being least preferred hosts for the mealybug.
- Do not grow bhindi, moong and arhar in the cotton crop and as border rows in order to reduce the incidence of American bollworm, spotted bollworms, jassid and whitefly. Bhindi, moong, dhaincha and castor are also the most preferred hosts of tobacco caterpillar, helping the pest to multiply and shift to cotton. The above pests on these crops grown in the vicinity of cotton fields, should be properly controlled in order to check their migration to the cotton crop.
- Do not throw the uprooted infested plants in cotton fields/water channels to check further spread of mealybug.
- Prevent movement of sticks from infested areas to new areas to avoid spread of mealybug.

- The trees/fruit plants near cotton fields harboring mealybug population should be sprayed with recommended insecticides.
- Egg masses and young larvae of tobacco caterpillar feeding gregariously should be collected along with leaves and destroyed.
- Terminate the crop as early as economically feasible. For this purpose give last irrigation by end of September. It would reduce bollworms damage and their carryover.
- After the final picking, infested cotton field should be shredded with the help of shredder to kill the larvae.
- Destroy all trash collected during the ginning process. Remove all seed from the ginneries by the end of March. Fumigate the seed left uncrushed in the mills before end of April with Celphos/Phostoxin/Delicia @ one 3-g tablet per cubic metre space, giving an exposure of 48 hours or use two tablets with an exposure of 24 hours. No un-fumigated seed should be retained or sold by the ginneries. Only cotton-seed cake (khal) should be fed to the cattle and no seed should be kept for this purpose.
- The seed meant for sowing must be acid-delinted in the ginneries before it is sold. The acid treatment kills the larvae of the pink bollworm. It also removes fuzz and thereby facilitates mechanical sowing.
- Even the apparently healthy seed-cotton (kapas) may be harbouring larvae of pink bollworm. Hence, kapas retained by the farmers should be ginned by the end of March and seed fed to cattle. If this seed is to be retained for sowing, it should be aciddelinted/fumigated or thoroughly dried the sun in a thin layer for 3-4 consecutive days in April.
- After the last picking, allow sheep, goats and other farm animals into cotton fields to feed on plant debris and un-opened bolls.
- Do not stalk the cotton sticks under shade or in the field. Beat the sticks on ground to dislodge the pink bollworm larvae surviving in the unopened bolls. Stalk the cotton sticks vertically.
- Prevent the movement of the cotton stalk from the infested areas to the new areas.

ii. Chemical Control

a. Insect Pests (Bt cotton)

Bt cotton provides effective protection against all cotton bollworms except pink bollworm. Regular monitoring should be done at weekly interval during reproductive phase. If at all American bollworms cross ETL level during late crop season, use insecticides as mentioned in Table 8.2. Bt cotton does not provide control of sucking pests.

Sucking Insect Pests: Among sucking pests, whitefly, jassid, mealybug, thrips and aphid are most serious on Bt cotton and they cause maximum damage during July-September.

Whitefly adults and nymphs suck sap from leaves and excrete honey dew on leaves which become sticky. Affected leaves and seed cotton turn black due to development of sooty mould.

Nymphs and adults of jassid suck sap from leaves and cause shedding in case of severe infestation. Aphids appear sporadically. The nymphs and adults of aphid suck sap and excrete honey dew on leaves on which black fungus develops. Petioles, internodes, flowers, buds, mature bolls and even leaves fall down and growth of the plant is retarded. Both nymph and adults of thrips first lacerate the leaf tissue and later feed on the oozing cell sap. Initially silver streaks appeared especially around the midrib and veins of the leaves. Later silvering got more severe and slight cupping of the leaves. Under severe infestation, leaves gave blasted appearance and extreme level of cupping observed. The decision regarding spray of insecticide should be taken based on Economic Threshold Level (ETL).

Spray against whitefly should be done when population reaches six adults per leaf in the upper canopy of plants before 10 AM or when honey dew appears on 50 per cent of the plants. Initiate spray against jassid whenever some of the fully formed leaves in the upper canopy show curling and yellowing at the margins on 50 per cent of the plants. Spray against thrips should be done when population of nymphs and adults reaches twelve per leaf in the upper canopy of plants. Spray against aphid should also be done on the appearance of honey dew on 50 per cent plants. Spray the crop as soon as the crawlers/adults of mealybug appear on the cotton plant. Use following insecticides for control of sucking insect pests given below in Table 8.1

Table 8.1: Insecticides for the management of sucking insect pests

Brand (Insecticides)	Dose/hectare
Whitefly	
Sefina 50 DC (afidopyropen)	1.0 litre
Osheen 20 SG (dinotefuran)	150 g
Polo/Craze/ Ruby/ Ludo/Shoku 50 WP (diafenthiuron)	500 g
Lano/Daita10 EC (pyriproxyfen)	1.25 litre
Oberon/ Voltage 22.9 SC (spiromesifen)	500 ml
Ulala 50 WG (flonicamid)	200 g
Dantotsu 50 WG (clothianidin)	50 g
Fosmite/E-mite/Volthion/ Gold Mit 50 EC (ethion)	2.0 litre
Nimbecidine or Achook (Neem based biopesticide)	2.5 litre

Jassid	
(a) Seed treatment: At the time of sowing, smear the seed with any of the following insecticides.	
Gaucho 70 WS (imidacloprid)	5 g/kg seed
Cruiser 30 FS (thiamethoxam)	7 g/kg seed
(b) Spray: Spray any of the following insecticides, if incidence is noticed in standing crop.	
Keefun 15 EC (tolfenpyrad)	750 ml
Osheen 20 SG (dinotefuran)	150 g
Neon 5 EC (fenpyroximate)	750 ml
Ulala 50 WG (flonicamid)	200 g
Actara/Extra super/Dotara/Thomson 25 WG (thiamethoxam)	100 g
Thrips	
Delegate 11.7 SC (spinetoram)	425 ml
Curacron/ Celcron 50 EC (profenophos)	1.25 litre
Polo 50 WP (diafenthiuron)	500 g
Mealybug	
Transform 21.8 SC (Sulfoxaflor)	375 ml

Note: In the beginning of crop season on appearance of whitefly, first spray of Nimbecidine or Achook @ 2.5 litre per hectare should be given.

- Do not spray any insecticide for the management of thrips on cotton up to 30 day old crop. In case thrips attack observed, irrigate the field immediately.
- In case of attack of aphid use the same insecticides as recommended for jassid.
- Use fix type solid cone nozzle. Thorough coverage of plants with insecticides is essential to check the multiplication of whitefly and mealybug.
- Spray insecticides before 12 PM or in the evening. Community approach should be adopted at village level for the application of insecticides
- Use only recommended insecticides with recommended dose and time. Avoid tank mixing and use of readymade insecticidal mixtures
- Mealybug is initially restricted to a few plants in a row, thus spot treatment with recommended insecticides is advocated. Spray of mealybug infested plants/rows of cotton after last picking

b. Insect Pests (Non-Bt cotton)

Sucking Insect Pests: See under Bt-cotton

Bollworms: Bollworms are the most harmful insects which attack cotton crop. Spotted bollworms damage growing points during May-June and cause heavy shedding of squares, buds, flowers and bolls during July to October. The American bollworm causes severe shedding of fruiting bodies during September-October especially on American cotton. The colour of its larvae greatly varies. They have one line on upper side and two wavy lines on lateral side of body. Their body also has sparse hairs. Pink bollworm does maximum damage from mid-July to mid-October. Due to severe attack of bollworms, the plants continue to grow without having adequate number of bolls. Spray any of the insecticides listed in Table 8.2.

Tobacco caterpillar: It is a polyphagous pest. The larvae cause serious damage to crop from August to October. The small larvae are black whereas grown up larvae are dark green with black triangular spots on body. Its' moths lay eggs in masses covered with brown hairs on the lower side of mature leaves. After hatching, first and second instar larvae feed gregariously and skeletonize the foliage. Later on grown up larvae disperse and feed singly. Besides leaves, they also damage the buds, flowers and green bolls. Control the tobacco caterpillar by spraying any of the insecticides given in Table 8.2. The larvae of leaf-roller, semi-loopers, hairy-caterpillars and bud moth may also appear sporadically and damage the crop during July-October. The monitoring of bollworms and tobacco caterpillar with sex pheromone should be done with the initiation of flowering stage of crop. Observations on moth catch should be recorded on every alternate day. This monitoring strategy will help in making decision for effective management of bollworms and tobacco caterpillar.

Pink bollworm: Use Sticka/Delta traps with at least 10 micro litre of gossyplure and place it at 15 cm above crop canopy. Replace the lure after 15 days and use 1 trap/ha.

Spotted/Spiny bollworms: Use Sleeve/Moth catch traps for spotted bollworms and replace the lure at 15 days interval. Place the trap at 15 cm above the crop canopy and use 2 traps/ha.

American bollworm: Use Sleeve/Moth catch traps with at least 2 mg of pheromone and place it at 15 cm above crop canopy. Replace the lure after 15 days and use 2 traps/ha.

Tobacco Caterpillar: Use Sleeve/Moth catch trap for tobacco caterpillar. Replace the lure after every 15 days. Place the trap 15 cm above crop canopy and use 2 traps/ha.

In order to control bollworms, conduct sprays on different varieties during their effective boll formation period based on ETL. Farmers should examine their fields twice a week in order to ensure that bollworms damage does not exceed 5 per cent among the freshly shed fruiting bodies (squares, buds and young bolls). For this purpose divide the field into four quarters and collect 25 freshly shed fruiting bodies at random in each quarter. The fruiting bodies damaged by bollworms will have feeding holes or their larvae. In case the damage exceeds 5 per cent, the crop should be sprayed immediately and thereafter spray as when need arises. (Table 8.2)

Table 8.2. Insecticides for the control of bollworms in cotton

Brands (Insecticides)	Dose/hectare
Pink and spotted bollworms	
A. Synthetic Pyrethroids	
Danitol/Meothrin 10 EC (fenpropathrin)	750 ml
Fastac/Alphaguard/Merit Alpha 10 EC (alphamethrin)	250 ml
Bulldock 0.25 SC (cyfluthrin)	750 ml

Ripcord/Bilcyp/Bullet/Ustad/Cyperguard 10 EC (cypermethrin)	500 ml
Cymbush/Cyperkill/Hillcyper/Colt/Basathrin/Agrocyper/Cypergaurd 25 EC (cypermethrin)	200 ml
Decis/Rukrain/Decicare 2.8 EC (deltamethrin)	400 ml
Sumicidin/Fenval/Agrofen/Fenlik/Triumph card/SB Fenvalerate/Milfen/ Markfenval 20 EC (fenvalerate)	250 ml
Pink, spotted and younger larvae of American bollworm	
A. Macrocycllc lactones or evermectins	
Prodaim 5 SG (emamectin benzoate)	250 g
B. Carbamates	
Larvin 75 WP (thiodicarb)	625 g
C. Organophosphatic	
Curacron/Carina/Profex/Celcron 50 EC (profenophos)	1.25 litre
Fosmite/E-mite/Volthion 50 EC (ethion)	2.0 litre
D. Miscellaneous group	
Fame 480 SC (flubendiamide)	100 ml
Grown up larvae of American bollworm	
A. Naturalyte	
Tracer 48 SC (spinosad)	150 ml
B. Oxadiazine	
Avaunt 15 SC/ Avaunt 15 EC (indoxacarb)	500 ml
C. Miscellaneous group	
Sumipleo 10 EC (pyridalyl)	750 ml
Coragen 18.5 SC (chlorantraniliprole)	150 ml
D. Organophosphates	
Coroban/Dursban/Durmet/Chlorguard/Radar/Lethal/Force/Markpyriphos 20 EC (chlorpyriphos)	5.0 litres
Orthene/Asataf/Starthene/Markphate 75 SP (acephate)	2.0 kg
Tobacco caterpillar	
A. Insect Growth Regulator	
Rimon 1O EC (novaluron)	375 ml
B. Miscellaneous group	
Coragen 18.5 SC (chlorantraniliprole)	150 ml

Note:

- Do not repeat the insecticide of same group in subsequent sprays.
- Do not use synthetic pyrethroids on cotton for the control of bollworm complex before September. • Repeat the spray immediately if it rains with in 24 hours after spray.
- If hairy caterpillars damage cotton crop during June-July use 1.25 litre quinalphos 25 EC in 250 litres of water per hectare.
- PRECAUTIONS: Cotton is highly sensitive to the 2, 4-D herbicide. Some farmers spray the ester form of 2, 4-D for controlling weeds in maize or sugarcane grown near the cotton fields. Owing to the volatile nature of 2, 4-D ester, its vapours cause serious injury to the cotton crop. Hence avoid the application of this herbicide in maize, if cotton is grown in the adjoining fields. - After using 2, 4-D on any crop, fill all spraying equipment as well as tubs, buckets, etc. with 0.5 per cent washing soda solution (500 g of washing soda in

100 litres of water) in the evening. Next morning, flush all equipment's thoroughly with fresh water. - To avoid the use of contaminated insecticides on cotton. It is advisable to test insecticide at least two weeks in advance on a few plants. If the insecticide is contaminated with 2, 4-D the tender leaves and shoots could become malformed within 10 days. Reject such an insecticide.

A. Diseases

Leaf curl: Disease is caused by whitefly transmitted virus. The diseased plants become stunted and have twisted internodes. Leaves remain small, show cupping and curling. Veins on the lower-side of the leaves become thickened with netted appearance. Small leaflets (enations) also develop on the underside of the leaves on the main as well as lateral veins. Number of fruiting bodies are reduced in the diseased plants. The disease can be reduced by adopting the following integrated measures:

- Avoid growing American cotton in and around citrus orchards and adjoining bhindi crop.
- Uproot and destroy the diseased plants from time to time.
- Protect the crop against whitefly vector by using recommended insecticides.
- Follow clean cultivation and destroy Kanghi buti (*Abutilon sp.*) and Peeli buti (*Sida* sp.) which act as collateral hosts.

Parawilt: Parawilt is a physiological disorder and no pathogen is involved. It generally occurs after droughts when the crop is heavily irrigated or there is heavy rain. Plants show sudden drooping of leaves which ultimately get wilted but the root system remains intact. The affected plants can be saved by spraying cobalt chloride @ 10 mg per litre of water (10 ppm) immediately after the appearance of symptoms. There would be no recovery if permanent wilting has already set in.

Root rot: This disease is caused by *Rhizoctonia solani* and *R. bataticola*. The main symptom are drying and shedding of leaves leading to complete wilting and death of the plant. The disease spreads in field in round patches. The affected plants can be pulled out very easily. The bark of the roots is broken into shreds.

Bacterial blight: It is caused by *Xanthomonas axonopodis* pv. malvacearum which survives in seed and plant debris. Lesions on the leaves appear as minute, water-soaked, angular spots, which subsequently turn brown and then are transformed into black angular dead lesions on both sides of the leaf. The bacterium also infects the young developing bolls and causes small, round, water soaked spots depressed in the centre. Use disease free seed.

Leaf spots: Foliar leaf spots are caused by different fungi. Leaf spots caused by Myrothecium roridum appear on leaves, bracts as well as on bolls. The disease is characterized by circular to semicircular brown coloured spots with broad violet margins. At later stage, shield shaped, small size fruiting bodies appear in the central necrotic portion of the spot. The pathogen is seed borne and also survives on the dead leaves. High humidity and intermittent rains are congenial for the development of the disease.

The fungus *Alternaria gossypina* also causes blightening of the leaves. In the early stages, the spots have a pale green area with irregular margins. As the spots enlarge, irregular concentric zones are formed. Sometimes severe shedding of leaves occur due to this disease. The plants with low vigour because of drought or deficiency of potash favour the development of this disease. The disease perpetuates through diseased debris.

Another type of leaf spots are caused by *Cercospora sp.* which generally appears towards the end of the season. It produces small, circular to irregular spots having whitish centre with dark brown margin. In advance stages, necrotic central portion may fall out 51 giving shot hole appearance. Low temperature (<25 degree C) and high relative humidity favour development of disease. Diseased debris is the main source of primary inoculum of the disease. Use disease free seed.

To control these fungal leaf spots, spray the crop with 500 ml Amistar Top 325 SC (azoxystrobin + difenconazole) in 500 litre of water per hectare immediately on the appearance of symptoms. If needed, repeat the spray at 15-20 days interval.

Tirak: It is a physiological disorder. It is characterized by the yellowing and reddening of leaves, followed by the bad opening of the bolls. It is particularly serious in pockets where cotton suffers from persistent drought, inadequate water supply, nutrient deficiency on light sandy soils, too early sowing or lack of plant protection measures. These factors may operate singly or in different combinations. Spells of high temperature prevailing during the flowering and fruiting further aggravate the intensity of this malady. Judicious fertilization and timely watering particularly during flowering and fruiting stages, and the adoption of recommended plant protection schedule help to mitigate the intensity of this disease.

Defoliation in cotton: Chemical defoliation with single spray of Ethrel 39% (Ethephon 39%) @ 5.0 ml/litre of water should be applied in last week of October. It leads to 85-90% defoliation after seven days of spray. Defoliation allows better sunlight penetration thereby resulting in early and uniform boll opening with increased productivity.

Picking: Cotton should be picked clean and dry to get a good price in the market. Picking should be done after every 15-20 days to avoid loss because of the Kapas falling to the ground. Do not keep the picked cotton in wet water channels in the field, as this practice impairs the quality of cotton. Store kapas in a dry godown. Keep produce of different varieties separately.

Removal of cotton sticks: Soon after the last picking, remove the cotton sticks alongwith the roots from the field and bury the remaining plant debris with furrow turning plough as sanitary measure against pests and diseases. Use or burn cotton sticks by the end of February at the latest. Use two-row tractor operated Cotton Stalk Uprooter for uprooting of cotton stalks. The Cotton Stalk Uprooter should be operated at a speed of 7 to 9 km/hr and at a depth of 12 to 15 cm with 45hp tractor for efficient field operation. This equipment will provide 10 to 15% more cotton sticks by weight than conventional manual stalk chopping method with a field capacity of 1.25 to 1.50 acre/hr. (0.4 hectare)

Marketing Hints

- Cotton should be picked dry, free from trash, with no dew on it.
- The first and the last pickings are usually of low quality and should not be mixed with rest of the produce. High-grade kapas mixed with low grade kapas sells at a relatively low price.
- Store cotton in damp proof and rat-free room.
- Store different varieties separately.

8.4. Jute (*Corchorus capsularis* and *C. olitorius*)

Jute (*Corchorus capsularis* and *C. olitorius*) ranks second in importance, next to cotton as a natural fibre and occupies important place in Indian economy. Jute fibre is extracted from phloem tissue in the stem of *Corchorus* as against seed fibre in the case of cotton. Of all the textile fibres, jute is the cheapest and is extensively used in the manufacture of packing material for agricultural and industrial products.

Origin and Distribution

The primary centre of origin of *Corchorus olitorius* is Africa and the secondary centre may be India or Indo-Burma and South China. Major jute growing countries are India, Pakistan and Bangladesh accounting for 90% of the global production.

Area and Production

India ranks first both in area (0.69 m ha) and production (1.81 mt) of jute. West Bengal accounts for more than 80% of total jute production of 9.94 million bales and about 75% area (more than 80%) in the country (FAO STAT, 2016).

Climatic Requirements

Jute thrives best under warm humid climate with temperature ranging from 24°C to 35°C and relative humidity 90%. Well distributed rainfall of 1500 mm of which about 250 mm from March to May is most ideal for jute cultivation. Alternate rainfall with bright sunshine is beneficial for a good crop of jute.

Soil Requirements

The new grey alluvial soils of good depth, receiving silt from annual floods are ideal for jute cultivation. However, it is grown in sandy and clay loams with varying soil management practices. Sandy soils and heavy soils are unsuitable. Well aerated sandy loams are best suited. Optimum pH range is 6.0 to 7.6. Below pH 5 or above 8 is not desirable. Jute requires a fine seedbed. Its tap roots extends to a depth around 50 cm and hence deep ploughing is necessary.

Jute Varieties released by ICAR 2020-21

Variety	Area of adoption	Characters
JROB 2	West Bengal, Odisha, Bihar, Assam, Parts of Uttar Pradesh, Tripura	A green-stem tossa jute variety specifically suited for biomass production beside its equal suitability for fibre production, average green biomass productivity 59 tonnes/ha with average fibre yield 3.2 tonnes/ha, fibre-crop maturity 110–120 days, biomass crop matures in 50–130 days depending upon the time of sowing, sown from March to August for biomass production and March to April as fibre crop, field resistance to stem rot, BHC and stem weevil.
JRCJ 11	West Bengal, Assam, Odisha, Bihar	Suitable to capsularis jute growing areas, optimum sowing period from 3rd week of March to 2nd week of April, average fibre yield 3.15 tonnes/ha, fibre maturity 110–120 days, days to seed maturity 135– 150 days, can tolerate water logging to some extent at the later stage of crop growth, highly tolerant to Bihar hairy caterpillar.

About 75% of the jute area is under *Capsularis*. It is highly adaptable and grown in both low lands and uplands. The *Colitorius* is basically an upland crop, which cannot withstand waterlogging. Some new high yielding *Colitorius* jute varieties

Seasons

Jute crop is confined to monsoon since low temperature affect the growth and yield during winter (*rabi*). Young seedlings are sensitive to water logging. As such, sowing in low lands begin in February, taking advantage of residual soil moisture for *Capsularis*. Generally, *C. olitorius* is sown in April.

Seed Rate and Spacing

A seed rate of 12 kg/ha is optimum by *C. capsularis* and 8 kg/ha for *C. olitorius*. Row distance for *C. capsularis* is 30 cm and for *C. olitorius* 20 cm. The final distance between the plants in the row should be around 10 cm to maintain optimum plant population around 2,50,000 and 5,00,000 plants per hectare for *C. capsularis* and *C. olitorius* respectively. Low plant population encourages branching leading to poor quality fibre.

Method of Seeding

Recommended seed rate is mixed with equal weight of sand or loose soil for uniform seed distribution. Broadcasting the seed is the usual practice, especially for complete rainfed crop. However, line sowing with seed drills to a depth of 3-5 cm is superior over broadcasting. Two thinnings are done, once when plants attain a height of 10 cm and another at 15 cm height. The final inter-plant distance should be 10 cm to maintain required plant population.

Nutrient Management

Judicious application of fertilizers provides one of the quickest means of increasing jute production. In jute, greater vegetative growth is conductive to higher fibre yield since it is obtained from bark of the plant.

On an average 2.6 to 3.6 t/ha of green matter is obtained from jute crop at harvest time. Generally, the average uptake of nitrogen varies from 40-80 kg/ha and potassium. 100-150 kg/ha. An *Colitorius* jute crop producing 3 t/ha of dry fibre absorbs 111 kg N, 64 kg P_2O_5, 199 kg K_2O, 175 kg Ca and 42 kg Mg/ha.

Recommended fertilizers

C. capsularis				
	Rainfed	40 kg N/ha 20 kg P_2O_2/ha 20 kg K_2O/ha	Supplemental Irrigation	80 kg N/ha 40 kg P2O2/ha 40 kg K2O/ha
C. alitorius				
	Rainfed	30 kg N/ha 15 kg P_2O_2/ha 15 kg K_2O/ha	Supplemental Irrigation	60 kg N/ha 30 kg P_2O_2 /ha 30 kg K_2O/ha

Nitrogen application is recommended in three splits: half at sowing, one-fourth 45 DAS and the remaining one-fourth 60 DAS by placement. Ammonium sulphate as a source of nitrogen, is ideal than other sources.

Irrigation

The ideal practice for February sown crops is to give a pre-sowing irrigation for seed germination and stand establishment. If water is available another irrigation should be given 30 DAS. Depending on climate and crop duration water requirement varies from 500 to 800 mm.

Weed Management

Weeds pose greatest problem in jute cultivation. In terms of expenditure, weed management constitutes one-third the total cost of cultivation. Weeds may reduce the fibre yield to the extent of 40-50%. Critical period of weed competition is 30-55 DAS. Two manual weedings, one each when the crop is 10 and 15 cm height, are considered optimum and economical for jute crop. Selective pre-emergence herbicides for jute crop are fluchloralin 0.5 to 1.0 kg/ha and thiobencarb 1.0 to 1.25 kg/ha and ideal post emergence herbicides is fluazifopbutyl 0.4-0.5 kg/ha.

Plant Protection

Jute aphids, semiloper and mealy bug among inserts and stem rot, black band and wilt among diseases cause considerable damage to the jute crop. Jute aphids can be controlled by the application of Acephate 75 SP at 1.5 g/litre or carbaryl 50 WP at 3 g/litre of water. Semiloper can be controlled with Acephate 75 SP at 1.5 g/litre or carbaryl 50 WP at 3 g/litre of water. Spraying Blitox 2-3 times at 15 days interval can control black band disease. Combination of Captan & Zineb is also equally effective

Harvesting

Jute may be harvested at any time between 120 and 150 days after seeding. Early harvesting gives finer fibre of good quality and late harvesting gives larger yield of coarser fibre. Harvesting is done by cutting the plants close to the ground level. In flooded lands, the plants are uprooted. After harvest, the plants are left in field for 2-3 days for the leaves to shed. The plants are then tied into bundles, 20-25 cm in dia and the branched tops looped off to rot in the field.

Retting

Retting is a microbial process by which the fibres are separated from the bark and woody core die to removal of pectins, gums and other mucilaginous substances by retting bacteria present in water. The processing of retting is completed in about

10 days after submergence in water. Over retting results in dazed weak fibre. Plant height and basal dia are the two major yield determinants of jute fibre.

Yield

As explained above plant height and basal dia are the two major yield determinants of jute fibre. Average yield of jute fibre is 1970 kg/ha. However, well maintained *C. olitorious* and *C. capsularis* crop can yield upto 2.75 and 2.0 t/ha respectively. Yield potential of these two has been assessed at 4.0 and 3.5 t/ha respectively. Seed crop of *C. capsularis* and *C. olitorius* on an average, give seed yields of 0.4 and 0.5 t/ha, respectively.

8.5. Sunhemp (*Crotalaria juncea*)

Sunhemp (*Crotalaria juncea*) Bombay hemp or Banaras hemp is grown for the best fibre. It is also a green manure crop and a minor fodder crop.

Origin and Distribution

Sunhemp is generally considered to have originated in India; where it has been cultivated since prehistoric times. It is cultivated commercially for fibre production in Russia, Italy, India, Austria, United States, China, Japan and France. In India, the crop can be grown all over the country but the major growing states are Rajasthan and Orissa.

Climate and Soil

Sunhemp is a tropical crop grown during *Kharif* in North India and throughout the year in South India. A minimum of 400 mm rainfall with equal distribution during the crop period is ideal. Sunhemp is drought resistant and is adapted to hot, semi-arid areas, yet can tolerate light frosts. Sunhemp has a wider range of adaptation to soil types. It grows on poor soils, but growth on such soils is improved by fertilization. For fibre production, light, loamy, well drained soils are preferred, however in low lying clay soils it makes vigorous growth, but then the fibre is coarser and yields are lower.

Improved varieties for Punjab

PAU 1691: It is quick growing variety with erect growth habit. It flowers in 60-62 days matures in 136 days. It attains the height of 160-220 cm and thus adds 10.0-16.0 tonnes green biomass per hectare to the soil when buried at 45-60 days after sowing. The average seed yield is 12 q/ha.

Narendra Sanai-1: It is also a quick growing variety with erect growth habit. The leaves are broad in size and green in colour. It flowers in 98-100 days and matures in about 152 days. Its average seed yield is about 9.75 q/ha.

Propagation and Planting

Sunhemp is established by seed. Seed rates of 40-45 kg/ha are used when it is sown as a forage crop or as green manure, but when it is sown as fibre crop, seed rates of 50-55 kg/ha are used. Can be sown with broadcasting or drill sowing method.

Growth and Development

Sunhemp is fast growing. Seedings emerge 3 days after sowing, and rapidly produce a thick ground cover that smoother weeds. Extensive cross-pollination occurs in sunhemp and self-pollination takes place after the stigmatic surface has been insect or mechanically stimulated.

Irrigation

The crop sown for green manure during summer period requires 2-3 irrigations depending upon the prevailing weather conditions. However, the grain crop should not suffer water stress at the time of flowering and grain development stages.

Weed Management

For seed production, the crop may require one hoeing after one month of sowing to keep the weeds under check. However, it is a quick growing crop and smothers weeds very effectively.

Fertilizer Management

Generally no fertilizer is required as the crop can fix atmospheric Nitrogen after the formation of nodules. However in poor soils starter dose of 10-15 kg N per hectare can be applied. In sunhemp, phosphorus and potassium are applied through fertilizers at 25 kg/ha each as basal dose at seeding. As sunhemp is a leguminous crop so there is no need to apply nitrogen.

Diseases and Pests

Sunhemp is attacked by many diseases and pests, including various fungi, insects and nematodes, but they usually cause little economic damage. In India, anthracnese caused by *Colletotrichum curvatum*, wilt caused by *Tusasium undam* can be serious. Damage from insects is more severe if crops are planted in the same area for more than 3 consecutive years. Fungicide seed treatments and crop rotations are the most recommended and practiced disease control measures.

Harvesting

Pod formation (120-140 DAS) is considered optimum stage of harvest for fibre. The plants are cut with sickles and the bundles are retained in the field for 2-3 days for the leaves to be shed. The process of retting and extraction of fibre are similar to that of jute.

Yield

Fibre yield is around 400 kg/ha. However, a well-managed crop can yield up to 1500 kg/ha. The fibre is dull yellow, somewhat cease, strong and durable. Due to high cellulose content, low lignin, the fibre is used in the manufacturing of tissue paper and paper currency. In rural areas, it is used for making ropes, twines and nets. Sunhemp fibre is not used for textile purposes, unlike jute and mesta.

9

Conservation Agriculture

9.1. Introduction

According to FAO guideline, "Conservation Agriculture (CA) is an approach to managing agricultural ecosystems for enhanced and sustained productivity, improved returns, and food security while preserving and enhancing the resource base and the environment."

The terms "conservation agriculture" (CA) and "resource-conserving technologies" (RCTs) are not synonyms, but sometimes both words are used for a similar meaning. Therefore, the practices which improve resource or input-use efficiency are termed as RCTs. RCT is a wider term than CA which is used for nitrogen-efficient varieties; fuel-, money-, and time-efficient techniques; and improvement in plot-level water productivity including CA.

Conservation agriculture (CA) is a resource-saving strategy encompassing three broad principles *viz.*, minimum soil disturbance, diversified crop rotation and adequate soil cover through residue retention. Nowadays organic farmers are encouraged to adopt conservation practices to ensure soil quality and sustainability along with food security.

Conservation agriculture, based on no-till, the permanent cover of soil surface by crops and their harvest residues and crop rotation, is an efficient system to protect the soil against erosion and to improve soil organic matter. Conservation agriculture has already contributed to the sustainable intensification of agriculture in several developed and developing regions of the world by improving soil quality and productivity but reducing the level of inputs (power, labour, fertilizers, agrochemicals, and water).

CA is a holistic approach to sustainability, which comprises the practical application of three interlinked principles, namely: no or minimum mechanical soil disturbance, biomass mulch soil cover, and crop species diversification, in conjunction with other complementary good agricultural practices of integrated crop and production management

The history of agriculture has essentially been a history of tillage in agriculture, and the culture of ploughing or tilling the soil to establish crops and to manage weeds has been a central part of agricultural development worldwide. After WWI, agriculture began to be intensified to achieve greater output. This was essentially based on the intensification of the use of tillage and of agrochemicals as part of crop nutrition and protection management of higher yielding crops under standardised mechanized systems. Initially, this change process began in North America, but after WWII, it spread to other industrialized countries in Europe and Eurasia as well as in Australia and New Zealand and in the independent countries of the tropics, with emerging economies such as in Latin America and South Asia as part of the Green Revolution drive from the West.

From 1960 to 2000, no-till production was being tested in all continents by researchers and larger-scale mechanized farmers, and limited scaling began in the 1980s and 1990s, mainly in countries such as the US, Canada, Brazil, Argentina, Paraguay, Uruguay, Bolivia, Venezuela, UK, Australia, New Zealand, Spain, Germany, Kazakhstan, Zambia, and South Africa. By 2000, these countries together covered some 65 M ha of no-till cropland systems. Prior to this period, soil and water conservation programmes in the US had led to the development of a range of soil and water management practices such as bunding, terracing, contour ploughing, reduced tillage as well as no-tillage, which were brought under a common term called 'conservation tillage'. During the period between 1970 and 1997, the no-till pioneers and champions, farmers, extension agronomists, and researchers had generated enough experience and expertise to be able to define the key components of a sustainable no-till system that was termed conservation agriculture (CA). This term was first proposed in Spanish at the IV RELACO (Latin-American Network for Conservation Tillage) meeting in Morelia, Michoacán, Mexico, in 1997 by the co-authors Rolf Derpsch and Theodor Friedrich. The term was also adopted in 1997 by the Food and Agriculture Organization of the UN (FAO) to describe sustainable production systems. This led FAO in 1998 to define the three interlinked principles of CA.

Principles of CA

Continuous no or minimum soil disturbance. In practice this means no-tillage. At least 30% of the soil must be covered after sowing to effectively protect it against erosion. However, it is recommended to leave more than 60% of the soil covered to have almost complete control over the soil degradation process.

Permanent soil mulch cover. In other words, this means maintaining crop biomass and stubble in annual crops and planting or preserving groundcovers between rows of trees in woody crops. In this way, soil organic matter and water infiltration into the soil increases, some weeds are inhibited, and water evaporation from the soil is limited.

Crop diversification through rotations and associations. In this way, pests and diseases are better controlled by breaking cycles that are maintained in monocultures, in addition to including crops that can improve the natural fertility of the soil and biodiversity.

Origin of CA

Soil ploughing or tilling began for the first time at Mesopotamia ~3000 BC (Hillel 1998). Tillage was an important component of agronomic practices in agricultural cultivation. It was used first time by Edward H. Faulkner in the 1930s in a manuscript called Plowman's Folly (Faulkner 1943). With time, the concept of conserving the soil with the help of minimum tillage or zero tillage and keeping the soil covered with crop residues gained popularity. This system of the soil and water conservation was called conservation tillage (Friedrich et al. 2012). The zero tillage (ZT) was introduced in Brazil in the early 1970s to reduce severe water erosion problems (Derpsch 2001). Moreover, escalating fuel prices and severe erosion problems during the 1970s forced the farmers to move in the direction of Soil ploughing or tilling began for the first time at Mesopotamia ~3000 BC. Tillage was an important component of agronomic practices in agricultural cultivation. It was used first time by Edward H. Faulkner in the 1930s. With time, the concept of conserving the soil with the help of minimum tillage or zero tillage and keeping the soil covered with crop residues gained popularity. This system of the soil and water conservation was called conservation tillage (Friedrich et al. 2012). The zero tillage (ZT) was introduced in Brazil in the early 1970s to reduce severe water erosion problems (Derpsch 2001). Moreover, escalating fuel prices and severe erosion problems during the 1970s forced the farmers to move in the direction of resource-saving agricultural systems. In this scenario, CA was adopted by commercial or large farmers to reduce the drought-driven soil degradation jointly with saving of fuel (Haggblade and Tenbo 2003).

9.2. Benefits and Challenges of CA

Zero-tillage farming with residue cover saves irrigation water, gradually increases soil organic matter and suppresses weeds, as well as reduces costs of machinery, fuel and time associated with tilling. Leaving the soil undisturbed increases water infiltration, holds soil moisture and helps to prevent topsoil erosion. Conservation agriculture enhances water intake that allows for more stable yields in the midst of weather extremes exacerbated by climate change.

While conservation agriculture provides many benefits for farmers and the environment, farmers can face constraints to adopt these practices. Wetlands or soils with poor drainage can make adoption challenging. When crop residues are limited, farmers tend to use them for fodder first, so there might not be enough residues for the soil cover. To initiate conservation agriculture, appropriate seeders are necessary, and these may not be available or affordable to all farmers. Conservation agriculture is also knowledge intensive and not all farmers may have access to the knowledge and training required on how to practice conservation agriculture. Finally, conservation agriculture increases yields over time but farmers may not see yield benefits immediately.

However, innovations, adapted research and new technologies are helping farmers to overcome these challenges and facilitate the adoption of conservation agriculture.

Although CA is a driving force in achieving sustainable farming, there are numerous challenges and constraints that affect CA adoption on farms around the world, in addition to the factors discussed above. The first and most prominent challenge that limits the adoption of CA cited in previous literature is small-scale farming. Small farmers lack social inclusion, institutional support, capital, and other resources, and they are less likely to implement CA on their farms. Moreover, farmers with owned animal traction are also less likely to adopt CA practices such as minimum tillage/zero tillage on their farms due to readily available tillage sources. Framers with animal traction typically replace mechanized soil management with animal-driven ploughing. Crop residual retention is also used as a CA practice in the fields by the farmers.

Crop cultivation and livestock rearing are complementary to each other for farmers, especially small farmers in developing countries, but farmers with crop and livestock interaction tend to have low crop residual retention on their farms. Furthermore, livestock is a more important source of traction and income security in an emergency, and crop residuals are a vital source of animal feed . Furthermore, managing crop residuals is more expensive than simply burning crop residuals. The cost of managing crop residuals is more than one-third higher than the cost of burning the residuals. Another major impediment to the adoption of CA practices such as crop rotation is a lack of timely seed availability, as well as dysfunctional markets for final farm outputs. The other challenge faced by farmers in the adoption of CA practices on farms is the unavailability of advanced equipment required for CA adoption. The equipment is either unavailable or its financial costs are high, especially in developing countries. Other than the above challenges, lack of awareness about different CA practices and their associated benefits among farmers in developing countries limits the adoption of CA on farms.

9.3. Global Adoption of CA and its Constraints

The historical chart of CA uptake at the global level is shown in Figure 9.1. The transformation of conventional tillage-based agriculture began in the 1930s after the 'Dust transformation of conventional tillage-based agriculture began in the 1930s after the 'Dust Bowl' that shook the farming communities in the midwestern US, causing the scientific Bowl' that shook the farming communities in the midwestern US, causing the scientific community to rethink what was not going right with farming, particularly with regards community to rethink what was not going right with farming, particularly with regards to soil conservation. Minimization of soil disturbance with stubble mulching was a major to soil conservation. Minimization of soil disturbance with stubble mulching was a major breakthrough in the understanding of how the objective of crop production intensification breakthrough in the understanding of how the objective of crop production intensification could be combined with the objective of soil and water conservation at the practical level could be combined with the objective of soil and water conservation at the practical level by farmers.

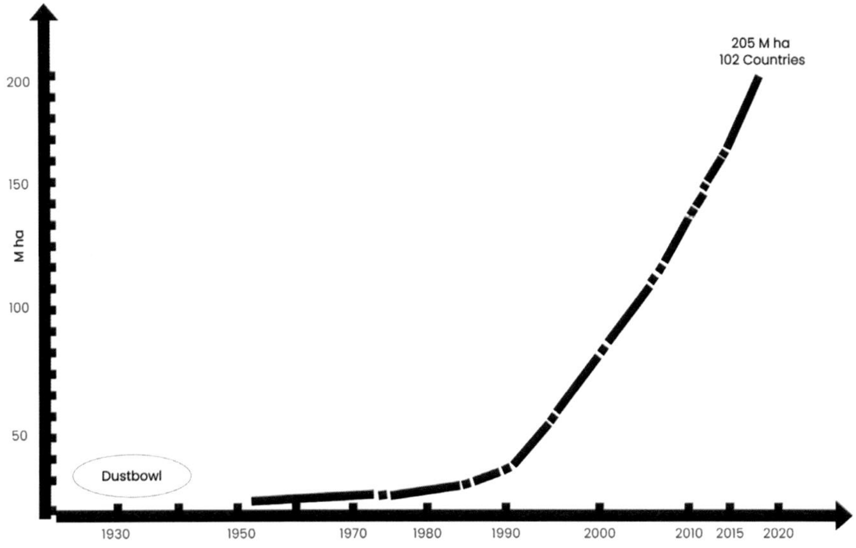

Fig. 9.1 Global increase in area under CA

The global total CA cropland area in 2018/2019 was approximately 205.4 M ha, corresponding to about 14.7% of the total global cropland. This represents an increase of 98.9 M ha or 92.9% from 106.5 M ha in 2008/2009, with the spread being more or less equally split between the Global South (50.5%) and the Global North (49.5%). The global CA cropland increased by some 48.6 M ha or 31.0% since 2013/2014 from 157 M ha, and some 25 M ha or 13.9% since 2015/2016, and from 180 M ha to 205.4 M ha in 2018/2019. Overall, the increase in the global CA cropland area since 2008/2009 has continued at an annual rate of approximately 10 M ha per year, from 106.5 M ha in 2008/2009 to 205.4 M ha in 2018/2019. Prior to that, the annual rate was about 5 M ha per year during the period from 1990 to 2008/2009. In 1990, the CA area of cropland was 11 M ha and in 2000, the CA area was 67 M ha.

The decade of the 1990s is considered as the decade when CA took off. During the first and the second decades of the new millennium, CA uptake by farmers spread out to Africa and Asia while it continued spreading in the Americas, Europe, and Australia. Smallholders had already been adopting CA systems, both manual and mechanized, in the tropics in South America in the late 1980s and 1990s, and smallholders in Africa and Asia also began to adopt CA systems during 1990s and the first decade in the new millennium.

The millennium opened with the first World Congress on Conservation Agriculture, which was held in Madrid, Spain. This helped to globalize the concept and principles of CA, and CA was promoted as part of sustainable production intensification by FAO and some donor agencies. Some centres of the CGIAR, particularly CIMMYT and ICARDA, began to conduct research on CA, and a number of national research systems also began to initiate CA research. More regional and national CA organizations and networks were established. Focus of attention also expanded to West Asia, South and South-East Asia, East Asia, and Africa, with countries such as Iran, Syria, India, Pakistan, China, South Africa, Mozambique, Zambia, Kenya, Ghana, and Morocco making significant progress in the expansion of CA area as well as in CA research to facilitate the effective application of the CA principles in specific contexts.

In July 2008, an international Consultation was organized by FAO in Rome to take stock and discuss the conditions that were necessary to achieve scaling of CA cropland systems. Experiences from all continents were discussed, and a global action plan was formulated to globalize the adoption and uptake of CA. To facilitate the implementation of the plan, a communication platform of Conservation Agriculture Community of Practice (CA-CoP) was established in early January 2009. The platform has enabled global CA stakeholders to be connected and exchange information on all aspects of CA from science and practice to sustainable agriculture development, conservation, and regeneration of natural resources and ecosystem functions.

By 2010, more than 105 M ha (7.0% of global cropland) were under CA cropland systems across 36 countries, covering all continents and most land-based agroecologies. During the 2010–2020 decade, the rate of global uptake increased to 10.5 M ha per year, reaching more than 205 M ha (14.7% of global cropland) in 2019 across 100 countries. CA area in Africa, Asia, and Europe expanded more rapidly as more attention and resources were directed to promoting and supporting the uptake of CA cropland systems. During this decade, three more World Congresses of CA were held, in Australia (2011), Canada (2014), and Argentina (2017).

The expansion of CA uptake continues to be largely farmer-driven, and an increasing number of governments are now providing policy and institutional support to the uptake of CA cropland systems in addition to private sector machine and service companies.

In 2008/2009, global CA area was spread over 36 countries. In 2013/2014, the number of countries with CA area had increased to 53 countries, and in 2015/2016, to 77 countries. In 2018/2019, the number of countries with CA area had increased to 102. The corresponding increases in the number of countries from 2008/2009 to 2018/2019 was from 8 to 13 in South and Central America, from 11 to 31 in Europe, from 2 to 26 countries in Asia, and from 9 to 25 in Africa. This shows that much greater interest is being shown for CA systems in recent years by farmers globally in every region. In addition, the extent of support from public sector institutions and governments, although modest, has been increasing steadily. This trend is expected to continue, benefitting an increasing number of smallholder farmers.

Historical development of no-till systems and the modern version of CA that has been promoted over the past three decades are documented . CA country areas for 2008/2009, 2013/2014, 2015/2016, and 2018/2019 are shown in Table 9.1

Table 9.1: Adoption trends of CA globally

S.No.	Country Name	CA Area 2008/2009	CA Area 2013/2014	CA Area 2015/2016	CA Area 2018/2019
1	USA	26,500.00	35,613.00	43,204.00	44,049.00
2	Brazil	25,502.00	31,811.00	32,000.00	43,000.00
3	Argentina	19,719.00	29,181.00	31,028.00	32,907.00
4	Canada	13,481.00	18,313.00	19,936.00	21,739.00
5	Australia	12,000.00	17,695.00	22,299.00	22,927.00
6	Paraguay	2400.00	3000.00	3000.00	3158.00
7	Kazakhstan	1300.00	200,000	250,000	3000.00
8	China	1330.00	6.670.00.00	9000.00	9000.00
9	Bolivia	706.00	706.00*	2000.00	1858.03
10	Uruguay	655.10	1072.00	1260.00	1278.00
11	Spain	650.00	792.00	900.00	1000.00
12	South Africa	368.00	368.00*	439.00	1607.08
13	Germany	354.00	200.00	146.00	352.89
14	Venezuela	300.00	300.00*	300.00#	300.00 +
15	France	200.00	200.00*	300.00	*no.oo*
16	Finland	200.00	200.00	200.00	120.00
17	Chile	180.00	180.00*	180.00 #	180.00 +
18	New Zealand	16,200.00	162.00*	366.00	366.00 +
19	Colombia	102.00	127.00	127.00#	127.00 +
20	Ukraine	100.00	700.00	700.00 #	900.00
21	Italy	80.00	380.00	283.92	432.00
22	Zambia	40.00	200.00	316.00	552.67
23	Kenya	33.10	33.10*	33.10 #	33.10+

S.No.	Country Name	CA Area 2008/2009	CA Area 2013/2014	CA Area 2015/2016	CA Area 2018/2019
24	UK	24.00	150.00	362.00	562.00
25	Portugal	25.00	32.00	32.00#	47.05
26	Mexico	22.80	41.00	41.00 #	149.22
27	Zimbabwe	15.00	90.00	100.00	100.00 +
28	Slovakia	10.00	35.00	35.00 #	365.00
29	Sudan	10.00	10.00*	10.00 #	10.00+
30	Mozambique	9.00	152.00	289.00	289.00 +
31	Switzerland	9.00	17.00	17.00#	11.02
32	Hungary	8.00	5.00	5.00#	24.29
33	Tunisia	6.00	8.00	12.00	14.00
34	Morocco	4.00	4.00	10.50	12.83
35	Lesotho	0.13	2.00	2.00	2.00 +
36	Ireland	0.10	0.20	0.20	13.66

Globally, the ten lead CA countries are: the US, Brazil, Argentina, Canada, Australia, China, Bolvia, Uruguay, Paraguay, and Kazakhstan. In South and Central America, the lead five countries are: Brazil, Argentina, Paraguay, Bolivia, and Uruguay; in Europe, Spain, France, Romania, United Kingdom, and Italy; in Africa, South Africa, Zambia, Mozambique, Ghana, and Malawi; and in Asia, China, India, Kazakhstan, Pakistan, and Iran.

Global CA Cropland Goal for 2050

The focus of the 8th World Congress on Conservation Agriculture in June 2021 was on the 'Future of Farming: Sustainable and Profitable Farming with Conservation Agriculture'. The Congress participants were convinced that CA must be the mainstay of the shift that the world has to make urgently towards sustainable farming and food systems, and that CA must become synonymous with sustainable farming.

The Congress set a notional goal for itself to increase the global CA cropland area to 50% of the total cropland by 2050, in particular to respond to the global challenge to mitigate advancing climate change and land degradation. This represents an area of 700 M ha. Achieving this goal would require increasing the rate of expansion in CA cropland area by 50%, to 15 M ha per annum. It would also require a massive boost to the momentum of the global CA community's activities with a concentration on the following six themes:

1. Catalysing the formation of additional farmer run CA groups in countries and regions in which they do not yet exist and enabling all groups to accelerate CA adoption and enhancement, maintaining high quality standards;

2. Greatly speeding up the innovation and mainstreaming of a growing array of truly sustainable CA-based technologies, including through engaging with other movements committed to sustainable farming;
3. Embedding the CA community in the main global efforts to shift to sustainable food and governance systems and replicating the arrangements at local levels;
4. Assuring that CA farmers are justly rewarded for their generation of public goods and environmental services;
5. Mobilizing recognition, institutional support, and additional funding from governments and international development institutions to support good quality CA programme expansion;
6. Building global public awareness of the steps being taken by the CA community to make food production and consumption sustainable.

Objectives of CA

CA practices aim to improve farm resource utilization by integrating natural resource management such as soil, water, and biological resources with the fewest external farm inputs. Therefore, CA is being adopted all over the world in response to the growing concerns of national and international institutions related to farm sustainability, and CA is one of the most important and rapidly expanding adoption systems in all regions of the world. Different types of CA practices are preferred on different farms and in different regions depending on the climate, the land type, the farmer's skills, and the farm's resources. Moreover, the CA practices adoption on farms is also majorly dependent on the purpose of adoption CA. Moreover, the CA practices adoption on farms is also majorly dependent on the purpose of adoption. As a result, CA practices used in one country or farm may be different from those used in another country due to the difference in intended objectives of CA adoption (Table 9.2) Different types of CA practices are preferred on different farms and in different regions depending on the climate, the land type, the farmer's skills, and the farm's resources. According to the literature, farmers in different countries adopted different CA practices on their farms, but zero tillage/no-tillage was one of the most widely adopted CA practices on farms. The minimum or zero-tillage CA practice is widely used around the world, but crop residuals retention in the field is more complicated . Moreover, cover crops and crop rotation are also commonly practiced CA strategies in the world. Crop rotation is underutilized in terms of pest control, disease cycle disruption, income risk reduction, and soil fertility.

Table 9.2: CA practices adopted along with influencing factors and objectives in different countries

Country/Region	CA practices	Influencing factors	Objective/s
Nigeria	Zero tillage, minimum tillage, contour stripping, not burning field, tree planting, cover crops, dead tree trunks, mulching	Age, education, innovativeness, attitude toward conservation, risk bearing, credit, farm income, input-output prices, off-farm occupation	Soil erosion controlling
Rwanda	Organic inputs	Monetary and physical factors, human capital, investment risk, wealth, and liquidity sources	Land conservation investment and organic inputs use
United States	Conservation tillage, contour farming, strip-cropping, grass waterways	Farm size, age, college education, program participation, land tenure, annual precipitation.	Adoption of CA practices and land tenure
Zimbabwe	Zero tillage, crop rotation, contour ridging technologies	Farm and farmer characteristics, institutional factors	Analysing the adoption of CA by small farmers
Spain	Not burning olive-de suckering debris, using shredded olive-pruning debris as soil cover, cover crops under mower control	Socio-economic characteristics of a farmer, social capital indicators, farm characteristics, farm management	Assessing the soil conservation practice as CA in olive groves
Zimbabwe	Winter weeding, digging planting basins, crop residues, manuring, basal fertilizer, topdressing, Timely weeding, crop rotation	Age, Education, own land, draught power, Extension services, labour, conservation agricultural experiences.	Assessing the adoption of CA among different clusters of farmers.
Australia	No-tillage, crop stubble, legumes rotation, controlled traffic farming,	Various socio-economic factors	Assessing the adoption of CA as climate change mitigation activity
Bangladesh	Conservation agriculture principles	Farm size, family size, farming experience, age and education of head, extension services, farm and off-farm income	Impact of CA adoption farms' economic viability

Country/Region	CA practices	Influencing factors	Objective/s
Kenya	Mulching, direct planting, shallow weeding, spraying herbicides	Attitudes, perceived norms and perceived behavioural control, farmer's perception Of the social norms towards CA, farmer's perceived behavioural control	Assessing the farmer's decision of selecting the CA over conventional agriculture.
South Africa	No-till conservation agriculture	Age, gender, education, experience, training, extension, credit access, land size, income	Assessing the adoption of no-till conservation by small-scale farmers
Malawi	Zero tillage, mulching of crop residual, and intercropping with legumes	Landholding, education, neighbour's adaption of CA, gender, no. of male and females in family	Decision-making analysis regarding CA adaptation
United States	Conservation tillage, cover crops, diverse crop rotation	Location and spatial variable, age, college education, area under operation, family network index, organization network index, perception of environmental benefits of the practices.	Assessing the adaption of CA practices

Excessive/meagre rainfall, relatively low temperature, poor and shallow soils, and soil erosion in terms of loss of fertile top soil are the major natural problems that plague the sustained agriculture in the hills. In the IHR, soil and water erosion leads to disastrous loss of soil and severe degradation of farmland. In response to these problems, CA is one of the best management practices (BMPs) to solve all these problems of hill agriculture. There are some important reasons for adoption of CA.

- It is very economical for farmers which helps in cost reduction of fuel and machinery and saving time for operations which allow development of other farming and non-farming harmonizing activities.
- It is very resilient to technical possibilities for sowing, fertilizer application, and weed control (helping in carrying out of operations within time).
- It is important to increase productivity and greater stabilization of productivity in long-term effect.
- It conserves the soil and water against water and wind erosion.
- It increases the agronomic efficiency (AE) and production efficiency (PE).
- It increases the nutrient use efficiency (NUE).

It increases water use efficiency (WUE) or water productivity and water economy in dryland areas where water scarcity is a serious problem.

9.4. Types of Agriculture

Conservation Agriculture vs Organic Agriculture

Conservation agriculture and organic farming both maintain a balance between agriculture and resources, use crop rotation, and protect the soil's organic matter. However, the main difference between these two types of farming is that organic farmers use a plow or soil tillage, while farmers who practice conservation agriculture use natural principles and do not till the soil. Organic farmers apply tillage to remove weeds without using inorganic fertilizers. Conservation agriculture farmers, on the other hand, use a permanent soil cover and plant seeds through this layer. They may initially use inorganic fertilizers to manage weeds, especially in soils with low fertility. Over time, the use of agrichemicals may be reduced or slowly phased out.

Conservation Agriculture vs Climate-Smart Agriculture

While conservation agriculture and climate-smart agriculture are similar, their purposes are different. Conservation agriculture aims to sustainably intensify smallholder farming systems and have a positive effect on the environment using natural processes. It helps farmers to adapt to and increase profits in spite of climate risks. Climate-smart agriculture aims to adapt to and mitigate the effects of climate change by sequestering soil carbon and reducing greenhouse gas emissions, and finally increase productivity and profitability of farming systems to ensure farmers' livelihoods and food security in a changing climate. Conservation agriculture systems can be considered climate-smart as they deliver on the objectives of climate-smart agriculture.

Conventional Agriculture vs Conservation Agriculture

In the conventional agriculture, management practices are extensively used of various tillage operations for plowing of the land for preparation of seedbed and to keep weed down, i.e., mouldboard or animal drawn plow or harrowing, drilling, cultivator, etc. These tillage operations are repeated many times; due to this conditions break down the soil structure and destroy pore and soil becomes prone to erosion and leads to heavy cost of time, fuel, and labour. However, conventional tillage exposed soil to air and sunlight which causes oxidation of organic matter and leads to low carbon content in soil which affects soil structure. The oxidation of organic matter releases CO_2 into environment which causes global warming or climate change (Grace et al. 2003). It was observed that soil organic matter decomposes more rapidly in the tropics compared to subtropical and moderate climates because of the higher temperature (Steiner 2002).

The conservation agriculture system involves specific agronomic field operations such as minimum soil disturbance or use of zero tillage or NT, soil cover with green manure or crop residue (mulching), and crop rotation; these management practices are highly beneficial for farm community because it saves fuel, time, and labour as well as it provides good soil structure, porosity, more accumulation of the organic matter in soil which provides better soil aggregation (Table 9.3), water-holding capacity, soil moisture for the long term, nutrient recycling, and transformation. Meanwhile, the conservation tillage improves soil fertility, water, and crop productivity; the no-tillage gives better soil protection than conventional tillage. This happens as the conventional tillage system leaves ~1–5 % of the soil surface covered with crop residues.

Table 9.3 Characteristic features between conventional and CA

Conventional agriculture	Conservation agriculture
Cultivation of land with the help of machinery using science and technology	Minimal distrubances to nature during crop grown processes
Excessive motorized tillage and leads to deteriorationof soil pores or structure	No-tillage or considerably reduced tillage (bio-tillage) increases porosity
Higher wind speed and water erosion	Lower wind speed and water erosion
Removal of crop resideue from field or burning or uncovered surface	Soil covered with crop residue permanently covered
Low water infiltration	High rate of infiltration of water
Decrease water-holding capacity	Enhance water-holding capacity
FYM/composts added from outside or green manuring (incorporated)	Use of in situ organics/composts or brown manuring/cover crops (stubble retention)
Due plowing eastablished weeds are kills but also stimulates more weed seed to germinate	Weeds create problem during early stage of implementation but reduce with time
Soil become very compacted due to more use of heavy traffic or machinery	Control of traffic, there is less compaction

Conservation Agriculture vs Sustainable Farming

Sustainable intensification is a process to increase agriculture yields without adverse impacts on the environment, taking the whole ecosystem into consideration. It aims for the same goals as conservation agriculture. Conservation agriculture practices lead to or enable sustainable intensification.

Sustainable development is defined as "the ability of the current generation to meet their needs without jeopardizing future generations' ability to meet their own needs". In order to apply the concept of sustainable development to farming, farming must be socially acceptable, economically viable, and environmentally friendly. Farming will therefore be sustainable if it is socially, economically, and environmentally sustainable. Conservation agriculture is a modern way of farming that helps farmers achieve their goals of sustainable development through sustainable farming. The impact of CA on crop yield can be used to explain CA's role in achieving economic sustainability. A high crop yield is one of the farmer's primary goals in order to enjoy a good economic return.

Crop yield is affected by numerous management (timely and proper application of off-farm inputs) and ecological factors (uneven rains, harsh weather, water deficiency, etc.). CA is critical in reducing the negative impact of these factors on crop yield. No-tillage, for example, improves soil fertility and structure while also softening the soil, which improves seed germination and crop growth. A well-grown crop with good germination results in a good crop yield, which increases the crop surplus. Farmers make a good profit from the high crop surplus, which lets farmers keep their high standard of living. CA also boosts crop yield and economic returns, which makes farming socially and economically more sustainable. Thus, the CA improves farmers' long-term welfare by increasing crop yield, high economic returns, and food security. Zheng et al. 2014, discovered that adoption of CA practices has a positive impact on crop yield in China. The study showed that adoption of CA on farms significantly increases crop yields. However, the impact of CA on crop yields is dependent on geographical location, climatic conditions, and the type of adopted CA practice. The study also found that conventional tillage with straw retention produced better crop yields than no-tillage with straw retention. Moreover, the study also reported that CA practices produce better results in geographical locations with annual precipitation of less than 600 mm and a mean temperature of greater than 5°C.

The potential for high crop yields with CA is greater in rain-fed areas than in conventional tillage systems. The CA is more effective in terms of yield and farm productivity when all three CA principles are implemented in combination than when they are implemented alone on farms. Even sometimes, CA principles implemented separately can have a negative impact on farm productivity. Pittelkow et al., 2015, stated that no-tillage has a negative impact on crop yield, but when combined with the other two principles (cover crops and crop rotation); it generates an equal or greater crop yield than conventional agriculture. Therefore, no-tillage, cover crop, and diversification (crop rotation) produce high crop yields ensuring the CA system's economic sustainability. The widespread adoption of three CA principles around the world has ushered in a new era of environmental control and mitigation for damages associated with conventional agriculture. No-tillage leaves the soil untouched, improving its physical properties, which is a major component of the environment. The organic carbon stock is three times more concentrated in the soil than in the atmosphere. Increased soil carbon level is highly associated with increased soil carbon level through improved mineralization processes, which reduces the negative effects of climate change on crop yield.

CA provides climate-smart sustainable farming systems that enable farmers to cope with the adverse impacts of climate change. Similarly, growing cover crops on fallow lands lowers the risk of soil erosion. By covering the field where no primary crop is grown, it can control weed germination and enhance N-input leaching. It also increases the soil's water-holding capacity, promotes microbial activity, maintains soil structure and porosity, and balances the nutrient cycle. Similarly, crop rotation reduces climate vulnerability and improves soil health by reducing herbivores, increasing yields, and generating high economic benefits. It also provides a more stable planting system in extreme weather conditions.

Likewise, CA improves soil fertility, reduces soil erosion, improves water filtration and retention, and reduces greenhouse gas emissions, all of which contribute to the farming system's high environmental sustainability. Therefore, the CA is an ideal solution for resolving the environmental problem in agriculture. CA's associated environmental benefit improves agriculture's environmental sustainability. Moreover, CA also contributes to social sustainability by increasing gender equality, labour participation, and farmer welfare, and it is expected that promoting the CA farming system will increase the participation of women in farming. For example, women's participation in Zimbabwe increased grain yield and improved food diversity and security. Furthermore, the use of CA has altered intra-household decision-making between males and females. Women's participation in decision-making, crop management, and improved agency were observed in Zimbabwe. Furthermore, crop residue retention can be used to describe the labor requirements of CA.

Women practicing CA are good time managers because they start clearing land on time to prepare it for early planting. Based on the preceding discussion, it is clear that CA adoption helps farmers cope with climate change while also increasing farm-level sustainability (social, economic, and environmental). Thus, adoption of CA works as a driver of sustainable farming. CA principles provide multifaceted benefits that lead to sustainable farming. Improving soil fertility, for example, improves farmers' economic conditions, which in turn affects farms' economic and social sustainability. Similarly, reducing the use of machinery for tillage reduces greenhouse gas emissions while also reducing costs, which helps to improve environmental, economic, and social sustainability. As an outcome, CA's minimal soil mechanical disturbance principle influences all three dimensions of sustainability. Similarly, the other two CA principles help farmers improve their farms' economic, social, and environmental sustainability. As a result, CA adoption can play a critical role in global farming sustainability.

9.5. Case Studies

1. By FAO - China

China started no-tillage studies in the 1950s, and no-tillage seeding experiments with two crops a year (wheat during winter /maize during summer) in a region of Northern China throughout the 1980s. In 1992, China began the systematic study on Conservation Agriculture (CA) technology, with emphasis on the development of no-tillage seeders. These studies generated convincing evidences that CA can benefit the dryland wheat and maize production ecologically and economically.

A series of CA equipment was developed based on local agricultural conditions. Typical CA equipment includes manual direct seeders, no/minimum tillage seeders for 2-wheel tractors, middle and small no/minimum tillage seeders (strip rotary hoe seeder, strip chop seeder, powered disc seeder, rice strip rotary hoe transplanter, etc.) for four-wheel tractor. In 1999, the Conservation Tillage Research Centre (CTRC) was set up at the China Agricultural University by Chinese Ministry of Agriculture (MOA), specifically for CA. CA studies were extended to the maize-wheat annual double cropping region, the Northeast ridge tillage region, the farming-pastoral region and the rice-wheat annual double cropping region.

All these laid the foundation for national CA extension. Since 2002, national and local CA demonstration projects have been launched successively – 226 national and 365 provincial demonstration counties – covering more than 3 Mha by the end of 2008. Within those counties, ten sites were established by CTRC to monitor CA benefits. Most of these sites have consistently shown that the synergetic application of the CA components including minimum soil disturbance, organic soil cover and crop diversity could enhance and sustain the agricultural production while promoting ecosystem services.

In China, farming systems based on CA principles were further studied and implemented to solve the specific regional problems. The Permanent Raised Bed (PRB) farming system, which has been shown to improve soil productivity and reduce water requirements worldwide, was used in Hexi Corridor, Gansu Province to solve the problems of soil degradation, limited water and subsequent yield decline. There on-farm demonstration sites in Hexi Corridor showed that PRB had more macro-aggregates (2.7%), a better distribution of pore size classes and higher crop yields (4.2%), and water use efficiency (21.3%) compared with traditional tillage.

Controlled Traffic Farming (CTF), which is also a CA based farming system, was studied in Linfen in the Chinese Loess Plateau. The over ten-year experimental results indicated significant improvements in soil physical and chemical property in cropping zones, as well as a winter wheat yield increase of 10.8% under CTF treatment.

After years of systematic research and demonstration, the CA farming system has been widely recognized as a sustainable way to cope with soil degradation, water shortage, as well as the changing climate in China. Given its noticeable ecological, economic and social contributions, many policies for CA have been formulated by Chinese government for rapid expansion of CA, such as the Central Document No. 1 which stressed the importance of CA over an eight-year period, and MOA and the Ministry of Finance (MOF) supported to extend the spread of CA from 2002 and beyond. Up to 2016, CA systems have been extensively implemented in China to reach an area averaging as much as 8 Mha.

2. By FAO – Malawi and Zambia

Where trees and shrubs cost less than fertilizer Leguminous trees and shrubs grown with maize provide high-quality, nitrogen-rich residues that improve soil fertility, boost yields and provide new sources of income. One of the main obstacles farmers face in increasing maize production is low soil fertility. Many maize farmers can neither afford mineral fertilizer nor obtain sufficient amounts of organic fertilizer, such as animal manure. Decades of intensive cultivation without fertilization have drained nutrients, particularly nitrogen, from the soil. To address the problem, the Zambia National Farmers' Union has explored ways of integrating nitrogen-fixing trees into maize production systems. The most promising candidate was found to be Faidherbia albida, an African acacia species, which has an unusual growth habit. The tree is dormant in the early rainy season and loses its leaves just as field crops are being established; the leaves only grow back at the end of the wet season. Maize can be grown directly under the leafless Faidherbia canopy, as the trees do not compete with the crop for light, nutrients or water while the maize is growing. Thanks to the decaying leaves, the soil under the trees contains up to twice as much organic matter and nitrogen as soil outside the canopy. There is also a marked increase in soil microbiological activity, and an increase in water holding capacity. Numerous studies have noted increases in yields when maize is grown in association with Faidherbia, and those increases tend to be higher where soil fertility is low. In Zambia, maize planted outside the tree canopy produced average yields of 1.9 tonnes per ha, compared to 4.7 tonnes when the crop was grown under the canopy; in Malawi, maize yields increased by 100 to 400 percent when the crop was grown with Faidherbia. Both countries promote Faidherbia as part of conservation agriculture systems that offer smallholder farmers a means of increasing maize productivity and earning higher incomes.

3. By FAO - Lesotho

Subsistence farmers in Lesotho have been able to boost agricultural yields and increase food production by adopting conservation agriculture. The locally adapted practice, known as likoti, also contributes to combating soil erosion and to enhancing fertility. Since the mid-1970s, maize yields have fallen from an average 1400 kg/ha to a current 450–500 kg/ha in most of the districts. In spite of its poor performance, agriculture remains a major source of livelihood for the vast majority of households engaged in subsistence farming. Under an FAO initiative, two sub-sample populations (117 CA and 112 'conventional' farmers, or a total of 229) were monitored through a household survey in the western lowlands and in the south-eastern highlands of Lesotho. The practice of likoti was first introduced in Tebellong – a harsh mountainous area in the south-eastern highlands of Qacha's Nek district – by the local NGO 'Growing Nations'. A planting basins system was developed and adapted to the local conditions and began to promote it with a Sesotho name, likoti, which means 'holes'. Since 2002, the practice has captured the interest of more non-governmental and international organizations, including, among others, German Red Cross, FAO and WFP, which have supported the dissemination of the practice throughout the country.

According to the likoti method, pits of about 15 × 30 cm in diameter and 15 to 20 cm deep (or smaller) are dug in a 75 × 75 cm grid. A small quantity of fertilizer (either inorganic or organic) and seeds (the number depends on the desired crop density) are placed in each basin and covered with soil. Additionally, farmers must leave enough crop residues on the field as mulch, and must also practice crop rotation and/or inter-cropping. The next season, farmers must plant in the same pits without turning the soil. Although likoti was originally deployed in the production of maize and beans, innovative farmers have used it to produce other crops such as sunflower, sorghum, potato and tomato. Analysis of the survey data showed that the adoption of likoti has brought about significant advantages compared to conventional tillage practices. The most important are: higher agricultural productivity, due to improved efficiency in the use of inputs and other resources; greater environmental sustainability, due to improved soil structure and enhanced fertility; and higher social sustainability, due to accessibility to the technology by all social categories, including the most vulnerable.

The socio-economic and environmental benefits help poor households to rehabilitate and strengthen their livelihood capital base and ultimately help rural communities to build system resilience in the face of widespread poverty and increasing vulnerability that affect the country. Comparing the value of the output with the costs of the inputs showed that farmers producing maize with likoti in Qacha's Nek returned a profit whereas those who ploughed incurred a loss. In Butha-Buthe, average profit of CA adopters was double that obtained by farmers who employed draught power, whereas it was almost four times that obtained by farmers who used tractor power. The results show that attending appropriate training is a crucial prerequisite for the correct adoption of likoti. However, training is more effective when trainers pursue true participation and when social capital among farmers is stronger. Further important determinants of adoption are the level of education and the economic incentives provided to vulnerable households. Stronger policy and institutional support in all these areas would thus help address the cultural and resource constraints that limit the full potential of likoti to be harnessed and ultimately hinder its further spread throughout the country.

By FAO - Kazakhstan

Farmers stop ploughing on Kazakhstani steppe Zero-tillage, soil cover and crop rotation would help many countries to reverse land degradation and produce more food. Kazakhstan's wheat growers are already well advanced in the transition to full conservation agriculture. Kazakhstan is one of the world's leading adopters of conservation agriculture. Direct-seeded, untilled land produces higher wheat yields than ploughed land, and carries lower production costs. Rotating wheat with other crops generates extra income and leaves residues that conserve soil moisture and suppress weeds. The area of land that is no longer ploughed at all rose from nil in 2000 to 1.4 million ha by 2008. That increase is attributed to very high adoption rates on large farming enterprises of more than 50 000 ha, where managers are striving to increase production while reducing costs. However, the approach has also been taken up on small to medium-sized farms, a category which, in sparsely populated Kazakhstan, ranges from 500 to 2 500 ha. The adoption rate has been particularly high on farms with rich black soils, where high returns provide the capital needed for investment in CA machinery.

In zero-tilled areas, weeds are often controlled with herbicides. However, many farmers have found that combining zero-tillage with permanent soil cover also helps to suppress weeds. Without tillage, the natural store of weed seeds in the soil diminishes over time, and decomposing residues release humic acids, which block the seeds' germination. While zero-tillage usually requires increased use of herbicides in the first few years of adoption, after four or five years the incidence of weeds – and herbicide use – decreases considerably. Another advantage of retaining crop residues in northern Kazakhstan is that it increases the availability of water to the wheat crop. Annual precipitation ranges from 250 to 350 mm, and winter snow accounts for around 40 percent of it; when the snow is blown away by wind, the soil surface is left bare and dry. Retaining the stubble of the previous wheat crop traps the snow which, when the weather warms, melts into the soil. That has two benefits: more moisture is available along the soil profile and erosion is reduced or even eliminated. On-farm research has found that the use of residues to capture snow, along with zero-tillage, can increase yields by 58 percent. Progress in the adoption of the third pillar of CA, diverse crop rotations, which would increase land productivity and help farmers to better manage wheat pests and diseases, has been slower. The vegetation period on the northern steppes in summer is short, with a high frequency of dry years. However, areas of traditional summer fallows are declining, as farmers take advantage of available – and sometimes abundant – rainfall to grow oats, sunflower and canola. Studies have shown the high potential of other rotational crops, including field peas, lentils, buckwheat and flax. A three-year study found that forage sorghum sown late in May and harvested in August provided not only fodder for sale or silage, but also left a durable post-harvest stubble that was very effective in trapping that precious snow.

10

Important Uses of Medicinal, Aromatic and Spice Crops

There is no plant in this world which is uneconomical/not useful. Medicinal plants have been considered as important therapeutic aid for alleviating aliments of human kind. The WHO estimates that 80% of people living in developing countries rely almost exclusively on traditional medicines for their primary health care needs. Aromatic plants are used for essential oils - volatile material extracted from plants or plant parts usually by stream distillation and used in flavours and fragrances for food, soap, detergents, perfumes, lotions etc. Spices are the dry part of a plant such as roots, leaves and seeds which impart to food a certain flavour and pungent stimuli. Spices contain variable amount of fat, protein, carbohydrates, fibres, minerals and vitamins.

10.1. Aloe vera (*Aloe barbadensis miller*)

In Ayurveda, Aloe vera is known as the 'King of medicinal plants.' It holds water in its fleshy leaves which is why it can sustain in extremely dry conditions as well. So, growing Aloe vera in your garden will not require you to take extra care of the plant every now and then. However, it surely can treat a wide variety of health problems. These include:

- Constipation
- Digestive distress
- Acne/Pimples
- Poor body immunity
- Skin problems
- Useful in fever, enlargement of liver, spleen and other glands
- Piles, jaundice, rheumatic diseases
- Treatment of burns and bruises

10.2. Tulsi *(Ocimum tenuiflorum)*

Tulsi is also called the Queen of medicinal plants. This plant holds immense significance in the Hindu religion. But it has so much more to itself than just religious significance. The strong aroma of Tulsi is good enough to keep bacterial growth at bay. It is known for its healing properties, and all it requires for growth is water. Here's a list of the uses and healing properties of Tulsi:

- It gives you strength to fight stress
- Promotes a longer life
- Treats cough
- Treats indigestion
- Anti-cancer
- Good for hair loss, heart diseases, diabetes, etc.

10.3. Mint (*Mentha* spps.)

This freshly fragrant medicinal plant serves a wide variety of purposes. From enhancing your mood to treating indigestion, mint can do it all. This plant requires a lot of water to grow. So you must sow its seeds and keep watering it. The best part about mint is its ability to repel pest and insects so your home will be a clean environment. The important uses of mint are

- Keeps the digestive system running
- Boosts immunity
- Enhances mood
- Expels cough from the body and helpful for chest pain
- Benefits respiratory health
- Keeps mosquitoes away
- Menthol and mint essential oil are used in medicine
- Mint is also used in shampoo products

10.4. Fenugreek *(Trigonella foenum-graecum)*

Metha or fenugreek is a wonder amongst other medicinal plants, and all because of its properties. It is an evergreen plant and both the leaves and the seeds are useful. Few important benefits of fenugreek are

- Controls cholesterol levels
- Curbs hair loss
- Increases appetite
- Boosts your excretory system
- Purifies blood
- Lowers blood pressure
- Beneficial for joint pains and diabetes

10.5. Fennel *(Foeniculum vulgare)*

Fennel or saunf is a flavourful and aromatic plant which is useful for a wide variety of health problems. In India, people are accustomed to chewing fennel seeds after every meal. It can be grown easily in your garden and does not require much care. Take a look at the many benefits of fennel seeds:

- Treats cough
- Controls cholesterol
- Improves eyesight
- Cure acidity
- Prevents bad breath
- Improves breast milk supply in lactating women

10.6. Coriander (*Coriandrum sativum*)

Coriander or dhania is an important ingredient of an Indian kitchen. Its leaves, seeds and powder of the seeds, everything is beneficial for your health. It adds a distinct flavour to your food and has a wide variety of medicinal properties. Here's how coriander keeps you healthy:

- Prevents the food from spoiling
- It is rich in antioxidants
- Cures urine retention
- Improves digestion
- It regulates your menstrual cycles
- Treats acne

10.7. Ginger (*Zingiber officinale*)

You name the problem and ginger is the answer. It is the root solution for a wide variety of health problems. All you need to do is sow the ginger root in your garden and it a few days it will grow it in a few days. It does not require much care. It is an important ingredient of Indian food due to its distinct flavor, and of course, its benefits for your overall health. Important benefits of ginger are:

- Treats indigestion
- Eases headaches
- Controls blood pressure
- Treats cold, cough, flu and asthma
- Relives menstrual pains and cramps

10.8. Ashwagandha *(Withania somnifera)*

Ashwagandha is the most popular medicinal plant. All the plant's parts (roots, leaves and fruits) exhibit medicinal properties.

- The roots hold special attention. They aid in treating a wide range of health complications.
- Ashwagandha helps to treat Insomnia
- Reduces stress and controls cholesterol levels.
- It also works as an immune booster
- Increases fertility and regulates testosterone.

10.9. Amla (*Phyllanthus emblica*)

Amla, also known as 'Indian gooseberry,' exhibits a sour taste.

- The rich source of vitamin C in amla acts as an antioxidant that promotes the function of the immune system.
- The carotene component in amla improves eyesight.
- Regular consumption of amla can increase overall eye health because the carotene component prevents eye diseases like redness, cataract and itchy and watery eyes.
- Amla also exhibits anti-inflammatory properties that can help to ease arthritis-related joints and help to treat wounds.
- Amla is widely used in hair oil as it reduces hair fall, improves scalp health and promotes hair growth

10.10. Turmeric *(Curcuma longa)*

Turmeric, a plant in the ginger family, is native to Southeast Asia and is grown commercially in that region, primarily in India. Its rhizome (underground stem) is used as a culinary spice and traditional medicine. Historically, turmeric was used in Ayurveda and other traditional Indian medical systems, as well as Eastern Asian medical systems such as traditional Chinese medicine.

- In India, it was traditionally used for disorders of the skin, upper respiratory tract, joints, and digestive system.
- Today, turmeric is promoted as a dietary supplement for a variety of conditions, including arthritis, digestive disorders, respiratory infections, allergies, liver disease, depression, and many others.
- Turmeric is a common spice and a major ingredient in curry powder.
- Curcumin is a major component of turmeric, and the activities of turmeric are commonly attributed to curcuminoids (curcumin and closely related substances). Curcumin gives turmeric its yellow color.
- Turmeric dietary supplements are made from the dried rhizome and typically contain a mixture of curcuminoids.
- Turmeric is also made into a paste for skin conditions.

10.11. Stevia *(Stevia rebaudiana)*

As stevia is 200 to 300 times sweeter than table sugar. It typically requires about 20 percent of the land and far less water to provide the same amount of sweetness as other mainstream sweeteners.

Here are some of the possible health benefits of stevia.

- **Diabetes:** Research has shown that stevia sweeteners do not contribute calories or carbohydrates to the diet. They have also demonstrated no effect on blood glucose or insulin response. This allows people with diabetes to eat a wider variety of foods and comply with a healthful meal plan. Studies related to type 2 diabetes showed that stevia triggered significant reductions in blood glucose and glucagon response after a meal. Glucagon is a hormone that regulates glucose levels in the blood, and the mechanism that secretes glucagon is often faulty in people with diabetes. Glucagon drops when blood glucose climbs. This regulates the glucose level.

- **Weight control**: Stevia can replace sugar in a diet to manage weight. Stevia contains no sugar and very few, if any, calories. It can be part of a well-balanced diet to help reduce energy intake without sacrificing taste.
- **Pancreatic cancer**: Stevia contains many sterols and antioxidant compounds, including kaempferol. Studies have found that kaempferol can reduce the risk of pancreatic cancer by 23 percent.
- **Blood pressure**: Certain glycosides in stevia extract have been found to dilate blood vessels. They can also increase sodium excretion and urine output. Cardiotonic actions normalize blood pressure and regulate the heartbeat.
- **Children's diets**: Foods and beverages containing stevia can play an important role in decreasing calories from unwanted sweeteners in the diets of children. Excessive sugars and calories are linked to obesity and cardiovascular disease.
- **Allergies**: In 2010, the European Food Safety Committee (EFSA) concluded that "steviol glycosides are not reactive and are not metabolized to reactive compounds, therefore, it is unlikely that the steviol glycosides under evaluation should cause by themselves allergic reactions when consumed in foods."

10.12. Air Potato / Air Yam (*Dioscorea* sp.)

- *Dioscorea* species and, being used in the formulation of pharmaceutical products, have a high medicinal, industrial and commercial importance.

- *Dioscorea* species account for the most important dietary supplements ingredients used in cosmetics and pharmaceutical industries.
- Tubers of different *Dioscorea* sps are used as a cure for different diseases and ailments (cough, cold, stomach ache, leprosy, burns, fungal infections, dysentery, skin diseases, rheumatism, arthritis, etc.) in several formulations, and even for birth control. It is well known the potential of this plant is related due to the different phytochemical compounds found in *Dioscorea.*
- Tubers and roots contain steroidal sapogenins, mostly diosgenin as well as volatile compounds. Chemical substances like diosgenin found in *Dioscorea* are commercially used in the pharmaceutical industry. There still a huge need for more research and information on the food, nutraceutical, and medicinal value worldwide of this plant species.

10.13. Periwinkle (*Vinca minor*)

- Periwinkle is an evergreen shrub that grows in Europe, northwestern Africa, central Asia, and some parts of North America. Periwinkle has likely been used for medicine for a long time; its Latin name, *Vinca,* is derived from the Latin word *vincere,* meaning "to overcome."
- European herbalists have used periwinkle for headaches, vertigo, and poor memory since medieval times. The flower and leaf of lesser periwinkle are used medicinally.

- The plant is sedative and tonic. It contains the alkaloid 'vincamine', which is used by the pharmaceutical industry as a cerebral stimulant and vasodilator.
- Since the discovery of vincamine in the leaves, the plant has been used herbally to treat arteriosclerosis and for dementia due to insufficient blood supply to the brain.
- The leaves are bitter, detergent and stomachic. Taken internally, they are used in the treatment of internal bleeding, heavy menstrual bleeding and nosebleeds. When crushed and applied to wounds they have astringent and healing properties.
- A mouthwash is used to treat gingivitis, sore throats and mouth ulcers
- The root is antispasmodic and hypotensive. It is used to lower the blood pressure.
- The fresh flowers are gently purgative, but lose their effect on drying.
- A homeopathic remedy is made from the fresh leaves. It is used in the treatment of haemorrhages.

10.14. Indian snake root *(Rauwolfia serpentine)*

- The root of Indian snakeroot plant is used to make medicine. Indian snakeroot is used for mild high blood pressure, nervousness, trouble sleeping (insomnia), and mental disorders such as agitated psychosis and insanity.

- Indian snakeroot is also used for snake and reptile bites, fever, constipation, feverish intestinal diseases, liver ailments, achy joints (rheumatism), fluid retention (edema), epilepsy, and as a tonic for general debilities.
- One of the chemicals in Indian snakeroot is the same as a prescription drug called reserpine. Reserpine is used to treat mild to moderate hypertension, schizophrenia, and some symptoms of poor circulation.
- Indian snakeroot contains chemicals such as reserpine that decrease heart rate and blood pressure.

10.15. Indian bdellium or Guggul (*Commiphora wightii*)

- **Balances cholesterol levels:** Guggul includes guggulsterone, ketonic steroid components that can lower blood levels of LDL cholesterol. These components also reduce plaque build-up on the artery walls, preventing atherosclerosis. In addition to encouraging healthy cholesterol levels, Guggul is highly effective at cleaning the blood and removing impurities from the circulatory system.
- **Assists in Weight Loss:** Guggul is an excellent natural treatment for weight loss and weight maintenance. The constituents present in Guggul are capable of activating the thyroid hormones that enhance the metabolism rate of the body. These hormones also play a crucial part in the regulation of normal metabolism, which aids in weight management. Additionally, it enhances digestion and thyroid function in addition to increasing metabolism.
- **Heals skin issues:** According to Ayurveda, when Kapha rises, the production of sebum surges, which clogs the pores. As a result, whiteheads and blackheads develop on the skin. Inflammation with pus and red blisters are also signs of Pitta aggravation. Tridosha-balancing characteristic of Guggul helps to balance Kapha-Pitta and lowers skin irritation and inflammation when consumed in moderation.

- **Relieves joint pain:** Joint pain, stiffness of the joints, swelling, and other symptoms of arthritis can be relieved with this ayurvedic herb. Guggul provides relief from pain and swelling in the joints. It also stimulates the regeneration of bones, nerve tissues and joints.
- **Treats inflammation, wounds and sores:** The anti-inflammatory characteristics of Guggul can also aid in healing issues like tonsillitis, mouth ulcers, sore throats and gum infections.
- Additionally, the characteristics of Guggul can increase the count of white blood cells that help fight against various infections. Overall, Guggul is an excellent immunity booster that assists in maintaining cholesterol levels, heals skin and wounds, and also aids in reducing body weight.

10.16. Common Coleus (*Coleus* Spps.)

Coleus is a plant from the mint family. It grows in tropical regions of Asia such as Nepal, India, and Thailand.

- The root is used as medicine. Coleus root contains a chemical called forskolin. Herbal product manufacturers often produce coleus extracts that contain high levels of forskolin. These preparations are being promoted for the same conditions for which forskolin has been traditionally used.
- Coleus or forskolin are used for high blood pressure, chest pain (angina), asthma, dry eye, and many other conditions, but there is no good scientific evidence to support any of these uses.

- Forskolin works on muscles in the heart and in the walls of the blood vessels. It produces a more powerful heartbeat and widening of the blood vessels, which lowers blood pressure.
- Folkloric plant is used to treat malarial fever, hepatopathy, renal and vesical calculi, cough, chronic asthma, hiccough, bronchitis, helminthiasis, colic, convulsions, and epilepsy.

10.17. Black nightshade (*Solanum nigrum*)

Solanum nigrum or black nightshade is native to Eurasia and introduced in the Americas, Australasia, and South Africa. Parts of this plant can be toxic to livestock and humans, and it is considered a weed.

- Ripe berries and cooked leaves of edible strains are used as food in some locales, and plant parts are used as a traditional medicine.
- In European traditional medicine, the plant has been used as a strong sudorific, analgesic, and sedative with powerful narcotic properties.
- In traditional Indian medicines, infusions are used in case of dysentery, stomach complaints, and fever and to treat tuberculosis.
- The juice of the plant is used on ulcers and other skin diseases.
- The fruits are used as a tonic, laxative, appetite stimulant, and for treating asthma.
- The juice from its roots is used against asthma and whooping cough.
- The plant is widely used in oriental medicine where it is considered to be antitumorigenic, antioxidant, antiinflammatory, hepatoprotective, diuretic, and antipyretic

10.18. *Senna (Senna alexandrina)*

- Senna is the fruit (pod) or leaf of the plant *Senna alexandrina*. It is approved in the US as a laxative for short-term treatment of constipation.
- Senna contains many chemicals called sennosides. Sennosides irritate the lining of the bowel, which causes a laxative effect.
- Senna is an FDA-approved over-the-counter (OTC) laxative.
- It is used to treat constipation and also to clear the bowel before procedures such as colonoscopy.
- People also use senna for irritable bowel syndrome (IBS), hemorrhoids, weight loss, and many other conditions.

10.19. Lemon grass *(Cymbopogon citratus)*

- Lemongrass is a plant whose leaves and the oil are used to make medicine.
- Lemongrass is used for treating digestive tract spasms, stomachache, high blood pressure, convulsions, pain, vomiting, cough, achy joints (rheumatism), fever, the common cold, and exhaustion. It is also used to kill germs and as a mild astringent.
- Some people apply lemongrass and its essential oil directly to the skin for headache, stomachache, abdominal pain, and muscle pain.
- By inhalation, the essential oil of lemongrass is used as aromatherapy for muscle pain.
- In food and beverages, lemongrass is used as a flavoring. For example, lemongrass leaves are commonly used as "lemon" flavoring in herbal teas.
- In manufacturing, lemongrass is used as a fragrance in soaps and cosmetics. Lemongrass is also used in making vitamin A and natural citral.

10.20. Citronella grass *(Cymbopogon nardus)*

- Citronella grasses are used for the production of citronella oil, which in turn is used in the production of soaps, insect repellents (especially mosquitoes) in insect sprays, candles, and in aromatherapy.
- Due to its various constituents like citronellol, geraniol, and citronellal, it has many medicinal properties like wound healing, antimicrobial, anticonvulsant, etc.
- Citronella has a naturally uplifting and happy smell, with research showing that it can be both uplifting and relaxing. It appears to work on both the

sympathetic and parasympathetic nervous system, providing natural stress relief.

- Whilst inflammation is a valid healing response by the body, persistent low level inflammation can lead to a whole host of problems and exacerbate existing ones such as osteoarthritis and rheumatoid arthritis.
- Citronellal is the chief compound found in Citronella oil, with research studies showing it exhibits a strong anti-inflammatory effect. It also contains potent antioxidant compounds that aid in the removal of free radicals, one of the major causes of persistent inflammation.
- This versatile oil can work wonders for the skin. It can help to heal dermatitis and eczema, slow down skin aging and treat fungal infections such as athlete's foot. As part of a beauty routine, its antibacterial and astringent qualities make Citronella essential oil great for use on oily skin and as a remedy for acne.
- The high limonene and methyl isoeugenol content in Citronella oil makes it effective in regulating and decreasing the amount of sebum oil produced by the scalp, combating greasy hair.

10.21. Celery *(Apium graveolens)*

- Widely recognized as one of the healthiest snacks available, celery is a wonderful source of fiber. Its fiber content is especially impressive given the vegetable's low calorie count. This makes it a great option for those concerned with losing weight or maintaining healthy digestion.

- Celery is loaded with antioxidants. These include well-known varieties such as flavonoids and vitamin C, as well as lunularin and bergapten. These and other antioxidants help to prevent the oxidative stress that contributes to cancer.
- Celery is rich in a phytochemical known as phthalides. This compound is thought to relax artery wall tissues to promote healthy blood flow. This increased blood flow may lead to lower blood pressure. Participants in a notable study involving celery seed extract experienced reductions in both systolic and diastolic blood pressure.
- Research suggests that a celery seed extract known as L-3-n-butylphthalide improves both cognition and memory. This extract could play a valuable role in both treating Alzheimer's and preventing it from developing in the first place.

10.22. Rose Geranium *(Pelargonium graveolens)*

- Rose geranium oil is an active ingredient in some cosmetic products, like lotions and fragrances.
- Antioxidant properties in rose geranium oil may help reduce the signs of aging. Antioxidants are well-established as natural agents for improving your skin's ability to heal itself from environmental toxins and exposure.
- Rose geranium oil could be the basis of new anti-inflammatory drugs that may have less harmful side effects than current medications. Rose geranium oil has strong antimicrobial, antifungal, and antiviral properties.
- The food service industry even uses rose geranium oil as a natural preservative in some food products. Rose geranium reduces bacterial, fungi,

and viruses that cause skin disease and infections.

- The scent of rose from the rose flower enhances relaxation, offer pain relief, and soothe anxiety in a clinical setting.
- Rose Geranium is reputed to balance the skin's natural oil production, minimize the appearance of enlarged pores, eliminate infection-causing or acne-causing bacteria, enhance circulation, reduce the appearance of varicose veins, cellulite, and dull skin, encourage the body's relaxation, promote the faster fading of scars and blemishes.

10.23. Vetiver or khus *(Chrysopogon zizanioides)*

- Vetiver roots are used to make medicine. People take vetiver for nerve and circulation problems and for stomach pain.
- Some women take vetiver to start their periods or to cause an abortion.
- Vetiver is sometimes applied directly to the skin for relieving stress, as well as for emotional traumas and shock, lice, and repelling insects. It is also used for arthritis, stings, and burns.
- Vetiver is sometimes inhaled as aromatherapy for nervousness, insomnia, and joint and muscle pain.
- In manufacturing, vetiver is used as a flavoring in alcoholic beverages.

10.24. Cardamom *(Elettaria cardamomum)*

- Cardamom may be helpful for people with high blood pressure.
- Contain cancer-fighting compounds.
- Protect from chronic diseases due to anti-inflammatory effects may help protect cells from damage and slow down and prevent inflammation in your body.
- Help with digestive problems, including ulcers
- Treat bad breath and prevent cavities
- May have antibacterial effects and treat infections
- Improve breathing and oxygen use

10.25. Black pepper *(Piper nigrum)*

- High in antioxidants. Free radicals are unstable molecules that can damage your cells
- Has anti-inflammatory properties
- May benefit the brain

- Improves blood sugar control
- Lower cholesterol levels
- May have cancer-fighting properties
- A versatile spice

10.26. White pepper *(Piper nigrum)*

- Packed with antioxidants and help the body combat free radicals.
- Anti-inflammatory properties
- Prevents gas
- Better digestion
- Beneficial for anorexia patients
- May prevent cancer
- Reduces blood pressure
- Heart-friendly

10.27. Dill Seed *(Anethum graveolens)*

- Dill is packed with flavonoids, which have been shown to help reduce the risk of heart disease and stroke. But that's not the only reason dill is thought to improve heart health. Research on animals shows that dill can also reduce LDL cholesterol levels.
- Improves digestion Dill acts as a laxative and improves the functioning of your digestive system, cleansing your bowels and improving its functioning
- Dill seeds are packed with flavonoids and monoterpenes that act as very potent anti bacterial agents; coupled with its digestive properties, dill seeds are a great remedy for an upset stomach. It not only helps get rid of the infection, but also aids in restoring the functioning of your digestive tract and strengthens your immune system.
- Beats bad breath. Due to its carmative properties, dill seeds are also great at beating bad breath. Not only does it help change the smell of your breath but also helps better your digestion, beating bad breath from the root.

10.28. Cumin *(Cuminum cyminum)*

Cumin contains compounds called flavonoids that work as antioxidants in the body. Antioxidants can help neutralize unstable particles called free radicals that cause cell damage. By neutralizing these particles, antioxidants can help prevent diseases like cancer, heart disease, and high blood pressure.

- Promotes digestion
- Is a rich source of iron
- Contains beneficial plant compounds
- Helps with diabetes
- Improves blood cholesterol
- Promote weight loss and fat reduction
- May prevent food-borne illnesses

- May help with drug dependence
- Fight Inflammation

10.29. Isabgol *(Plantago ovata)*

- Removes toxins from the body.
- Adds a protective lining to the stomach that helps in curing acidity.
- Adds volume to the stool while softening it for easy passage.
- Promotes the growth of healthy gut bacteria.
- Boosts digestion.
- Absorbs extra water from the stool.
- Regulates cholesterol levels.

10.30. Clove *(Syzygium aromaticum)*

- Contain important nutrients.
- High in antioxidants.
- May help protect against cancer.

- Can kill bacteria.
- May improve liver health.
- May help regulate blood sugar.
- May promote bone health.
- May reduce stomach ulcers.

10.31. Cinnamom *(Cinnamomum verum)*

Cinnamon, from the bark of the cinnamon tree, has long been used as both a spice and a traditional medicine.

- Lower blood sugar.
- Boost metabolism.
- Great skin.
- Help treat cancer.
- Lower blood pressure
- Protect the brain.
- Reduce inflammation.

10.32. Liquorice / Mulethi *(Glycyrrhiza glabra)*

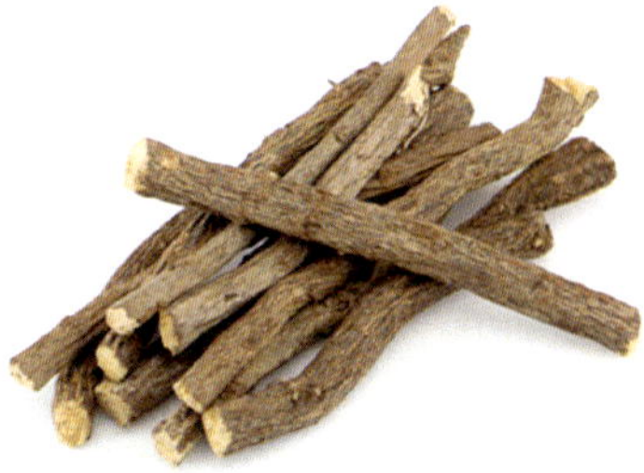

There are more than 300 different compounds in licorice, some of which have anti-inflammatory, antiviral, and antimicrobial properties.

- Many compounds in licorice root may help to reduce inflammation of the skin and other parts of the body. These compounds can help to treat several skin conditions.
- Glycyrrhizin found in licorice may help treat hepatitis C, a virus that infects the liver.
- Some research suggests that licorice extract may help kill bacteria in the mouth that cause tooth decay.
- Licorice root extract can help reduce the growth of Streptococcus mutans bacteria in the mouth.
- Different forms of licorice may help treat sore throats and other upper respiratory disorders.

10.33. Dry Ginger Powder/Sonth *(Zingiber officinale)*

- It improves digestion, used in bloating, colic pain, acidity - add it in vegetables, in buttermilk or have it with amla powder.
- Relives swelling, pain in joints, stiffness, rheumatoid arthritis - consume it, massage with til oil + shunthi.
- It is good for your heart.
- Good for liver.
- Useful in anemia, jaundice - have shunthi with jaggery.
- Helps in relieving pain during periods- take it with ajwain and jaggery.
- Piles- shunthi with jaggery
- Toothache - apply paste of shunthi.
- Obesity
- It is useful in respiratory disorders.
- Headaches -apply its paste on forehead.

10.34. Harad/Haritaki *(Terminalia chebula)*

- Constipation.
- Weight loss.
- Cough and cold.
- Weak immunity.
- Skin disease.
- Arthritis.
- Alzheimer's disease
- Improves weak digestion.

10.35. Baheda (*Terminalia bellirica*)

- Enhances digestion
- Relieves constipation
- Boosts immunity
- Promotes cardiac functioning
- Treats wounds and ulcers
- Remedies pain and inflammation
- Regulates diabetes
- Improves cognitive functioning
- Cures respiratory troubles
- Baheda for fever and anorexia nervosa
- Baheda for oral health
- Baheda for skin quality
- Baheda for hair problems

10.36. Kalonji /Nigella Seeds (*Nigella sativa)*

- Packed With antioxidants.
- Lower cholesterol.
- Have cancer-fighting properties.
- Help kill off bacteria.
- May alleviate inflammation.
- Could help protect the liver.
- Aid in blood sugar regulation.
- Prevent stomach ulcers

10.37. Linseed/Flax (*Linum usitatissimum)*

- Loaded with nutrients. Flaxseed is one of the world's oldest crops.
- High in omega-3 fatty acids.
- Help protect against cancer.
- Rich in fiber.
- May lower cholesterol levels.
- May reduce blood pressure.
- May stabilize your blood sugar levels.
- May help you manage your weight.

10.38. Methi Seeds (*Trigonella foenum-graecum*)

- It has an antioxidant property.
- It may show liver protection property
- It helps in reducing inflammation
- Shows antiulcer property
- It may show anticancer property
- It has antibacterial property
- It may show neuroprotective (protective of neurons) properties.

10.39. Garlic (*Allium sativum*)

- Garlic helps boost your body's immune system.
- Garlic helps reduce high blood pressure.
- Garlic helps reduce cholesterol levels.
- Garlic can help with prevention of cancer.

- Garlic has antibiotic properties.
- Garlic may prevent alzheimer's and dementia.
- Garlic can improve athletic performance.

10.40. Onion (*Allium cepa*)

- Gave many antioxidants.
- Help with blood sugar levels.
- Improved digestive health.
- Loaded with nutrients.
- Anti-inflammatory properties.
- May keep certain cancers at bay.
- Reduces risks of alzheimer's disease.
- Great source of vitamin c.

10.41. Star Anise *(Illicium verum)*

Boosts brain power

- Reduces anxiety and depression
- Good for digestion
- Beneficial during cold and flu
- May balance blood sugar levels
- Relaxes mind and muscles
- Reduces headache
- Helps in managing healthy weight
- Supports breastfeeding mothers

10.42. Mace/Japatri/Javitri/Nutmeg *(Myristica fragrans)*

- Japatri powder good for digestion
- Mace spice acts as appetite enhancer
- Mace spice benefits blood circulation
- Dental health benefits of mace, relieves toothache
- Mace spice protects kidneys
- Improves the heart functioning
- It acts as a stress buster
- Mace effectively treats cold and cough
- Mace effectively treats inflammatory conditions like joint pains, arthritis, and lumbago.

- Mace has proven to stimulate and enhance brain functionality.

10.43. Bay leaf/Tejpatta *(Cinnamomum tamala)*

- *Helps In improving digestion*
- Heart friendly
- A herb perfect for diabetics
- Acts as a stress-buster
- Protects the body from inflammation
- Natural wound healer

10.44. Stone Flower / Phattar Phool *(Parmotrema perlatum)*

- It has wound healing properties.
- *It* has antioxidants, anti-inflammatory, antimicrobial, and kapha-pitta balancing abilities.
- It is an excellent spice for treating kidney stones
- It helps in stimulating the digestive enzymes, which reduces the number of stomach related problems like diarrhoea

- It has analgesic activity (pain relieving) for the liver, stomach, and uterus.
- It helps with its anthelmintic activity against worms.
- It has antimicrobial properties.
- It may show carminative (relieving flatulence) activity.

10.45. Fox nut / Makhana *(Euryale ferox)*

- Maintains kidney health by controlling the blood flow and regulating urination.
- Healthy heart.
- Detoxifies liver.
- Good For diabetic patients.
- Make bones strong.
- Weight loss.
- Hormonal balance.
- Keeps digestive system healthy.
- Good for fertility
- Prevents inflammation
- Prevents ageing

10.46. Asafoetida/Hing (*Ferula assa-foetida*)

- Helps reduce bloating and other stomach problems
- May lower blood pressure
- May relieve menstrual pain
- Reduces headaches
- Can heal insect bites and stings
- It may help reduce acne
- Acts as a good hair conditioner

10.47. Saffron / Kesar *(Crocus sativus)*

- Providing antioxidants that protect cells against free radicals and oxidative stress

- Preventing nervous system disorders
- Boosting mood
- Promoting libido
- Reducing PMS symptoms
- Promoting weight loss.
- May reduce appetite and aid weight loss
- May have cancer-fighting properties
- Improve mood and treat depressive symptoms
- Reduce heart disease risk factors.
- Lowers blood sugar levels.
- May improve eyesight in adults with age-related macular degeneration (AMD).
- Improve memory in adults with Alzheimer's disease.

10.48. Safed musli *(Chlorophytum borivilianum)*

- Safed Musli is given to people suffering from malnourishment because of its weight and muscle gain properties.
- It has aphrodisiac properties that help improve sex drive, virility, and sexual stamina in men.
- It has the ability to boost stamina, energy levels, and immunity in men and women.

- It contains anti-arthritic and anti-inflammatory properties that can provide relief from inflammation and pain caused by arthritis.
- Safed Musli contains anti-bacterial and anti-diarrhoea properties that can treat and prevent diarrhoea.
- It has anti-stress properties that manage serotonin levels to combat anxiety and depression.
- Safed Musli is a natural cardio-protective herb because of its antioxidant-rich makeup.
- It contains antioxidants that protect the skin from free radical damage which can improve skin health.
- It has hypoglycaemic properties that boost insulin production to lower blood glucose levels.
- It can treat urinary disorders like pain or burning sensation during urination as well as urinary incontinence.
- It is fibre-rich, which helps reduce digestive problems like flatulence, bloating, constipation, and stomach distension.
- Safed Musli is said to be a powerful galactagogic that helps increase milk secretion from the mammary glands.

10.49. White til / Sesame seeds (Sesamum indicum)

- Triggers hair growth.
- Anti-ageing properties.
- Boosts skin health.
- Boosts dental health.

- Boosts digestion.
- Good source of energy.
- Stabilizes your blood pressure.
- Boosts bone health.

10.50. Black til / Sesame seeds *(Sesamum indicum)*

- Rich in antioxidants and prevents cell damage in body.
- May improve blood pressure
- May have anticancer properties
- May promote healthy hair and skin
- Good for digestion
- Great for nursing mothers

10.51. Moringa / Sohanjana *(Moringa oleifera)*

- Rich in vitamins and minerals.
- Rich in amino acids.

- Fight inflammation.
- Rich in antioxidants.
- Lower blood sugar levels.
- Lowers cholesterol.
- Protects the liver.
- Protects against arsenic toxicity.

11

Crop Residue Management Techniques

11.1. Biomass

Globally, all agricultural countries have promoted the use of biomass sources for many household applications and biomass energy generation. Biomass is defined as the bio residue available from the sum of crop production, agricultural waste, vegetation, forestry waste and organic waste. Biomass is a renewable source of energy and contains a complex mixture of carbon, nitrogen, hydrogen, and oxygen. Various biomass resources are available in different parts of countries and are classified based on their availability. The broad classification of biomass energy is in three formats; residue of agricultural crops, energy plantations, and municipal and industrial waste. Agriculture residues are organic materials produced as the by-product from processing and harvesting of agriculture crops. Energy plantations are wastes from agro-based industries, road-side shrubs, plantations, vegetable markets, etc. The municipal and industrial wastes include those from water and industrial, food industry, solid and liquid waste, municipal, sewage and animal waste.

These biomass resources can be used as fuel to avoid burning, burying, storage or inefficient usage. The carbon neutrality of biomass sources has made it an ideal renewable for usage in domestic, industrial, and power generation applications. With this chief property of biomass sources, the possibility of another biomass fuel named biomass briquettes was explored. Biomass briquettes are a combination of two or more biomass sources or fuels, purposely combined to augment the properties of biomass materials for efficient use as fuel. The technology followed for the preparation of briquettes is a compaction process in which raw biomass is compressed under high pressure, which causes the lignin in the biomass to be liberated so that it binds the material into a firm briquette. Briquetting requires the collection of combustible unused biomass and its compression into a solid block to use as a fuel, that is similar to wood or charcoal. The chief idea behind briquetting is to prepare a fuel with high bulk density, uniform shape and size, low moisture content, and above all else, with good burning properties.

Surplus biomass production in India was found to be 44.86 MT in rice, 41.46 MT. in sugarcane, 29.74 MT in cotton and 25.07 MT in wheat and minimum of 1.76

MT in tur (Table 11.1). The bioethanol potential (BL) was recorded in sugarcane (14.63), rice (12.02), cotton (7.41) and wheat (6.92) and minimum in tur (0.45 BL) and groundnut (0.85 BL). In India surplus biomass production by all crops was 178.74 MT with total bioethanol potential of 51.35 BL (Table 11.1)

Table 11.1: Crop wise area, total dry biomass, surplus biomass, and bioethanol production potential of selected crops in India

Crop across all seasons	Area (Mha)	Dry Biomass (MT)	Surplus Biomass (MT)	Bio Ethanol Potential (BL)
Rice	44.36	225.49	44.86	12.02
Wheat	30.84	145.45	25.07	6.92
Maize	8.78	27.88	6.04	1.35
Sugarcane	5.04	119.17	41.46	14.63
Gram	8.48	26.52	8.72	2.17
Tur	4.11	9.17	1.76	0.45
Soybean	10.69	27.78	9.95	2.94
Rapeseed & Mustard	5.87	17.09	5.16	1.50
Cotton	12.16	66.58	29.74	7.41
Groundnut	5.47	12.90	3.87	0.85
Castor	1.19	4.60	3.02	1.13
Grand Total	136.99	682.62	178.74	51.35

Source: Estimation of Surplus Crop Residues in India for Biofuel Production (Joint report of TIFAC and IARI, October 2018)

Biomass Energy

Biomass is organic, meaning it is made of material that comes from living organisms, such as plants and animals. The most common biomass materials used for energy are plants, wood, and waste. These are called biomass feedstocks. Biomass energy can also be a non-renewable energy source.

Biomass contains energy first derived from the sun: Plants absorb the sun's energy through photosynthesis, and convert carbon dioxide and water into nutrients (carbohydrates).

The energy from these organisms can be transformed into usable energy through direct and indirect means. Biomass can be burned to create heat (direct), converted into electricity (direct), or processed into biofuel (indirect).

Sources for Biomass Energy

Wood and wood processing wastes—firewood, wood pellets, and wood chips, lumber and furniture mill sawdust and waste, and black liquor from pulp and paper mills

Agricultural crops and waste materials—corn, rice, soybeans, sugar cane, switchgrass, woody plants, and algae, and crop and food processing residues, mostly to produce biofuels

Biogenic materials in municipal solid waste—paper, cotton, and wool products, and food, yard, and wood wastes

Animal manure and human sewage for producing biogas/renewable natural gas

Benefits of biomass energy

- Biomass is widely available in many forms.
- Biomass energy is less expensive to produce than fossil fuels.
- Using biomass as an energy source reduces garbage in landfills.
- Since biomass only releases the C_2O that it captures, it's 100% carbon neutral.
- Biomass energy helps mitigate the effects of climate change.

Thermal Conversion

Biomass can be burned by thermal conversion and used for energy. Thermal conversion involves heating the biomass feedstock in order to burn, dehydrate, or stabilize it. The most familiar biomass feedstocks for thermal conversion are raw materials such as municipal solid waste (MSW) and scraps from paper or lumber mills.

Different types of energy are created through direct firing, co-firing, pyrolysis, gasification, and anaerobic decomposition.

Before biomass can be burned, however, it must be dried. This chemical process is called torrefaction. During torrefaction, biomass is heated to about 200° to 320° Celsius (390° to 610° Fahrenheit). The biomass dries out so completely that it loses the ability to absorb moisture, or rot. It loses about 20% of its original mass, but retains 90% of its energy. The lost energy and mass can be used to fuel the torrefaction process.

During torrefaction, biomass becomes a dry, blackened material. It is then compressed into briquettes. Biomass briquettes are very hydrophobic, meaning they repel water. This makes it possible to store them in moist areas. The briquettes have high energy density and are easy to burn during direct or co-firing.

11.2. Process for Converting Biomass To Energy

1. Direct Firing and Co-Firing

Most briquettes are burned directly. The steam produced during the firing process powers a turbine, which turns a generator and produces electricity. This electricity can be used for manufacturing or to heat buildings.

Biomass can also be co-fired, or burned with a fossil fuel. Biomass is most often co-fired in coal plants. Co-firing eliminates the need for new factories for processing biomass. Co-firing also eases the demand for coal. This reduces the amount of carbon dioxide and other greenhouse gases released by burning fossil fuels.

2. Pyrolysis

Pyrolysis is a related method of heating biomass. During pyrolysis, biomass is heated to 200° to 300° C (390° to 570° F) without the presence of oxygen. This keeps it from combusting and causes the biomass to be chemically altered.

Pyrolysis produces a dark liquid called pyrolysis oil, a synthetic gas called syngas, and a solid residue called biochar. All of these components can be used for energy. Pyrolysis oil, sometimes called bio-oil or biocrude, is a type of tar. It can be combusted to generate electricity and is also used as a component in other fuels and plastics. Scientists and engineers are studying pyrolysis oil as a possible alternative to petroleum.

Syngas can be converted into fuel (such as synthetic natural gas). It can also be converted into methane and used as a replacement for natural gas.

Biochar is a type of charcoal. Biochar is a carbon-rich solid that is particularly useful in agriculture. Biochar enriches soil and prevents it from leaching pesticides and other nutrients into runoff. Biochar is also an excellent carbon sink. Carbon sinks are reservoirs for carbon-containing chemicals, including greenhouse gases.

3. Gasification

Biomass can also be directly converted to energy through gasification. During the gasification process, a biomass feedstock (usually MSW) is heated to more than 700° C (1,300° F) with a controlled amount of oxygen. The molecules break down, and produce syngas and slag.

Syngas is a combination of hydrogen and carbon monoxide. During gasification, syngas is cleaned of sulfur, particulates, mercury, and other pollutants. The clean syngas can be combusted for heat or electricity, or processed into transportation biofuels, chemicals, and fertilizers. Slag forms as a glassy, molten liquid. It can be used to make shingles, cement, or asphalt.

Industrial gasification plants are being built all over the world. Asia and Australia are constructing and operating the most plants, although one of the largest gasification plants in the world is currently under construction in Stockton-on-Tees, England. This plant will eventually be able to convert more than 350,000 tons of MSW into enough energy to power 50,000 homes.

4. Anaerobic Decomposition

Anaerobic decomposition is the process where microorganisms, usually bacteria, break down material in the absence of oxygen. Anaerobic decomposition is an important process in landfills, where biomass is crushed and compressed, creating an anaerobic (or oxygen-poor) environment.

In an anaerobic environment, biomass decays and produces methane, which is a valuable energy source. This methane can replace fossil fuels. In addition to landfills, anaerobic decomposition can also be implemented on ranches and livestock farms. Manure and other animal waste can be converted to sustainably meet the energy needs of the farm.

Biomass at the farm can be used for following purposes.

Biomass is the only renewable energy source that can be converted into liquid biofuels such as ethanol and biodiesel. Biofuel is used to power vehicles, and is being produced by gasification in countries such as Sweden, Austria, and the United States. Ethanol is made by fermenting biomass that is high in carbohydrates, such as sugar cane, wheat, or corn. Biodiesel is made from combining ethanol with animal fat, recycled cooking fat, or vegetable oil.

Biofuels do not operate as efficiently as gasoline. However, they can be blended with gasoline to efficiently power vehicles and machinery, and do not release the emissions associated with fossil fuels. Ethanol requires acres of farmland to grow biocrops (usually corn). About 1,515 liters (400 gallons) of ethanol is produced by an acre of corn. But this acreage is then unavailable for growing crops for food or other uses. Growing enough corn for ethanol also creates a strain on the environment because of the lack of variation in planting, and the high use of pesticides.

Ethanol has become a popular substitute for wood in residential fireplaces. When it is burned, it gives off heat in the form of flames, and water vapor instead of smoke.

Biochar

Biochar, produced during pyrolysis, is valuable in agricultural and environmental use. When biomass rots or burns (naturally or by human activity), it releases high amounts of methane and carbon dioxide into the atmosphere. However, when biomass is charred, it sequesters, or stores, its carbon content. When biochar is added back to the soil, it can continue to absorb carbon and form large underground stores of sequestered carbon—carbon sinks—that can lead to negative carbon emissions and healthier soil. Biochar also helps enrich the soil. It is porous. When added back to the soil, biochar absorbs and retains water and nutrients.

Biochar is used in Brazil's Amazon rain forest in a process called slash-and-char. Slash-and-char agriculture replaces slash-and-burn, which temporarily increases the soil nutrients but causes it to lose 97% of its carbon content. During slash-and-char, the charred plants (biochar) are returned to the soil, and the soil retains 50% of its carbon. This enhances the soil and leads to significantly higher plant growth.

Black Liquor

When wood is processed into paper, it produces a high-energy, toxic substance called black liquor. Until the 1930s, black liquor from paper mills was considered a waste product and dumped into nearby water sources.

However, black liquor retains more than 50% of the wood's biomass energy. With the invention of the recovery boiler in the 1930s, black liquor could be recycled and used to power the mill. In the U.S., paper mills use nearly all their black liquor to run their mills, and the forest industry is one of the most energy-efficient in the nation as a result. More recently, Sweden has experimented in gasifying black liquor to produce syngas, which can then be used to generate electricity.

Hydrogen Fuel Cells

Biomass is rich in hydrogen, which can be chemically extracted and used to generate power and to fuel vehicles. Stationary fuel cells are used to generate electricity in remote locations, such as spacecraft and wilderness areas. Yosemite National Park in the U.S. state of California, for example, uses hydrogen fuel cells to provide electricity and hot water to its administration building.

Hydrogen fuel cells may hold even more potential as an alternative energy source for vehicles. The U.S. Department of Energy estimates that biomass has the potential to produce 40 million tons of hydrogen per year. This would be enough to fuel 150 million vehicles.

Currently, hydrogen fuel cells are used to power buses, forklifts, boats, and submarines, and are being tested on airplanes and other vehicles.

However, there is a debate as to whether this technology will become sustainable or economically possible. The energy that it takes to isolate, compress, package, and transport the hydrogen does not leave a high quantity of energy for practical use.

11.3. Paddy Straw Management Techniques

Rice straw is a lignocellulosic material essentially constituted of cellulose (24.0%), hemicelluloses (27.8%), and lignin (13.5%), with important amounts of ash (17%) that correspond mostly to silica. Due to the significant amounts of carbohydrates and lignin, and to its low price and high availability, rice straw has been considered a suitable feedstock for the production of biofuels, biochemicals, and bio based materials in the context of lignocellulosic biorefineries. In addition, rice straw also presents significant amounts of lipophilic compounds (3.4%) that can also be used to obtain valuable phytochemicals of industrial interest. Most of the functionally important plant-based compounds derive from food processing as by-products. Nevertheless, lignocellulosic wastes (cereal by-products such as wheat straw, rice husks, or maize fibers, among others) that are easily available at a very low cost are considered alternative sources for obtaining valuable phytochemicals

Paddy straw is available in huge amounts and at present there is a big problem of its management both or farmers and state government. Farmers should be provided with alternate options like incorporation, surface mulching or mechanical collection in order to avoid burning. Burning of paddy straw not only creates environmental pollution but nutrients and microbial losses also takes place with this practice. At present there are different options of paddy straw management which are as follow.

1. Burning
2. Incorporation
3. Surface mulching
4. Mechanical removal

1. Burning

It is very common, cheap and quick method for management of paddy straw. After combine harvest, farmers burn rice straw as such or cutting it into small pieces (after shredding). This method is costless and very quick for preparing the field for wheat sowing because of less availability of time between rice harvesting and wheat sowing with the farmers. Also alternative host plants/ materials for some insects or diseases are eliminated with burning. On the other hand this practice is very harmful for our environment resulting in accidents, human health problems, increase in toxic grass in the environment, decreases in nutrients and microbial levels, burning of economical adjoining crops etc.

The rice–wheat system is cultivated in the nearly 13.5 million hectares in the Indo- Gangetic Plains of South-Asia and plays a key and crucial role in the food security of millions of people; India alone shares 10.5 m ha. In Punjab, about 91% and 82% of the area under rice and wheat is harvested by combines, respectively, leaving a large amount of residue in the field that creates problems in the smooth sowing of succeeding wheat crops. While more than 75% of wheat straw is used as fodder, rice straw has no economic use. In India particularly in Punjab, more than 90% of rice straw and 25% of wheat straw are burnt annually by the farmers few years back. Nowadays due to the introduction of heavy machinery both for incorporation of paddy straw into the soil or its removal, there is a considerable decrease in burning. It leads to loss of nutrients, about 80% of nitrogen (N), 25% of phosphorus (P), 21% of potassium (K), and up to 60% of sulfur (S) along with the emission of 18% of black carbon, which is the second largest contributor to global warming. Burning or removing a straw from the system leads to the loss of a substantial amount of plant nutrients which need to be replenished in the soil through costly energy-expensive inorganic fertilizers, resulting in reduced profits for the farmer

2. Incorporation of Rice Straw to Enhance Soil Fertility

Incorporation of rice straw into paddy soil can maintain and enhance soil fertility with a proper management. However, ineffective management of straw incorporation can result in a decrease in production efficiency and increased greenhouse gas emissions. It should be noted that slow decomposition is often the reason why farmers do not want to incorporate crop residues. Some research has been undertaken on speeding this up, for example, by using fungal inoculums.

Straw incorporation with MB plough: After harvesting of paddy, 'Tractor operated Straw mulcher/Straw cutter cum spreader' is used for chopping of paddy straw and then straw incorporated into the soil with 'Tractor mounted Mould board plough'. After incorporation various tractor drafted implements like disc harrow, cultivator, rotavator, etc. can be used for the tillage. The sowing should be done with seed-cum fertilizer drill.

Straw incorporation with Rotary Mulcher: Rotary Mulcher shreds the crop residues/grass into small pieces and spread equally on the field. It is highly compatible to paddy straw. It can shred entire plant after harvesting. Bio mulch serves dual proposes, benefits of mulching and improve overall organic structure and soil fertility after decomposition

Ping etal 2020, reported that a novel deep-buried straw incorporation management was conducted to improve organic matter and over-compacted subsoil structure in Huang-Huai-Hai Plain of China. Four straw incorporation rates (CK, 0 kg ha−1; T1, 6 000 kg ha−1; T2, 12 000 kg ha−1; T3, 18 000 kg ha−1) were set up in this three-year field positioning test. It was found that soil water content (SWC), soil penetration resistance (SPR), soil organic carbon (SOC) and total nitrogen (TN) were improved with straw incorporation rate increasing. Especially in the subsoil layer, compared with CK, T3 created the highest average increase in SWC (61.3%), the lowest SPR (59.4%–39.6%) and the highest increase in SOC (25.9%)

and TN (19.2%) during the three years. However, T2 had the highest average SOC conversion rate (41.1%) over the three years and the highest C:N (7.1%) in the third year. Treatments of T2 and T3 showed similar significant increases in wheat and maize yield (7% and 5%, respectively), and root biomass (approximately 19%), which were positively correlated with SWC and SOC. We concluded increasing straw incorporation rates showed more positive effects on improving subsoil SOC and structure, in which T3 had a higher SOC increase while T2 had a higher C-conversion efficiently.

With incorporation of rice straw into the soil, a lot of changes in physical and chemical properties of soil takes place. This practice improves water holding capacity as well as improve soil water infiltration rate. Also availability of oxygen to plant root improves. With passages of time there is a drastic increase in organic matters content of soil of the fields where paddy straw is being incorporated. However, sometimes wheat crop initially turns yellow due to immobilisation of nitrogen because of decomposition of paddy straw. This practice helps in improving soil fertility when it is adopted for a long period over the same piece of land. The physical and chemical properties are improved with continuous adoption of this practice resulting in increased yield of not only rice but succeeding crops like potato, peas, wheat, sarson etc. So, this practice is environmentally friendly and improves the income of farmers due to less use of inorganic fertilisers as well as irrigation water. This practice can be done with the use of super seeders which are run with heavy tractors with more than 65 HP capacity.

3. Surface Mulch

Surface mulching method of paddy straw management has a lot of direct and indirect benefits to wheat as well as on following rice crop. Surface mulching has lot of benefits which are listed below.

0. Weed population in wheat is reduced due to less germination of weed seeds and whatever the weed plants emerge, these remain pale green due to unavailability of sun light and their growth is limited.

i. During wheat season, paddy straw helps in modernizations of soil temperature resulting in good growth of crops.

ii. It also helps in conserving soil moisture with better root growth .

iii. Paddy straw mulch helps to increase organic carbon in soil so, these will be more availability of nutrients like N,P_2,O_5,K_2O and sulphur. So in the long run paddy straw mulching has a lot of advantages. However, under short term, this practice can boost attack of insect pests and diseases as well as attack rodents on wheat crop.

Generally surface mulching of paddy straw is done both in vertical or horizontal form. In first method the lose straw of combine is spread uniformly in the field and wheat is sown in the standing stubbles of rice with the use of happy seeder or smart seeder. Smart seeder is modified form of happy seeder in which the crop row slit is cultivated and sowing is done.

Happy seeder sow the crop in zero till conditions after cutting paddy straw from crop row area only. In the second method, the straw of paddy after combine harvest is chopped into small pieces and in the presence of rice straw. Sowing of wheat is done with suitable machine mostly smart seeder.

The straw of basmati rice is low in silicon and of better quality than coarse rice. Hence the farmers collect straw of basmati rice after sickle harvest or combine harvest and feed to their cattle may be after mixing in green fodder. Basmati rice straw is more safe to farm animals especially milch cattle than of coarse rice.

Sowing of wheat in anchored rice stubbles

Wheat in standing rice stubbles

On-farm straw management options include surface retention as mulch and soil incorporation. Gupta etal, 2022 reported that in situ incorporation of rice straw is generally not preferred by the farmers due to the narrow window of 20–25 days between harvesting of rice and sowing of wheat and the high cost incurred on incorporation which require heavy machinery. Generally, rice is harvested in the first week of October and wheat is sown in the last week of October or the first week of November. In paddy-based systems, the management of paddy straw in fields is a serious problem due to scarcity of labour availability. The farmers generally follow the legally banned practice of burning paddy straw in their fields, and major portion of rice straw produced is being burnt annually during last weeks of October and beginning of November. The problem is more severe under irrigated conditions particularly in the mechanized rice–wheat system of northwestern India. With the recent development of the zero-till seed drill (known as Happy Seeder), wheat can be planted in combine harvested rice fields, leaving rice straw as surface mulch. The surface- placed residue presented a slow decomposition, which generally does not contribute to the N nutrition of wheat over a short period and might even immobilize soil N.

On the other hand, a recent study showed that the long-term incorporation of rice straw increased N use efficiency in both rice and wheat in the rice–wheat system. Mulching wheat crop with rice straw enhances wheat yield from 11 to 22%, water use efficiency up to 25%, and root length density up to 40% in mulch-treated plots because of higher soil water retention. A large number of short-term and medium-term studies on the effect of rice residue management on productivity and soil properties in the rice–wheat system are available in the literature, but more research is needed to study the long-term effect of crop residue management with varying N levels on wheat yield, N use efficiency, grain quality and soil properties in north-western conditions of India. There are several scenarios for residue management in the rice–wheat cropping system. Every management scenario has its advantages as well as disadvantages, since crop residue management practices are affected by the climatic and soil conditions of an area and every crop residue management practice is not suitable under all conditions. Therefore, the testing of different management practices is imperative before their adoption in the area. We hypothesize that the continuous retention of rice residues on the surface as mulch or its incorporation into the soil for more than three years and its subsequent decomposition, besides improving soil health, will release nutrients for wheat crop and thereby help save fertilizer input, increase N use efficiency, improve wheat grain yield and quality, maintain or improve soil health and decrease the environmental pollution.

Iqbal et al 2017 reported that grain yield differ significantly from each other during both the years of study. The highest grain yield 3030 kg/ha and 3920 kg/ha were obtained in happy seeder zero till (HSZT) followed by broadcasting (2836 and 3478 kg/ha). This increase in yield may be attributed due to good soil tilth and better resource utilization in better uptake of nutrients due to better root development. Sidhu et al., (2007) reported that sowing of wheat with Happy Seeder tillage without burning of previous crop residues, eliminating air pollution and loss of nutrients and organic carbon due to burning, at the same as maintaining or increasing yield upto 10% compared to conventional method.

Table 11.2: Showing yield (kg/ha) by HSZT and CM of sowing techniques during 2014-15 and 2015-16

Tillage Equipment's	Two year Mean Yield (kg/ha)	Total Income (Rs/ha)	Total Cost (Rs/ha)	Net Benefit (Rs/ha)	CBR
Conventional Method	3157	102602	77290	25312	1:1.33
Happy Seeder Zero Tillage	3475	112938	75000	37938	1:1.51

Paddy straw management practices followed in wheat cultivation

Before going into the details of input-output pattern of wheat crop, it is prerequisite to know the paddy straw management technologies adopted by the sampled farm households. As per Singh et al, 2019, the following straw management techniques were practised on the selected farms:

Wet mixing: After harvesting of paddy using combine, straw was chopped with straw reaper. Then the field was irrigated and powered by tractor using various implements such as disc harrow, rotavator etc. to plough and mix the paddy straw into the soil in wet conditions. After the field reached at the adequate moisture level for sowing, wheat was sown using rota-seeder or seed drill.

Dry mixing: After harvesting of paddy, straw was chopped with a straw reaper. The straw was incorporated into the soil by using various tractor driven implements such as; disc harrow, rotavator, cultivator, etc. However, difference in this technique is that no pre-sowing irrigation was applied to the field and only available moisture of last irrigation given to paddy crop was used for straw incorporation. Thereafter, wheat was sown using seed-drill or rota-seeder.

Happy Seeder: Paddy straw was cut using a straw reaper and wheat was sown in the standing stubbles using 'Happy seeder' run by tractor. No preparatory tillage is required in this practice.

Conventional tillage post burning straw: As per common practice, the paddy straw was burnt after cutting it using a straw reaper. Afterwards, field was prepared by ploughing using tractor driven disc harrow, cultivator, etc. and wheat was sown using seed-cum- fertiliser drill.

Singh et al 2019 reported that the estimated gross returns did not vary much in different straw management practices and were the highest at Rs.103985/ha on the farms using Happy seeder and lowest (Rs. 98120/ha) on farms where wet mixing of paddy straw was followed. Moreover, farmers using Happy seeder, dry mixing and conventional practice received almost equal gross returns with slight variation. Returns over variable cost (ROVC) were the highest (Rs.73930/ha) on the farms where wheat was sown using Happy seeder and were the least on the farms following wet mixing practice (Rs.61285). In case of wet mixing practice, ROVC were nearly 18 per cent lower than conventional practice while it was about 4 per cent lower in dry mixing. In case of Happy seeder technology, ROVC were about two per cent higher than conventional practice of wheat sowing.

Nutrient Cycling

The incorporation of rice straw can improve soil quality through enhanced nutrient cycling and soil organic C sequestration. Straw incorporation has been shown to enhance nutrient recycling and provide soil fertility benefits. The straw is the main organic material that is available for most rice farmers and serves as an important source of K. At harvest, the straw contains 0.57% N, 0.07% P_2O_5, 1.5% K_2O, 0.1% (sulfur) S, and 5% silicon (Si). Rice straw at harvest can contain 0.5–0.8% N, 0.07–0.12% P_2O_5, 1.16–1.66% K_2O, 0.05–0.1% S, and 4–7% Si. This translates to about 40%, 30–35%, 80–85%, 40–50%, and 80%, respectively, of the N, P_2O_5, K_2O, S, and Si taken up by the plant. The addition of rice straw was shown to improve soil pH, soil organic C, and nutrient content compared to the initial conditions. The incorporation of rice straw resulted in greater increases in soil organic C, pH, and nutrient contents compared to addition of ash from burned straw, although the increase in N was small. Due to the low N content in straw, large quantities would be needed to supply adequate amounts of N. However, the straw has to decompose before the nutrients can become available for uptake and the rate of decomposition and supply of nutrients depends on soil type and season. Additionally, only a proportion of the nutrients become available in the season of application.

11.4. Mechanized Collection of Paddy Straw and Processing

Collection of rice straw spread in the field by combine harvesters is only economically feasible with mechanized collection. It plays an important role in the rice straw supply chain to prepare feedstock for further uses. The rice straw balers are currently used.

I. Mechanized Straw Collection Technologies

Mechanized collection of straw scattered in the field involves three main operations:

(a) picking up the straw from the field,

(b) compressing it into bales, and

(c) transporting the bales to the bunds.

In some areas, there are also some machines that just pick up the straw in loose form and transport it to the side of the field for further densification and transport.

Rice straw balers can be classified according to their mobility, the technical operation of the compacting unit, the manner of straw collection in the field, and/or the bulk density of produced bales. A mobile baler moves on the field to collect and compress straw into bales; it can be self-propelled or pulled by a tractor. A stationary baler, on the other hand, can be used to compress rice straw disposed by stationary threshers, which are still quite common in Asia.

Balers can also be classified according to compression density (high or low), shape of bales produced (round or square), and scale (large, at least with a 100-HP tractor or engine or small, with a less than a 60-HP engine). It is pulled by a tractor connected via three points with hydraulic ports to control pick-up height. It has a series of rollers that form the round bales. The pressure for baling is delivered through the tractor's power take-off (PTO) and the bale expeller consists of a built-in, independent hydraulic mechanism. It cannot work continuously as the operation must stop periodically so that the bales can be tied with a rope and then unloaded.

II. High-Density Straw Compacting, Briquetting, and Pelletizing

Rice straw densification—through compacting, briquetting, or pelletizing—results in better handling and storage of the by-product, which, in turn, reduces transportation costs and makes efficient use of storage facilities. The technologies now available, such as briquette presses and pelletizers, also provide options for other uses of rice straw, such as animal feed, fuel, and feedstock for energy generation

Transporting bales after collection from the field has become feasible and costs less than transporting loose straw. However, for high-end markets, such as industrial cattle farms, large amounts of rice straw must be transported long distances (sometimes more than 500 km) and stored for from 3 to 6 months. Round bales should be compacted into larger and higher-density square bales to reduce transportation and storage costs. Compacting the bales utilizes technologies that apply high pressure, such as screw or piston presses. A few compacting machines that use the piston press are found in Asia. Two common variations are the vertical and horizontal compacting systems.

III. Briquetting

Briquettes are produced by compressing chopped straw into a cylindrical form through a briquetting press. The hydraulic operation pressure of the briquetting press is set to 15 MPa and the press works in three stages. First, the feedstock passes through the pre-compression unit, which presses it inside the pressing block in a Y-direction. When a resistance of 8 MPa is achieved, the piston starts to operate and densifies the feedstock in an X-direction. At the open end of the pressing block, a pressing clamp is installed that opens to eject the produced briquette. The pressure at which the pressing clamp opens, which varies from 4 to 8 MPa, is related to the compaction density of the briquettes. Briquettes are also produced through:

(a) the press-chamber principle, which consists of two parts: a heated die that acts as a press and a punch that fits in tight;

(b) the screw principle, based on continuous extrusion of the feedstock by a screw through a heated tape die; and

(c) a piston press, where a reciprocating ram presses the straw biomass in a die. The finished products would have varying energy density depending on the technology used

IV. Pelletizing

Pellets are produced based on the principle of compressing ground straw. The compression unit is composed of a horizontal or vertical ring die and rollers that put pressure on the inner surface of the ring die. During the pelletizing process, the ring die and rollers rotate and the raw materials fall in the clearance between the ring die and the roller, which are pressed into the holes on the ring die. Pellets are cut at the outer surface of the ring die and collected. Compared with other compacting processes, such as briquetting and tumble agglomeration, pellets are generally regarded as the most durable because they are placed under the highest amount of pressure during formation

Pelletized rice straw can be used as fuel, animal feedstock, or material for anaerobic digestion. The pelletizer die hole size is known to have an important effect on the moisture content of the pellet, while the temperature reached during pelletization can also influence pellet quality

V. Rice Straw Mushroom Production

Rice straw mushrooms, *Volvariella volvacea*, are considered to be one of the easiest mushrooms to cultivate because of their short incubation period of 14 days. This tropical species thrives best at 30–35 °C for mycelia development and at 28–30 °C for fruiting body production. Outdoor mushroom production is a common practice as it has low investment cost, however its yield is just 0.8 kg of mushroom per 10 kg of dried straw. Indoor mushroom growing with higher investment cost produces about 2 kg higher yield per 10 kg of dried straw.

VI. Rice Straw Silage for Cattle Feed

Rice straw has too few nutrients to be used as the only source of food for cattle but is a good feed additive and can also be treated to increase the supply of energy and protein. In general, the daily recommended maximum intake of rice straw by ruminants is 1.0 to 1.5 kg per 100 kg live-weight per day. Urea-treated straw (rice straw is ensilaged with 2-3 % urea) can increase intake and digestability of this rice straw based feed.

Among agricultural by-products, rice straw is most abundant, low in cost, and a practical source of fodder for ruminants. Its utilization as a livestock feed is limited due to problems in collection, hauling, and storage. Rice straw has low nutritional value (low protein content and poor digestibility) compared to grasses, thus it cannot support the nutrients required by high-yielding milk cows and buffaloes. There are technologies that have been developed to increase the nutritive value, nutrient digestibility, and utilization of rice straw, such as physical processing, pre-treatment using chemicals, and/or biological treatment. However, adoption of these developed technologies is still low due to farmers' limited skills and inputs (e.g., farm equipment) and their doubts regarding applicability to the farm situation and the benefits for the animals and livestock producers. To maximize the utilization of rice straw as fodder for ruminants, mechanization is most important to facilitate collection, hauling, and stacking or processing all available rice straw from the field.

VII. Rice Straw Composting

a) Vermi-composting

Vermi-composting is a biological process of bio-oxidation and stabilization of organic material involving the joint action of earthworms and microorganisms. While microbes are responsible for the biochemical degradation of organic matter, earthworms serve as the important drivers of the process in conditioning the substrate and alteration of biological activity. The end product, or vermicompost, is a finely divided peat-like material with high porosity and water-holding capacity that contains most nutrients in forms that are readily taken up by the plants. The invertebrates have indirect effects on the structure and activities of bacterial and fungal communities through inoculum dispersal, grazing, litter combination, gut passage, and aggregate formation. The use of earthworms in compost heaps, beds or boxes makes the process faster for the breakdown of organic waste and its decomposition, i.e., composting

b) **Mechanized Windrow Composting**

The windrow-composting method consists of linear rows of compost materials (rice straw and cow manure), which are placed layer by layer and mechanically turned periodically. The air contained in the interspaces of the composting mass varies in composition. The C_2O content gradually increases and the O2 level falls during composting process. The concentration of O_2 for composting varies from 15 to 20%. (Diaz et al. 2007). Thus, the turning process helps improve aeration and mixing of compost constituents. The windrow composting method relies on mechanical aeration, typically with a compost windrow turner, to optimize the composting process. During turning, microbial inoculum is mixed with water and sprayed in the windrow to speed up the composting process and obtain the required moisture content. The interval between turnings is usually 10–14 days but 15-day intervals have also been reported. After turning, the windrow is covered using a plastic sheet to maintain the proper moisture content and temperature. The height and width of the windrows are typically set to fit the size of the turner

Mechanized composting can significantly improves the bio-physical processes of composting. The compost product can be used as medium for growing vegetables and other crops or spread on the rice field as soil amendment. It enhances nutrient (i.e., nitrogen and carbon contents) and organic matter content of the soil.

12

Integrated Farming System and Allied Enterprises

Agriculture is not only the main source of employment, income and food for over 70% of the population but also the main culture of the rural families. But the linear growth in population and unplanned colonization lead to rapid fragmentation of land holdings and shrinkage in fertile cultivated land. In India, more than 70% of the total population of the country still live in villages and mainly depend on agriculture and/or related enterprises. Marginal and small farmers constitute more than 84% of the 115 million operational holdings in India which are cultivating only 29% of the arable land (Singh et al., 2011). The livelihoods of the small and marginal household families are the major concern. The small land holders are better contributors to the total production (78%) but weak in terms of generating adequate income and sustaining their own livelihood. Small holding (below 0.8 ha) does not generate enough income to keep a farm family out of poverty despite high productivity (Chand et al., 2011). Growth in human and livestock populations has led to an expansion of cultivated land and shortened fallow periods. High population growth has also resulted in some area which was used for agricultural purpose being converted to residential area. This therefore has resulted in great pressure on agricultural land, leading to intensification of land use. Widespread occurrence of ill-effects of green revolution technologies in all intensively cultivated areas is threatening the sustainability of the important agricultural production systems and national food security. A mismatch between the national food grain production and requirement has already crept into the system, which is further widening.

The prosperity of any country depends upon the prosperity of its people. Food, clothing, shelter, health, education, security, roads, electricity and drinking water are the today's basic needs. Farmers produce food. About 70% of Indian population is engaged in farming. Therefore, prosperity of India would depend upon the prosperity of its farmers. This in turn depends upon the adoption of improved technology and judicious allocation of resources (land, labour, capital, machinery etc). Human race depends more on farm products for their existence than anything else since food and clothing – the prime necessaries are products of farming. Even for industrial prosperity, farming forms the basic raw material.

Farming system takes into account the combined needs of the family- the economic factors like relative profitability of the technically feasible enterprises, availability of farm resources, infrastructure and institutions such as irrigation, marketing facilities including storage and transportation and credit besides the agro-biological consideration-namely-interdependence, if any among various technically feasible enterprises and the performance of individual farmers.

Farming is defined as the way in which the farm resources are allocated to the needs and priorities of the farmers in his local circumstances which include

- Agro climatic condition such as the quantity, distribution and reliability of rainfall, temperature, etc.
- Soil type and topography and
- Economic and institutional circumstances like market opportunities, prices, institutional and infrastructure facilities and technology.

Integrated farming is becoming popular in India as part of sustainable agriculture system. It is being promoted and is gaining acceptance all over the world as part of the latest efforts to encourage agriculture systems that are socially, ecologically and economically sustainable. As it is a judicious mix of agricultural enterprises like dairy, poultry, piggery, fishery, sericulture etc. suited to the given agro-climatic conditions and socio-economic status of the farmers would bring prosperity in the farming. Integrated farming system (IFS) is considered as one of the best option towards intensification of small holder farm income to ensure sustainable livelihood. Integration of resources is made through a combination of land, water and animal resources of a farm through careful planning including recycling of bio-resources.

Governments and development agencies have designed projects/programmes in promoting IFS through demonstration of successful models and other means. Integration of resources is a process which involves great thought in investing in one of the resource combined and gaining benefits. Integrated farming system approach not only fulfils the household needs but enrich diet of human being and animals both for nutritional security. Further, diversified nature of the model provides employment opportunity for unemployed rural youth. Economic and livelihood analysis of the system revealed that beside household food, feed, fodder and fuel security, the system generates a sizable amount of savings which will assist to meet other liabilities of the family including education, health and social obligations and overall improvement in livelihood of small farm holders. Over two decades extension agencies have been encouraging farmers to adopt ways of integrating resources for better efficiency and to reduce dependency on adopting practices with high input cost. Any planning in this regard to be ecologically sound, economically viable, adaptable, socially acceptable and humane should be based on the needs of the targeted population and take into account the "6-M Kits" which consists of Manpower, Money, Material, Market, Motivation and Management aspects with Knowledge, Information, Technology and Skill of both extension worker and beneficiaries for its' successful promotion and propagation.

IFS is a judicious mix of one or more enterprises along with cropping having complementary effect through effective recycling of wastes and crop residues and encompasses additional source of income to the farmer. It could be crop-fish integration, livestock-fish integration, crop-fish-livestock integration or combinations of crop, livestock, fish and other enterprises. IFS activity is inter-dependent, inter-related and inter-linked production system. Farming systems consist of several enterprises like cropping system, dairying, piggery, poultry, fishery, beekeeping etc. Integrated Farming System is a combination of one or more enterprises with cropping when carefully chosen planned and executed gives greater dividends than a single enterprise, especially for small and marginal farmers.

Key Principles

Cyclic: The farming system is essentially cyclic (organic resources – livestock – land – crops). Therefore, management decisions related to one component may affect the others.

Rational: Using crop residues more rationally is an important route out of poverty. For resource-poor farmers, the correct management of crop residues, together with an optimal allocation of scarce resources, leads to sustainable production.

Ecologically sustainable: Combining ecological sustainability and economic viability, the integrated livestock-farming system maintains and improves agricultural productivity while also reducing negative environmental impacts.

Components of Integrated Farming Systems

1. Crops, livestock, birds and trees are the major components of any Integrated Farming System.
2. Crops may have subsystem like monocrop, mixed/intercrop, multi-tier crops of cereals, pulses, oilseeds, fodder, vegetables, etc.
3. Livestock component may be milch cows, buffaloes, goats and sheep.
4. Tree component may include timber, fuel, fodder and fruit trees.
5. Other enterprises like apiculture, sericulture, poultry, etc.

12.1. Importance of Integrated Farming Systems

1. **Deteriorating Resource Base:** During post-green revolution period, our attempt to solve food problem and attain self-sufficiency in food production through excess use of agrochemicals, inevitable dependence on irrigation and high cropping intensity has led to contamination of food with harmful chemicals, pollution of ground water, degradation of soil quality and damage to agriculturally beneficial microorganisms. In many regions both surface and ground water are already becoming unfit for human and animal consumption due to high concentration of pesticide residues. Intensified agriculture, coupled with indiscriminate use of irrigation water and non-judicious fertilizer application, especially in irrigated areas of the country has led to various kinds of physical and chemical degradation of the soil.

2. **Climate Change:** The increasing green-house gases has resulted in global warming. The Intergovernmental Panel for Climate Change (IPCC) projections on temperature predicts an increase of 1.8 to 4.0 °C by the end of this century. Temperature and sea level changes will affect agriculture through their direct and indirect effects on crops, soils, livestock, fisheries and pests. The brunt of environmental changes is expected to be very high in India due to greater dependence on agriculture, limited natural resources, alarming increase in human and livestock population, changing pattern in land use and socioeconomic factors that pose a great threat in meeting the food, fiber, fuel and fodder requirement. Recent studies done at the Indian Agricultural Research Institute indicated the possibility of loss of 4-5 million tonnes in wheat production in future with every rise of 1°C temperature throughout the growing period.

3. **Narrowed Biodiversity:** The narrowing of genetic biodiversity occurs as traditional crop varieties and local animal breeds are being replaced by modern ones. These new varieties/breeds are certainly better matched to modern intensive agriculture, but rarely any consideration is given to preserving the bio-diversity of an agricultural ecosystem. In addition, the increased farming density tends to erode the biodiversity of flora and fauna in the agricultural ecosystems. For example, extensive adoption of rice-wheat monoculture in the Indo-Gangetic plains has replaced the other traditional crops. Soil micro-flora is also adversely influenced on account of large-scale use of agro-chemicals and lack of recycling of crop residues in the region.

4. **Multiplicity of Integrated Farming Systems:** Very often, almost all Indian farmers, in pursuit of supplementing their needs of food, fodder, fuel, fiber and finance resort to adopt integrated farming systems, majority of them revolving around the crops + livestock components. Livelihood

of small and marginal farmers, comprising more than 80.00% of total farmers, depends mainly on crops and livestock, which is often affected by weather aberrations. Under present scenario, in the absence of scientifically designed, economically profitable and socially acceptable appropriate integrated farming systems models, they are unable to harness the benefits of integration. An important consequence of this has been that their farming activities remain, by and large, subsistent in nature rather than commercial and many a times uneconomical.

5. **Low Rate of Farm Resource Recycling:** In the absence of adequate knowledge among farmers about techniques and benefits of recycling of farm, industrial and municipal organic wastes in agriculture, these remain unutilized. A vast untapped potential exists to recycle these solid and liquid organic wastes of farm origin. Recycling of crop residues may be a potential organic source to sustain the soil health. Incorporation of crop residues of either rice or wheat increases the yield of rice and nutrient uptake and also improves the physico-chemical properties of the soil, ensuring better soil environment for crop growth.
6. **Technology Adoption Gaps:** In our effort to develop and improve upon existing technologies, involvement of people in conceptualization and extension of technologies would appear very important. The farm family had never been the focal point of our investigations. This top down approach had given a poor perception of the problems that they tried to solve. Due to poor extension mechanisms at national as well as state levels, many farmers, especially those at lower strata of social structure, remain uninformed about many of the development schemes and the desired impact of such schemes is not obtained. One of the reasons for poor rate of transfer of agricultural technologies is poor linkages between the different clientele groups of agriculture. Practically linkages among farmers, service providers, technological and financial institutions are either weak or non-existent (NAAS, 2009).

The basic aim of IFS is to derive a set of resource development and utilization practices, which lead to substantial and sustained increase in agricultural production (Kumar and Jain, 2005). Hence, integrated farming systems are often viewed as a sustainable alternative to commercial farming systems particularly on marginal lands with the objective of reversing resource degradation and stabilizing farm incomes. Survey on farming systems in the country as a whole also revealed that milch animals (cows and buffaloes), irrespective of breed and productivity, is the first choice of the farmers as an integral part of their farming system. However, from economic point of view, vegetables and fruits (mango and banana in many parts of the country) followed by fish cultivation were the most enterprising components of any of the farming systems prevailed in the country. A number of success stories on integrated farming system models in different parts of the country suggested that farmers' income can be increased manifold by way of integration of enterprises in a farming system mode.

12.2. Benefits of Integrated Farming Systems

1. **Productivity:** Integrated farming system provides an opportunity to increase economic yield per unit area per unit time by virtue of intensification of crop and allied enterprises.
2. **Profitability:** Use waste material of one enterprise at the least cost as input for other enterprise. Thus, reduction in cost of production, form the utilization linkage of waste material and elimination of middleman interference in most input used.
3. **Sustainability:** Organic supplementation through effective utilization of by-products of linked component is done thus providing an opportunity to sustain the potentiality of production base for much longer period.
4. **Balanced food:** Integrated Farming System links components of varied nature enabling to produce different sources of nutrition for farm families.
5. **Environmental safety:** In Integrated Farming System, waste materials are effectively recycled by linking appropriate components, thus minimize environment pollution.
6. **Income round the year:** Interaction of enterprises with crops, eggs, milk, mushroom, honey, fish, cocoons, etc. provides flow of money to the farmers round the year.
7. **Adoption of new technologies:** Money flow round the year due to Integrated Farming System gives an inducement to the small and marginal farmers to go for the adoption of new technologies.
8. **Saving energy:** Effective recycling technique the organic wastes available in the system can be utilized to generate biogas. Energy crisis can be postponed to the later period.
9. **Meeting fodder crisis:** Every piece of land is effectively utilized. Plantation of perennial fodder trees on field borders. These practices will greatly relieve the problem of non – availability of quality fodder to the animal component linked.
10. **Solving fuel and timber crisis:** Linking agro- forestry appropriately the production level of fuel and industrial wood can be enhanced without determining effect on crop. This will also greatly reduce deforestation, preserving our natural ecosystem.

11. **Employment generation:** Combing crop with livestock enterprises would increase the labour requirement significantly and would help in reducing the problems of under employment to a great extent. Integrated Farming System provide enough scope to employ family members round the year.

12. **Agro-industries:** When one of produce linked in Integrated Farming System are increased to commercial level there is surplus value addition leading to development of allied agro–industries.

13. **Increasing input efficiency**: Integrated farming system provide good scope to use inputs in different components efficiently and effectively.

12.3. Farming System Concept

Farming system is a complex inter related matrix of soil, plants, animals, implements, power, labour, capital and other inputs controlled in parts by farming families and influenced to varying degree by political, economic, institutional and socials forces that operate at many levels. Thus farming system is the result of a complex interaction among a number of interdependent components. Farm activities interact with market forces (socio-economic) and ecosystem (biophysical) for purchasing inputs and disposing outputs by utilizing and degrading natural resources (land, water, air, sunshine etc.).

A farm is a system which has INPUTS, PROCESSES and OUTPUTS

INPUTS - these are things that go into the farm and may be split into physical inputs (e.g. amount of rain, soil) and human inputs (e.g. labour, money etc.)

PROCESSES -these are things which take place on the farm in order to convert the inputs to outputs (e.g. sowing, weeding, harvesting etc.)

OUTPUTS - these are the products from the farm (i.e. wheat, barley, cattle)

Depending on the type of farming e.g. arable/pastoral, commercial/subsistence, the type and amount of inputs, processes and outputs will vary. You need to make sure you are able to define and give examples of inputs, processes and outputs in farming systems.

Income through arable farming alone is insufficient for bulk of the marginal farmers. The other activities such as dairying, poultry, sericulture, apiculture, fisheries etc. assume critical importance in supplementing their farm income.

The overall objective is to evolve technically feasible and economically viable farming system models by integrating cropping with allied enterprises for irrigated, rained, hilly and coastal areas with a view to generate income and employment from the farm.

The Goals of Integrated Farming Systems (IFS) are to:

- **Provide a steady and stable income** through rejuvenation/amelioration of the system's productivity and
- **Achieve agro-ecological equilibrium** through the reduction in the build-up of pests and diseases, through natural cropping system management and the reduction in the use of chemicals (inorganic fertilizers and pesticides).

Principles involved

The FSR approach includes:

- Viewing the farm as a whole
- Identifying the farming system, the interacting component and delineating boundaries
- Systematic investigation of the nature and extent of interdependence among the enterprises and identifying constraint
- Applying modern technical know-how to the system so as to make it yield optimum results
- Studying the equity gender income, employment and resources use efficiency, and
- Dealing with the issue at integration level through analysis and solution of problems towards sustainable farming system development.

Farming system research is a 'research method' designated to understand farmer's priorities, strategies and resources allocation decisions. It is most often used in conjunction with on-farm highly location specific research with multi and interdisciplinary in nature and uses a whole farm approach for improved technologies to enhance and stabilize agriculture production.

Characteristics of Farming System Research

- It is holistic or system-oriented,
- It is problems solving: involvement of farmers in problem identification and solving process,
- It is farmer participatory,
- It envisages location specific technology solutions,
- It is for specific client group–small/ marginal farmer,
- It adopts bottom up approach,

- It compasses extensive on farm activities, collaboration between farmer and scientist,
- It is gender sensitive,
- It ultimate objective is sustainability,
- It focuses on actual adoption,
- It recognizes interdependence among multiple clients.

Components of IFS

- Agriculture
- Fish farming
- Horticulture
- Duck rearing
- Forestry
- Pigeon rearing
- Apiary
- Mushroom cultivation
- Sericulture
- Azolla farming
- Dairy
- Kitchen gardening
- Poultry
- Fodder production
- Goat rearing
- Nursery
- Sheep rearing
- Seed production
- Piggery
- Vermiculture
- Rabbitory
- Value addition

Elements of Integrated Farming System

Following elements may be included in Integrated Farming System demonstrations depending upon the individual farmers' resources, interest and opportunities.

- Watershed
- Farm ponds
- Bio-pesticides
- Bio-fertilizers
- Plant products as pesticides
- Bio-gas
- Solar energy
- Compost making (Vermi, Japanese, Improved etc.)
- Green manuring
- Rain water harvesting.

Cropping systems based on climate, soil and water availability have to be evolved for realizing the potential production levels through efficient use of available resources. The cropping system should provide enough food for the family, fodder to the cattle and generate sufficient cash income for domestic and cultivation expenses.

12.4 Allied Enterprises

Cattle rearing in India is carried out under a variety of climatic and environmental conditions. Draught cattle broadly classified into following groups.

- **Draught breeds:** The bullocks of these breeds are good draught animals, but the cows are poor milkers. e.g. Nagore, Hallikar, Kangeyan, Mali.
- **Dairy breeds:** The cows are high milk yielders but the bullocks are of poor draught quality e.g., Sahiwal, Sindhi, Gir.
- **Dual purpose:** The cows are fairly good milkers and the bullocks are with good draught work capacity e.g. Hariana, Ongole and Kankrej.
- **Exotic breeds:** The exotic breeds are high milk yielders. e.g. Jersey, Holstein Friesian, Aryshire, Brown Swiss and Guernsey.
- **Buffaloes:** Important dairy breeds of buffalo are Murrah, Nili Ravi (which has its home tract in Pakistan), Mehsana, Suti, Zafrabadi, Godavari and Bhadawari. Of these Godavari has been evolved through crossing local buffaloes in coastal regions of AP with Murrah.

- **Housing:** Each cow requires 12-18 sq. m space and buffaloes need 12-15 sq. m. it is important to provide good ventilation and an open shed of housing is always preferable. Dairy building should be located at an elevated place to facilitate easy drainage. The floor should be rough and have gradient of 2.5 cm for every 25 cm length.

Breeding and maintenance: The cows remain in milk for 9-10 months, the average calving interval being 16-18 months. A cow does not require more than 6-8 weeks of dry period. From the economic point of view, cows should ordinarily be bred during the second and third months after calving. In weak animals and heavy milkers, breeding may be delayed by 1 or 2 months. Cattle come to heat in more or less regular cycles of about 21 days which lasts for about a day. The best time to serve a cow is during the last stage of heat. If artificially inseminated, a second insemination may be given within 6-8 hours after the first insemination. The gestation period varies with individual cows and normally it is about 280 days. In the case of buffaloes, the lactation period last for 7-9 months. She buffaloes come to heat every 21 to 23 days. The gestation period is 310 days. Calf rearing is very important in the case of buffalo maintenance. Since they require abundant water, wallowing is required. Regular deworming is needed for buffalo maintenance. Under Indian conditions, cattle commonly mature at the age of about three years. This period can, however, be reduced by six months under well managed herd.

Feeding: Cattle feed generally contains fibrous, coarse, low nutrient straw material called roughage and concentrates.

Roughages: Dairy cattle are efficient use of roughages and convert large quantities of relatively inexpensive roughage into milk. Roughages are basic for cattle ration and include legumes, non-legume hays, straw and silage of legume and grasses.

Concentrates: grains and by-products of grains and oilseeds constitute the concentrates. They are extensively used in dairy cattle ration. These include cereals (maize, sorghum, oats, barley), cotton seeds, industrial wastes (bran of wheat, rice, gram husk) and cakes of oilseeds (groundnut, sesame, rapeseed, soybean, linseed)

- **Vitamins and mineral mixtures:** It is advisable to feed a supplement containing vitamins A and D. Mineral mixtures containing salt, Ca and P should also be provided in the ration. The ration per animal per day normally includes concentrates at 1 kg for 2 litres of milk yield, green fodder at 20-30 kg, straw 5-7 kg and water 32 litres.

12.5. Goat Rearing

In India, the activity of goat rearing is sustained in different kinds of environments, including dry, hot, wet and cold, high mountains or low lying plains. The activity is also associated with different systems such as crop or animal based, pastoral or sedentary, single animal or mixed herd, small or large scale. Goats are mainly reared for meat, milk and hide. Goat meat is the most preferred meat in India. A goat on hooves fetches a better price than a sheep on hooves.

Breeds: There are nearly 102 breeds of goats in the world, of which 20 breeds are alone found in India.

Following are the important breeds in India.

Milk and meat (dual purpose) –Osmanabadi, Barbari, Jamnapari, Sangamneri, Mehasa and Zalwadi, Beetal, Ajmeri and Kachhi.

Meat–Assam, Kali Bengali, Brown Bengali and Marwari.

Wool–Angora, Gaddhi and Pashmina.

Housing: Goats can be maintained under stall fed conditions. Successful goat rearing depends on the selection of site. Goats do not thrive on marshy or swampy ground. Goats are to be provided with a dry, comfortable, safe and secure place, free from worms and affording protection from excessive heat and inclement weather. The kids are kept under large inverted baskets until they are old enough to run along with their mothers. Males and females are generally kept together. The space requirement for a goat is 4.5 to 5.4 sq. m.

Breeding and maintenance: Goats mature in about 6-7 months. Breeding is allowed for buck (male goat) at one year and does /nanny (female goat) after 10 months of age. Gestation period is 145- 155 days. It gives birth to 1-3 kids at a time. Once in five years, the buck can be changed to avoid deterioration due to inbreeding. When the young ones attain a body weight of about 25-30 kg in about nine months, they can be sold.

Feeding: The per head requirement of nutrients in respect of goats is relatively low. Hence, they are suitable for resource poor farmers with marginal grazing lands. They are essentially browsers and eat which any other animals won't touch. Goats eat 4-5 times that of their body weight. Since the profit depends on weight addition, adequate protein and calorie should be given to goats. They eat more of tree fodder and the rest with other grass species. Goats should be fed with concentrates of maize, wheat, horse gram, groundnut cake, fish meal and wheat bran. Common salt and vitamin mixtures should also be added. Abundant clean fresh water should be made available to the goats. Water should be changed every morning and evening. Fresh water is required for digestion, blood circulation and removal of waste from the body. Water is also required for regulation of the body temperature.

12.6. Sheep Rearing

Sheep are well adapted to many areas. They are excellent gleaners and make use of much of the waste feed. They consume large quantities of roughage, converting a relatively cheap food into a good cash product. Housing need not be elaborate or expensive. However, to protect the flock from predatory animals, the height of the fencing should be raised to 2 m.

Breeds of Indian sheep

There are three types of sheep in India based on the geographical division of the country.

(i) **The dry temperate Himalayan region:** Gurez, Karanah, Bhakarwal, Gaddi, Rampur-Bushiar.

(ii) **Dry western region:** Lohi, Bikaneri, Marwahi, Kutchi, Kathiawari.

(iii) **Southern region:** Deccani, Nellore, Bellary, Mandya, Bandur.

- **Breeding and maintenance:** One ram can be maintained for 40-50 ewes. Rams are liable to fight when two or more of them are put in a pen. Unlike other farm animals, ewes in general do not come in heat at regular intervals throughout the year but are seasonal in this respect. The duration of heat period will range from 1-3 days and 75% of ewes remain in heat for 21-39 hours. The optimum time of service is towards the end of heat period. Average heat interval is 18 days during the breeding season. The gestation period will vary from 142-152 days with an average of 147 days. A normal ram is in full vigour for breeding from the age of 2.5-5 years. Sheep grow fully at two years of age when the ewe is ready for breeding, under average range condition; ewes may be expected to produce about five crops of lambs.
- **Feeding:** A sheep requires about 1-2 kg of leguminous hay per day depending on the age of the sheep and its body weight. Proteins may be supplied through concentrates such as groundnut cake, sesame cake or safflower cake when the pastures are poor in legumes or when scarcity conditions prevail. Normally 110-225 g of cake is sufficient to maintain an average sheep in good condition. Feeding a mixture of common salt, ground limestone and sterilized bone meal in equal parts is required to alleviate deficiency of minerals in the feed.

12.7. Poultry

Poultry is one of the fastest growing food industries in the world. Poultry meat accounts for about 27% of the total meat consumed worldwide and its consumption is growing at an average of 5% annually. Poultry industry in India is relatively a new agricultural industry. Till 1950, it was considered a backyard profession in India.

Breeds: Specific poultry stocks for egg and broilers production are available.

A majority of the stocks used for egg production are crosses involving the strains or inbred lines of white Leghorn. To a limited extent, other breeds like Rhode Island Red, California Grey and Australop are used.

Heavy breeds such as white Plymoth Rock, White Cornish and New Hampshire are used for cross-bred broiler chickens. Hence, it is essential to consider the strain within the breed at the time of purchase. Several commercial poultry breeders are selling day old chicks in India. It is best to start with the day old chicks.

Housing: Adequate space should be provided for the birds. Floor area of about 0.2 m^2 per adult bird is adequate for light breeds such as white Leghorn. About 0.3-0.4 m^2 per bird is required for heavy breeds. The house should have good ventilation and be reasonably cool in summer and warm during winter, it should be located on well-drained ground from flood waters.

- **Feed:** The feed conversion efficiency of the bird is far superior to other animals. About 60-70% of the total expenditure on poultry farming is spent on the poultry feed. Hence, use of cheap and efficient ration will give maximum profit. Ration should be balanced containing carbohydrates, fats, minerals and vitamins. Some of the common feed stuffs used for making poultry ration in India are:
- **Cereals:** Maize, barley, oats, wheat, pearl millet, sorghum, rice-broken.
- **Cakes/meal:** oil cakes, maize-gluten-meal, fish meal, meat meal, blood meal.
- **Minerals/salt:** Limestone, oyster shell, salt, manganese. From the day old to 4 weeks of age, birds are fed on starter ration and thereafter finisher ration, which contain more energy and 18-20% protein. Feed may be given 2-3 times. In addition antibiotics and drugs may also be added to the poultry ration. Laying hens are provided with oyster shell or ground limestone. Riboflavin is particularly needed.
- **Maintenance:** The chicks must be vaccinated against Ranikhet diseases with F1 Strain vaccine within the first 6-7 days of age. One drop of vaccine may be administered in the eye and nostril. When chicks get the optimum body weight of 1.0-1.5 kg around six weeks, they can be marketed for broiler. Hens may be retained for one year for production i.e. upto the age of 1.5 years. After that they are disposed off for table purpose. It may not be economical to keep the hens beyond 1.5 years since egg production would get reduced. One hen is capable of laying 180-230 eggs in a year starting from the sixth month. In addition, a laying hen produces about 230 g of fresh droppings (75% moisture) daily.

12.8 Piggery

Pigs are maintained for production of pork. They are fed with inedible feeds, forages, certain grain by-products obtained from mills, meat by-products, damage feed and garbage. Most of these feeds are either not edible or not very palatable to human beings. Pigs grows fast and is a prolific breeder, farrowing 10 to 12 pigs at a time. It is capable of producing two litters per year under good management conditions. The carcass return is high at 65-70% of the live weight.

- **Breeds:** Imported breeds of large white Yorkshire and Landrace are being used widely. Yorkshire is the most extensively used exotic bred in India. It is a prolific breed having carcass quality, growth rate and feed conversion ability. For a small breeding farm of a unit, the selection of the herd boar is extremely important. A good boar weighs 90 kg in about 5-6 months and is strong on feet and legs. The mother of pig is to be selected should have large litter of eight piglets or more.

- **Housing:** Housing should provide maximum comfort to pigs so that their growth is optimum. There should not be dampness, draft and overheating. Locally available materials can be used for housing. One pig requires 2.7 sq. m with a wall of 1.2 m height. Eight boars can be kept in 2.7-4.5 sq. m area with 2.4-6.0 sq. m open space.

- **Feeding:** Feed plays an important role in successful pig production. Pigs are the moist rapidly growing livestock and suffer more than nutritional deficiencies than the ruminants. Proteins, fats, carbohydrates, minerals, vitamins and ample good water form a complete diet for pig. Pigs have a simple stomach; therefore they must be fed with maximum of concentrates and minimum of roughage. The main ingredients of swine ration are cereals and millets and their by-products. The fiber contents in swine ration should be very low (around 5-6%) for better feed utilization efficiency. Mixed ration should be also containing 0.5% of added salt. Swine requires comparatively higher %age of Ca and P than do cattle or sheep. When pigs are maintained with agricultural/kitchen wastes or fish and slaughter house waste, the cost of production remains low. On an average, the consumption of feed is 3.5% of total weight. Feed allowance is calculated as 2.5-3.0 kg/100 kg body weight +@ 0.25 kg per piglet with the lactating mothers.

- **Management:** As a general rule, well developed gilts weighing about 100 kg, when 12-14 months old, may be used for breeding. The body weight is more important than age at breeding. Sow with low body weight show higher rate of fetal and pre-weaning mortality and have been proved with mothers with poor nursing ability. The gestation period is on an average 114 days. Litter size at birth may be 1-16 numbers with the body weight of 1-25 kg. Normal period between birth of piglets is 10-20 minutes. Time taken for the whole process of farrowing ranges from 1.5 to 4.0 hours. Sows are weaned after 40 days. It is advantageous to cull the sow after the fifth or sixth litter in a commercial herd. Weaned sows come into heat in 3-10 days after the weaning and may be allowed to breed. The boar-sow ratio should be 1:15. It is profitable to raise two litters from each sow each year.

Mortality in piglets is an important cause of heavy economic loss leading to failure of pig industry. In general, one fourth of piglets farrowed die before they are weaned.

Another one-tenth are also categorized into stunt or unprofitable group due to disease or parasite infection. Thus about 60-65% of the piglets farrowed perform as healthy piglets at slaughter age. Death rate is high during farrowing and the first week after farrowing. The farrowing season also determines mortality rate. Mortality in new-born piglets is maximum when farrowing takes place in acute cold or hot climate. Therefore, mating should be planned in such a way that farrowing could be avoided in such seasons.

12.9. Rabbit Rearing

Rabbit is reared all over the world in various climatic conditions. Rabbit rearing in India is of recent origin, though hunting of wild rabbits for meat is not uncommon. Rabbit can easily be reared with relatively less concentrate feed and with very high reproduction rates. Rabbit meat is pure white and is low in fat and cholesterol content. Further rabbit meat does not have any religious taboo, unlike pork and meat from cow.

- **Breeds:** Rabbit can be classified into two categories as meat and wool producing ones. Important meat breeds are New Zealand White, Soviet Chinchilla and Kandinrex. Breeds for wool are Angora, Polish Bolamina, Dan and Flirorida.
- **Housing:** Rabbits can be kept in hutches outside or in cages inside. Cages should have doors (30 × 30 cm) with welded wire mesh at floor and on sides. The area required by a rabbit depends on the body weight. For one kg weight, 1.8 sq m area is needed. Angora rabbits cannot be kept at high temperature and should be restricted to hills. Breeding bucks and adult does are housed single. Castrated males can be kept in groups. Young rabbits also live in groups till the age of 4 months and thereafter they are separated. Rabbit keeping require calm and quiet areas. Sudden noise can cause death of rabbits. Disturbance by dogs, cats and humans should be kept to the minimum.
- **Feeding:** Rabbit is a non-ruminant herbivorous animal having a simple stomach. Like any other animal, rabbit requires protein, carbohydrates, fats, minerals, and vitamins in its feed. Rabbits digest the fibre fraction very

poorly and diet having 5-7% crude fibre is adequate. Alfalfa and other forages are well utilized by rabbits. The feed ingredients may constitute maize (20%), alfalfa (25%), wheat bran (40%), groundnut cake (12%), mineral mixture (2%) and salt (1%). The feed mixture should be given twice a day, in the morning and evening. Other grasses and vegetables can be given intermittently in small quantities. Water should always be provided.

- **Maintenance:** Rabbits of 5-6 months age are ready for breeding and the body weight should not be less than 2.0 kg. The female rabbit is taken to the cage of male for mating. At the time of mating, the male produces a particular noise. After that the female rabbit is taken out of the cage of the male. Early morning is the best time for mating. Rabbits mostly breed from March to July. Gestation period is 28-32 days. The eyes of the young ones (pups) open around two weeks and at 3 weeks, they feed themselves. During this period, the pups should not be often handled. Around four weeks, the young ones are separated from the mother. After two weeks of weaning, the female can be allowed for mating. In a year, it gives birth five times, each time producing about six numbers. Mortality in the kindlings is the most important component that determines the economics of rabbit rearing. In general, mortality in adult rabbits is very low ranging from 3-5%. However, it ranges from 10-15% in kindlings up to the age of 30-42 days. Mating during extreme hot and cold season causes high mortality of kindlings.

12.10. Beekeeping

Beekeeping is one of the most important agro-based industries which does not require raw material from the artisan like other industries. Nectar and pollen from flowers are the raw materials which are available in plenty in nature. Beekeeping can even be started with a single colony.

- **Species:** There are two species, *A. piscerana* and *A. mellifera* are complementary to each other but have different adaptations. *A. cerana* is better acclimatized to higher altitudes of the Himalayan region. *A. mellifera* is more profitable in the plains.
- **Management:** The beekeeper should be familiar with the source of nectar and pollen within his locality. The most important source of nectar and pollen are maize, great millet, bulrush, sunflower and palm. The beginner should start with 2 and not more than 5 colonies. A minimum of two colonies is recommended because in the event of some mishap, such as the loss of the queen occurring in one, advantage may be taken with the other. The hive consists of a floor board, brood chamber, top cover, frames and entrance rod. These parts can easily be separated. The hive may be double walled or single walled. The single walled hive is light and cheap. The most suitable time for commencing beekeeping in a locality is the arrival of the

swarming season. Swarming is a natural tendency of bees to divide their colonies under conditions that are generally favourable for the survival of both parent colony and the swarm. This occurs during the late spring or early summer.

- **Honey collection:** Honey should be of good quality to meet the national and international standards. Qualities such as aroma, colour, consistency and floral sources are important. Proper honey straining and processing are needed to improve the quality of the produce. Honey varies in the proportion of its constituents owing to the differences in the nectar produced by different plants. The nectar collected by bees is processed and placed in comb cells for ripening. During the ripening, sucrose is converted into glucose and fructose by an enzyme called invertase which is added to it by the bees. Honey is an excellent energy food with an average of about 3500 calories per kg. It is directly absorbed into the human blood stream, requiring no digestion.

12.11. Azolla Cultivation

Presently, due to low rainfall and decline in cultivable land if alternate fodder was used as feed it can reduce fodder scarcity and cost. Azolla is a chemical fertilizer-free alternate fodder, which is a balanced feed for all varieties of cattles. Ingredients added in livestock feed grain forages such as maize, cumbu, sorghum and ingredients for the preparation of concentrates like oilseeds price hike caused increasing livestock and poultry feed cost increasing day by day. The protein rich Azolla can be used as regular alternate feed for livestock and poultry we can significantly reduce the production. It has been used as best natural fertilizer for rice.

Nutrients in Azolla: Azolla contains 25-30% protein, 14-15% fiber, 3-4% fat, 45-50% carbohydrates, minerals and many micronutrients.

Production Methods

- Shallow standing water ponds, paddy / rice nursery
- Cement tanks.
- Silpanulinsheet spreaded cavity method

In paddy field, if Azolla added @ one kg per cent and stagnate the water to the level of 5 cm, within two weeks it grows in the entire place.

Cement tank method: 10 feet long, 2 feet wide and 1 feet deep concrete tank spread with 25-30 kg of soil with this mix 5 kg of decomposed dung, with this soils of gravel quarry or well soil @ 100g could be added. Keep the water level up to 5 cm. in this add 5 kg of Azolla within two weeks we can collect 35-40 kg Azolla. Once in ten days pour the cow dung slurry.

Azollaas bio-fertilizer: It has ability to fix atmospheric nitrogen. It contains 4.5% nitrogen. For paddy it can be added @ 500 kg per hectare. It gives nitrogen source to the crop and 15-20% yield also increased.

Feeding cattle: Production of 1 kg Azolla costs about 0.75 paise only. When Azolla is used as feed it could be mixed 1:1 proportion with concentrates and fed to the cattles. After familiar with this, mulching cows will started to consume Azolla alone. One kg of Azolla equal to 1 kg of oil cake. For mulching cows 1-1.5 kg, white pigs 1-1.5 kg, Goat 300-500 grams, Rabbits 100 grams and 25-30 grams of Azolla for chickens might be given to these animals.

Approximately 15-20% increase in milk production in dairy cows, in addition to this increase in cholesterol and also cholesterol free solids enhanced the milk quality. Azolla feeding chickens showed increase in body weight and size of the egg. Further, egg yolk seems to be dark in colour. So, production of Azolla under Integrated Farming System not only gives income and also protects the environment.

12.12. Aquaculture

Ponds serve various useful purposes, viz. domestic requirement of water, supplementary irrigation source to adjoining crop fields and pisciculture. With the traditional management, farmers obtain hardly 300-400 kg of wild and culture fish per ha annually. However, poly-fish culture with the stocking density of 7500 fingerlings and supplementary feeding will boost the total biomass production.

Pond: the depth of pond should be 1.5 to 2.0 m. this depth will help for effective photosynthesis and temperature maintenance for the growth of zoo and phytoplanktons. Clay soils have high water retention capacity and hence are best suited for fish rearing. Pond water should have appropriate proportion of nutrients, phosphate (0.2 -0.4 ppm), nitrate (0.06-0.1 ppm) and dissolved oxygen (5.0 - 7.0). Water should be slightly alkaline (pH 7.5-8.5). if the pH is less than 6.5, it can be adjusted with the addition of lime at an interval of 2-3 days. Higher pH (>8.5) can be reduced with addition of gypsum. Application of fresh dung may also reduce high pH in the water. Soil of the pond should be tested for N and P content. If the nutrient content is less, nitrogenous fertilizers like ammonium sulphate and urea and phosphate fertilizers like super phosphate can be added. Organic manures such as FYM and poultry droppings may also be applied to promote the growth of phyto and zooplanktons.

Species of fish

- Among the Indian major carps, Catla (*Catlacatla*) is a fast growing fish. It consumes a lot of vegetation and decomposing higher plants. It is mainly a surface and column feeder.

- Rohu (*Labeorohita*) is a column feeder and feeds on growing plants, decomposing vegetation, large colonial algae, zooplanktons and detritus to a small extent.
- Calbasu (*Labiocalbasu*) is a common feeder on detritus.
- Mrigal (*Cirrhimusmrigale*) is a bottom feeder, taking detritus to a large content, diatoms, filamentous and other algae and higher plants.
- Common carp (*Cyprinuscario*) is a bottom feeder and omnivorous.
- Silver carp (*Hypophthalmichlthysmolitrix*) is mainly a surface and phytoplankton-feeder and also feeds on microplants.
- Grass carp (*Cyernuscarpia*) is a specialized feeder on aquatic plants, cut grass and vegetable matter. It is also a fast growing exotic fish.

Poly fish culture: The phytophagous fish (Catla, rohu and mrigal) can be combined with omnivorous (Common Carp), plankton- feeder (Silver Carp) and mud eaters (Mrigal and Calbasu) in a composite fish culture system.

Management: The fish are to be nourished with supplementary feeding with rice brands and oilseed cakes. This will enable faster growth and better yield. Each variety of carps could be stocked to 500 fingerlings with the total of 5000-8000 per ha. This stocking density will enable to get a maximum yield of 2000-5000 kg/ha of fish annually.

12.13. Sericulture

Sericulture is defined as the practice of combining mulberry cultivation, silkworm rearing and silk reeling. Sericulture is a recognized practice in India. It plays an important role in socio-economic development of rural poor in some areas. The climatic conditions in India are favourable for luxuriant growth of mulberry and rearing of silkworms throughout the year. The temperature in Karnataka state, major silk producing site in India, ranges from 21.2 to 30°C. Climatic conditions in Kashmir are favourable to rear silkworm from May to October.

Moriculture: Cultivation of mulberry plants is called moriculture. There are about 20 species of mulberry, of which four are commonly cultivated. They are *Morusalba*, *M indica*, *M. serrata* and *M. latifolia*. The crop can yield well for 12 years, after which they are pulled out and fresh planting is done. Yield of mulberry leaves is 30-40 t/ha/ year.

Silkworm rearing: There are four types of silkworms:

- Mulberry silkworm (*Bombax mori*)
- Eri silkworm (*Philosamia ricini*)

- Tussar silkworm (*Antheraea paphia*)
- Muga silkworm (*Antheraea assamensis*)

Rearing and maintenance: The fertilized moth is covered with an inverted funnel or cellule and eggs are allowed to be laid over a cardboard. Parasites may be removed by brushing the egg masses with a fine brush. This will also enable to obtain an uniform hatch. In a bamboo tray rice husk is spread. Tender chopped mulberry leaves are added to the tray. The hatched out larvae are transferred to the leaves. It is important to change the leaves every 2-3 hours during the first 2-3 days. The cocoon is constructed with a single reelable thread of silk. If the moths are allowed to emerge from the cocoons, the silk thread is cut into pieces. Hence, the pupa are killed 2-3 days before the emergence of the moth and processed. The cocoons required for further rearing are kept separately and moths are allowed to emerge from them. A family of three can manage 0.75 acre without engaging extra labour. Under irrigated conditions one hectare of mulberry garden gives employment to 13 people throughout the year. It requires 607 man days and 827 woman days.

12.14. Mushroom Cultivation

Mushroom is an edible fungi with great diversity in shape, size and color. Essentially mushroom is a vegetable that is cultivated in protected farms in a high sanitized atmosphere. Just like other vegetables, mushroom contains 90% moisture with high quality protein. Mushrooms are fairly good sources of vitamin C and B complex. The protein have 60-70% digestibility and contain all essential amino acids. It is also rich source of minerals like Ca, P, K, Cu. They contain less of fat (0.3%) and CHO and are considered good for diabetic and blood pressure patients. Fibre content in mushrooms is very high which helps in excretion of waste and prevention of constipation.

Types of Mushroom

- Oyster mushroom-*Pleurotus ostreatns*
- Milky mushroom-*Calocybe indica*
- Button mushroom-*Agricus bisporus*

Oyster mushroom basically occurs in dead woods in nature and resembles the shape of oyster shell. It can be grown on a wide range of agricultural wastes. It can be grown on a wide range on a wide range of temperature and with low investment. The production of this mushroom is being done on a small scale in different parts of the country and is most popular one. Milky mushroom is new of its kind and was introduced for commercial cultivation in the year 1998 from TNAU. The mushrooms are umbrella shaped, robust, attractive and milky white in nature. This variety is tropical in nature and can be grown when button and oyster mushroom cannot be grown due to high temperature. Button mushroom is particularly suitable for winter season, in the northern states. Cultivation of this mushroom requires more investment. This mushroom cannot be cultivated in paddy or wheat straw.

Method of Production

Oyster mushroom: Take fresh paddy straw and cut into small pieces of 3-5 cm length. Soak them in water for 4-6 hours and then boil for half an hour. Drain the water and dry the straw in shade till it is neither too dry nor wet. Take polythene bags of 60 x 30 cm of size and make two holes of one cm diameter in center of the bag such that they face opposite sides. Tie the bottom of the bag with a thread to make a flat bottom. Fill the bag with paddy straw to 10 cm height. Then inoculate with the spawn. Likewise prepare 4-5 layers of straw and spawn alternately. The last layer ends up in the straw of 10 cm height. Keep this in a spawn running room maintained a temperature of about 20-25 °C. and with RH 85 - 90%. After 15-20 days when the spawn running is completed, cut open the polythene bag and take it to the cropping room and allow it to grow for 7 days and harvest the mushroom. Yield: 0.5–1.0 kg/bed.

12.15. Duck Rearing

Ducks account for about 7% of the poultry population in India. They are popular in coastal and water logged states like West Bengal, Orissa, Andhra Pradesh, Tamil Nadu, Kerala, Tripura and Jammu & Kashmir. Ducks are predominantly of indigenous type and reared for egg production on natural foraging. They have a production potential of about 130-140 eggs/bird/year. Ducks are quite hardy, more easily brooded and resistant to common avian diseases. In places like marshy river side, wetland and barren moors where chicken or any other type of stock do not flourish, duck farming can be a better alternative.

- **Breeds:** The important Indian breeds are Sylhet Mete and Nageswari which are mostly found in the eastern region of the country. Their annual production 150 eggs/bird/year. Improved breeds for eggs and meat production are available. Khaki Campbell and Indian Runner are the most popular breeds for egg laying. Khaki Campbell has a production potential of 300 eggs/bird/year. Indian Runner is the second best producer. White Pekin, Muscovy and Aylesbury are known for meat production. White Pekin is the most popular duck in the world. It is fast growing and has low feed consumption with fine quality of meat. It attains about 3 kg of body weight in 40 days. Indigenous types, however, still continue to dominate in duck farming. Desi ducks are robust, well adapted to local conditions and free of diseases.

- **Housing:** Ducks prefer to stay outside day and night even during winter or rains. In mild climate, it is possible to raise ducks without artificial shelter. A free fence of atleast 1.2 m high enclosing the yard is enough to stop any predator. One nest of size 0.3 x 0.3 x 0.45 m to every 3 ducks is sufficient. In case of laying birds, a mating ratio of 1 drake: 6-7 ducks and in meat type 1: 4-5 is allowed. The duck house should be well ventilated, dry, leak and rat proof. The roof may be thatched or asbestos sheeted. A water channel of 0.5 m wide and 0.2 m deep is constructed at the far end of both sides parallel to the night shelter in the rearing or layer house.
- **Feeding:** Ducks normally require lesser attention. They supplement their feed by foraging, eating fallen grains in harvested paddy fields, small fishes and other aquatic materials in lakes and ponds. However, for intensive rearing, pellet feeding may be given. Ducks prefer wet mash due to difficulties in swallowing the dry mash. Hence, ducks should never have access to feed without water. During the first 8 weeks, birds should always have an access to feed. Later on they may be fed twice a day in the morning and late afternoon.
- **Maintenance:** The general management of ducks is similar to that of chickens. The incubation period is 28 days. A broody duck or hen may be used for small-scale hatching and incubator for large-scale hatching. During the early part of the life, newly hatched ducklings require warm temperature under the natural or village conditions. A duck or broody hen can take care of 10-15 ducklings. Artificial brooding may be resorted for large number of ducklings. High egg-laying strains of duck come into production at 16-18 weeks of age. Ducks are resistant to common avian diseases. Some of the common diseases in ducks are duck plague, ducks virus hepatitis, duck cholera and aflatoxicosis.

12.16. Turkey Rearing

Turkey is a robust bird and can be reared in humid tropics. It actively feeds on a variety of crop residues and insects in the farm and hence can be raised with much ease.

- **Physical features:** There are a few peculiar features in the physical appearance of a turkey. Male has black beard attached to the upper breast regions. This beard is small in females of white turkeys and absent in those of coloured varieties. A fleshy protuberance called snood is present near the base of the beak in both males and females. The head bears caruncles and a large flap of skin below the chin is known as dewlap. Broad Breasted Bronze breed are ordinarily seen in India. Norfolks and the Cambridge are the other important breeds.

- **Housing:** Housing for turkeys may be of range system, confinement or a combination of both. There is a growing trend towards confinement housing. Breeds require a floor space of 0.4 m2/bird in confinement and 32.5 m2/bird in the range rearing.
- **Feeding:** Turkeys have much higher growth rate compared to chickens. Hence they have relatively higher feed and nutrient requirement. Turkey can be fed with all types of dry feed comprising pellets, wetmesh, grain and green feed in well balanced combination. They also need plenty of clean water.
- **Maintenance:** Mating ratio in natural flock is 14 hens: 1 tom for small and 12:1 for medium sized turkeys. The birds should be carefully segregated when they reach the mating stage. Mating takes place during summer. Particular attention should be bestowed in regulating mating. The males should be kept away from their mates in the morning and selectively mated with other birds by turn. Mating in the evening increases the number of fertile eggs. Generally turkeys lay eggs on alternate days from six onwards. In a year, they lay about 150 eggs. The bird attains a weight of 5-8 kg in a year. The birds should not be allowed to fatten too much as it would make them vulnerable for diseases. Changing the breeding stock periodically would be gainful for the farmer.

12.17. Pigeons for Meat

Domestic pigeons (*Columba livia*) belonging to the sub-family Columbia can be reared for its meat. Females are smaller than the males and finer in the head. They hold their tails higher than males and their pelvic pones are spaced wide apart. The males are large, aggressive and make a loud cooing noise. They strut around the females during the breeding season. There is difference in the growth rate among the males and females. The former grows faster than the latter. Thus, even at an early stage, the gender difference can be easily made based on their body weight.

Feeding: Adults feed the young ones (squabs) with pigeon milk, a fluid that is secreted in their crop. Pigeon milk looks like thick crusted and has high protein content. Pigeons require water for drinking and for bathing to remove external parasites. A pair of breeds drinks about 500 ml of water per day. Pigeons are not fussy eaters, but do best on diets having whole grains. Nutrient requirement of pigeons is crude protein (13.5%), carbohydrates (65%), crude fibre (3.5%) and fat (3.0%). The daily feed intake of an adult pair is about 120 g.

Maintenance: Each pair normally produces 14 squabs in a year. The incubation period of the eggs is 18 days. Both partners build their nest and take turns to sit on the eggs. The female starts laying eggs again when the squabs are two weeks old. The young ones become independent in 35 days. Squabs of 30-35 days are best suited for meat purpose since recovery is high with the body weight of 350 g. In addition to the meat, pigeon droppings (25 g per bird day) can be used as fish feed or enriched manure.

12.18. Japanese Quail

Quail belongs to class Aves, family Phasianidae and genus *Coturnir*. Quails, introduced initially as experimental birds, have become popular for their eggs and meat. Efforts are being made by Central Avian Research Institute, Izatnagar, to develop the guinea fowl for meat production. Males are lighter than females and are identified by the cinnamon coloured feathers on the upper throat and lower breast region. Females are similar to males in colour except that the feathers on the throat and upper breast are pointed and much lighter cinnamon in colour.

Maintenance: Quails need small space and are highly suitable for cage rearing. Therefore, it is possible to rear them even in urban areas where space is a major constraint. They are socially and commercially more adaptable due to their small body size, shorter generation interval, highly prolific nature and simplicity in rearing and are more economical. Females lay eggs as early as at the age of 5 weeks. The average age of the first egg production is 40 days. They are in full production by 50 days. They have a capacity to lay 300 or more eggs each per annum. The shell of a typical quail's egg is a mixture of greenish black and brownish speckles or patches of varying sizes. An egg weighs about 10-12 g on an average and is comparatively thin and delicate in nature. Albumen content is 56 to 59%, yolk 28-32% and shell 12-14%. The yolk content in relation to albumen is more in quail's egg than in chicken's egg. Special care should be taken for collecting and handling the egg. The rapid growth rate of the Japanese quail renders it an ideal broiler. At six weeks, quails attain a body weight of 125-150 g with a feed efficiency of 3.0. Japanese quails are fairly resistant to most of the common diseases affecting chicken. However, they are sensitive to abrupt environmental changes during first and second week of their life.

12.19. Vermicompost

Vermicompost is a one of the important natural fertilizers. In the past, agriculture was done using only natural fertilizers. Because of this increased production and also soil fertility was preserved. But the current situation due to growing population, because of Green Revolution, by using chemical fertilizers high yielding varieties is created. Excessive use of chemical fertilizers and pesticides caused reduction in soil fertility and also reduced the number of beneficial microorganisms. Therefore, conversion of this degraded land into fertile lands, organic manures play a major role for the transformation. Naturally available organic residues are ingested by worms. It was digested by the intestine microorganisms and enzymes and then tiny pellets form forced out through anal of the earthworm is called vermi composting.

Earthworm is called as “farmer’s friend”. According to living place, earthworm is categorized as top, intermediate and grassroots worms. Top and intermediate type of earthworm’s are best suited to produce compost. Earthworms takes food as 2 to 5 times organic matter of its body weight, out of this 5 -10% of them spend as food and the rest of the organic matter excreted as fertilizer.

Selection of earthworm: Despite, nearly 3,000 varieties of earthworms, available worms such as Eudrilus and Eisenia of African species are suitable for commercial production of manure, as they are the top feeders in the soil. The species should have higher reproduction rate, adjusting to climatic conditions and economic importance has to be selected. And also reproduce quickly, suitable for temperatures and with the economic importance of the species should be selected. *Perionyx excavatus, Eudrilus eugeniae*, and *Eisenia fetida* are such kind of earthworm, much of it has been used to produce vermicompost.

Collecting earthworms: Using local earthworm helps to improve the fertility of the soil and environment. Choose a shady place in the garden. The place of tree shade where chemical fertilizers has not been applied or select a place where organic waste from the house pass out would be right choice. Choose a small place and spread cow dung and place on straw or sprinkle dust and cover with old gunny bags and frequently sprinkle water based on need. After two weeks we can see top and middle layer earthworms. While collecting earthworms soil also to be taken together. Culturing of earthworms and manufacturing fertilizers, worms are fed continuously. Nitrogen rich cow dung, goat dung and pig dung may be used as feed.

Earthworm production systems: Vermicompost production methods may change according to prevalent situation in the place of worm culturing area. Some of the methods which include are cavity, basin system and bed methods etc. Cavity method is a very simple method for preparation of the earthworm manure. In this method large-scale vermicompost production is possible on commercial scale. In this system, long beds are prepared and grow worms and produced fertilizer.

In this method, dig a pit in the dimension of 10” × 3” × 2” on a shaded and elevated surface and even extend the length up to 20 feet. Release 2000 – 2500 earthworms per square metre.

Thus, inner side of the excavated pit is lined with bricks. Bottom of the pit has to be filled with coir waste or agriculture waste. Finally add cow dung or decomposed agriculture waste or leaves has to be spread over there. Finally cover the pit with biogas slurry or cowdung mixture and release the earthworms. After this pits are to be covered with straw or stakes to preserve the moisture.

Separation method of earthworm excreta

Once decided to collect the excreta, stop sprinkling water two days in advance. Earthworms move to moisturized portion in the bottom. This is directly used for cultivated land or long term usage, bagged it without loss in moisture and it was used later. If maintained the compost without loss in moisture, it can be used for six to eight months.

Uses

- Increasing the microorganism populations in the soil.
- Increase the soil aeration.
- Play a role in increasing the growth of roots.
- Environmental protection.

12.20. Biogas Plant

Biogas is a clean, unpolluted and cheap source of energy, which can be obtained by a simple mechanism and little investment. The gas is generated from cowdung during anaerobic decomposition. Biogas generation is a complex biochemical process. The celluloitic materials are broken down to methane and carbon dioxide by different groups of microorganisms. It can be used for cooking purpose, burning lamps, running pumps, etc.

Types of biogas plant

1. **(i) Float dome type:** Different models are available in this category, e.g. KVIC vertical and horizontal, Pragati model and Ganesh model.
2. **(ii) Fixed dome type:** The gas plant is a dome shaped underground construction. The masonry gas holder is an integral part of the digester called dome. The gas produced in a digester is collected in the dome at vertical pressure by displacement of slurry in inlet or outlet. The entire construction is made of bricks and cement. The models available in this category are Janata and Deen-bandhu.

The selection of particular type depends on technical, climatologically, geographical and economic factors.

Size: The size of a biogas plant is decided by the number of family members and the availability of cattle. One cubic meter capacity plant will need 2-3 animals and 25 kg of dung. The gas produced will meet the requirement of the family of 4-6 members. It would suffice to have a 2 cubic meter plant to cater to the needs of a family of 6-10 members.

Site selection and management: The site should be close to the kitchen or the place of the use. It will reduce the cost of gas distribution system. It should also be nearer to the cattle shed to reduce the cost of transportation of cattle dung. Land should be levelled and slightly above the ground level to avoid inflow or run off of water. Plant should get clear sunshine during most part of the day. Generation of dung has directed bearing on the quantity of gas generated. The amount of gas production is considerably higher in summer followed by rainy and winter seasons. Gas production would be maximum at a temperature between 30 and 35° C. if the ambient temperature falls below 10°C, gas production is reduced drastically. Biogas slurry is obtained after the production of biogas. It is enriched manure. Another positive aspect of this manure is that even after weeks of exposure to the atmosphere, the slurry does not attract flies and worms.

13

Role of Organic Farming in Agriculture

As per United States Department of Agriculture (USDA), organic farming is defined as a system which avoids or largely excludes the use of synthetic inputs (such as fertilizers, pesticides, hormones, feed additives etc.) to a maximum extent and feasible rely upon crop rotations, crop residues, animal manure, off farm organic waste, mineral grade rock additives and biological system of nutrient mobilization and plant protection. FAO suggested that organic agriculture is a unique production management system which promotes and entrance agro-ecosystem health, including biodiversity, biological cycles and soil biological activity and this is accomplished by using on-farm agronomic, biological and mechanical methods is exclusion of all synthetic off-farm inputs.

Organic farming is a specified form of diversified agriculture that not only aims to produce nutritious food of high quality but also lead to the generation of more income and employment opportunities for rural population. In true sense, organic farming is a production system which avoids or largely excludes the use of synthetic fertilizers, growth regulators, livestock feed additives and pesticides. Moreover, it relies on agronomic management practices like crop rotation, crop residues, animal manures, legumes, green manures, off -farming organic wastes and aspect of biological pest control to maintain soil productivity.

Organic farming has entirely different perspective between developed and developing countries. Developed countries are endowed with surplus resources, but limited population with high purchasing power. Hence, a consumer in a developed country expects enough food of highest possible quality that has been produced without any serious ecological impact and harmful residues. Thus organic farming may be the best answer to fulfil such expectations in rich countries (Marwaha and Jat, 2004). However, on the other hand, for a resource poor, highly populous country like India, with ever growing demand for food, the primary concerns may be enough availability of food at affordable prices with reasonable quality. In this backdrop, considerable efforts have been taken to promote the organic farming in India without compromising the food grain production to feed the ever growing population. Organic farming is also called biological agriculture/ ecological agriculture/ bio-dynamic or natural agriculture.

13.1. Area Under Organic Farming

As per Mahapatra*etal* (2009), organic agriculture is developing rapidly and is now practiced in more than 120 countries of the world. Its share in agricultural land continues to grow in many countries and globally 30.4 m ha is under organic agriculture, which constitutes 0.65% of agricultural land. According to the report of International Federation of Organic Agriculture Movement, Germany (IFOAM, 2016), the area under organic agriculture rose from 15.0 m ha during 2000 to 74.9 m ha during 2020 (Fig. 13.1). Cultivation of permanent grasses, arable crops and perm-anent crops contributes major share in organic agriculture globally.

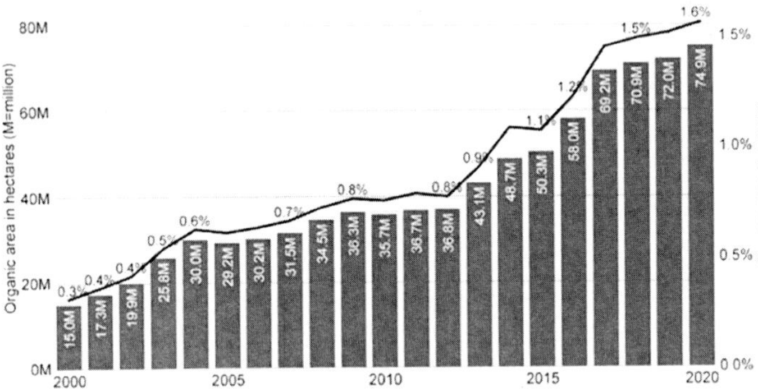

Fig. 13.1. World: Growth of the organic agricultural land and organic share 2000-2020
Source: FiBL-IFOAM-SOEL Surveys 2001-2022

Among different countries, Australia leads in organic farming with total area of 35.69 m ha during 2020 (FIBL Survey 2022) (Fig. 13.2). Argentina is at second position with area of 4.45 m ha and it was followed by Uruguay (2.74 m ha), India (2.66 m ha), France (2.55 mha) and Spain (2.44 m ha).

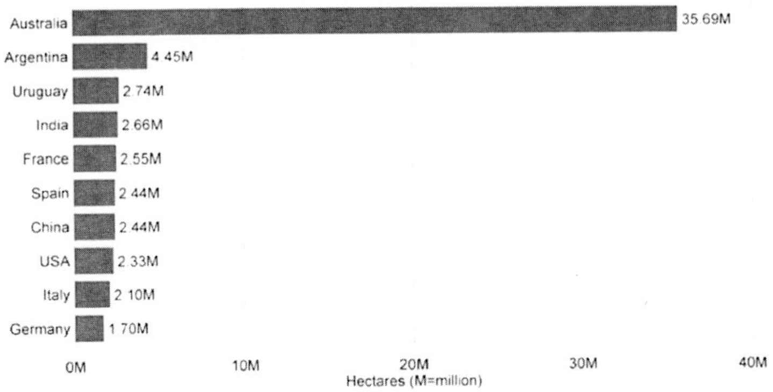

Fig. 13.2: Top 10 countries under organic agriculture (2020) *Source*: FIBL - IFOAM Surveys 2022

On global level, the area under key crops under organic agriculture is presented in Fig. 13.3. As per FIBL Survey 2022, the area under cereal crops is highest (50.88 lac hectare) which is followed by green fodder (32.17 lac ha), Oilseeds (17.95 m ha), dry pulses (7.48 lac ha), textile crops (6.17 lac ha) etc. (Fig. 13.3) FIBL Survey 2022.

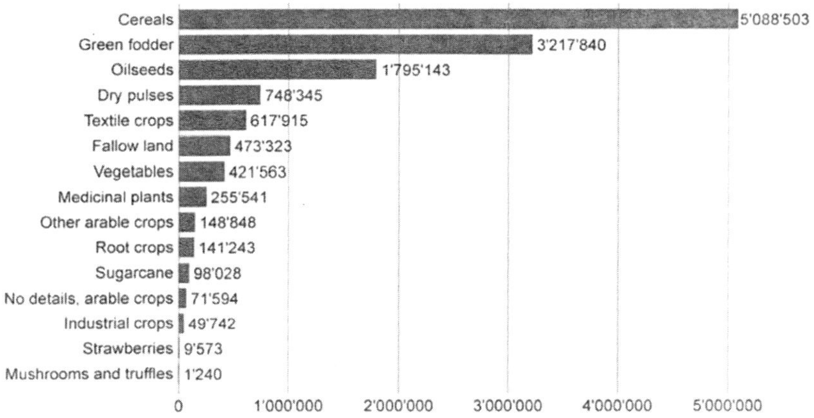

Fig. 13.3: World: Use of arable cropland by crop group 2020 *Source*: FiBL Surveys 2022

In India, organic agriculture is done is nearly all the states. However, Chhattisgarh leads with organic agriculture area of 30.09 lac ha which is followed by Madhya Pradesh (23.71 lac ha), Maharashtra (11.34 lac ha), Rajasthan (6.86 lac ha), Gujarat (6.02 lac ha), Himachal Pradesh (2.03 lac ha). (Figure 13.4)

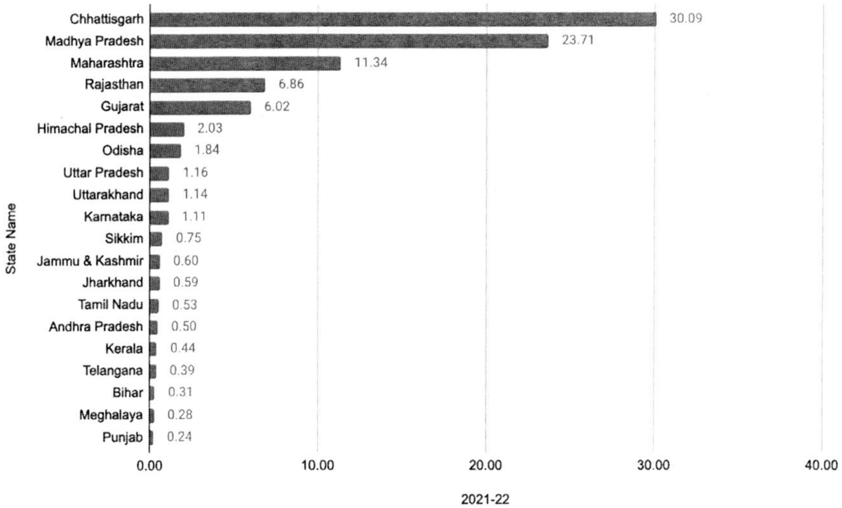

Fig. 13.4: Organic Agriculture area in India by State in ha 2021-2022. *Source*: APEDA

Organic agriculture is becoming very popular in India with highest production of fibre crops (1140.60 thousand ton) which is followed by Oil Seeds (464.82 thousand ton), Sugar (336.88 thousand ton), Cereals & Millets (242.42 thousand ton) (Figure 13.5)

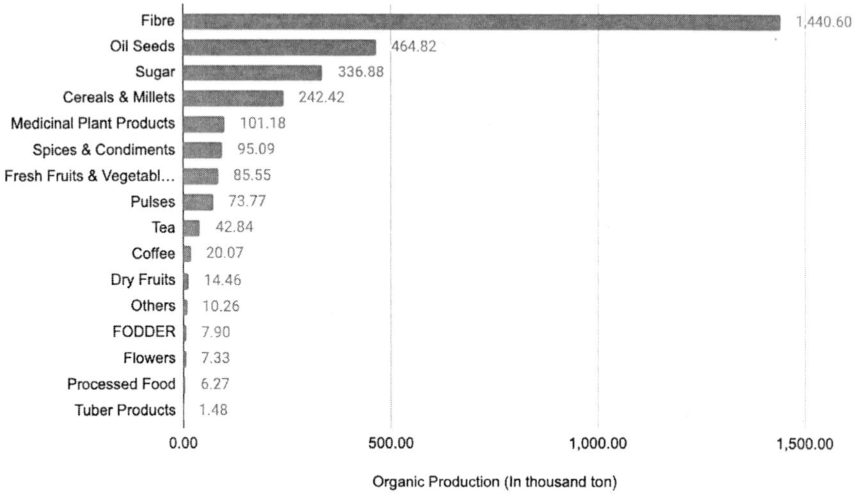

Fig. 13.5: Production of organic crops in India by category in thousand MT for 2021-22
Source: APEDA

13.2. Movement of Organic Farming

Global

At global level, the roots of organic farming can be traced back to the first quarter of the early 20th century in Europe (Stockdale *et al.,* 2001). In 1924, the Austrian philosopher Dr. Rudolf Steiner conceptualized and advocated organic agriculture and in 1927 a trademark 'Demeter' was introduced for organically produced food. The formation of the International Federation of Organic Agriculture Movements in 1972 gave framework for the discussion and codification of internationally recognized principles of organic farming.

The main principles of organic farming are:

(i) to produce food of high quality in sufficient quantity,

(ii) to interact in a constructive and life-enhancing way with natural systems and cycles,

(iii) to consider the wider social and ecological impact of the organic production and processing systems,

(iv) to encourage and enhance biological cycles within the farming system involving micro-organisms, soil flora and fauna, plants and animals,

(v) to maintain and increase the long-term fertility of soils,

(vi) to maintain the genetic diversity of the production system and its surroundings, including the protection of wildlife habitats,

(vii) to promote the healthy use and proper care of water, water resources and all life therein,

(viii) to use, as far as possible, renewable resources in locally organized production systems,

(ix) to give all livestock conditions of life with due consideration for the basic aspects of their innate behaviour,

(x) to minimize all forms of pollution and

(xi) to progress towards an entire production, processing, and distribution chain which is both socially justifiable and ecologically responsible.

In India

Organic farming was practised in India since thousands of years. The great Indian civilization thrived on organic farming and was one of the most prosperous countries in the world. In traditional India, the entire agriculture was practised using organic techniques, where the fertilizers, pesticides etc., were obtained from plant and animal products. At that time organic farming was the backbone of the Indian economy. The country had witnessed a tragic Bengal famine in 1942-43 that claimed millions of lives due to starvation. India's annual food grain production in 1950-51 was only about 50 m t. In this backdrop, modern agriculture played a vital role in alleviating the hunger through green revolution. India's achievement in agricultural production after the green revolution has been exemplary and mainly due to the increased use of inputs like fertilizers, pesticides and farm machinery. The food grain production in India itself more than doubled during the post-green revolution period with virtually no increase in net cultivated area. The intensive modern agriculture for the past five decades has led to several environmental and health issues. The land is losing its fertility and is demanding larger quantities of fertilizers. Pests are becoming resistant requiring the farmers to use stronger and costlier pesticides. The over-use of pesticides resulted in residues much above the safety levels and this brought to the attention of ill-effects of modern agriculture, and paved the way for organic farming. According to the Agricultural and Processed Food Products Export Development Authority (APEDA), around 2.8 m ha of land was under organic farming in India in 2007. Further, there are over 15,000 certified organic farms in India. India therefore is one of the most important suppliers of organic food to the developed nations. No doubt, the organic movement has again started in India.

13.3 Status and Scope of Organic Farming in India

According to Marwaha and Jat (2004), the organic farming probably is seldom practised and there is dearth of data on comparative yield and economics of crops with pure organic farming versus intensive farming. Whatever little organic farming that is being practised in India seems to be taken up mainly as a part of contract farming. In India, the area under organic farming is estimated to be only 0.3% of total cultivated area. Even in Europe and other developed countries of the world, despite serious efforts and incentives, the area under organic farming is generally less than 5%. Organic farming may be far from reality for 3 reasons firstly, the limited availability of organic sources; secondly, for getting optimum yields, crop plants require nutrients in sufficient quantities and in readily available form. This could be best possible through fertilisers, and hardly through organic sources. Thirdly, for plant protection, the organic sources and biological methods that have been tried so far have not met with considerable success in the country.

Marwaha and Jat (2004) were of the opinion that, organic farming may be promoted in selected rainfed areas and for export oriented crops in India. In India, only about 38% of the cultivated land is irrigated and in rest of the rainfed where there is little or no use of fertilisers and other agro-chemicals on account of associated risk in farming and poor resources. So, almost of such areas of the country may be regarded as akin to organic farming. Similarly, some areas under tea, coffee, spices and cashew may be easily brought under organic farming with a thrust on export of organic produce. Thus, rather than promoting organic farming enmasse and universally, it would be desirable to carefully delineate areas for organic agriculture, which provides better profitability and boost to the export.

Regulation of Organic Production: Global Perspective

Organic agriculture is based on the commitment of farmers and processors to work according to set standards and regulations, which define the organic production system. Further, organic agriculture is based on transparency to make the production system comprehensive and reliable. Finally, organic agriculture needs the consumer confidence for market development. Certification and accreditation systems primarily serve as tools to enhance trade, market development and foster confidence among the consumers (Herrmarm and Rundgren, 2007). The first international standards for organic products were published by International Federation of Organic Agriculture Movements (IFOAM) in 1980. The first legislative initiatives were developed by some European countries (e.g. Austria, France) in 1980s. In 1991, the European Countries (EC) passed the organic regulation and set standards with major implications for international trade, and included not only production standards, but also standards for labeling and inspection. Various countries in Europe, Latin America and Asia introduced legislation in the 1990s. In 1999, Codex Alimentarius approved the first guidelines for organic plant production. In the new millennium, most major economies have implemented legislation on organic production. Currently more than 60 countries have a regulation. In India the regulations for organic agriculture is implemented only for export commodities (Huber *et al.*, 2007).

As of 2006, around 395 organizations world-wide offer organic certification services. Most certification bodies are in Europe (160), followed by Asia (93) and North America (80). Many of the certification organizations also operate outside of their home country. The certification bodies 40% are approved by the European Union, 32% have ISO 65 accreditation, and 28% are accredited under the US National Organic Program (Herrmann and Rundgren, 2007).

Need of Organic Farming

With the increase in population our compulsion would be not only to stabilize agricultural production but to increase it further in sustainable manner. The scientists have realized that the 'Green Revolution' with high input use has reached a plateau and is now sustained with diminishing return of falling dividends. Thus, a natural balance needs to be maintained at all cost for existence of life and property. The obvious choice for that would be more relevant in the present era, when these agrochemicals which are produced from fossil fuel and are not renewable and are diminishing in availability. It may also cost heavily on our foreign exchange in future.

Key Characteristics of Organic Farming

- Protecting the long term fertility of soils by maintaining organic matter levels, encouraging soil biological activity, and careful mechanical intervention.
- Providing crop nutrients indirectly using relatively insoluble nutrient sources which are made available to the plant by the action of soil micro-organisms.
- Nitrogen self-sufficiency through the use of legumes and biological nitrogen fixation, as well as effective recycling of organic materials including crop residues and livestock manures.
- Weed, disease and pest control relying primarily on crop rotations, natural predators, diversity, organic manuring, resistant varieties and limited (preferably minimal) thermal, biological and chemical intervention.

- The extensive management of livestock, paying full regard to their evolutionary adaptations, behavioural needs and animal welfare issues with respect to nutrition, housing, health, breeding and rearing.
- Careful attention to the impact of the farming system on the wider environment and the conservation of wildlife and natural habitats.

13.4 Principles of Organic Farming

The four principles of organic agriculture are as follows:

1. Principle of Health

Organic Agriculture should sustain and enhance the health of soil, plant, animal, human and planet as one and indivisible. This principle points out that the health of individuals and communities cannot be separated from the health of ecosystems - healthy soils produce healthy crops that foster the health of animals and people. Health is the wholeness and integrity of living systems. Immunity, resilience and regeneration are key characteristics of health.

The role of organic agriculture, whether in farming, processing, distribution, or consumption, is to sustain and enhance the health of ecosystems and organisms from the smallest in the soil to human beings. In particular, organic agriculture is intended to produce high quality, nutritious food that contributes to preventive health care and well-being. In view of this it should avoid the use of fertilizers, pesticides, animal drugs and food additives that may have adverse health effects.

2. Principle of Ecology

Organic Agriculture should be based on living ecological systems and cycles, work with them, emulate them and help sustain them. This principle roots organic agriculture within living ecological systems. It states that production is to be based on ecological processes, and recycling. Nourishment and well-being are achieved through the ecology of the specific production environment. For example, in the case of crops this is the living soil; for animals it is the farm ecosystem; for fish and marine organisms, the aquatic environment.

Organic farming, pastoral and wild harvest systems should fit the cycles and ecological balances in nature. These cycles are universal but their operation is site-specific. Organic management must be adapted to local conditions, ecology, culture and scale. Inputs should be reduced by reuse, recycling and efficient management of materials and energy in order to maintain and improve environmental quality and conserve resources. Organic agriculture should attain ecological balance through the design of farming systems, establishment of habitats and maintenance of genetic and agricultural diversity.

3. Principle of Fairness

Organic Agriculture should build on relationships that ensure fairness with regard to the common environment and life opportunities. Fairness is characterized by equity, respect, justice and stewardship of the shared world, both among people and in their relations to other living beings.

This principle emphasizes that those involved in organic agriculture should conduct human relationships in a manner that ensures fairness at all levels and to all parties - farmers, workers, processors, distributors, traders and consumers. Organic agriculture should provide everyone involved with a good quality of life, and contribute to food sovereignty and reduction of poverty. It aims to produce a sufficient supply of good quality food and other products.

Natural and environmental resources that are used for production and consumption should be managed in a way that is socially and ecologically just and should be held in trust for future generations. Fairness requires systems of production, distribution and trade that are open and equitable and account for real environmental and social costs.

4. Principle of Care

Organic Agriculture should be managed in a precautionary and responsible manner to protect the health and well-being of current and future generations and the environment. Organic agriculture is a living and dynamic system that responds to internal and external demands and conditions. Practitioners of organic agriculture can enhance efficiency and increase productivity, but this should not be at the risk of jeopardizing health and well-being. Consequently, new technologies need to be assessed and existing methods reviewed. Given the incomplete understanding of ecosystems and agriculture, care must be taken.

This principle states that precaution and responsibility are the key concerns in management, development and technology choices in organic agriculture. Science is necessary to ensure that organic agriculture is healthy, safe and ecologically sound. However, scientific knowledge alone is not sufficient. Practical experience, accumulated wisdom and traditional and indigenous knowledge offer valid solutions, tested by time. Organic agriculture should prevent significant risks by adopting appropriate technologies and rejecting unpredictable ones, such as genetic engineering. Decisions should reflect the values and needs of all who might be affected, through transparent and participatory processes.

Basic Steps of Organic Farming

Organic farming approach involves following **five** steps:

- Conversion of land from conventional management to organic management.

- Management of the entire surrounding system to ensure biodiversity and sustainability of the system.
- Crop production with the use of alternative sources of nutrients such as crop rotation, residue management, organic manures and biological inputs.
- Management of weeds and pests by better management practices, physical and cultural means and by biological control system.
- Maintenance of live-stock in tandem with organic concept and make them an integral part of the entire system

13.5 Current Status of Regulatory Mechanism in India

National Programme for Organic Production

There is tremendous scope for cultivation and export of wide range of organic produce in India. Looking into its importance, Ministry of Commerce, Government of India launched National Programme for Organic Production (NPOP) in 2000. Declaring standards for organic production, policy on accreditation system, inspection and certification process and institution of logo were the major objectives of this programme at the initiation. A national organic logo "India Organic" was also developed that has been in use for certified organic products. During July, 2002, Ministry of Agriculture, Government of India has also circulated guidelines for promotion of organic farming (Bhattacharyya and Gehlot, 2004).

Accreditation Agencies

As of 2004, Agricultural and Processed Food Product Export Development Authority, Coffee Board, Spices Board, Tea Board, Coconut Development Board, and Cocoa and Cashewnut Board were authorised by the Ministry of Commerce, Government of India as accreditation agencies for organic agriculture (Bhattacharyya and Gehlot, 2004).

Certifying Agencies

At present there are 16 accredited certification bodies (www.apeda.com) in India. In 2001, a group of organizations and corporate bodies took the initiative to set up the Indian Organic Certification Agency (INDOCERT). It has become an important element of the organic movement in India and mobilizes new forces and partnerships. In 2003, together with other partners, INDOCERT created the International Competence Center for Organic Agriculture (ICCOA), a service provider for networking, capacity building and market development in the organic sector in India. Among other activities, ICCOA implements the Indian Organic Market Development Project (2005-2007), the activities of which were focussed on awareness raising, market intelligence, developing organic market initiatives, and the India Organic Trade Fair (Kilcher, 2007). An overview of organic agriculture in India (2004-05) Organic agriculture Accountability area under certified crops 2.5 m ha - Total certified product (tonnes) 1,15,238 Number of processing units -158. Accredited inspection and certifying agencies 11 Number of products exported-35. States involved in organic export - Kerala, West Bengal, Karnataka, Tamil Nadu, Punjab, Himachal Pradesh, Maharashtra etc. Premium collected against organic Rs 80-90 export crores (tentative) *Source:* Bhattacharyya and Chakraborty (2005).

National Institute of Organic Farming

The Indian Central Government set up a National Institute of Organic Farming (NIOF) in October 2003 in Ghaziabad, Uttar Pradesh. The National Bio-fertiliser Development Centre was renamed as the NIOF. The purpose of this institute is to formulate rules, regulations and certification of organic farm products in conformity with international standards.

India has now become a leading supplier of organic herbs, organic spices, organic basmati rice, etc. India produced around 58,597 MT of certified organic products which includes all varieties of food products, viz. cereals, pulses, honey, tea, spices, coffee, oilseeds, fruits, vegetables and their value added products. The production is not limited to the edible sector but also produces organic cotton fibre, garments, cosmetics, functional food products, body care products, etc. India exported 35 items last year (2007-08) with the total volume of 19,456 MT. The export realization was around 78 million US $ registering a 200 % growth over the previous year. As many as 98 agencies are involving in the export of organic products in India. Organic products are mainly exported to European Union, United States of America, Australia, Japan, Switzerland, and Middle East. Basmati rice leads among the products exported (5,139 t) (www.apeda.com). In February 2009, the export duty on export of Basmati rice has been withdrawn by the Government of India (Notification No. 10/2009-Customs) with a view to promote the organic products. It is concluded that rainfed areas provide ample scope for pure organic farming. Moreover, there is a lacuna to show the comparative crop performance under pure organic farming and intensive agriculture in crop productivity and food quality. Hence, efforts need to be taken to address these issues with a view to promote the organic farming with great confidence in India. During the last four decades of the 20th century, the global population doubled from 3 to 6 billion and is estimated that by 2020, it will reach the 8 billion mark. Food and nutritional security is therefore, a global concern. Neither conventional farming with inorganic alone nor the organic farming only with the use of organic inputs can face this challenge. The combination of both organic and inorganic is undoubtedly the best option as on today for developing country like India. Above all organic agriculture must be discussed with an open mind, with advantages and disadvantages being clearly consider- ed to create positive mindset among the decision makers, because the most important factor that will enable organic agriculture to usefully con-tribute to food security is the attitude of decision-makers (Wynen, 1998).

Marketing

Marketing is a process of communicating the value of a product or service to customers. Marketing might sometimes be interpreted as the art of selling products, but sales is only one part of marketing.

From a societal point of view, marketing is the link between the society's material requirements and its economic patterns of response. Marketing satisfies these needs and wants through exchange processes and building long term relationships. Marketing is a science of choosing target markets through market analysis and market segmentation, as well as understanding consumer buying behaviour and providing superior customer value.

Export Potential

This term is derived from the conceptual meaning as to ship the goods and services out of the port of a country. The seller of such goods and services is referred to as an 'exporter' who is based in the country of export whereas the overseas based buyer is referred to as an 'importer'. In international trade, 'export' refers to selling goods and services produced in the home country to other markets. Export should be high then only our foreign currency will increase.

Export of commercial quantities of goods normally requires involvement of the customs authorities in both the country of export and the country of import.

Traditional Export Crops

Beyond these basic food crops, the agriculture and often the whole economy of many developing countries depend to a high degree on the production of one or a few commodities destined principally for export. Various organic food products exported from India are:

- **Organic cereals:** Wheat, rice, maize or corn
- **Organic pulses:** Red gram, black gram
- **Organic fruits:** Banana, mango, orange, pineapple, cashew nut, walnut
- **Organic oil seeds and oils:** Soybean, sunflower, mustard, cotton seed, groundnut castor
- **Organic vegetables:** Brinjal, garlic, potato, tomato onion
- **Organic herbs and spices:** Chilli, peppermint, cardamom, turmeric, black pepper, white pepper, tamarind, ginger, vanilla, clove, cinnamon, nutmeg, mace;

- **Others:** Jaggery, sugar, tea, coffee, cotton, textiles.
- **Persp:** The economics of countries dependent on exports of these commodities are subject to changing conditions in the world market. Slow growth in the world demand, combined with increasing supplies from the main producing and exporting countries, which compete with one another, have led to declining and widely fluctuating prices in the markets for several commodities.

Organic Standards

The organic standards describe the specific requirements that must be verified by a USDA- accreditation certifying agent before products can be labelled USDA organic. Overall, organic operations must demonstrate that they are protecting natural resources, conserving biodiversity, and using only approved substances. A brief summary is provided below:

- **Organic crops:** The USDA organic seal verifies that irradiation, sewage sludge, synthetic fertilizers, prohibited pesticides, and genetically modified organisms were not used.
- **Organic livestock:** The USDA organic seal verifies that producers that met animal health and welfare standards, did not use antibiotics or growth hormones, used 100% organic feed.
- **Organic multi-ingredient foods:** The USDA organic seal verifies that the product has 95% or more certified organic content. If the label claims that it was made with specified organic ingredients, you can be sure that those specific ingredients are certifies organic and provided animals with access to the outdoors.

Certification Process

Certified organics refers to agricultural products that have been grown and processed according to uniform standards, verified by independent state or private organizations accreted by standards setting organization.

The certification processes actually start with conversion process. The steps to be followed

- Fulfil the requirement of conversion period
- Develop a comprehensive operations plan for the farm including an eco-system management plan
- Convert the farm completely or comply with strict regulation in the cases of split and parallel operations
- Farm follow up inspections by certifier

- Continuously update the chosen certifier on major changes.
- Keep track of inspection certificates issued by certifier

Procedure of Certification

- The certification process start with conversion process i.e requirement of conversion process must be fulfilled
- The farmer makes contact with agency. Certification agency provide information on standards, fees, application inspection, certification and appeal procedure
- Then producer then submit application along with field history, farm map, record keeping system etc

The cost of certification depends upon size of farm, type of production system, group of farmer, extent of animal; husbandry, location of unit, travel time to reached site, food and accommodation during inspection

Inspection

Inspector verify the organic practice such as long term soil management, buffering between organic farms and neighbouring conventional farmland land record keeping are being followed. The inspection procedure review the facility's cleaning and pest control method, ingredient transportation and storage and land record and audit control. He can take sample for laboratory testing and may conduct unannounced inspection. The inspector transmits his finding to the certification body as a written report.

Certification

Certification includes annual submission of an organic system plan and inspection of farm fields and processing facilities. The certification body cross the results of the inspector with the requirements of the of the organic standards. A certification committee decides whether certification may be granted or not and then the agency issue approval or denial of certificate. The certificate is given for current year's harvest only and hence annual certification is required.

Labelling

The label convey clear accurate information on the organic status of the product (i.e. conversion in progress or organic). The label for the organic and conversion in progress products should be distinguishable by different coloured labels. The details like name of product, quantity of the product, name and address of the producer, name of certification agency, certification, lot number etc, are to be given on label.

Lot number is helpful in tracking back the product particularly the field number which it is grown in case of contamination. For packing, recycling and reusable material like clean jute bag should be used. The biodegradable material should be used. Unnecessary packing material should be avoided. Organic and non-organic products should not be stored and transport together except when labelled.

Accreditation

In order to make sure that the certification programme is competent to carry out inspection and certification, a third level of quality control is needed. Accreditation is a process in which certification of competency, authority, or credibility is presented. Authorized bodies regularly evaluate certification programme and check their proper functioning according to certain criteria. In case of the certification body complies with the criteria, they accredit the certification programme.

Accreditation Agencies of India

In India 6 accreditation agencies approved by central government's Ministry of Commerce.

I. APEDA,

II. Spice Board,

III. Coffee Board

IV. Coconut Board,

V. Tea Board,

VI. Cashew nut Board

13.6 Benefits of Organic Farming

Organic farming tries to bridge the widening gap between man and nature with the following broader aims. However, the relative importance, for an individual or a community of farmers, may vary. Few benefits of organic farming as reported by Prasad and Gill (2009) which are presented below:

1. Sustainability of Natural Resources

Organic agriculture is a holistic way of farming, and besides production of goods of high quality, it primarily aims at conservation of the natural resources (soil, water, climate, bio-diversity and non-renewable energy) for sustainable productivity in agriculture. In the context of organic farming the term 'sustainability' is used in a wider sense, to encompass not just conservation of non-renewable resources (soil, water, minerals, energy, bio-diversity), but also the issues of environmental and social sustainability. The very basic approach to organic farming envisages to:

(i) improve and avoid over-exploitation and pollution of natural resources;

(ii) minimize the consumption of non-renewable energy sources;

(iii) exploit synergies that exist in natural ecosystems;

(iv) maintain and improve soil health by stimulating activity of soil organisms with organic manures and avoid harming them with pesticides;

(v) obtain optimum economic returns, within a safe, secure and healthy working environment; and

(vi) acknowledge the virtues of indigenous knowledge and traditional farming systems (ITKs).

2. Minimizing Cost of Cultivation

Organic farming is one of the environment-friendly approach of reducing dependency on external inputs and achieving the optimum productivity by making the best use of ecological principles and processes, leading to reduced costs of cultivation. This is very important for resource-poor farmers, especially for those who are operating in high risk-prone areas of dryland and rainfed agriculture.

3. Healthy Food

Healthy food means a food that is free from harmful chemicals and heavy elements, and is tasteful and nutritious. Nevertheless, the organic agriculture practices cannot ensure that products are 'completely free' of harmful residues, as they may possibly trespass into the organic production systems through general environmental pollution also. But this is one of the major aims of organic farming and all feasible methods are used to minimize pollution of not only farm products but also of surrounding environment, including air, soil and water.

4. Augmentation of Profits

Organically produced foods have a great demand in export markets, especially those of European and North American countries, and they fetch a sizeable premium as compared with conventionally grown farm products. In domestic sector also, the demand for organic food is increasing tremendously, especially among mid-and high-income segments, which has become more cautious about harmful effects of food grown with the use of pesticides and synthetic fertilizers, and potential hazards of environmental pollution caused due to modem practices in agriculture.

5. Improvement in Soil Health

As such the objective of soil health and production sustainability lies at the heart of organic farming and is one of the major considerations determining its acceptability. In organic production systems the soil health is maintained and improved through stimulation of activity of soil organisms with organic manures and by avoiding harming them with the use of synthetic pesticides as well as fertilizers. Some of the important features of organic farming in the context of soil health and environment protection are given below:

(i) Organically managed soils have a high potential to counter soil degradation, erosion and desertification, as they are more resilient both to water stress and nutrient loss.

(ii) Organic farming protects the long-term fertility of soils by maintaining organic matter levels, fostering soil biological activity, giving due importance to the basic principles of crop rotation and intercropping, and providing crop nutrients indirectly by using sources that are not readily soluble.

(iii) In organic farming, the supply of the nutrients is more balanced, which helps keep the plants healthy and improve the biological activity of soil, which in turn improves nutrient mobilization from organic matter and native soil reserves and minimizes the losses of nutrients, thus protecting the environment.

(iv) Micro-organisms have a good feeding base and create a stable soil structure.

(v) Inclusion of legumes or cover crops, mulching, intercropping and agro-forestry, plays an important role in protection against soil erosion and degradation.

(vi) Organic-agriculture technologies increase the organic matter content of the soil, which has a positive effect on soil aggregation and water-holding capacity.

(vii) Animal manure, green-manure and composts favour the composition processes and can replenish nutrients required by the crops and supply the soil with essential organic matter.

(viii) Legume crops are a highly valuable source of nitrogen. Closed nutrient cycles and efficient use of local resources (e.g. compost, cow dung or seeds) are especially important for subsistence farmers depending on few and limited assets.

(ix) Water-retention capacity of the soil is increased due to their higher level of organic matter content and permanent soil cover. Due to the resulting capacity for higher moisture retention, the amount of water needed for irrigation is reduced.

(x) Organic agriculture helps mitigate the climate change, because it reduces emission of greenhouse gases, especially nitrous oxide, because no chemical nitrogen fertilizers are used and nutrient losses are minimized.

(xi) Minimizes energy consumption by 30-70% per unit of land by eliminating the energy required to manufacture synthetic fertilizers and pesticides, and by using internal farm inputs, thus reducing the amount of fuel used for transportation.

(xii) Organic agriculture helps farmers adapt to climate change because it prevents nutrient and water loss through high organic matter content and soil covers thus making soils more resilient to droughts and land-degradation processes.

(xiii) Preserves seed and crop diversity, which promote natural cycles in the production systems. Maintenance of diversity also helps farmers evolve contingent cropping systems to adapt to climatic changes.

(xiv) Organic agriculture encourages natural habitats of agriculturally beneficial flora and fauna and forbids the clearances of primary ecosystems.

13.7 Major Limitations

Presently the organic food is produced to very limited extent in the State. Some of the factors identified for its lower adoption are as under:

1. **Bulky nature of organic manure:** Since the content of major plant nutrients in organic manure is very less (less than 1% N, 0.25% P_2O_5 and 0.5% K_2O) this leads to voluminous material needed. This also increases the transport and application expenses.
2. **Divergent nutrition value:** The divergent nutrition values of different organic manures confuse the farmers to replace the NPK requirements from different sources. The NPK content in FYM, green manure, crop residue, vermicompost are not uniform and depend largely on the method of storage/ preparation/ incorporation.
3. **Small farm holding:** In Punjab nearly 49% of farmers are small and marginal with low education status. Their farming is just subsistence type. They do not participate or come forward for following organic farming approach on account of its complex nature and resource endowed technology and risk of low yield especially during the initial years.
4. **Lack of awareness among farmers:** The farmers are not well acquainted about this concept of organic farming as well as organic manures therefore compost/FYM are improperly prepared which is the breeding place for many insect-pests. There is a need to ask the farmers about preparing enriched compost as well as to store FYM efficiently and as scientifically desired. Most of the work being got done from the hired (migratory) labour.

5. **Marketing of organic produce:** This is an important limitation because the consumer wants the authenticity of organic produce for paying the price premium. The farmers avoid the registration on account of high registration fee, complex and skill oriented registration procedure. This does not allow them to fetch the price premium. Moreover, there is no retail marketing of organic produce.

6. **Non-availability of produce certification Labs. for analysis:** No laboratory is available in the State to get the produce tested about its genuineness.

7. **Labour and cost intensive:** The enriched compost, vermicompost and green manuring are the labour and cost intensive prepositions which also discourage the farmers to make more investments on them.

8. **Low incentives from Govt:** The farmers are not aware about any scheme formulated by the State Govt. for extending monetary help for the growers while in the developed countries this facility is very common and easily accessible.

9. **Non availability of bio-pesticides:** With the extension of rice-wheat system on large area, the attack of insect-pests, diseases and weed bio-types (both traditional and newly emerging ones) have become very common and no specific bio-pesticide is available for controlling such serious maladies which make the farmers further averse to its adoption.

10. **Dramatic and large benefits in terms of yields and returns:** Such benefits cannot be expected in short-time from organic farming practices. The increase in the yield and returns commensurate with the resource structure can be gradually build up, stabilized and then sustained over a longer period.

11. **Cultivation of high nutrient responsive cultivars:** Most high yielding varieties/hybrids of cultivated crops developed in the past 3-4 decades are tuned to their high responsiveness to application of fertilizer. There nutrient demands are so high that meeting their requirement by organic sources alone is economically not feasible.

12. **Promotion of biological agents:** Many more efforts are needed to identify natural enemies, predators of some pests. The pests have to be managed by cultural and breeding methods which may involve more time and labour and above all the urgency of meeting food-security concerns which impel on realization of high yields without any let down which in effect means growing crop with use of high fertilizer inputs. Lot of insects and disease organisms attack field crops and at present very limited biopesticides are available and moreover there is a need to find out the effective biopesticides to make organic farming a success.

13.8 Different Biopesticides and Biofertilizers

Biopesticides

I. Delfin WG (*Bacillus thuringiensis*)- To control American bollworm (*Helicoverpa armigera*)

II. Spodocide 0.50% AS (Aqueous suspension) - S-NPV (Nuclear polyhedrosis virus) - To control Tobacco caterpillar (*Spodoptera litura*)

III. Helicide 0.43% AS - H-NPV (Nuclear polyhedrosis virus) - To control American Bollworm (*Helicoverpa armigera*)

IV. Niprot 0.50% WS (*Trichoderma viride*)

- For seed treatment@10 g/kg seed.
- For nursery bed treatment 5g/litre of water/m^2 area
- For soil treatment 1 kg/100 kg well decomposed FYM and keep for 5-10 days under polythene cover.

V. Neemjal 300 ml/ha- For sucking insects like white fly, aphid and jassid.

VI. Trichocards@ 100/ha (containing eggs of *Corcyra cephelonica* parasitized by *Trichogramma chilonis*) Used in rice, cotton and maize.

Pest Management

Bio pesticides obtained from plants like neem (*Azadirachata indica)*, pongamia (*Pongamia glabra*), tulsi (*Ocimum sanctum*), sorghum, marigold and bougainvillaea have been found effective against a number of diseases and insect-pests. Many neem based formulations like Neemark, Neemguard, Neemazal, Nimbicidin, Wellgro, Agricef, Neoconeem and Limnonol are available in market and are in great demand (Singh, 1993). Bio-pesticide agents are abundantly available in nature. Several pathogens including viruses such as nuclear polyhedrosis virus (NPV) and granuosis virus (GV), bacteria like *Bacillus thurienglensis*, *Veticillium* and protoaoan like *Schizogregrine* cause diseases in insects and destroys them. Similarly, several insect parasitoids (parasites thriving on insects) are also known in nature. *Trichoderma* is an egg parasitoid of several pests (Singh *et al.,* 1997). Selective microbial pesticides offer particular promise, of which strains of *bacillus thuringiensis* is an example (Palaniappan and Annadurai 1999).

Formulations (biopesticides) prepared from neem have shown satisfactory results against 300 pests-insects of agricultural, horticultural, vegetable crops, house hold pests, locusts, grass hoppers, caterpillars, stem borers, pod borers, beetle, jassids, aphids, mealy bugs, whiteflies, plant hoppers, fruit flies and mosquitoes. Neem formulations are environment friendly and also save farmer friendly insects

(Ghosh, 2000). *Goniozus, Elamus, Eriborous, Bracon, Trichopirillus, Tetratichus* and *Chelonus* are some of other parasitoids attacking insect pests. Insect predators such as *Crytorihinus, Pharoscymnus* and predatory nematodes like species of *Neoplectana* and *Heterorhabditis* destroy insects by causing diseases in insects and by predation. Fungi namely, *Trichoderma, Aspergillus, Chaetomium* and *Agrobacterium* are effective against different diseases (Kler *et al.*, 2001).

The use of organic source of nutrients (vermicompost and FYM) proved significantly superior in controlling the fruit borer in *bhindi* over chemical fertilizers (Surekha and Rao, 2000). These findings were in accordance with the reports of Choudhary and Kashyap (1987) on cotton boll worm and Venketeswara Rao (1989) on chilli pod borer and it was mainly attributed on account of regulated growth of crop plants which discouraged shelter to the insect-pest and encouraged predator with more perched places to feed on insect-pest because the canopy is not thick and avoid over shaded micro environment beneath the plants. It also avoids oviposition to the pests and as a result pest population remains less.

Biofertilizers

I. **Farm Yard Manure (FYM) :** FYM is the basic source of nutrients in organic agriculture which contains nearly all the macro as well as micro nutrients. Apart from supply of nutrients, FYM improves water holding capacity, infiltration and other physical, chemical and biological properties of soil. The FYM occupies prominent position among the bulky organic manures available in India with a potential supply of 13.99 million tonnes N, P_2O_5 and K_2O per annum (Manna and Ganguly, 2003) besides it improves the soil properties. Apart from FYM, many other sources such as vermicompost, neem cake, press mud etc. can also be used.

II. **Green Manure :** Green manuring has lot of scope in organic farming in order to supply nutrition to the crops in desired amounts. It is cheaper alternative to mounting price of fertilizer nitrogen and has become an effective technology in economizing the agricultural production system, ensuring productive capacity of soil without causing any impoverishment and combating many ecological and environmental problems.

III. **Crop Residues:** In organic farming, incorporation of crop residues is again of great significance for its success. The residues of all crops except grains should be incorporated to improve soil health. Beri *et al.* (2002) reported that 462.93 million tons of different crop residues were available in India which amounts to 2.5, 1.1 and 2.0 million tons of N, P_2O_5 and K_2O, respectively.

IV. **Jeev Amrit:** It can be prepared with the following ingredients.

Cow Dung: 10 kg, Gram flour: 2 kg, Cow Urine: 10 L, Soil : 1 kg, Jaggery : 2 kg

Make final volume 200 litres with water and for fermentation keep under shade for 3-5 days with stirring 2-3 times a day. In soil and foliar application treatment, four sprays of Jeev Amrit will be done along with soil application.

V. **Panchagavya:** In Sanskrit, Panchagavya means the blend of five products obtained from cow (all these five products are individually called 'Gavya' and collectively termed as 'Panchagavya'). It contains ghee, milk, curd, cow dung and cow's urine. Panchagavya had reference in the scripts of Vedas (divine scripts of Indian wisdom) and Vrkshyurveda (Vrksha means plants and ayurveda means health system). The texts on Vrkshayurveda are systematizations of the practices the farmers followed at field level, placed in a theoretical framework and it defined certain plant growth stimulants; among them Panchagavya was an important one that enhanced the biological efficiency of crop plants and the quality of fruits and vegetables (Natarajan, 2002). For making approx. 20 litres of panchagavya you need the following:

Fresh cow dung	5kg
Fresh cow's urine	3 litres
Cow's milk	2 litres
Cow's curd	2 litres
Cow's ghee	500 gms
Jaggery	500 gms
Water or sugarcane juice	3 litres
Ripe banana fruit	1 bunch (12 nos.)
Tender coconut water	3 litres
Toddy	2 litres

(If toddy is not available, you can ferment 3 litres of tender coconut water by keeping it in a pot for 1 week. That will become toddy. In case both these items are not available then 100 g brewing yeast can be taken).

Method of Preparation

- Take 5 kgs of fresh cow dung and mix it thoroughly with 500 gms of cow's ghee and keep it in a plastic drum or a mud pot. It should not be kept in a metal container because it will react with the metal. Keep the ghee and the cow dung mixture for 4 days, mixing it twice a day.
- On the 5th day add cow's urine, cow's milk (cow's milk can be boiled, cooled and then added) and cow's curd, then jaggery with water or sugarcane juice and banana fruit (which has to be mashed and mixed thoroughly). Then tender coconut water has to be added.

- Wait for another 15 days, stirring twice daily. Stirring the contents for about 20 minutes each time facilitates aerobic microbial activity.
- On the 19th day the panchagavya solution will be ready. This solution must be kept under a net, i.e. it must be covered with a muslin or fine cloth so that the common fly cannot sit on it and lay eggs. For use after 19th day, stir the mixture at least once a day to aerate it.

Application

200ml of this solution can be diluted with 10 litres of water with proper stirring for spraying on plants. For soil application, dilute 1000 ml in 10 litres of water. Periodicity of use once in 15 days, it can be used for all crops. In winter crops, 1.5-2% usage is sufficient but for all other crops 3% should be used.

13.9. Weed Management

Weed management techniques plays very important role in organic agriculture. So, for higher crop yields, weeds should be controlled very effectively. Since, no herbicide is to be used in organic farming and hence weeds create big problems for the success of organic crops. Following methods can be adopted for controlling weeds from organically cultivated crops.

1. **Use of Well Rotten Farm Yard Manure:** Farm yard manure (FYM) especially in raw form, contains huge amount of weed seeds. Since, the required amount of FYM is very high resulting in big problem of weeds in the fields where it is applied. The seed bearing weed plants which are fed to animals along with green fodder, the seeds of different weeds comes in intact form along with cow dung and when raw cow dung is applied in field all weed seeds germinate. On the other hand when cow dung is stored in proper deep pits, all weed seeds lose their viability due to development of high temperature in the pit (due to fermentation). Farm yard manure stored in open heaps is a big source of weed seeds and it is a wrong method of storing FYM. So, always use well rotten farm yard manure especially in organic agriculture in order to reduce the incidence of weed problem.
2. **Prefer Green Manuring Crops:** When 40 to 60 days period (interval) between harvesting of one crop and sowing of other crop is available, prefer cultivation of green manuring crops as compared to use of FYM. Green manuring crops stimulates the germination of weed seeds due to frequent irrigation and grown up weed plants are buried in soil before flowering along with green manuring crops. So, this technique in addition to supplying plant nutrients also reduces weed seed bank in soil as compared to FYM especially in raw form. Green manuring crops especially Sunhemp, Dhaincha, Cowpea etc can be sown after the harvest of *Rabi* season crops in

the month of April and are buried in soil before sowing of *Kharif* crops. A seed rate of 50 kg/ha can be used for Sunhemp and Dhaincha. For Cowpea a seed rate of 30-50 kg/ha can be used depending upon varieties. Any crop can be selected for green manuring purposes. Since, the availability of FYM is limited, so green manuring crops have high scope in organic farming.

3. **Dab System:** It is a technique in which sufficient time is provided for the germination of weed seeds and the germinated / germinating weed seedlings are uprooted before sowing the crop. Generally two pre-sowing irrigations are given before seed bed preparations, which stimulate all the weed seeds, present in the plough layer to germinate. Second option which holds good during winter season only is to leave the prepared field (after planking) as such for 8 to 10 days and after this period either plough up the field or do planking twice/thrice in order to kill germinated weed seedlings and to create soil mulch. Then sow the crop in moist zone. Adoption of this technique helps in significant reduction in weed population as compared to normal practice of crop cultivation.

4. **Use of Clean Crop Seeds:** Always use weed seed free crop seed in order to reduce weed infestations. Before sowing, sieve out the weed seeds, shrivelled and broken seeds with a proper siever. The seed crop should be properly maintained by uprooting weed plants and other off type plants. It is a preventive method of controlling weeds.

5. **Crop Rotation:** Crop weed association is very common and few weeds are associated with a particular crop year after year. Crop rotation is a non-monetary method of weed control. For instance *Phalaris minor* which is a big problem in wheat can be effectively controlled by rotating wheat with berseem crop. The plants of *Phalaris minor* before flowering and seed setting will be removed along with each cutting of berseem crop. The weed seed bank in soil will be exhausted within 2 to 3 years. Similarly during *Kharif* season, cultivation of rice / basmati rice will eliminate all perennial as well as *Kharif* season weeds of all crops except of paddy. Crop rotation has lot of scope for reducing weed intensities without any extra effort.

6. **Planting Pattern:** It has been observed that proper distribution of crop plants in the field have more smothering potential on weeds. Cross sowing or closer sowing with the recommended seed rate helps in maintaining proper distribution of crop plants in the field. For example, a significant reduction in dry matter of weeds can be achieved by cross sowing (22.5 cm x 22.5 cm) or closer sowing (15 cm) as compared to normal sowing technique (22.5 cm) with constant seed rate. Similarly in tillering crops as well as in fodder crops, weeds can be smothered by using 25% extra seed rate than the recommended level.

7. **Bed Planting:** Bed planting technique have lot of advantages over flat planting such as saving in irrigation water, no crop mortality due to excessive rains, efficient interculture with tractor, reduced crop lodging etc. Apart from these benefits the infestation of weeds is comparatively less in bed planting as compared to flat planting technique because majority of weed seeds are buried beneath the bed which are not able to germinate. Secondly, mechanical hoeing is possible without any crop injury in bed planted crops. This technique is very successful for the cultivation of wheat, maize, soybean, moong, mash, cotton, sugarcane, vegetable crops etc. Another advantage of bed planted crops is that hoeing and earthing up operations can be done side by side resulting in healthy plants with deep root system. Bed planted crop is less liable to attack of insect pests and diseases as compared to flat planted crops due to proper air circulation.

8. **Mulching:** Mulches are known for regularization of soil temperature, conservation of soil moisture, proper growth of soil microorganisms, efficient weed control, addition of organic matter, proper and quick crop germination etc. Residues of any crop can be used as soil mulch. Plastic mulching has big scope in orchards and wider row sown crops. Mulching is very common in sugarcane, turmeric, maize, soyabean, cotton, potato, chillies, brinjal etc. About 20-25 q/ha of mulch can be used. Mulching is very effective for controlling annual grassy and broad leaf weeds and their effect on perennial weeds is partial. The availability of mulches is a big problem, however this practice must be adopted on maximum possible area. In organic farming mulching have lot of scope for controlling weeds and improving soil fertility.

9. **Soil Solarization:** Soil solarization is possible on limited area. In this technique field is irrigated and is covered with black plastic sheet in such a way that no air circulation can take place. The weed seeds present in top soil layer will germinate and will die due to suffocation created with high temperature. This practice should be adopted before crop sowing and during hot months only. Crop sown without tillage (zero till) will have less weed infestation as compared to conventional tillage sown crop. However, this practice has limited scope.

10. **Hand Hoeing:** Hand hoeing is the only popular method of controlling weeds in organically sown crops. Weed problem is very serious in organic system of crop cultivation and generally 3 to 4 hoeings are required for controlling weeds depending upon their intensity. For hand hoeing khurpa, kasola, wheel hoe, spade etc can be used. However, in wider row sown crops or in bed planting, use of tractor mounted hoeing implements can be done. This method is very costly due to high wages of labour but farmers have no alternative for controlling weeds except hand hoeing. This practice must be adopted at proper time i.e. before critical period of crop weed completion.

11. **Uprooting of Weed Plants before Flowering:** The left over weed plants should be uprooted before seed settings so that future weed problems can be minimized. In organic farming, weeds are a big problem and effort must be made to reduce soil weed seed bank which is only possible by uprooting left over weed plants. In bed planted crops, weed plants can be easily uprooted without any damage to crop plants. The uprooted plants should not be left in the field but can be used as fodder or can be used for composed making.

Weed control is very expensive and difficult in organic farming system. Farmers should not depend on one method but must adopt integrated weed management approach as per their situation. Preventive methods such as use of well rotten FYM, Dab system and use of clean seed, cultural methods such as crop rotation and planting techniques as well as physical methods such as mulching, soil solarization, bed planting, hoeing, uprooting of weed plants etc. have their own significance. Weeds should be controlled properly so that good crop yield can be obtained

Transition Period and Crop Yield

The wheat yield initially decreased with residue incorporation (35.8 q/ha) in comparison to burning (43.7 q/ha) and removal (42.8 q/ha) of crop residues. However, after 4 years of gestation period, deleterious effect of previous residues incorporation on wheat yield was recuperated and highest wheat yield (49.2 q/ ha) was obtained under crop residues incorporation followed by burning (44.7 q/ha) and removal (43.0 q/ ha) of residues (Dhiman *et al.*, 2000). Change from conventional to organic farming takes 2-3 years for the establishment of same yield magnitude, which might be due to dynamics in soil fauna and flora. While working on a permanent site having organic versus chemical farming system at Punjab Agricultural University, Ludhiana, Kler *et al.* (2002b) reported that chemical farming gave higher yields of *Kharif* (maize/ soybean) as well as *Rabi* crops (wheat/*gobhi sarson*) during first three years of study as compared to organic farming. In the fourth year, the *Kharif* crops showed increasing trend in productivity (4.6%) towards organic farming than chemical farming, while *Rabi* crops did not show any difference. The maize crop raised with chemical sources of nutrition produced 78.4 % higher grain yield over organic farming (receiving 1/3 N each from farmyard manure, vermicompost and non-edible cake) during first year. The corresponding values during second and third year were 10.9 and 3.2 per cent, respectively. However, during fourth and fifth year the organic maize grown with organic sources of nutrition produced 7.8 and 20.9 per cent more grain yield as compared to chemical sources of nutrition.

13.10. Organic Farming and Soil Health

Use of organic sources of nutrients helps to conserve the soil health by maintaining the equilibrium of organic matter and soil micro flora ultimately helping to improve the physical, chemical and biological properties of the soil. Still better the soil can conserve the water more efficiently and ensure its availability as per the need of the plant. Recycling of crop residue will check the pollution hazards. Intensive cropping and imbalanced use of agrochemicals have resulted in severe depletion of soil fertility and environmental pollution. Organic materials improve soil properties after decomposition.

Physical Properties

Addition of organic manures, improves soil physical properties such as

1. Bulk Density and Soil Strength

Organic manuring with FYM significantly reduced bulk density (1.41 Mg/m^3) as compared with no manure (1.52 Mg/m^3) (Mishra and Sharma 1997). In Bangladesh, Hussain *et al* (2001) reported that soil bulk density was significantly lower in fields under ecological farming compared to those under conventional management. At Dapoli, Maharashtra, on clay loam soil bulk density, water holding capacity and water stable aggregates were significantly improved by FYM (5 t/ha), paddy straw (5 t/ ha) compared to poultry manure (5t/ha) and control (Ramteke *et al.*, 2001). Parkash *et al* (2002) also reported that soil bulk density undergoes more reduction with application of nutrients through FYM than through chemical fertilizers. The effect of FYM was however, more pronounced after second year due to accumulation of FYM.

Soil bulk density tended to decrease as organic carbon concentration increased in all (coarse, medium and fine) soil textural groups. Bauer and Black (1992) observed a linear relationship between soil bulk density and organic carbon content. More crop residues, higher organic matter content and better aggregation decreased bulk density. The physical properties like bulk density, soil strength, water holding capacity, aggregate stability were greatly influenced by the use of organic manures (Sur *et al.*, 1993). These properties make the rhizosphere congenial for the availability of plant nutrients.

Penetration resistance, a measure of soil strength tends to decrease with use of organic sources of plant nutrients, as was the case with soil bulk density. At Pusa, Mishra and Sharma (1997) observed that application of FYM at 10 t/ha before rice transplanting significantly reduced bulk density (1.41 Mg/m^3) and soil strength (2.32 M Pa) than no manure application (1.52 Mg/m^3 and 2.52 M Pa). The decrease in bulk density and soil strength has been ascribed to better organic carbon content and improved soil aggregation. Penetration resistance decreased with application of FYM in 0-5, 5-10 and 10-15 cm layers due to lower bulk density (Tiwari *et al.*, 2000). Ray and Gupta (2001) at New Delhi studied the effect of green manuring (*Sesbania aculeata*) before rice on coarse loamy soil (Typic Ustrochrept) and reported a significant reduction in soil strength (18.5 M Pa) over no green manure (19.2 M Pa).

2. Infiltration Rate

Higher infiltration rate with straw incorporation and farmyard manure treatments was observed than straw burning (Bhagat and Verma, 1991). More (1994) observed improvement in infiltration rate and soil aggregation due to application of different organic materials. Application of 100% NPK along with FYM increased the infiltration rate. This was an outcome of improved aggregation and subsequent changes in pore size distribution. Infiltration rate was increased with FYM (0, 5,15, 30 t/ha) application in the tune of 1.42, 1.57, 2.19, 2.61 cm/hr (Patil *et al.*, 2003).

It may be concluded that the long term use of organic manures, green manures and crop residues tends to decrease bulk density, soil strength and improves cumulative infiltration rate as compared to chemical fertilizers.

Chemical Properties

Continuous application of nitrogenous fertilizers has depleted soil organic matter, resulting into inherent loss of native soil N, available P, available K and lower production (Srivastava, 1998).

1. Organic Matter and Nitrogen

Incorporation of any green manure crop along with wheat or rice straw improved the organic matter content and physical properties of soil. The organic farming significantly ameliorated soil structure with an increased horizon depth, organic matter content, porosity, earthworm abundance and activity with coarser and better developed aggregates than the conventional farming. Dhiman *et al.* (2000) reported that organic carbon, available P and K were highest where an additional FYM application was made. Soil organic matter and soil total nitrogen levels were increased by about 120 per cent over 150 years in the manured plots. The organic carbon in soil can be improved by growing legumes for grain or green manure. Bhandari and Walia (2000) reported that burying of crop residues increased the organic carbon content of soil. However, the notable increase in organic carbon content, P and K values were observed in the presence of FYM followed by green manuring.

The organic manures results in gradual (slow) mineralization rates during the crop period with synchronization with crop uptake and results in reduced mineral N losses from the rooting zone during the growing season. In contrast, application of inorganic N led to substantial losses of soil mineral N during the growing season. Green manured plots will show longer periods of high mineral N levels compared with crop supplied with inorganic source of nitrogen. Rice residue incorporation may also depress levels of mineral N initially followed by sharp increase.

2. Phosphorus

Bhagat and Verma (1991) reported increased available P in straw incorporation + animal manure treated plots followed by animal manure alone and it was minimal under straw burnt and control plots. Ohno and Erich (1997) suggested that the management systems that return crop residues back into system, the soil might increase availability of P by decreasing the adsorption of P. The addition of rapeseed and soybean residues lowered water soluble P because of microbial immobilization and adsorption and precipitation process in soil (Beri *et al.,* 1995).

3. Potassium

Straw of cereals, particularly rice, generally contains large amounts of K (about 2.0%), about 75% of which is water soluble and its incorporation increase potassium content of soil (Bhat *et al.,* 1991). Addition of organic residues increased exchangeable and available K in the surface soil layers.

4. Micronutrients

The organic manures enhance the availability of the native soil Zn and are also a good source of other micronutrients which are effectively utilized by the crop plants. Use of organic manures alone or in combination with $FeSO_4.7H_2O$ has been found to be effective in correcting Fe deficiency in different crops. This was evident from the significant increase in available Fe status of soil with incorporation of *Sesbania* green manure before transplanting rice in coarse textured soils (Nayyar and Chhibba, 2000).

They also reported substantial increase in the DTPA-extractable Mn as a result of regular incorporation of *Sesbania* green manure over years in rice-wheat system on coarse textured soils of Punjab, owing to formation of complexes of this metal with organic acids provided during the anaerobic decomposition of these materials.

The combined use of organic materials (FYM/ wheat cut straw/GM) to compensate 50 to 75 per cent of N along with 25 to 50 per cent NPK dose through fertilizers increased the content of DTPA extractable Fe, Mn, Cu and Zn over application of 100 per cent NPK dose through fertilizers. The depletion of micronutrients in the soil (Fe, Cu, Zn and Mn) was more in the mineral fertilizer plots than in plots having both mineral and organic sources (Kumar and Yadav, 2003).

Organic farming treatments also resulted in substantial increase in labile C. The microbial biomass C was more with the addition of organic manures due to higher activity of dehydrogenase and phosphatase enzymes.

The after effects of green revolution advocated the necessity of organic farming due to over exploitation of the natural resources (land, water and vegetation), appearance of contaminants in food products and disturbed agro-ecosystem.

- The crop yield during first three years of organic farming remained low and started to increase thereafter. The long term studies clearly showed the significant edge of organic farming over chemical farming.
- The application of organic sources helped to improve the physico-chemical and biological properties of the soil.
- The synchronous nutrient supply at slow pace in the growing season decreased the insect-pests incidence.
- The quality of organic produce was better than that grown inorganically.

It is therefore, imperative to take up the organic farming on atleast 5-10 per cent of the cultivated area in the Punjab. It will create opportunities of organic food availability at premium prices and serve dual purpose of healthy food with conservation of resources.

14

Bioherbicides and Agricultural Biotechnology: New Approaches in Weed Management

According to Watson, bio-herbicides are living entities (natural enemies) used deliberately to suppress the growth or reduce the population of a weed species. This may include an insect; a microbe or a parasite nematode. These organisms operate directly or indirectly (by producing toxins) and their mode of action proves deleterious to the weed species. The latter toxins are termed bio rationals, phytotoxins or natural herbicides and are excluded in the definition.

Bio-herbicides are formulations containing plant pathogens capable of mass production *in vitro* that are applied directly to target weeds uniformly to kill or suppress their growth. This method employs the annual release of a bio-agent into specific weed infested fields to infest and kill susceptible weeds. At present only the potential use of fungal pathogens as mycoherbicide has been studied in depth. Exploitation of bacteria seems promising where use of viruses as biocontrol agents are problematic since they are not host specific and require vector for their transmission.

The utilization of biological weed control agents in weed management offers many advantages including

i. a high degree of specificity for the target weed

ii. no effect on non-target and beneficial plants or man

iii. absence of weed resistance development

iv. absence of residue builds up in the environment and

v. potential impact from biotechnological research and development.

Some drawbacks associated with the production and use of bioherbicides include:

i. Bioherbicides have to be registered with the Environmental Protection Agency (E.P.A.) and the registration process may be lengthy.

ii. Suppression or killing of weeds by bioherbicides may be a slow process.

iii. Stability of bioherbicides under field conditions is highly dependent on environmental conditions.

iv. Production of a bioherbicide for large-scale application may be an expensive process.

v. Numerous fungi need to be discovered and developed as bioherbicides because of the high degree of specificity of these agents.

Commercial bioherbicide first appeared in the market in the USA in early 1980 with the release of the product Devine in 1981 (Kenny, 1986) and of Collego in 1982 (Bowers,1982) and Biomal in Canada (Makowski and Mortenson, 1992). Success stories of these products and expectation of getting perfect analogues of chemical herbicides have opened a new vista of weed management.

14.1. Mycoherbicde Development

Mycoherbicide

A type of biological control of weeds when formulations containing plant pathogenic fungi are used at high inoculum level in the same way as chemical herbicide to control the target weeds.

Process of development

The process of development of an effective bio-herbicide has been found to have three steps (Prasad, 1993).

The discovery phase: The usual scenario began with the discovery of diseased weed species in the field; the causal agent was then cultured on artificial media, identified and re-inoculated back into the host weed. If the disease was reproduced, the same agent was re-isolated, the fungus was further tested as a potential mycoherbicide.

The developmental phase: It included testing of infectivity and pathogenicity under different environmental conditions, small scale field tests, host range studies, studies of fungus survival in the field, test of compatibility with chemicals and other treatments and evolution of herbicidal efficacy. Based on these indications, efforts were made for development of mass culture techniques, devising suitable formulation and application technology, testing for allergic or toxic effects on animals and further tests of efficacy, non-target effects and environmental fate.

The commercialization phase: During this phase a successful registration usually in co-operation with a private industry leading to mass production and marketing is done.

Criteria of selection of potential mycoherbicide

General requirements for selection of potential bio-herbicide are that pathogens must be able to produce abundant and durable inoculum (spores) in artificial culture, be genetically stable and specific for the target weed and be able to infect and kill the weed in an environment of reasonably wide latitude. While selecting potential bio-herbicide, high parasitism between saprophytes and obligate parasitism was also given importance.

Common restraints on pathogen

Although till date a good many products fungus has been reported as potential mycoherbicides, only a few were released as commercial products. Many of these fungi failed a few tests and were discarded. Common problems associated with pathogen were reported to be poor over wintering capacity, poor inoculum dissemination, requirement of long incubation period, low virulence and host resistance, inadequate host density and fastidious environmental parameters or sequences required by the pathogen for its efficacy.

14.2. Constraints in the Development of Bioherbicides

Bioherbicides have certain advantages. They are highly host specific, eco-friendly and beneficial to non-target species, free from chance of resistance development, non-persistent, cost effective, very promising in controlling resistant weed, mimic weeds, parasitic weeds, but certain factors have been reported to limit the development of bio-herbicides into a commercial product. Bruce and Louise (1995) have classified those factors under four categories.

1. Biological constraints

Plant populations: Host variability and host range: Within a population of host weed species, there will usually be a range of genetically diverse biotypes which may include some resistant types. It is also common to have different bio-types of micro-organism with slightly different host ranges (Auld et al., 1992). There is some evidence of a lack of clear relationships between plant phylogeny and pathogen specificity (Walker et al., 1991) making choice of species used in complex host range tests difficult. Besides, concerns have been raised in recent years regarding the potential for sexual and asexual gene exchange between bio-herbicide strains and strains attacking distantly related crop plants and natural exchange of genetic material between pathogen (Tebeest et al., 1992). Successful mating in culture and plant tissue of *Colletotrichum gleosporioides* f. sp. *aeschynemone*, the active ingredient of Collego, with strains of the fungus infecting winged water primrose (*Ludwigia decemens*) and *Cary illinoensis*.

Individual plants: Resistance mechanisms and assessment: There are myriad barriers to infection on individual plants, some inherent some induced including physical barriers such as hair and cuticle, biological barriers such as antagonistic phylloplane microorganisms and chemical barriers such as phytoalxins.

2. Environmental constraints

Epidemiology was also dependent on optimum environmental conditions. In the field, a high number of initial infection sites were generally essential for weed control (Yang and Tebeest, 1993).

Aerial environment: The prevailing environmental conditions were found to be limiting factors for biological control agents especially for fungi, for example requirement of particular dew period and temperature.

Soil environment: Moisture and nutrient status of the soil were reported to influence bioherbicide efficiency which affects the physiology of target plants and their interaction with aerially applied bioherbicide. Soil borne pathogens applied at pre-emergence stage of weeds were reported to tolerate environmental extremes and became persistent and effective to residual control but they usually had high virulence on a wide range of plant species (Jones and Honocko, 1990).

Aquatic environment: Charudattan *et al* (1990) have discussed the technical and environmental limitations regarding application of bioherbicides and maintaining their efficacy in water. They have listed the limiting factors to be oxygen concentration, temperature, salinity and diffusion of spore in water. Pathogens of aquatic weeds that have been tested as microbial pesticides include species of *Fusarium* and *Macrophomina*, on hydrilla (*Hydrilla verticillata* (L.F.) Royle), and species of *Acremonium, Colletotrichum, Fusarium, Pythium,* and *Phytophthora* for control of eurasian watermilfoil (*Myriophyllum spicatum* L.) but no promising control agents have been found among these isolates.

3. Technical constraints

Mass production: Submerged fermentation technology used in case of Devine, Collego and Biomal were not cost effective. Large scale solid substrate fermentation technology which has been available in industrialized countries is unfortunately not available for mycoherbicide en masse production. Selection or design of the growth in medium improvement of efficiency of submerged fermentation and stabilization process for propagules, and selection of type of fermentation systems were also suggested to be considered.

Formulation: Inadequate formulation has been one of the major constraints to the development of reliable and efficacious bioherbicides. Liquid formulation of Devine has a very short shelf-life of 6 weeks only limiting market area. Most challenging aspect of formulation has been to overcome the dew requirement which has also been reported to reduce the dosage of inoculum and cost. In this direction invert emulsion was found to have certain potential. However, some non-target damage has been reported due to its high viscosity.

4. Commercial limitations

Market size: Several efficient bioherbicide candidates remained underdeveloped for commercial use by industry because of their low market potential. Bioherbicides for economically important weeds, opportunities of cost recovery, size of foreseeable markets, availability of alternative chemical herbicides had been the determining factor of market size.

Patent protection and secrecy: The aim of the patent system is to promote science by recognizing and rewarding inventors but it has been found to delay the exchange of ideas and results among scientists.

Regulation: Increasingly restrictive and often unrealistic regulations for bioherbicide in North America have raised the cost of registration and discouraged investors. There is growing concerns amongst regulators regarding the possible production of mycotoxins by fungus e.g. *Phomopsis emieis, Fusarium spp.* used as bioherbicide.

14.3. Prominent Mycoherbicides

DeVine

Devine is a submerged fermentation product containing chlamydospores of the fungus *Phytophthora palmivora* MWV pathotype was registered during 1981. The liquid formulation is available in one-pint containers which must be kept refrigerated (2 to 8°C) until use. Devine is a mycoherbicide for the specific control of *Morrenia odorata,* strangler or milkweed vine, in citrus groves. This fungus will initiate a root infection in milkweed vine plants that kill the vine in two to ten weeks following application, depending on the size and maturity of the vine. The surface of the soil must be wet at the time of application. De Vine may be applied in any type of citrus grove from May to September after the weed has germinated or is actively growing. Apply Devine with a herbicide boom sprayer to achieve uniform coverage of the soil under the tree canopy. Use at least 124 lt./ ha of spray water.

Phytophthora palmivora infection of milkweed vine shows typical *Phytophthora* rot symptoms. Dying plants are girdled at the soil line and up to an inch above it. The infection, initially occurring at the soil line, progresses until it encompasses all plant roots. Infected plant roots slough the cortex, leaving only the stele when the plant is pulled from the soil. The root rot induces leaf wilt, the leaves wither and eventually fall from the plant. The fungus can be consistently isolated from the diseased plant roots. *Phytophthora palmivora* is not a mammalian pathogen and presents no imminent hazard upon human exposure. Liquid suspension formulated with shelf life of only six weeks was marketed on 'fresh milk' and 'made to order' concept. This soil applied mycoherbicide proved to be very effective (Kenny, 1986). The formulated material has a shelf-life of only six weeks and must be handled much like fresh milk.

Collego

The active ingredient of Collego is living spores of the fungus *Colletotrichum gloeosporioides* f. sp. *aeschynomene*. Collego is a two-compound product. Component A consists of a water-soluble spore dehydrating agent and component B is a wettable powder formulation of living fungal spores of *C. gloeosporioides* f. sp. *aeschynomene*. It is a selective post-emergent mycoherbicide for the control of northern jointvetch (*Aeschynomene virginica* (L.) B.S.P.) in rice and soybean. It was registered for commercial use during 1982 with a shelf life of 18 months. Collego should be applied to emerged jointvetch plants that are from 20 to 30 cm tall and have not reached the bloom stage. Rice fields should be flooded before application. Soybean fields should be irrigated just prior to application. Free moisture or relative humidity about 80% and air temperatures of approximately 26°C for at least 12 hours are necessary for development of the highest degree of infection.

Colletotrichum gloeosporioides f. sp. *aeschynomene* is an anthracnose disease that forms lesions on the above ground parts of the northern jointvetch. Lesions occur principally on the stems and once the stems are girdled, plant parts above the girdle collapse and die. Death of plants may not occur for 4 to 5 weeks after Collego is applied (Smith, 1986). No known hazards to humans or to the environment are attributable to Collego. Nontoxic to mallard duck, bobwhite quail, bluegill fish, channel catfish, crayfish and earthworm. Acute oral LD50, in rat 5000 mg/kg.

Biomal

The successful development of Collego, led to the discovery of another *Colletotrichum* based mycoherbicide Biomal by Philom Bios Inc., Sarkatoon, Canada. Biomal contain spores of *Colletotrichum gloeosporioides* (Penz.) Sacc. f. sp. *malvae*. It is used as post-emergence to control *Malva pusilla* (round-leaved mallow) in field crops of Canada and USA. It is applied with any field sprayer to the actively growing round-leaved mallow plants after the two-leaf stage and preferably before the weeds are 15 cm tall. Although Biomal is effective when applied at any stage of weed growth, control occurs at a slower rate in older and the more mature plants. The most effective stage of application is at an early seedling stage. The suspension should be agitated constantly during spraying to ensure that the spores stay in uniform suspension. The suspension should be applied within 3 h of preparation. High humidity for approximately 18-24 h after application ensures successful spore germination and infection. These conditions usually occur on overcast, humid days; during late afternoon or early evening; or when rain is imminent. Rainfall during or immediately after application facilitates spore germination. If available, sprinkler irrigation can be used to create more humid conditions after application.

Biomal infested round-leaved mallow plants show typical anthracnose disease symptoms. Lesions will form on the leaves, petioles and stems of infested plants within 2-4 h after application. As the disease progresses, the stems are girdled by lesions, resulting in plant mortality. The commercial Biomal formulation does not appear to have adverse toxicological effects on humans and the environment.

Agricultural Biotechnology

Agricultural biotechnology is a collection of scientific techniques used to improve plants, animals and microorganisms. Based on an understanding of DNA, scientists have developed solutions to increase agricultural productivity. Starting from the ability to identify genes that may confer advantages on certain crops, and the ability to work with such characteristics very precisely, biotechnology enhances breeders' ability to make improvements in crops and livestock. Biotechnology enables improvements that are not possible with traditional crossing of related species alone.

Uses of Agricultural Biotechnology

Genetic engineering: Scientists have learned how to move genes from one organism to another. This has been called genetic modification (GM), genetic engineering (GE) or genetic improvement (GI). Regardless of the name, the process allows the transfer of useful characteristics (such as resistance to a disease) into a plant, animal or microorganism by inserting genes (DNA) from another organism. Virtually all crops improved with transferred DNA (often called GM crops or GMOs) to date have been developed to aid farmers to increase productivity by reducing crop damage from weeds, diseases or insects.

Molecular markers: Traditional breeding involves selection of individual plants or animals based on visible or measurable traits. By examining the DNA of an organism, scientists can use molecular markers to select plants or animals that possess a desirable gene, even in the absence of a visible trait. Thus, breeding is more precise and efficient. For example, the International Institute of Tropical Agriculture has used molecular markers to obtain cowpea resistant to bruchid (a beetle), disease-resistant white yam and cassava resistant to Cassava Mosaic Disease, among others. Another use of molecular markers is to identify undesirable genes that can be eliminated in future generations.

Molecular diagnostics: Molecular diagnostics are methods to detect genes or gene products that are very precise and specific. Molecular diagnostics are used in agriculture to more accurately diagnose crop/livestock diseases.

Vaccines: Biotechnology-derived vaccines are used in livestock and humans. They may be cheaper, better and/or safer than traditional vaccines. They are also stable at room temperature, and do not need refrigerated storage; this is an important advantage for smallholders in tropical countries. Some are new vaccines, which offer protection for the first time against some infectious illnesses. For example, in the Philippines, biotechnology has been used to develop an improved vaccine to protect cattle and water buffalo against hemorrhagic septicemia, a leading cause of death for both species.

Tissue culture: Tissue culture is the regeneration of plants in the laboratory from disease-free plant parts. This technique allows for the reproduction of disease-free planting material for crops. Examples of crops produced using tissue culture include citrus, pineapples, avocados, mangoes, bananas, coffee and papaya.

14.4. Global Status of Biotech Crops

Introduction of biotech crops can lead to increase in productivity with decreased cost of plant protection measures. During 1996, the global area of biotech crops was 1.7 million hectares which increased to more than 125 million hectares during 2008 and this increase is about 74 folds as compared to 1996. The area under biotech crops further increased to 190.4 million hectares during 2019. (Table 14.1) Highest area of biotech crops (71.5 m ha) was in the USA which was followed by Brazil (52.8 m ha), Argentina (24.0 m ha), Canada (12.5 m ha), which is 37.6, 27.7, 12.6 and 6.6 percent of global area respectively. In India only biotech cotton is cultivated which occupies 11.9 m ha.

Table 14.1: Global Area of Biotech Crops in 2019: By Country (Million Hectares)*

Rank	Country	Area (million hectares)	Biotech Crops
1.	USA*	71.5	Maize, soybeans, cotton, alfalfa, canola, sugar beets, potatoes, papaya, squash, apples
2.	Brazil*	52.8	Soybeans, maize, cotton, sugarcane
3.	Argentina*	24.0	Soybean, maize, cotton, alfalfa
4.	Canada*	12.5	Canola, soybeans, maize sugar beets, alfalfa, potatoes
5.	India*	11.9	Cotton
6.	Paraguay*	4.1	Soybeans, maize, cotton
7.	China*	3.2	Cotton, papaya
8.	South Africa*	2.7	Maize, soybeans, cotton
9.	Pakistan*	2.5	Cotton
10.	Bolivia*	1.4	Soybeans

Rank	Country	Area (million hectares)	Biotech Crops
11.	Uruguay*	1.2	Soybeans, maize
12.	Philippines*	0.9	Maize
13.	Australia*	0.6	Cotton, canola, safflower
14.	Myanmar*	0.3	Cotton
15.	Sudan*	0.2	Cotton
16.	Mexico*	0.2	Cotton
17.	Spain*	0.1	Maize
18.	Colombia*	0.1	Maize, Cotton
19.	Vietnam*	0.1	Maize
20.	Honduras**	<0.1	Maize
21.	Chile**	<0.1	Maize, Canola
22.	Malawi**	<0.1	Cotton
23.	Portugal**	<0.1	Maize
24.	Indonesia**	<0.1	Sugarcane
25.	Bangladesh**	<0.1	Brinjal/Eggplant
26.	Nigeria**	<0.1	Cotton
27.	Eswatini**	<0.1	Cotton
28.	Ethiopia**	<0.1	Cotton
29.	Costa Rica**	<0.1	Cotton, pineapple
	Total	190.4	

Source: ISAAA 2019

*19 biotech mega-countries growing 50,000 hectares, or more, of biotech crops.

**Rounded-off to the nearest hundred thousand.

Among the Biotech/GM crops, soybeans occupied the highest area i.e. 95.9 mha during 2018 as compared to 94.1 mha during 2017 showing an increase of about 2 percent (Table 14.2). The area under maize was 58.9 mha during 2018 which is 1 percent less than 2017. The area under cotton and canola during 2018 was 24.9 and 10.1 mha which was 3% higher and 1.0% less than the previous year.

Table 14.2: Global Area of Biotech Crops, 2017 and 2018 by Crop Million Hectares

Crops	2017	%	2018	%	+/-	%
Soybeans	94.1	50	95.9	50	1.8	2
Maize	59.7	31	58.9	31	–0.8	–1
Cotton	24.1	13	24.9	13	0.8	3
Canola	10.2	5	10.1	5	–0.1	–1
Alfalfa	1.2	<1	1.3	1	0.1	5
Sugar beets	0.5	<1	0.5	<1	0.0	<1
Papaya	<1	<1	<1	<1	<1	<1
Others*	<1	<1	<1	<1	<1	<1
Total	189.8	100	191.7	100	1.9	1.0

Source: ISAAA 2019

*Others include biotech squash, potato, eggplant, sugarcane and apples.

15

Dryland Agriculture and Its Farming Systems

Dryland agriculture refers to cultivation of crops entirely under natural rainfall without irrigation. It is a form of subsistence farming in the regions where deficit of the soil moisture retards the growth of water consuming crops like rice *(Oryza sativa)*, sugarcane etc. Dryland areas are characterized by low and erratic rainfall and no assured irrigation facilities. Dryland agriculture is important for the economy as most of the coarse grain crops, pulses, oilseeds, and raw cotton are grown on these lands. Dryland areas receive annual rainfall between 500 and 1200 mm.

Types of Dryland Agriculture

Depending on the amount of rainfall received, dryland agriculture has been grouped into three categories:

Dry farming: It is production of crops without irrigation in areas where annual rainfall is less than 750 mm. Crop failures are more frequent under dry farming condition owing to prolonged dry spells during crop period. The growing season is less than 200 days. It is generally practiced in arid regions of the country

- **Dryland farming:** Cultivation of crops in areas receiving rainfall above 750 mm is known as dryland farming. Dry spell during crop duration occurs, but crop failures are less frequent. Semi-arid regions are included under this category.
- **Rainfed farming:** It is practice of crop cultivation without irrigation in areas receiving 1150 mm rainfall, mostly in sub-humid and humid areas. Here chances of crop failure and water stress are very less.

15.1. Area Under Drylands

Globally the area under drylands is about 6150 mha. In India out of the total cultivated area of 143 mha, the area under drylands is about 85 mha, which comes to 60% . It is estimated that even after creating entire irrigation potential for irrigation about 55% of total cultivated area remains as rainfed. Except in the

States of Punjab, Haryana and Pondicherry the percentage of area under drylands is high in all other. Among Indian regions, highest rainfall area is of north eastern region followed by semi-arid to arid central and southern region, cold and northern region. (Table 15.1)

Table 15.1: Dryland area in different regions of India

Region	States	Per cent of rainfed area
Cold and northern region	Jammu and Kashmir, Uttaranchal and Himachal Pradesh	60 to 81
Semi-arid to arid central and southern region	Madhya Pradesh, Maharashtra, Andhra Pradesh, Andhra Pradesh, Karnataka, Tamil Nadu	76 to 82
Sub humid to humid eastern region	Eastern Uttar Pradesh, Bihar, Jharkhand, Orissa, West Bengal	33 to 73
Humid to per humid north eastern region	Assam and north eastern hill States	Up to 90

Distribution of Drylands

Our country has fertile cultivable land and receives the highest rainfall on per unit area basis anywhere in the world due to short duration of rainfall in a year. One hundred and twenty eight districts in India have been recognized as dryland farming areas. Of these, 91 districts are spread in the states of Madhya Pradesh, Chhattisgarh, Uttar Pradesh and Tamil Nadu, representing typical dry farming tracts. Rest of the districts belongs to Central Rajasthan, Saurashtra region of Gujarat and rain shadow region of the Western Ghats.

India has about 108 million hectares of rainfed area which constitutes nearly 75% of the total 143 million hectares of arable land. In such areas crop production becomes relatively difficult as it mainly depends upon intensity and frequency of rainfall. The crop production, therefore, in such areas is called rainfed farming as there is no facility to give any irrigation, and even protective or life-saving irrigation is not possible.

Major dry farming crops are millets such as *jowar, bajra, ragi*, oilseeds like mustard, rapeseed, and pulse crops like pigeon pea , gram and lentil. Almost 80% of maize and *Jowar*, 90 per cent of *Bajra* and approximately 95% of pulses and 75% of oilseeds are obtained from dryland agriculture. In addition to these, 70% of cotton is produced through dryland agriculture. Dryland areas also contribute significantly to wheat and rice production. Thirty three per cent of wheat and 66% of rice are still rainfed.

Prospects of Dryland Areas

More than 75% of the peasants involved in dry farming are small and marginal. Therefore, improvement in dry farming would raise the economic status of farmers thus helping in poverty elimination. Dryland farming holds immense significance especially in the context of fluctuating food grain production and expanding population in our country. The biggest employer in our country, the cotton mills are fed by raw cotton grown mostly in dryland areas. Increasing production of cotton subsequently leads to increase in exports of cotton good. The expanding import of oilseeds is a cause of concern to Indian nation. The improvement of production of oilseeds in these regions will save valuable foreign exchange reserves. By enhancing the productivity of crops like *jowar, bajra* and *ragi* which are mainly grown in dryland farming would increase the nutrient consumption levels of our nation.

Marginal lands in the semi-arid regions offer potential for fodder production to feed the cattle population which is an integral component of farming practice of this region. Providing importance to these areas can solve the problems of pulses, oilseeds and cotton. The dryland areas have also tremendous potentiality of increased food grain production. Thus enhanced agricultural production in these areas would boost the agriculture dependent economy of India. Moreover it would also be helpful in eliminating the problem of hunger and malnutrition prevailed in below poverty line society of the country.

Constraints of Drylands

Drylands are characterized by low and uncertain rainfall therefore, crop failure is common feature. The various constraints of drylands includes:

1. In dryland areas in general, the rainfall is low and highly variable which results in uncertain crop yields. The distribution of rainfall during the crop period is uneven, receiving high amount of rain when it is not required and lack of it when crop need it.
2. Generally in dryland areas when the monsoon sets in late, the sowing of crops are delayed resulting in poor yields. At times, the rains may cease very early in season exposing the crop to drought and during flowering and maturity stages which reduces the crop yields considerably.
3. Soils of the drylands are not only dry but also deficient in macronutrients like nitrogen and phosphorous. Thus in other words drylands are not only thirsty but they are hungry too.
4. The temperature in dryland varies greatly. During the period of moisture stress and drought, the temperatures accelerate the crop development resulting into forced maturity. Chilling or frost injury at flowering results in poor grain setting and deteriorates the grain quality.

5. Dryland areas suffer from various process of soil degradation especially soil erosion.
6. Small size of land holdings (less than 2 hectares) usually fragmented and scattered, lack of market facilities, frequent crop failure, poor economic condition and other socio-economic problem related to drylands.
7. Extremely poor condition of farmers, lack of infrastructure to boost production.

Major areas of concern

Major areas of concern in dryland agriculture are:

- Proper marketing and price policy to cover crops and animal products.
- Conservation of soil and water resources.
- Need to evolve high yielding and drought resistant crop varieties.
- Low cost and locally suited agricultural implements.
- Judicious and balance use of costly chemicals.
- Proper financial availability to purchase inputs; and extension education.

15.2. Dryland Technologies

Reducing Evaporation Losses

Soil moisture is the most limiting factor in dryland agriculture. It is lost as evaporation from the soil surface and as transpiration from the plant surfaces. Evaporation has to be arrested as it is not directly related to productivity whereas transpiration can be reduced to some extent without affecting productivity of plants. The evaporation losses can be reduced by:

1. Mulches
2. Antitranspirants
3. Wind breaks
4. Weed control
5. Crop planning

1. Mulches

About 60 to 75% of the rainfall is lost through evaporation. These evaporation losses can be reduced by applying mulches. Mulch is any material applied on the soil surface to check evaporation and improve soil water. Application of mulches results in additional benefits like soil conservation, moderation of temperature, reduction in soil salinity, weed control and improvement of soil structure. Following types of mulches can be used.

(i) **Soil mulch or dust mulch:** If the surface of the soil is loosened, it acts as a mulch for reducing evaporation. This loose surface soil is called soil mulch or dust mulch. Intercultivation creates soil mulch in a growing crop.

(ii) **Stubble mulch:** Crop residues like wheat straw or cotton stalks etc., are left on the soil surface as a stubble mulch. The advantages of stubble mulch farming are protection of soil from erosion and reduction of evaporation losses.

(iii) **Straw mulch:** If straw is used as mulch, it is called as straw mulch. Mostly, by- products / waste materials of crops are used as straw mulch.

(iv) **Plastic mulch:** Plastic materials like polyethylene, polyvinyl chloride are also used as mulching materials.

(v) **Vertical mulching:** To improve infiltration and storage of rainwater in these soils, vertical mulches are formed. It consists of digging narrow trenches across the slope at intervals and placing the straw or crop residues in these trenches. The pruned plant material is placed in contour trenches formed between rows or in trenches around the plants in concentric circles each year in one circle.

2. Antitranspirants

About 99 per cent of the water absorbed by the plants is lost in transpiration. If transpiration is controlled, it may help in maintenance of favourable water balance.

Antitranspirant is any material applied to transpiring plant surfaces for reducing water loss from the plant. These are of four types:

- Stomatal closing
- Film forming
- Reflective
- Growth retardant

(i) Stomatal closing type: Most of the transpiration occurs through the stomata on the leaf surface.

Fungicides like phenyl mercuric acetate (PMA) and herbicides like atrazine in low concentrations serve as antitranspirants by inducing stomatal closing.

These might reduce the photosynthesis also simultaneously. PMA was found to decrease transpiration to a greater degree than photosynthesis in a number of plants.

(ii) Film forming type: Plastic and waxy materials which form a thin film on the leaf surface retard the escape of water due to formation of physical barrier.

Mobileaf, hexadeconol, silicone are some of the film forming type of antitranspirants. The success of these chemicals is limited since they also reduce photosynthesis.

The desirable characteristics of film forming type of antitranspirants are:

- They should form a thin layer,
- They should be more resistant to the passage of water vapour than carbon dioxide and the film should maintain continuity and should not break

(iii) Reflectant Type: These are white materials which form a coating on the leaves and increase the leaf reflectance (albedo). By reflecting the radiation, they reduce leaf temperatures and vapour pressure gradient from leaf to atmosphere and thus reduce transpiration.

- Application of 5% kaolin spray reduces transpiration losses.
- A diatomaceous earth product (celite) also increases reflection of solar radiation from crop canopy

Growth Retardants: These chemicals reduce shoot growth and increase root growth and thus enable the plants to resist drought. They may also, induce stomatal closure.

- Cycocel is one such chemical useful for improving water status of the plant
- Antitranspirants generally reduce photosynthesis. Therefore, their use is limited to save the crop from death under severe moisture stress. If the crop survives, it can utilize the rainfall that is received subsequently.

Antitranspirants are also useful for reducing the transplantation shock of nursery plants. They have some practical use in nurseries and horticultural crops.

3. Wind Breaks and Shelterbelts

(i) **Wind breaks** are any structures that obstruct wind flow and reduce wind speed while shelterbelts are rows of trees planted for protection of crops against wind. The direction from which wind is blowing is called windward side and direction to which wind is blowing is called leeward side.

(ii) **Shelterbelts** are planted across the direction of wind. They do not obstruct the wind flow completely. Depending upon their porosity, certain amount of wind passes through the shelterbelts while the rest deflects and crosses over the shelterbelts. It thus reduces wind speed without causing turbulence. The protection offered by the shelterbelts is dependent on the height of central tree row in the shelterbelts. Generally, shelterbelts give protection from desiccating winds to the extent of 5 to 10 times their height on windward side and up to 30 times on leeward side. Due to reduction in wind speed, evaporation losses are reduced and more water is available for plants. The beneficial effect of shelterbelts is seen more clearly in drought years. In addition, shelterbelts reduce wind erosion.

4. Weed Control

- Prompt weed control eliminates the competition of weeds with crops for limited soil moisture.
- Transpiration rate from weeds is more compared to crops.
- Effective weed control in dryland agriculture leads to increasing availability of soil moisture to crops. This is the most useful measure to reduce transpiration losses.

5. Crop Planning

Crop varieties for dryland areas should be of short duration through resistant tolerant and high yielding which can be harvested within rainfall periods and have sufficient residual moisture in soil profile for post-monsoon cropping.

(i) **Planning for weather:** Variation in yields and output of the dryland agriculture is due to the observation in weather conditions especially rainfall. An aberrant weather can be categorised in three types viz.

- Delayed onset of monsoon.
- Long gaps or breaks in rainfall and
- Early stoppage of rains towards the end of monsoon season.
- Farmers should make some changes in normal cropping schedule for getting some production in place of total crop failure.

(ii) **Crop substitution:** Traditional crops/varieties which are inefficient utilizer of soil moisture, less responsive to production input and potentially low producers should be substituted by more efficient ones.

(iii) **Cropping systems:** Increasing the cropping intensities by using the practice of intercropping and multiple cropping is the way of more efficient utilization of resources. The cropping intensity would depend on the length of growing season, which in turn depends on rainfall pattern and the soil moisture storage capacity of the soil.

(iv) **Fertilizer use:** The availability of nutrients is limited in drylands due to the limiting soil moisture. Therefore, application of the fertilizers should be done in furrows below the seed. The use of fertilizers is not only helpful in providing nutrients to crop but also helpful in efficient use of soil moisture. A proper mixture of organic and inorganic fertilisers improves moisture holding capacity of soil and increase during tolerance.

Nutrient solution spray is recommended in the event of revival of rain and release of moisture stress. Urea or DAP spray (2% solution) is useful for quicker regeneration of crops like legumes and castor after rain.

(v) **Rain water management:** Efficient rainwater management can increase agricultural production from dryland areas. Application of compost and farm yard manure and raising legumes add the organic matter to the soil and increase the water holding capacity. The water, which is not retained by the soil, flows out as surface runoff. This excess runoff water can be harvested in storing dugout ponds and recycled to donar areas in the server stress during rainy season or for raising crops during winter.

(vi) **Watershed management:** Watershed management is a approach to optimize the use of land, water and vegetation in a area and thus, to provide solution drought, moderate floods, prevent soil erosion, improve water availability and increase fuel, fodder and agricultural production on a sustained basis.

(vii) **Alternate land use:** All drylands are not suitable for crop production. Same land may be suitable for range/ pasture management and for tree farming and let farming, dryland horticulture, agro-forestry systems including alley cropping. All these systems which are alternative to crop production are called as alternate land use systems. This system helps to generate off-season employment in mono-cropped dryland and also, minimizes risk, utilizes off-season rains, prevents degradation of soils and restores balance in the ecosystem. The different alternate land use systems are alley cropping, agri-horticultural systems and silvi-pastoral systems, which utilizes the resources in a better way for increased and stabilized production from drylands.

To summarize following farming technologies are needed to enhance agricultural production in dryland areas.

- Timely preparatory and seeding operations including conservation of stored soil moistures.
- The use of improved crop varieties should be done which can withstand stress. For moisture conservation in the soil, deep tillage, surface cultivation and stubble mulching need to be practiced. Deep tillage is required to break plough soles and layers because repeated ploughing over centuries has resulted in the growth of hard compacted layers which restrict infiltration and movement of water and penetration of water.
- Conjunctive use of rainfall, surface and groundwater.

- Harvesting of water for use in dry periods. Watershed a natural hydrological unit is a good device for water harvesting. Proper watershed management can stop not only further degradation of ecosystem, but degraded lands can also be restored.
- Soil conservation by contour bunding, terracing, land sloping and land levelling and also by practicing conservational tillage (zero tillage and minimum tillage).
- Practice of drip irrigation to save water.
- Lining of canals to minimize water loss.
- Agronomic practices like mixed cropping and crop rotation which increase the yield of crops need to be practiced.
- Integrated nutrient management need to be practiced with special emphasis on use of bio-fertilizers to maintain the soil fertility.
- Integrated weed management and integrated pest management need to be adopted to control weeds and pests, respectively.
- For the non-farm operation dryland areas have to be supplemented with non- form occupation like animal husbandry, fisheries, poultry, social forestry and cottage for the development of these areas.
- Alley cropping, pasture management, tree farming, silvi-pastoral management systems and agro-horticultural system which are more relevant to dryland situations have to be adopted for successful dryland farming system.

Dryland areas constituting more than two-third of total arable lands in India are the chief contributor of pulses, oilseeds, coarse grain crops and cotton. Drylands also contribute significantly to wheat and rice production. Therefore, it is the need of the hour to adopt and practice the available dryland technology to maximum extent for the enhancement of agricultural production in these areas which would not only boost the food grain production of the country but would also improve the economic status of farmers in these areas.

15.3. Problems of Crop Production in Dryland

Dry farming crops are characterized by very low and highly variable and uncertain yields. Crop failures are quite common. These are mainly due to the following causes.

- **Inadequate and uneven distribution of rainfall:** In general, the rainfall is low and highly variable which results in uncertain crop yields. Besides its uncertainty, the distribution of rainfall during the crop period is uneven, receiving high amount of rain, when it is not needed and lack of it when crop needs it.

- **Late onset and early cessation of rains:** Due to late onset of monsoon, the sowing of crop are delayed resulting in poor yields. Sometimes the rain may cease very early in the season exposing the crop to drought during flowering and maturity stages which reduces the crop yields considerably
- **Prolonged dry spells during the crop period:** Long breaks in the rainy season is an important feature of Indian monsoon. These intervening dry spells when prolonged during crop period reduces crop growth and yield and when unduly prolonged crops fail.
- **Low moisture retention capacity:** The crops raised on red soils, and coarse textured soil suffer due to lack of moisture whenever prolonged dry spells occur due to their low moisture holding capacity. Loss of rain occurs as runoff due to undulating and sloppy soils.
- **Low fertility of soils:** Drylands are not only thirsty, but also hungry too. Soil fertility has to be increased, but there is limited scope for extensive use of chemical fertilizers due to lack of adequate soil moisture.

Importance of dryland farming in India

In India, the importance of dryland farming is increasing year by year. With the continuous growth in the size of population in India the gap between the requirements and the supply of agricultural output is increasing gradually leading to a crisis in respect of both food and non-food crops.

The World Bank in a study had observed that at the turn of the century, the total requirements of food would increase to about 235 million tonnes which would again require an additional production of food grains to the force of 65 million tonnes over the present level of production.

Prospects of Dryland Farming

With the present initiatives, the production of food grains would increase at best by another 35 million tonnes, leaving a gap of 30 million tonnes. Under this situation, the only way open to the country is to introduce and promote better dry-land farming which has a huge potential throughout the country.

Until recently, the development of technology for dry-land farming remained totally neglected. Considering the potential and also its importance, the Indian Council of Agricultural Research (ICAR) very recently evolved dry farming technology for its application in those areas which are experiencing inadequate rainfall even in the rainy season.

In India, a huge area of land remained fallow for want of assured water supply required for agriculture. Recently, the Planning Commission has made an estimate of such arid lands where dryland farming technology can be adopted.

As per this estimate, a total of about 37 million hectares of such arid lands, spread over 92 districts in different regions of the country can be brought under dry-land farming technology which could serve the twin purpose of controlling soil erosion, as well as increasing agriculture production and productivity.

In the meantime, the Eighth Plan has given much thrust on the improvement of dry- land farming more particularly for the production of pulses and oilseeds in which the country is deficient. Accordingly about 5,000 micro watersheds have been identified by the Government for adopting dry-farming technology.

In order to improve dry-land farming in the country, the Government proposed to follow an integrated and multi- disciplinary approach for the adoption and development of dry-land farming in the country. NABARD and cooperative banks have already expressed their willingness to extend financial help for running those viable dry-land farming projects in the country.

15.4 Status of Dryland Farming

Out of 14.3 million ha of net sown area in the country, rainfed agriculture is practiced in 85 million ha (67%). Nearly 60 m ha of rainfed area falls in the mean annual precipitation range of 500-1500 mm.

The average annual rainfall of the country is 1200 mm amounting to 400 mm of rain water over the country's geographical area (329 m ha). However, the distribution across the country varies from less than 100 mm in extreme arid areas of western Rajasthan to greater than 3600 mm in NE states and 1100 mm from east coast 2500-3000 mm in the west coast. The broad area of the summer monsoon activity extends between 30^0 N to 30^0 S and from 30^0 W to 16.5^0 E. the detail information on rainfall and monsoonal pattern in India has been summarized in the following Table 15.1

Table 15.1: Rainfall pattern in India

Season/Period	mm	Percent
Winter (Jan-Feb)	12	3
Pre-monsoon (Mar-May)	52	13
South-west monsoon (Jun-Sept)	296	74
North-east monsoon (Oct-Dec)	40	10
Total for the year	400	100

Rainfed farming comprises about 91% area of coarse cereals (sorghum, pearl millet, maize and finger millet), 91% pulses (chickpea and pigeon pea), 80% of oilseeds (groundnut, rape seed, mustard and soybean), and 65% of cotton. Also, about 50% area under rice and 19% area under wheat is rainfed.

During the last 25 years there had occurred significant changes in the area and yield of imported crops of rainfed farming areas. The area under coarse cereals decreased by about 10.7 million ha and most of this was under sorghum. The area under oilseeds increased by 9.2 million ha and most of this increase was due to irrigated rapeseed and mustard and soybean. The total area under pulses and cotton remained constant but more of cotton became irrigated and shifts in the area under occurred from one agro-ecological region to others. Area under chickpea in the northern belt decreased but increase in the central belt. This change occurred due to increase in area under rice-wheat cropping system which displaced chickpea and also pearl millet to a great extent and maize to a small extent.

According to the present concept, there are 128 districts in the country which face the problems of dryland, of these 25 districts covering 18 m. ha of net area sown with 10 % irrigation receive 375-750 mm rainfall annually spread over Central Rajasthan, Saurashtra region of Gujarat and rain shadow region of Western Ghats in Maharashtra and Karnataka. Twelve districts have irrigation covering 30-50% of the cropped area and do not pose serious problems. The remaining 91 districts covering mainly Madhya Pradesh, Gujarat, Maharashtra, Andhra Pradesh, Karnataka, Uttar Pradesh, parts of Haryana, Tamil Nadu etc., represent typical dryland area. The total net sown area in these districts is estimated to be 42 million hectares of which 5 m ha are irrigated. Rainfall in these districts varies from 375 to 1125 mm. therefore, more and more efforts are to be made for enhanced and stable production in these areas so that the recurring droughts do not stand in the way of meeting the growing food demands.

It is not that no attention has been paid in the country towards the development of dryland farming. Efforts were made right from 1923 to improve crop yields with the establishment of a research projects at Manjari in Maharashtra and later at Solapur, Bijapur, Raichur and Hagari in Deccan and Rohtak (Haryana) in the north. An All India Coordinated Research Project for Dryland Agriculture was launched by ICAR in 1970 in collaboration with Government of Canada and later Central Research Institute for Dryland Agriculture (CRIDA) was established in 1985 at Hyderabad. These projects generated technology, which, if followed, can bring marked improvement in cropping intensity, productivity and stability in production.

Limitations of Dryland Farming

In dryland agriculture, scarcity of water is the main problem. Apart from the low and erratic behaviour of rainfall, high evaporative demand and limited water holding capacity of the soil constitute the principle constraint in the crop production in dryland area. Yield fluctuations are high mainly due to vagaries of weather, often much behind the risk bearing capacity of the farmers. It is surprising to a layman that even humid areas with 2000 mm of annual rainfall not only suffer from moisture stress, but also face drinking water scarcity. Monsoon starts in the month of June and ends in last week of September or sometimes in the first week of October. Most of the rainfall is received during this period. With undulating topography and low moisture retention capacity of the soil, major portion of the rain water is lost through runoff, causing erosion and adding to the water logging of low lying areas. After the rain stops, very little moisture is left in the profile to support plant growth and grain production.

In dryland area deficiency and uncertainty in rainfall of high intensity causes excessive loss of soil through erosion which leaves the soil infertile. Owing to erratic behaviour and improper distribution of rainfall, agriculture is risky, farmers lack resources, tools become inefficient and ultimately productivity is low.

15.5. Improved Agronomic Practices for Dryland

The improved techniques and practices, which have so far been generated and recommended for achieving the objective of increased and stable crop production in dryland areas, have been summarized as follows.

Crop Planning

The farmers of the dryland areas, prior to the development of dryland techniques, were growing a crop either on rainwater in *kharif* or on conserved soil moisture during the winter. The crop varieties grown when moisture is sufficiently available. Such varieties have low genetic potential for yield. Selecting suitable crops and varieties capable of maturing with in actual rainfall periods will not only help in enhancing production of a single crop but in intensifying the cropping intensity. Many criteria have been laid out for selecting a crop variety for drylands. The capacity to produce a fairly good yield under limited soil moisture conditions is the most desirable criteria. The duration of *kharif* crops/varieties should not normally exceed the number of rainy days. In other words, crop varieties for dryland areas should be of short duration, through resistant tolerant and high yielding which can be harvested with in rainfall periods and have sufficient residual moisture in soil profile for post-monsoon cropping.

Under dryland agriculture determination of length of growing period (LPG) i.e., moisture availability of a given soil type, provides better index than total rainfall based crop planning. LPG is defined as the period when the moisture and temperature regions are suitable for crop growth and the period is determined by the FAO method. The LPG is computed as the sum of the period when P is more than 0.5 PET plus time taken to utilize stored soil moisture (assured 100 mm) after P falls short of PET. For example 'Nagpur' and Ratnagiri in Maharashtra receive mean annual rainfall of 1120 mm and 2500 mm, respectively but LPG determination indicates that both the places have LPG of 210 days in deep black soils. Therefore, both the places are suitable for single long duration on a short duration crop with a relay *rabi* crop.

Planning for Aberrant Weather

Dryland agriculture is subject to high variability in areas sown, yields and output. These variations are the results of aberrations in weather conditions, especially rainfalls. Delays in normal monsoonal pattern cause problems of timing and the organization of preparatory tillage and other initial activities for commencing cultivation processes for the season. Such monsoonal delays have repercussions on the programme of activities for the entire agricultural year. Even after the onset of monsoon and the commencement of planting, there may be monsoonal withdrawal causing moisture stress on plants and creating difficulties in the adoption and timing of approval cultural practices ultimately causing reactions in yields and outputs. Some crops are highly susceptible to such mid-season variations in moisture availability such as at the flowering stage in rice. Major crops like rice and maize get seriously affected if monsoonal rains cease early.

The need for modifying and introducing new technology for increasing and sustaining yield in dryland areas can hardly be overemphasized. Equally urgent is the need to decelerate and ultimately eliminate the process of damage to agricultural assets which are proceeding unbated in dryland areas. Erratic rainfall results in fluctuating production. This in turn leads to frequent scarcities, like the ones experienced in Indonesia and Vietnam in 1977 which created severe food shortages. Droughts in China in 1972, 1974 and 1985 brought depression of food grain production by up to about 25 million tonnes. Frequent droughts in India during 1966, 1968, 1972, 1974, 1979, 1982 and 1987 seriously affected the food and fodder production in the country. Hence, it is necessary to understand the distribution of South-West monsoon within the season to determine the extent to which the crop productions are likely to be affected by the vagaries of monsoon.

Several attempts have been made to understand the behaviour of South-west monsoon rainfall in different agro-climatic regions on the basis of historical rainfall records. These studies (Ramanna Rao, 1988) have brought out that

(i) there is large variation in dates of commencement of South-west monsoon from year to year in different parts of the country,

(ii) the monsoon rainfall is of sequential nature with long dry spells or breaks extends sometimes to the period of even one month or more,

(iii) there is large year to year variation in dates of withdrawal of South-west monsoon,

(iv) there is variation in quantum of rainfall received from year to year and

(v) high intensity rainfall occurs in association with movement of cyclones or depression resulting in sizeable loss of rainwater through run-off and deep drainage.

Thus, crop production in drylands fluctuates widely from year to year due to vagaries of weather. An aberrant weather can be categorized under three heads i.e.

(i) delayed onset of monsoon,

(ii) long gaps or breaks in rainfall, and

(iii) early stoppage of rains towards the end of monsoon season. Therefore, to mitigate such weather situations, farmers should make some changes in normal cropping schedule for getting some production in place of total crop failure.

Crop Substitution

Alternate crop strategies have been worked out for important regions of the country for *vertisols, alfisols, entisols, submontane* and *sierozemic* soils. Strategy has also been evolved for normal onset of rains, breaks in rains, early withdrawal, its uneven distribution; through selection of uneven crops/varieties which are inefficient utilize of the soil moisture, less responsive to production input and potentially low producers should be substituted by more efficient ones. Appropriate crops, suiting varying rainfall situations, have been identified for most of the dryland regions of India which are presented below in Table 15.2

Table 15.2: Traditional and alternate efficient crops in different dryland regions of India

S.No	Region	Traditional Crop	Alternate efficient crop
1	Deccan *Rabi* season	Cotton, wheat	Safflower
2	Malwa Plateau	Wheat	Safflower, Chick pea
3	Uplands of Bihar Plateau and Orissa	Rice	Ragi, Blackgram, Groundnut
4	South-east Rajasthan	Maize	Sorghum
5	North Madhya Pradesh	Maize	Soybean
6	Eastern UP	Kalitur	Chick pea
7	Sierozems of North-west India	Wheat	Mustard, Taramira

Crops which do not produce yield under normal rainfall years may not do so under abnormal years. Studies conducted in agro climatic conditions of Varanasi (eastern U.P.) revealed that under normal monsoon crops like short duration upland rice, maize, pearl millet, blackgram, greengram, sesame, pigeon pea etc. should be taken up on the basis of needs. These crops should be followed by chickpea, lentil, barley, mustard, safflower, linseed etc. on residual moisture during winter season.

If monsoon sets in as late as second week of July, short duration upland rice (variety - NDR-97 and NDR-118) may be included in place of Akashi and Cauvery. If the rains are delayed beyond the period but start somewhere in last week of July or first week of August and growing season is reduced to 60-70 days, then cultivation of hybrid pearl millet (NHB 3-4, B.J. 104), blackgram (Type 9), greengram (var-

Jagriti and Jyoti) may be included in place of T-44 and K-851 etc. should be grown. Yet another alternative could be to harvest a fodder of either pearl millet, maize, sorghum or a mixture of cowpea, blackgram and one of the above fodder crops.

In case monsoon rains stop early towards the end of season, normal sowing of short duration upland rice, blackgram and sesame may be taken up. If the rain stops very early, i.e. by the end of August or first week of September, only fodder crops or grain legumes could be harvested. Depending upon the soil moisture condition, relay sowing of crops like chickpea, lentil, mustard, linseed and barley could be done in rabi season.

Efficient Cropping System

Besides putting various measures to increase the productivity levels of dryland crops, efforts would also be needed to increase the cropping intensity in dryland areas which was generally 100%, implying that a single crop was taken during the year.

Cropping intensities of these areas could be increased by practice of inter cropping and multi cropping (sequential) by way of more efficient utilization of resources. The cropping intensity would depend on the length of growing season which in turn depends on rainfall pattern and the soil moisture storage capacity of the soil. For example in Indore region, receiving 1000 mm annual rainfall, only single crop can be taken on shallow soils, inter cropping in medium depth soils and double cropping on deep soils. Similar crop combinations have been identified for different regions of the country. In dryland of Varanasi region upland rice-chickpea/lentil sequence can be practices with advantage.

Inter cropping of vegetables with grain crops was pursued vigorously in some centers such as Varanasi and Phulbani. At both the places long duration pigeon pea was inter cropped with vegetables such as okra, radish and chilli. Such inter cropping systems would be very useful to get maximum returns from rainfall agriculture. Even at Solapur, leafy vegetables and some short duration beans were grown as intercrops during the rainy season.

Fertilizer Use

Soils of drylands in the country are not only thirsty but hungry also because these soils are severely eroded horizontally as well as vertically. Whenever efforts are made towards bunding and levelling of the fields in dryland areas, it is the surface soil which is removed. The resultant effect is that the fields are rendered shallow in depth and completely deprived of plant nutrients, particularly nitrogen, phosphorus and potassium. It is, therefore, necessary to apply all the three major nutrients in adequate amounts. Since soil moisture is limiting in drylands, the availability of nutrients becomes limited, attempt should always be made to apply fertilizers in furrows below the seed. If seed-cum-fertilizer drills drawn by bullocks or tractors are available, this objective can be fulfilled. There has been a belief among the farmers of dryland areas that use of fertilizer increases the chances of crop failure but recent findings have shown that the use of fertilizer is not only helpful in providing nutrients to crop but also helpful in efficient use of profile soil moisture (Table 15.3). If dryland farmers are shown such results, they will be convinced to use more and more fertilizers.

Table 15.3: Effect of N-levels on yield and moisture use efficiency (MUE) of barley and wheat (Varanasi Centres)

Nitrogen levels (kg/ha)	Grain yield (q/ha)	Total moisture use (mm)	MUE (kg/mm)
Barley			
0	14.05	133.7	10.5
30	20.45	136.3	15.0
60	30.00	142.3	21.0
90	37.20	141.6	26.3
Wheat			
0	9.55	145.5	6.6
30	13.55	144.4	9.3
60	18.35	153.6	11.9
90	24.15	155.1	13.6

In dryland areas, a proper mixing of organic and inorganic would be desirable. Organics have low nutrient content, but help to improve the moisture holding capacity of soils. In addition to yield advantage, nutrients like potassium help to increase drought tolerance by affecting plant-soil relationship. Transpiration losses are reduced and productivity per unit water increases.

Rainwater Management

Efficient management of rain water can boost agricultural production from drylands. The broad bed and furrow system of the International Crop Research Institute for the Semi-Arid Tropics (ICRISAT) for managing rainwater in vertisols made it possible to increase crop yields four to five times as compared to normal practice. However, this method could not be adopted widely by the farmers in India because it is costly and labour intensive. The vertical mulching developed at Bellary centres increases the infiltration of water in soil profiles and improves in situ moisture conservation. The scope for managing profile moisture is limited in alfisols but the surface run off in such soils can be reduced by ridge-and-furrow technique. Alternatively, application of compost and farm yard manure as well as raising legumes will add the organic matter to the soil and increase the water holding capacity.

The water which is not retained by the soil flows out as surface runoff. The run-off-recycling holds immense prospects in deep black soils where the seepage losses are much less. This runoff water, if not permitted to drain out safely, causes erosion. Therefore, safe disposal of excess water from the field drains to the disposal system should be planned properly. This excess runoff water can also be harvested in storing in dug out ponds and recycled and donated as in the event of severe moisture stress during rainy season or for raising crops during the winter.

Watershed Approach for Resource Improvement and Utilization

Watershed management is a holistic approach arrived at optimizing the use of land, water and vegetations in an area and thus, providing solution to alleviate drought, moderate folds, prevents soil erosion, improves water availability and increases fuel, fodder and agricultural production on a sustained basis. On the basis of the experiences of ICAR Operational Research Projects, which attracted the attention of our farmers, State departments, administrators and scientists, 47 model watersheds were established during the year 1983 for development, jointly by the Ministry of Agriculture, ICAR and various State Government Departments and Agricultural Universities, in 16 states and then the Department of Agriculture and Co-operation launched the National Watershed Development Project for Rainfall Areas (NWDPRA) covering almost the same states. Out of these 47 model watersheds, the Central research Institute for dryland Agriculture (CRIDA). Hyderabad has been entrusted with 30 watersheds. These activities were in micro and mini-watersheds covering 500-2000 ha. Major components in these model watersheds are:

(i) Improvement of water resources,

(ii) In situ soil and water conservation : rain water harvesting for safe disposal of surface runoff,

(iii) increase in cropping intensity and

(iv) alternate land use system for efficient use of lands as per land capability to provide stability in productivity.

15.6. Cropping Systems of Dryland

All drylands are not suitable for crop production. Some lands may be suitable for range/ pasture management, while others for tree farming, ley farming, dryland horticulture, agroforestry systems including alley cropping. All these systems which are alternatives to crop production are called as alternate land use systems. This system not only helps in generating much needed off-season employment in mono crop dryland but also minimizes risk, utilizes off season rains which may otherwise go waste as runoff, prevents degradation of soils and restores balance in the ecosystem.

Crop production may be disastrous in the years of drought, whereas drought resistant grasses and trees could be remunerative. Scientists of dryland have developed many alternate land use systems which may suit different agro ecological situations. These are alley cropping, agri-horticultural system and silvi-pastoral system which utilize the resources in better way for increased and stabilized production from drylands.

Alley Cropping: For imparting stability and providing sustainability to the farming system, a tree-cum-crop system will be one most appropriate for such situations. One such system called 'alley cropping' - a version of agro-forestry system, could meet the multiple requirements of food, fodder, fuel, fertilizer etc. Alley cropping is a system in which food crops are grown in alleys formed by hedge rows of trees or shrubs. The essential feature of the system is that hedgerows are cut back at planting and kept pruned during cropping to prevent shading and to reduce competition with food crops.

For example, fast growing leguminous trees such as *Leucaena leucocephala* or *Liliricidia* spp. are planted in rows. During the cropping season, trees are lopped at about 0.5 metre height. These loppings are used as much to reduce moisture loss and improve the nutrient status of soil. Arable crops like maize, rice, pearl millet, legumes, oilseeds etc. are planted in the alleys formed by the two rows of threes. This is also known as agrisilviculture. Alley cropping is also a form of conservation farming which enhances soil fertility and prevents erosion.

One very strong argument in favour of alley cropping is its ability to produce usable material even in years of severe drought. At Rajkot in 1985, rainfall received during the season was only 30% of the normal. There was total failure of grain production of the three legume crops tried in the system. In sole crop plots production was limited to 5.0 q/ha to 17.0 q/ha of green fodder. However, in alley cropped plots, *Leucaena* hedge-rows produced over 50.0 q/ha of green fodder.

Agri-horticultural System: Agri-horticultural system plays an important role in dryland areas, especially in semi-arid regions where production of annual crops is not only low but also highly unstable. Fruit trees if suitably integrated in dryland farming system could add significantly to overall agricultural production including food, fuel and fodder, conservation of soil and water and stability in production and income. Dryland fruit trees being deep rooted and hardy, can better tolerated monsoonal aberrations than short duration seasonal crops. Hence, in drought year when annual crops usually fail or their production is highly depressed, fruit trees species yield considerable food, fodder and fuel.

A suitable example of agri-horti-system is growing of cow pea/green gram/horse gram in inter space of *ber (Zizyphus mauritiaria)* at 6 x 6 m spacing at Hyderabad. Phalsa (*Grewia asiatica*) may be planted in between two ber plants in a row with a view to intensify the system. A well-managed dryland orchard of *ber* should give 50 kg fruits per tree/year. There should be 250 plants/ha for optimized production. The grow income would touch around Rs. 50,000/ha (250 x 50 x 4), assuming that one kg *ber* fetches Rs. 4. One could get an additional income of Rs. 800-Rs. 1000 from green gram/cow pea (2.5-3.0 q/ha).

Silvi-Pastoral System: This system is suitable for marginal drylands and is most preferable where fodder shortages are experienced frequently. The system essentially consists of a top feed tree species carrying grasses on legumes (preferable perennial) as understorey crops. Dryland farmers having larger holdings and keeping a land follow for a longer period for one reason or the other, should go in for this system which could provide both fodder and fuel. In a survey carried out in Andhra Pradesh, Karnataka and Maharashtra by CRIDA scientists, it was revealed that after food it is fodder which is of paramount importance for sustaining animal wealth in rural areas. In years to come, fuel will assume greater importance.

In August 1981 *Leucaena leucocephalla* was planted in contour trenches 7.5 m apart, the plant to plant spacing being maintained at 2.0 m at CRIDA. Four strips at upper reaches of plot (2% slope) were put under *Cenchrus ciliaris,* while lower four strips were seeded with *Stylosanthes hamata.* The system has come up very well.

Efficient Implements

In order to take full advantage of annual precipitation in dryland agriculture, higher doses of energy input is essential. Farmers in drylands have been using traditional and outdated farm equipment's which not only perform poorly but also demand a lot of energy and time and post-harvest operations. Farm implements can help to conserve as much rain water in situ as possible and to harvest rain water. Shallow off season tillage with pre-monsoon showers ensures better moisture conservation and lesser weed intensity. It has resulted in 20% yield increase in sorghum in Andhra Pradesh. Deep tillage helps in increasing water in soils having textural profiles and hard pan. This has resulted in 10% yield increase in sorghum and 9% yield increase in case of caster. For in-situ moisture conservation, land has to be opened so that it can cause hurdle to flow of rainwater. Tillage machines of appropriate size and type matching the power sources need to be used. Location specific seeders have been developed for dryland areas and these have shown good prospects and promise. A feature of these machines is that the seeds and fertilizers are placed in the moist zone of the soil resulting in a high percentage of seed germination and good crop vigour. In deciding farm mechanization in dryland areas, where farmers are generally poor, and their socio-economic condition should always be kept in mind.

The foregoing discussions show that technology of crop production in dryland areas have been generated to a great extent. What is important now is to view it in socio- economic context of the farmers. Once the technology is adopted by the farmers, the contribution of dryland areas to the total production can be sizably improved and the living standards of the farmers of these areas can be improved. This has been clearly shown in selected watershed areas and what is needed is to have more watersheds identified, proper technology to be developed and implemented.

Choice of Crops and Varieties in Drylands

Traditional cropping pattern in the dry farming areas is dominated by food grains viz., millets and pulses. In a predominantly subsistence type of farming system, such dominance of food crops is natural. The choice of crops for drylands is affected by

- Rainfall quantity and distribution
- Time of onset of rainy season
- Duration of rainy reason
- Soil characters including amount of rain water stored in the soil
- Farmer's requirements

The major focus of research under AICRPDA has been on the identification of most efficient crops for each dry farming region. The criteria for choice of crops for dry farming regions comprise the following

- Tolerance to drought
- Fast growth during initial period to withstand harsh environment
- Genetic potential for high yield
- Short or medium duration to escape terminal drought
- Adaptability to wide climatic variations
- Responsive to fertilizers.

For many dry farming regions of India, more suitable crops than existing ones have been identified. However, the acceptance and adoption of the practice of crop substitution by dryland farmers is poor since in most instances the new crops replace food crops (Table 15.4).

Table 15.4: Suggested crops in place of traditional crops

Region	Traditional crop	Yield (q/ha)	More suitable crop	Yield (q/ha)
Agra	Wheat	10.3	Mustard	20.4
Bellary	Cotton	2.0	Sorghum	26.7
Bijapur	Wheat	9.4	Safflower	18.8
Varanasi	Upland rice	28.0	Maize	33.8

Monocropping

In general a single crop is being taken in dry farming areas where the annual rainfall is below 500 mm. Rainy season crop is taken in light soils like alfisols, inceptisols and oxisols. Crops are being grown on residual moisture in vertisols in the post rainy season in winter season. The most common monocropping systems are groundnut in alfisols and associated soils in Rayalaseema region of Andhra Pradesh, sorghum or chickpea or wheat in black soils of Madhya Pradesh, Andhra Pradesh, and Karnataka

Intercropping

Intercropping refers to growing two or more crops in the same field with specific row arrangement during the same season. In areas where the annual rainfall is in the range of 600-850 mm, intercropping is being recommended and practiced. In such areas, at least one of the component crops succeeds in drought years and there is every chance to get two crops in good rainfall years. The different intercropping systems are followed in different regions of the country based on soil, category effective cropping season(weeks) in different areas Potential system < 20 weeks Bellary (8), Hissar (17), Jodhpur (11), Rajkot (17), Anantapur (13), Bijapur (17), Sole cropping 20-30 weeks Jhansi (21), Udaipur (22), Anand (25), Kovilpatti (21), Sholapur (23), Akola (27), Hyderabad (22), Agra (24), Intercropping > 30 weeks Bhubaneswar (32), Bangalore (36), Rewa (36), Samba (44), Varanasi (32), Hoshiarpur (35), Ranchi (45), Indore (36), Dehradun (44), Sequence cropping rainfall, market price and marketing facilities. Intercropping systems are subject to change due to change in market price of component crops. Intercropping is widely practiced in dry farming since it offers many advantages as outlined below.

- Intercropping is a risk minimization strategy and provides an insurance against complete crop failure due to rainfall abnormalities.
- It provides more yield and income per unit area per unit time than sole cropping
- Stability in production is achieved
- Multiple products for home consumption as well as for marketing are made available

- When legumes are included in intercropping, soil fertility is enriched
- Intercrop canopy suppresses weed growth
- Resource use efficiency is increased viz., light, water and nutrients efficiently used are
- One crop provide physical support to other crop.

Disadvantages

- Yield may be decreased due to competition effect
- Allelopathy effect
- Build-up of certain diseases and pests
- May not be suitable for large farms with adequate resources

15.7. Drought Management Strategies

1. Short-term Strategies

I. Ground-water utilisation

(a) Efficiency of the electric pumps is higher than the diesel pumps. However, because of the subsidised or free supply of electricity, the farmers do not care for the efficiency of the motors or pump-sets and look for cheaper options in the market.

(b) Uninterrupted supply of electricity: Frequent tripping of the supply was complained to almost all the teams who visited various States and interacted with the farmers. Frequent tripping leads to repeated irrigation of the same spot whereas rest of the field remained uncovered.

(c) Proper maintenance of the motors and pumping sets to reduce friction by way of greasing and other maintenance should be advised for efficient pumping.

(d) In case of rice, continuous standing of water is required only in the initial 15- 20 days so as to suppress growth of weeds. However, later on, irrigating one day after disappearance of water is the most economical and efficient way of scheduling irrigation.

(e) Sowing of cotton, soyabean, maize, etc. on the ridges and furrows and letting water in alternate furrows can save 20-30% water.

(f) Sprinklers for cereal crops like wheat and drip system for widely spaced crops sown in lines like sugarcane, cotton, maize etc. can give an efficiency of 80- 90%.

II. Rain water harvesting: About 11-37% run-off is generated even by the delayed monsoon and should be stored in the farm ponds or tanks. These will recharge ground water during normal or excessive rainfall year. Rain water stored in self-sealing or lined ponds can be used for irrigation if there is long break in the rainfall or for pre- sowing of the *rabi* crops to ensure proper germination.

III. Use of poor quality water

(a) Mixed and/or alternate use of limited good quality and underground poor quality waters for irrigation.

(b) Cultivation of salt tolerant varieties should be preferred.

(c) Farmers having residual sodium carbonate (RSC) rich waters should be encouraged to use this for irrigation judiciously after amending with gypsum or as soil amendment.

(d) Resorting to dry sowing of mustard followed by irrigation with saline water especially in light textured soils.

IV. Contingent cropping : Selection of crops, cropping sequences and agronomic practices are very important. Relatively more drought tolerant, deep rooted and short duration crops, varieties and cultivars are available for different agro-ecological and rainfall situations. If the rain is excessively delayed or main crop has failed cultivation or re-sowing with fodder is the best option. Fodders can be harvested at any stage keeping in view sowing of the next *rabi* season crop.

(a) Date bound contingency crop plan for rainfed areas

(b) Suitable early maturing varieties of different *kharif* crops

(c) Crop group/crop-wise contingency plan for *kharif*

(d) Contingency plan relevant to all crops and cropping systems

(e) Specific contingent plan for Boro or Summer Cropping

(f) Crops and cropping system-wise agronomic practices.

(g) Conserve water and soil

V. Arrangements of quality fodder

(a) Decrease area under paddy and increase area under maize, sorghum (jowar), pearl millet.

(b) As the sowing of main *rabi* fodder crops will start in October-November, catch crop of maize, bajra, sorghum, cowpea, bajra + cowpea, maize + cowpea and toria may be taken up after light showers during August-September.

(c) Dual purpose crops like barley may be sown in October. One cutting may be taken for fodder at 50-60 days after sowing and subsequent regenerated crop left for grain production.

(d) Oats may be grown in October as multicut fodder to ensure availability of green fodder for longer period.

2. Long and Medium Term Strategies

I. Securing good quality water in drought prone areas

(a) Networking of rivers, reservoirs, lakes and other water bodies existing in high rainfall areas which are prone to periodic flooding. Transferred water could be used for ground water recharging and to fill up dried lakes, water storage structures in dry areas whenever such necessity arises. This seems a viable option because the country receives more than one metre average rainfall in a year with lot of inter-regional variation.

(b) Less exploitation of ground water by resorting to low water demanding crops, introduction of precision micro-irrigation techniques such as drip and sprinkler methods in overexploited/ critical areas and matching water application schedules with critical growth stage concept may be prioritized.

(c) Diversification into less water demanding cropping systems. Vast ranges of options are available to make preferred choices.

II. Fine tuning of crops, cropping and farming systems: The strategy should include:

(i) relooking and upgradation of our knowledge about mixed cropping, intercropping, catch cropping, mixed farming and multi-strata cropping concepts;

(ii) promotion of agro-forestry, silvipasture, horti-pasture and silvi-horti-pasture systems etc. through large pilot scale demonstrations in farmer's participatory mode;

(iii) agronomic manipulations such as zero tillage, bed furrow irrigation, fertilization, adjusting spacing, soil and water conservation through mulching, use of anti- transpirants/Jal shakti etc.

III. Exploiting under-exploited and under-utilized plant resources: Large number of trees, bushes, shrubs and grasses are naturally growing as wild plants in one or the other rainfed region of the world. Such plants are adapted to rainfed situation and have tremendous potential to be exploited as a food, forage, fuel and/ or as industrial crop. Some of these plants have already been exploited in different parts of the world. For example, edible cactus (*Opunita spps*) is in countries like Mexico, Argentina, France, Brazil, Italy, South Africa and even in south western USA.

16

Efficient Irrigation Managament Practices

Water is indispensable for human, animal and plant life. It is a major component of all organisms some of which contain more than 90% of water. Water is a constituent of protoplasm. It maintains turgidity of cells which is essential for stomatal opening, expansion of cells and for different physiological processes. Water is also required for translocation of nutrients and dissipation of heat.

16.1. Irrigation

Irrigation is generally defined as the artificial application of water to soil for the purpose of supplying the moisture essential for plant growth.

However, a broader and more inclusive definition of the term irrigation is the application of water to the soil for any number of the following six purposes —

I. To add water to soil to supply the moisture essential for plant growth.

II. To provide crop insurance against short duration drought.

III. To cool the soil and atmosphere, thereby making more favourable environment for crop growth.

IV. To wash out or dilute salts in the soil.

V. To reduce the hazard of the soil piping.

VI. To soften tillage pans (big clods are soften for sowing).

In short irrigation is the science of harnessing and controlling various natural sources of water for the benefit of agriculture.

Sources of Water Supply in Agriculture

I. Main source - The basic source of surface water supply is precipitation. This is in form of rainfall, snowfall/hails/sleet.

II. Minor sources

(a) *Dew* - The water is absorbed by plant leaves.

- Reduce the loss of water due to transpiration.

(b) *Fog:* Water is absorbed through leaves of plants

(c) *High humidity*- Reduce loss of water and lower the demand of evapotranspiration.

III. Ground water - It is the water below the soil surface. It is available to plant by the phenomenon of capillarity action. The water table should not be low and also should not be too high so that supply of moisture continue and doesn't affect aeration.

IV. Flood water:-Only a small quantity of flood water can be utilized by the crop but it higher the underground water table.

Necessity of Irrigation

I. It controls soil moisture and overcome the drought situation.

II. It makes multiple cropping system possible.

III. It increases the crop yields because if irrigation facilities are available, we can grow long duration high yielding crop varieties instead of ordinary ones

IV. Degree of famines are reduced

V. Addition of manures and fertilizers are possible if water is available which ultimately increase efficient use of inputs.

VI. Labour will be engaged throughout the year.

VII. It is possible to grow crops like sugarcane, cotton, potato and vegetables.

VIII. It ensures the returns all around the year.

IX. It softens the soil clods and dissolve the plant nutrients.

X. It makes tillage operations possible (pre-sowing – for soil preparation, post-sowing- for intercultural practices).

XI. It helps in leaching down the excess salts.

XII. It improves the efficiency of soil applied pesticides and herbicides. Movement of herbicide is fast and reaches the lower layers of soil and is protected from sun.

16.2. Role of Water In Crop Production

Role of water in crop production can be discussed under two main heads.

1. Ecological importance
2. Physiological importance

1. Ecological importance

The distribution of vegetation over the surface of earth is controlled more by the availability of water than by any other single factor. Regions where rainfall is abundant and fairly distributed over the growing season, have dense vegetation e.g. forest in hills. Where summer droughts are frequent and severe, forests are replaced by grassland. Further decrease in the rainfall results is semi-desert with scattered shrubs and finally deserts. Ecological significance of water can be discussed as -

I. Water as a factor affecting plant characters: In-spite of the limitations imposed by heredity, the morphological and physiological characters of plant are markedly affected by water balance in the plant system. The plant have developed varying characters to fit in varying water conditions. On one hand, plants have their habitat in wet conditions as Hydrophytes while on other hand, have their habitat in regions of extreme dryness, are called xerophytes and those having their habitate in moderately wet conditions are called mesophytes. While heliophytes are those plants which are capable of growing or surviving in water or water logged habitats.

Trophophytic plants are such deciduous trees which are alternately mesophytic and xerophytic. But none of the plants are as versatile as many species of algae, Lichens and Mosses which can survive very prolonged state of intense desiccation and again resume activity when water becomes available.

Characteristics of xerophytic plants

- Thick cuticle. (the outer corky or waxy covering of the plant)
- Rolling or folding of leaves of narrow or prickly leaves.
- Pubescent (hairy) leaves and stems.
- Extensive root system.
- Reduced number of stomata or sunken or covered stomata.
- Resistance to protoplasmic dehydration under high moisture stress conditions.

Hydrophytes have the following characters

- Poorly developed and less extensive root system.
- Large number of air cavities and more of the mesophyll cell
- Cuticle (soft cellular tissue) outside the epidermis is absent
- More number of stomata (an opening in the epidermis for the passage of gasses and water vapours).

II. Water as a factor affecting the soil properties: Water plays a major role in the geological processes leading to soil formation, its nature, properties and processes that govern their agricultural usefulness. Soils of dry regions are generally of light type as compared with those of humid and sub-humid regions.

Soil reaction is also controlled by water. Soils of hilly areas are generally acidic because of excessive leaching of salts by high rainfall. Soils of arid areas are generally alkaline. The availability of nutrients in a soil is also affected by water.

III. Ecological distribution of plants - Each species varies in its essential requirements for successful reproduction and growth. In respect of every habitat, however, there are optimum conditions and limits within which a plants species can exist and these are referred to as their 'range of tolerance'.

2. Physiological Importance of Water

Almost every plant process is affected directly or indirectly by the water supply. The importance of water in the physiological process of plants can be discussed as under:

I. **Constituent of protoplasm:** Water constitutes 80-89% of the fresh weight of the herbaceous plants and 50% of the woody plants. It is an important part of the protoplasm. Reduction of water content below some level changes structure and ultimately causes the death of protoplasm. The death is due to the fact that most of the organic substances in protoplasm including proteins and carbohydrates are hydrated in the natural state and if the water is removed, their physical and chemical properties are affected. Protoplasm of very few cells can survive if their water content drops below 10% of their saturation level. Most plants are killed if it is 30-50% below the saturation level.

II. **Water act as solvent:** Water has high dissolving power in which gases, minerals and other solutes are dissolved enter plant cells and move from cell to cell and organ to organ, on account of permeability of most cell walls and membrane in plant system; Water acts as a medium for movement of dissolved substance in plant system.

III. **Water act as reagent:** Water participates in many chemical reactions occurring in protoplasm. Hydrolysis and condensation reactions in which water is added, and removed from molecules are important in metabolic processes. Water plays important role in the interconversion of carbohydrates in the process of photosynthesis and respiration.

IV. **Water act as a buffer against high and low temperature:** By transpiration water cools down the vegetation. This is because water has high value of latent heat of vaporization.

At 100°C – 540 calories

By transpiration water evaporate from plant surface, much of the solar energy is utilized which otherwise would have given much heat to the plant and may not be tolerated sometimes. Because of the high latent heat of fusion, 1gm water at 0 °C releases 80 calories of heat during freezing. Therefore, it releases some heat during freezing because of this it protects the vegetation from low temperature injury.

V. **Water act as maintenance of turgidity:** Turgidity is essential for cell enlargement and growth and maintaining the form of herbrebaceous plants. The shape of cell, is dependent upon water. In turgid cells, the water accumulates in the vacuoles of the cell and builds up a pressure, which forced the protoplasm against the cell walls.

VI. **Water helps in germination:** Seeds absorb water for swelling and initiation of growth activities. Generally the quantity absorbed is double their weight. Most of the seeds remain dormant till adequate water is not absorbed. Seed germination is rapid when soil is at field capacity. Most seeds, however, germinate even when the water content of the soil is much lower, provided the other conditions are satisfactory. Seeds can absorb water from humid air and will germinate in saturated air.

Seed absorbs water by imbibition process-gets swollen and seed coat becomes week, permeable to gases-enzymes get activated, dissolved in water and diffused to stored food material (cotyledons). By enzymes, starch, fats and proteins are converted to sugar, amino acids and fatty acids translocated and dissolved-cells gain maximum size development.

16.3. Scheduling Irrigation

Irrigation scheduling is defined as frequency with which water is applied based on needs of the crop and nature of soil. It is total number of irrigations and their frequency required to meet the crop water requirement.

It can also be defined as scientific management technique of allocating irrigation water based on the indication of crop water requirement with an aim to achieve maximum crop production per unit of water applied over a unit area in a unit time. In irrigation scheduling, there are three major question:

- How much irrigation to apply?
- When to apply irrigation?
- How to apply irrigation?

Three major considerations which influence these questions are:

- Water needs of the crop
- Availability of water with which to irrigate
- Capacity of the root zone soil to store the water

Time of irrigation is most important concept of irrigation

Basic Principles of Scheduling Irrigation

The broad basic principle governing irrigation is that water should be applied at longest possible interval but before the crop receives a set-back in terms of economic yield. The quality of water should be such that there is no wastage of water as well as entire root zone is sufficiently supplied with soil water. Irrigation scheduling is a effective management of supplying water in accordance with crop needs. Criteria most suitable for scheduling irrigation vary from one situation to another. For example where water availability is less and expensive, irrigation should be applied at most critical crop stage to maximum crop production per unit of applied water. The criteria selected for irrigation should permit favourable crop yields and optimum use of water and proper attention to other factors involved. The scheduling of irrigation can be done on following basis:

(1) Plant basis

(2) Soil factors

(3) Climatological approval

1. Plant Basis

It is now widely recognised that plant growth is directly related to the water balance in plant tissues.

General appearance of plant

(a) **Colour of leaves:** A dark green/bluish green colour in gram, berseem and cotton show the deficiency of moisture while light green colour is generally indication of the adequate moisture and satisfactory growth. In wheat yellow colour show need for watering. The change in colour, however, may also be due to deficiency of certain nutrient especially Nitrogen or due to diseases etc.

(b) **Angle or posture of leaves or degree of wilting or rolling of leaves:** Grain crops indicate need for water by temporary wilting particularly during the warmest part of the day. Sugarbeet and clovers readily indicate need for water by temporary wilting.

(c) **Degree of opening and closing of stomata:** The loss of water from the intercellular spaces to the atmosphere through the stomata reduces the turgor of guard cells decrease the stomata opening. So the degree of stomata may be taken as basis of scheduling irrigation.

(d) **Exudation:** Exudation from topped plants depends directly on root pressure. Exudation from crop plants stops at soil moisture content above wilting point. eg: It is suggested by some workers that irrigation to cotton crop before the exudation from topped plants ceases permanently.

(e) **Use of indicator plant:** Growing indicator plant in association with main crop that exhibit water stress symptoms earlier than the main crop plants (e.g. Sunflower plant)

(f) **Plant growth– Leaf growth:** Thickness of sugar beet leaves, stem or trunk growth – shortening of internodes of sugarcane, cotton plants etc.

(g) **Leaf reflection and temperature:** Turgid cells reflect more of infra-red light and have low temperature as compared to fluccid cells.

Limitations

i. In fruit trees, it is difficult to detect need for irrigation on appearance basis.

ii. When there is a lag of water absorption to transpiration, plant develop water stress. If this lag further increases the leaves loose turgor and droop. Wilting is an expression of internal water deficit. Drooping is an indicator of deficit. Drooping occurs in some plants having parenchyma cells whereas plants having schernchyma cells do not exhibit drooping at all although they are wilted.

iii. At noon evaporation is more than absorption and wilting appears even if there is sufficient moisture in the soil and this basis thus frequency of irrigation will be high.

iv. If we schedule irrigation on basis of appearance of wilting, then the crop might have already been adversely affected.

Important stages of different crops

Rice – Tillering - 10-20 days after transplanting (DAT)

Maize – Tasselling - (55 ± 3 DAS)

Silking -(74 ± 3 DAS)

Wheat – CRI (20-30 DAS)

Gram – Pre-flowering (45 ± 3 DAS)

Flowering (70 ± 3 DAS)

Pigeon Pea – Flower initiation (75 ± 5 DAS)

2. Soil Factors

i. **Soil appearance and feel method:** Determination of soil moisture in the absence of more sophisticated method. Feel the soil with fingers. Examine whether bolls forms or not etc. Though long experience is required in this case.

ii. **Soil water content:** A common criteria. Allows depletion of various fractions of available water in the crop rooting depth. i.e. 25%, 50% and 75% prior to irrigating a crop. Water content can be determined by direct and indirect methods.

Direct methods: Thermo gravimetric method, Volumetric method and Neutron Scattering Probe.

Indirect methods: Tensiometer, Gypsum Block method.

3. Climatological Approach

The moisture in the soil is regarded as being a balance between what enters in it as a result of precipitation or irrigation and what amount is lost through evaporation and transpiration. Precipitation can be measured by means of rain gauge. It is not easy to measure the ET but the climatological approach provides a way to determine it. Consequently we can determine the daily moisture loss from the soil through ET and can compare it with daily rainfall. An irrigation schedule thus can be set up in a book keeping procedure. The moisture is the soil as a bank account ET will drain from it. We surely need to keep pace of the ET and restore by irrigation whatever, is not promptly returned by precipitation.

Day degree concept (Dass, 1936): In Hawii, he used this concept in sugarcane. He accumulated day degrees above 70°F daily and when the total day degrees exceeds 350°F give irrigation. In this case maximum temperature was to be taken. Infact it is not temperature only but its duration, radiation effects etc also counts.

I. Use of evapormetry: ET depends primarily on climatic factors such as temperature, radiation, wind velocity, relative humidity and so on. Each of these elements can be correlated to ET by crop plants. An evaporimeter is an instrument which integrates the effect of all the different climatic factors. It is relatively simple, cheaper and thus more useful in irrigation practices but require a good knowledge of soil moisture relationship.

a. US Weather Bureau Class I Pan – for measuring Evaporation

b. Sunken screened Pan Evaporimeter – for estimating ET

These evaporimeters give close co-relation of the values of consumptive use of water by field crop and evaporation from the evaporimeter. An even simpler evaporimeter can be constructed for field use which is called Pan Evaporimeter. This could be 1 kg capacity or more with a pointer inside at 1.5 cm below the brim to facilitate recording of water levels.

16.4. Methods of Irrigation

Irrigation method refers to the manner in which irrigation water is applied to the field. There are four principal methods of irrigation:

1. Surface Irrigation

Water is applied from the field ditches to the field without bunds to guide the flow, very inefficient method as only small per cent of water is used by the plants. The rest goes as waste in the form of run-off, seepage and evaporation.

Advantages

- Installation or operation cost is low.
- No interference with the use of farm implements.
- No risk of damage by live stock.

Disadvantages

- Excessive loss of water.
- Uneven distribution of water.
- Soil erosion losses.
- Fertilizer and farm yard manure gets eroded.
- Loss of N by deep percolation.
- Harmful to crops sensitive to water logging.

2. Border Strip Method

It consists of dividing the field into a number of strips separated by low ridges. Water turned into the upper end of each strip flows slowly towards the lower end in a thin sheet guided by the ridges. Depending upon the situation, strips may be straight or contour. The strip has a uniform gentle slopes (0.05–5%) in direction of irrigation.

Width: It depends upon size of stream and varies from 3–15 meters.

Length: It depends upon

- Size of stream.
- Slope.

Advantages

- The ridges can be constructed cheaply with simple farm implements.
- Large area can be covered as compared to check method.
- Uniform distribution of water.
- High water application efficiency.
- Large irrigation streams can be used.
- Operation of the system easy and simple.
- Surface drainage can be provided.

Disadvantages

- Requires, more precise leveling than other methods.
- Large irrigation streams are required.

3. Check Basin Method

It consists of running water into relatively level plots surrounded by small ridges. It is the most commonly used method of irrigation in India. The check may be rectangular or ring or contour. In the rectangular check basin method, the field is divided into series of small rectangular plots surrounded by ridges to hold water within the compartments. The method is suited to nearly level areas. In this method there is usually one field distribution ditch for every two lines of check basins. It supplies water to the plots on either side.

The size of check basins depends upon

- Flow of water source
- Infiltration rate of the soil.
- The slope of field.

As the infiltration rate increases, as in light sandy soil, the stream size must be increased or the size of basin reduced in order to cover the area in a short period. If the size of stream is small, the basin size may also be reduced.

The size of check basins may vary from 1 meter square for vegetable growing to as large as one acre for growing rice under wet land farming.

Advantages

- It is familiar to an Indian farmer.
- It can be used for small discharges.
- Uniform application of water and no run-off losses.
- Efficient use of water.

Disadvantages

- High initial cost on labour in leveling.
- Much area is wasted in bunds and channels.
- Forbids the use of bullock drawn implements.
- Difficult to practice on the soil with slopes.
- Not suitable for crops sensitive to wet conditions.
- Impedes surface drainage.

4. Ring Basin Irrigation

Ring basin irrigation is suitable for orchards and other high value crop where size of the plot to be irrigated is very small. Variable ring basins are commonly used for irrigating fruit trees depending upon the tree spacing and land leveling.

Advantages

- No water is lost by run-off
- Rapid irrigation possible
- Almost no loss of applied fertilizer and farmyard manure through run-off.

Disadvantages

- Initial cost is high.

5. Furrow Irrigation

The water is applied by running small streams in furrows between the crop rows. The water soaks into the soil and spreads out into the crop root zone laterally to irrigate areas in between the furrows. Duration will depend upon the deficit and infiltration rates. The method is suitable for crops like maize, potato, sorghum, sugarcane, cotton, tobacco, potato, pulses, oilseeds and vegetables. The method is good where crops are sensitive to water ponding like maize, potato and beans. The furrows can be

- Graded furrows
- Level furrows
- Contour furrows
- Corrugations
- Alternate furrows which require limited water and check salt accumulation.

Length: Depends upon

- Permeability
- Slope of furrow
- Size of stream

On a porous sandy soil the length is shorter than on clay soil. The length can be as short as 45 m on soils with infiltration rate of more than 5 cm / hour or as much as 300 m or more on heavy soils having infiltration rates less than 3 cm / hour.

6. Corrugation

It is commonly used for close growing crops such as small grains, vegetables as carrots and for pasture growing on steep slopes. Corrugations are very shallow furrows and are generally combined with border irrigations. This may be straight or contour.

7. Sub–surface Irrigation

In sub-surface irrigation, water is applied beneath the ground surface usually by creating and maintaining an artificial water table at some pre-determined depth (10– 100 cm) depending upon root characteristics of the crop, texture and layering of soil.

Moisture reaches plant roots through capillary movement.

I. **Natural:** Impervious layer, water is led into series of ditches dug upto the impervious layer which then moves laterally and wets the root zone.

II. **Artificial:** Raising water table by laying porous tubes where artificial water table could be maintained. It is suitable for irrigation of vegetable, crops and most of field crops, small grains and forage crops.

Advantages

- Maintenance is cheap.
- Efficient use of water.
- No evaporation losses.
- Soil ever fit for cultivation and weed removal.
- No wastage of area in bunds and channels.
- No Interference to farm machinery.
- Very suitable for ornamental gardens and high value crops.
- Weed seeds are not carried with water.
- Channels serve as drainage channels also.

Limitations

- Costly.
- Risk of soil getting saline.
- Neighbouring land damaged due to heavy seepage.
- Deep rooted crops difficult to grow.

In India, this method is used in areas around Dal Lake in Kashmir for vegetable production and in soils of Kuttanad (Kerala) for irrigation of coconut palms but on a limited scale.

8. Sprinkler Irrigation

Water sprayed into air and allowed to fall on land surface in a uniform pattern with rates lesser than the infiltration rate of the soil.

Advantages

- Applicable under odd conditions where other methods cannot be used.
- Very efficient in areas with excessive slopes.

- Suitable for irregular topography.
- Very effective method for erosive soil.
- Suitable method for excessively permeable soil.
- No harm to crop in excessively impermeable soil.
- Land preparation cost less.
- No wastage of land in channels.
- Quantity of water can be controlled.
- Rate of application according to infiltration rate.
- No deep percolation and run off losses.
- Efficient use of water.

Disadvantages

- High initial cost.
- High degree of technical skill.
- More evaporation losses on hot windy days.
- Water is required under pressure (0.5 to 10 kg/cm^2) and that increases the cost of pumping.
- Little scope under ordinary farming conditions, scope only under high value crops.
- Interference with use of inter–cultivation operations
- Not useful on ripening stage of plants having soft fruits
- Dirty water cannot be used.

Types of sprinklers

(i) Rotating Head System. (ii) Perforated Pipe System.

9. Drip Irrigation

This is also called trickle irrigation. The irrigation is given by using small diameter PVC tubing with devices called emitters or drippers at selected places near the base of plants. The losses by deep percolation and evaporation are minimum. The amount of water applied daily is equal to consumptive use and it keeps the soil at low tension. It was originally developed by Simca Blass a water engineer in Israel in 1959. The method is good for economizing on water use, where poor quality water is used and is gaining popularity in arid areas.

It has been found successful in musk melon, cucumber, vines and other orchard plants, sugarcane, maize, tomato etc. Salt distribution is better than sprinkler when poor quality waters are used.

Advantages

- No deep percolation losses.
- We can maintain good moisture.
- Economizing water use saving.
- We can maintain good quality waters.
- Fertilizer can also be applied along with water.

Disadvantages

- High cost.
- Rats and other animals may damage the system.
- Dirty water cannot be used.
- Does not work under very low pressure.

16.5. Irrigation Requirements of Crops

Wheat (*Triticum aestivum*)

Water requirement of wheat crop is 50-55 cm and irrigation requirement is 35-40 cm. Sow wheat after a heavy pre-sowing irrigation (10 cm) except when follows rice. In case wheat sowing is delayed due to late harvest of rice, the pre-sowing irrigation of wheat can be given to the standing rice 5-10 day (depending upon soil type) before its harvest except where the crop is to be harvested with combine. This practice advances the sowing of wheat by about a weak.

The first irrigation should be relatively light and be given about three weeks after the date of sowing to the normal sown crop and after four weeks of sowing to late sown crop. The subsequent irrigations are determined by the prevailing weather soil conditions. In general 4 to 5 irrigations at CRI, jointing, ear head initiation, milk and dough stages are required. However, irrigation at CRI (crown root initiation) stage is very important for getting higher yields.

Gram (*Cicer arietinum*)

Water requirement of gram is 25-30 cm and irrigation requirement is 17.5-25 cm. Where irrigation facilities are available give a heavy pre-sowing irrigation (*rauni*). It will ensure deep rooting for proper utilization of soil moisture. Afterwards, give one more irrigation between mid-December and end January depending upon the date of sowing and the rainfall. This irrigation reduces the incidence of wilt disease. In no case should this irrigation be given earlier than four weeks after sowing. If early rains are received delay the irrigation. Excess of irrigation results excessive vegetative growth and depresses grain yield.

Raya (*Brassica juncea*)

Water requirement = 25-32 cm. Irrigation requirement = 17.5-25 cm. Crop sown after a heavy pre-sowing irrigation (10-12 cm) should be irrigated 3-4 weeks after sowing. Irrigation at this stage promotes deeper rooting which helps in the use of water stored in the lower soil layers. If necessary, another irrigation at flowering stage may be given. If the crop is threatened by frost damage, the second irrigation may be given earlier. Taramira (*Eruca sativa*) crop can be raised without any irrigation.

Gobhi Sarson (*Brassica napus*)

First irrigation like raya. Second irrigation may be given at the end of December or beginning of January. The third and last irrigation be given during second fortnight of February. The crop should not be applied any irrigation, thereafter, as it lodges.

Sunflower (*Helianthus anus*)

Water requirement = 50-55 cm, irrigation requirement 45-50 cm. This crop generally requires 6-9 irrigations depending upon the soil type, rain and weather. Apply the first irrigation about a month after sowing and reduce the intervals during the subsequent months to 3 to 2 weeks during March and 8-10 days during hot summer months of April-May. Stop irrigations about 12-14 days before harvest. Ensure that crop does not suffer moisture stress at flower initiation and seed development stage.

Potato (*Solanum tuberosum*)

The first irrigation should be given immediately after sowing, as it ensures better germination. The potato crop, responds better to light and repeated irrigations. While applying irrigation, avoid the over-flooding of the ridges to avoid consequent hardening of the soil surface which interfere with germination and the development of tubers. Water requirement 70-80 cm, irrigation requirement 60-70 cm.

Berseem (*Trifolium alexandrinum*)

Water requirement = 120-125 cm, irrigation requirement =100-115 cm. Berseem is considered to be rice of winter. The first irrigation is important and should be applied early before the soil surface gets dried so as to obtain a good stand. On light soils, the first irrigation may be given within 3-5 days after sowing. Whereas on heavy soils, it may be delayed for 8-10 days. Afterwards, irrigate the crop at 15-20 days interval depending upon the weather consideration. For seed crop, apply frequent irrigation during seed formation and ripening.

Sugarcane (*Saccharum officinarum*)

Water requirement = 150-160 cm. Irrigation requirement = 120-130 cm. Because of dry hot season, April to June is the most critical period for the growth of sugarcane. During this period, irrigate the crop at 7-12 days intervals. During the rainy season, adjust the frequency of irrigation according to rainfall. Drain away excess water from the sugarcane fields if they got flooded during the rainy season.

During winter (November to January), irrigate the crop at monthly intervals. To save the crop from frost, apply one irrigation in mid-December and another in the first week of January.

Autumn Sugarcane

First irrigation one month after planting, followed by three irrigations upto February and subsequent irrigations as per the recommendations for the spring crop. Other recommendations are the same as for the spring crop.

Cotton (*Gossypium arboreum*)

Water requirement = 75 cm. Irrigation requirement = 40-55 cm. Cotton requires four to six irrigations, depending upon the seasonal rainfall. The first irrigation should be given about one month after sowing and the subsequent ones at intervals of two to three weeks. The crop must not be allowed to suffer for want of water during the flowering and fruiting stages. Otherwise a lot of shedding of flowers and bolls will take place, resulting in low yields. Cotton during its early growth is very sensitive to water stagnation. To hasten boll opening, give the last irrigation to *Desi* cotton and American cotton by the end of September.

Rice (*Oryza sativa*)

Water requirement = 150-180 cm. Irrigation requirement 120-130 cm. Keep the water standing continuously in the crop for two weeks after transplanting so that the seedling establish properly. Afterwards apply irrigation two days after the pounded water has infiltrated into the soil. In this way, irrigation water can be saved without causing any reduction in yield. The depth of standing water should not exceed 10 cm. Drain away excess water before applying fertilizers, interculturing or weeding and irrigate the field after these operations. Stop irrigation about a fortnight before maturity to facilitate easy harvesting and the timely sowing of the succeeding *rabi* crops.

Maize (*Zea mays*)

Water requirement = 75 cm, irrigation requirement = 40-55 cm. Generally 4-6 irrigations are required, depending on the rainfall. Adequate water supply is essential throughout the crop season, particularly during the pre-tasseling and silking stages. Maize can tolerate heavy rains, provided water does not stand in the field for long period. Therefore, drain away excess water by making a drain of adequate capacity at the lower end of the field. Do not grow maize in fields from which it is not possible to drain away excess water.

16.6. Quality of Irrigation Water

The quality of irrigation water is determined by composition and concentration of salts in water. The different types of hazards are :

1. Salinity hazard
2. Sodium hazard
3. Boron hazard

1. Salinity Hazard

The relative tolerance of crops is as under:

Semi-salt tolerant crops : Rice, wheat, oats, jowar, maize, castor, turnip, senji, metha, berseem, cowpeas, tomato, cabbage, cauliflower, potato, onion, carrot, peas, cucumber, bitter gourd.

Salt-tolerant crops: Dhaincha, sugarcane, barley, tobacco, cotton, taramira, dubgrass, sugarbeet, spinach.

2. Sodium Hazard

Sodium is harmful when present in soil in exchangeable form, even at low concentrations. It causes adverse chemical and physical conditions. Sodium makes the moist soil impermeable to air and water and on drying it makes it hard and difficult to cultivate. The crust interferes with germination and emergence. Excessive concentration of sodium is directly toxic to plants. Sodium hazard is measured in different terms:

I. Sodium adsorption ratio (SAR)
II. Sodium percentage
III. Residual sodium carbonate (RSC)

3. Boron Hazard

All natural waters contain boron in minor concentrations. Though boron is an essential nutrient for plant growth but is toxic if present in soil solution more than a few ppm. In Punjab, 2 ppm limit is given for all soils. General classification of crops in respect of boron tolerance is as under:

Sensitive up to 1 ppm : Plum, grape, orange.

Semi-tolerant up to 2 ppm : Potato, sunflower, cotton, barley, wheat, maize, oats

Tolerant up to 4 ppm : Sugarbeat, alfalfa, onion, carrot, cabbage.

Improving poor quality water

i. Dilution with good quality water.

ii. Alternate irrigation with good and poor quality water in different ratios.

iii. Treatment with amendments: Gypsum is either mixed with water or gypsum powder is sprinkled on soil surface. It is a source of calcium. The added Ca would tend to correct the SAR or reduce the HCO_3 by precipitation as $CaCO_3$. It reduces pH. All waters testing more than 2.5 RSC would bring forth the need of applying gypsum to the soil or in water.

iv. Growing salt tolerant crops and crop varieties.

16.7. Preventing Wastage of Irrigation Water

Following losses during conveyance of water are common.

Canals - 17%

Distributaries – 8%

Field channels – 30%

Different methods for preventing irrigation water losses are:

i. **Lining of water courses:** Helps in reducing/eliminating water losses.

ii. **Discharge:** It should be maximum, but easy to control.

iii. **Maintenance of water courses:** Water courses should be properly maintained. Properly cleaned from weeds and silt.

iv. **Size and shape of plot**: Not that big that there is wastage of water. No that small that wastage of land in bunds and channels. It should be rectangular rather than square.

v. **Land preparation for irrigation**: Land leveling is very important. There is need for use of scrapper and float. Float does micro-levelling. Planker does not perform that operation so efficiently as it just follows contours.

vi. **Use of water control equipment:** For regulation of water at the farm. The following equipment needs to be used.

(a) Head gate: For opening and closing of water supply

(b) Diversion boxes: For diverting water in different directions. They can be 2,3 or 4 ways.

(c) Check dams: Canvas check dams, iron check demands are useful

(d) Turn out boxes: To divert water from supply channel to field.

(e) Syphones: For conveying water from channels to beds or furrows.

(f) Spill tubes: Spill tubes made of bamboo, iron, can be used to take out water from secondary ditch to field.

(g) Drops A drop is used to discharge water in a channel from one land to a lower one.

vii. **Scheduling irrigation properly:** Adequate water leads to maximum yield and limited water leads to maximum water use efficiency.

viii. **Optimum quantity of water to be applied:** Different soils hold different quantities of water. Water applied in excess of that a soil can hold will go as deep percolation loss.

ix. **Tapping shallow water table:** Irrigations can be reduced under these conditions.

x. **Irrigation in relation to manuring:** The response to fertilizer is of higher order under irrigated conditions than unirrigated. Water use efficiency increases with fertilizer.

Crop Growth Under Excess Water Conditions

Causes of excess water

(i) Over irrigation

(ii) Seepage from canals, water courses

(iii) Excessive rains. The movement of excess water being restricted by canals, railway lines

(iv) Geological formations as rising of springs and

(v) Water coming from higher areas.

Effects on crop raised in excess water conditions

i. Soil aeration: Less oxygen as small amount of oxygen dissolved in water and extremely slow rate of gaseous diffusion. Gaseous exchange is confined only to top fraction of an inch of soil.

ii. Increase in CO_2 concentration.

iii. Decrease infiltration rate: With continuous standing water due to blocking of channels with decomposition of products from anerobic microbiological activity.

iv. Formation of incompletely oxidized and reduced organic compounds like methane, marsh gas, methyl compounds, complex aldehydes.

Oxidized form	Reduced form
Nitrate ions	Ammonia, nitrite, nitrous oxide and nitrogen gas
Sulphate ions	Sulphide ions or Hydrogen sulphate
Ferric ions	Ferrous ions
Manganic	Manganus
Phosphate	Phosphine

v. Accumulation of O.C. - N remain locked up.

vi. Lack of nitrogen - Fixation

vii. pH increases and decreases with submergence and drainage respectively due to sulphide and sulphate ions respectively.

viii. Toxic concentration of certain nutrients. Mn, manganous- toxic Ferric – Ferrous-toxic

ix. Leaching of nutrients

x. Degradation of soils to leaching of salts under prolonged submergence.

xi. Poor soil structure as conditions alternate wetting and drying do not exists.

xii. Accumulation of salts under water logged conditions.

xiii. Transpiration, absorption, photosynthesis, respiration and nutrient uptake are ultimately reduced.

xiv. Low temperature conditions.

Plant characters adopted to water logged conditions

i. Develop superficial rooting system by specialized roots functional in water. They are massed just below soil surface and supply sufficient O_2 because of their large absorbing surface.

ii. Deep penetrating roots on stems containing large intercellular spaces. Oxygen of shoot released in photo synthesis is used by the roots. Rice roots maintain oxidizing conditions around them due to O_2 excretion.

iii. Development of adventious roots – observed in beans, tomato.

iv. Many crops have roots with large intercellular spaces like bar-ley, oat, corn, tomato. These spaces increase by collapse of cells due to lack of O_2.

v. Inherent characteristics of the plant.

16.8. Agricultural Drainage

Agricultural drainage is the removal of the excess water, known as free water or gravitational water, from the surface or below the surface of farm land so as to create favourable soil conditions for plant growth.

Types of drainage

I. Surface drainage

The process of removing the excess water from the land surface is known as surface drainage. If the excess water saturates the pore spaces of the soil, the process of its removal by downward flow through the soil, is known as sub-surface drainage or internal drainage. The excess water that is drained away is the gravitational water which is not available for plant growth. Drainage does not remove the capillary water and there is no danger of over-draining the soil by placing the drains too close together.

Soil may need artificial drainage for either or both of the following reasons.

(a) Where there is a high water table and should be lowered.

(b) When excess surface water cannot more downward through the soil or over the surface of the soil.

A permeable soil seldom needs artificial drainage, except when there is high water table. Slowly permeable soils often need artificial drainage especially when land surface is level and rainfall is high.

Particle size affects the pore size distribution and drainage property of the soil. The finer the texture, smaller the pore size or greater the need for drainage. Conversely the need for removal of excess water by artificial drainage is reduced as particle size increase from clay to silt to loam to sand.

Platy structure of the soil indicates poor drainage whereas the granular structural offers favourable conditions for drainage.

Benefits of drainage

I. Drainage provides a better environment for plant growth. Excess water fills the pore spaces and air is excluded. The air is essential for root respiration, growth of beneficial soil bacterial etc.

II. Drainage improves structure, infiltration capacity of soil. Higher infiltration capacity reduces soil erosion.

III. Drainage promotes leaching of salts and prevents their accumulation in the soil.

Surface drainage is required in (a) heavy soils (b) shallow soils (c) No outlet available (d) Larger quantity rain water to be drained in short time.

The surface drainage is done through open ditches drainage system. They are like open channels, however, the capacity of drainage channel should increase downstream as against irrigation channel. The size of the drain is fixed in relation to the quantity of 'runoff' which varies with rainfall, soil type and of 'drainage coefficient'. The drainage coefficient is defined as the rate of drainage expressed as the depth in cm of water drained off from a given area in 24 hours.

II. Sub Surface drainage

Methods

i. Pole type

ii. Stone drain

iii, Tile drain.

Water management in saline soils

Points to be considered for water management in saline soils

i. Method of irrigation

ii. Amount and frequency of irrigation

iii. Planting practices in relation to irrigation system.

1. Methods of Irrigation

The method of irrigation depends upon many factors like:-

- Salinity status
- Types of salts
- Topography

- Soil characteristics
- Availability of water

Points to remember

- Flood irrigation method may be used where salinity is a serious problem for leaching, soluble salts like Ca, Mg are present.
- When sodium is in excess flooding to be avoided unless gypsum is added.
- Sub surface, method of irrigation is not suitable as it leads to further rise of salts.
- Levelling for uniform application of water so that leaching from whole field is also uniform.
- Higher salt content water should not be used. This may aggravate the problem.

2. Amount and Frequency of Irrigation

Under irrigation is harmful. Frequency should be more to keep the salts under check i.e. to reduce osmotic pressure. Total quantity of water should be more as some quantity is needed for leaching salts and some to meet the water needs of the crops.

Increasing soil permeability

When evaporation rate is more than soil permeability salt conc. increases. If low soil permeability is due to hard pan it may be broken with deep ploughing/ chiselling. If the permeability is low due to high exchangeable sodium, it can be improved by replacing excessive exchangeable sodium content with calcium by the application of gypsum.

3. Planting Practices in Relation to Irrigation

i. Selection of salt tolerant crops and varieties.

ii. Keeping soil moist at emergence stage. At germination stage special care is needed. Pre-emergence irrigation for better imbibitions of water. This will make surface soil moist, reduce soil strength and emergence with improved sprinkling irrigation at this stage will also be useful. This will keep down the salt concentration. If moisture content is less the salt concentration increases.

iii. Planting on the sides of ridges is better than on the top. As the salts move inside with movement of water.

iv. Planting on northern side. Utilization of solar energy for desalinization. Row direction evaporation will take place from southern side due to sun's rays and salts will move to that side. The northern side will be free of salts.

v. Alternate row irrigation and planting in those rows. The salts with irrigation water will move to the other side.

vi. Puddling- in rice - In alkali soil puddling is not needed.

vii. Increasing soil permeability.

16.9. Frost Damage

When the temperature of air falls below 0 °C the water vapours change into solid form i.e. frost.

Conditions favourable for frost

i. A prevailing stable air mass with cool surface temperature.

ii. Clear sky to allow loss of heat by radiation.

iii. Little or no surface wind to mix the cold air near the ground with the warmer air above.

iv. Lengthy nights.

v. High pressure area-air comes down from above.

Measures

i. Irrigation – temperature of air and soil is increased. 0°C → 80 Cal released

ii. Wind machines Big fans- air is put into circulation which does not allow the temperature to fall below 0°C.

iii. Haters or smookers - Paddy straw → air become hot

iv. Sprinklers - directly increase the temperature of air.

v. Polythene or Straw covers - in such a way they cover NW direction and exposed to SE direction (NW Cold wind) covers from NW e.g. tomato

Irrigation for Frost Protection

Frost can cause serious damage to plants. Frost occurs if the temp. falls to 32°F (0°C) or below after dew has formed. Dew formation depends on the R.H., which is the %age of moisture in the air at any temperature compared with the maximum moisture that the air could hold at that temperature. Remember that, as air temperature drops, the absolute amount of moisture the air can hold decreases when R.H. reaches 100%, dew is formed, if this dew freezes white frost forms.

Sometimes the air is so dry that dew is not formed at 32°F (0°C), however, plants still suffer frost damage (usually burning of leaves or flower damage), this type of dry frost damage is referred to as 'black frost'.

Winter burn is another form of cold damage. It occurs if the above ground parts of plants start active growth (break dormancy) while the ground is still frozen. The new growth requires water, but because the plants cannot remove water from frozen soil, a drought like condition is established that causes leaf burn, bud death and death of the entire plant if the conditions persist for a long enough period. Water absorption by plants reduces sharply below 10°C of soil temperature. Because of : (i) reduced root growth/activity, increased viscosity of water (twice viscous at 0°C than at 25°C), decreased permeability of cell membrane of root cells resulting into increased resistance to the entry of water and decreased metabolic activities of root cells.

Effect of frost on plant cells

If the temperature drops rapidly to below freezing, cytoplasm freezes and cellular organization is disrupted; enzymes may be precipitated and cell die.

Measures for frost protection

There are a few practices a crop producer can rely on to control temperature within a limited range in order to reduce damage from frost.

i. The use of large fans installed permanently on poles (wind mills) above orchards and vineyards. By stirring the air, they force higher, warmer air down into the crops atmosphere. The higher air can be as much as 5°F (2.8°C) warmer than the air surrounding crop. Such fans can increase temperature around plants from 2-5°F (1-3°C).

ii. Smudging (outdoor fire with dense smoke) can be used especially to warm air in orchards. Smudging requires some type of petroleum burner to serve as a heater. The heaters are lighted as the temperature approaches freezing (32°F or 0°C). Smudging can raise temperature in an orchard more than 7°F (4°C). In some areas, the burning of oil-soaked rags and old tires has been used in place of commercial heaters. Such smudge pots create some heat at tree level and form a dense pall of smoke which spreads over the surrounding area and reduces the loss of heat by radiation in much the same way that a cloud cover does. But during day a smoke cover retards warming at the surface of earth and also creates air pollution. More efficient heaters which give off less smoke are in common use. Crop residue (Paddy straw burning) such as crop straw can also be used to heat the air.

iii. Use of polythene or straw mulch can reduce the loss of heat from soil (advection) during night but it will also check gaining of heat by the soil during day. Sensitive vegetable and plantation crops can be protected from frost by covering them from N-W direction and keeping them exposed to S-W direction.

iv. Use of irrigations (light and frequent protective irrigation).

Frost protection by irrigation

Overhead irrigation can be used in two ways to guard crops against damage by frosts: by wetting the surface soil in advance, and thus releasing warmth from the ground to combat the frost and by continuously forming a protective ice coating on tender leaves and buds, so, that their temperature cannot falls to damaging level.

Advance anti-frost irrigation

This method makes use of the fact that a wet, compact mineral soil is a better conductor of heat than a dry, loose one. Hence, irrigated soil absorbs more heat on a warm day and releases more of the ground's heat on a cold night than dry soil. Soil temperature at a depth of 5 cm is observed to be higher by 2.5-4.0°C than air temperature, when the soil is at field capacity. These effects make the air immediately above wet soil warmer by about 1–2°F on a frosty night than it would be if left in a dry state. Hence, irrigation affords limited protection against night frosts for young crops. This pre- wetting technique is less applicable to organic soils, because their thermal conductivity is very low and wetting may not make an effective difference to temperature above the soil. Using conventional irrigation, an hour's application of 1/3 inch of water per acre every 3 or 4 days, while the risk of frost persists, should give useful protection in marginal frosts, but success depends upon the day time weather being sunny or warm, so as to add warmth to the soil. When frost occurs after a cool, overcast day, the pre- wetting technique is less likely to afford effective protection.

Continuous anti-frost sprinkling

This method of frost protection depends on keeping the plants warm by means of the latent heat released when water changes into ice (In the reverse process, when ice melts the same amount of heat has to be added to bring about the change). Fine misty sprays are desirable and the water has to be applied continuously and fast enough to make sure that it is continuously freezing on the trees and bushes. So long this is going on, their temperature cannot fall below freezing point (32°F). The sap in blossoms, buds and young tender leaves does not freeze until the temperature is several degrees below this, so they remain undamaged as long as fresh ice continuously to form outside them.

Sprinkler systems have been used for frost protection for several years on barriers, almonds, citrus, vegetables, fruits and flowers. Crops have been saved in temperature as low as -8°C by sprinkler irrigation. Sprays should be started at or above 30°F for frost protection. Water should be applied until the ice is melted.

17

Role of Plant Nutrition in Crop Production

Essential elements for plant growth

A large number of mineral elements are present in plants, but they do not require all these for growth. Out of the 107 elements in the periodic table, 90 elements are naturally occurring in plants and out of which only 17 elements are essential for plant growth.

Although common sense goes a long way in defining the concept of an essential element, Arnon and Stout in 1939, stated that an essential element should fulfil following criteria:

1. A deficiency in absence of the elements should make it impossible for the plant to complete its lifecycle.
2. Must not be replaceable by another element.
3. Must be directly involved in plant metabolism, that is, it must be required for a specific physiological function.

   To Arnon and Stout's three requirements for essential elements should be added a fourth:

4. The element must be required by a substantial number of plant species, not just a single species or two.

17.1. Mineral Elements

There are actually 20 mineral elements necessary or beneficial for plant growth. Carbon (C), hydrogen (H), and oxygen (O) are supplied by air and water. The six macronutrients, nitrogen (N), phosphorus (P), potassium (K), calcium (Ca), magnesium (Mg), and sulphur (S) are required by plants in large amounts. The rest of the elements are required in trace amounts (micronutrients). Essential trace elements include boron (B), chlorine (Cl), copper (Cu), iron (Fe), manganese (Mn), sodium (Na), zinc (Zn), molybdenum (Mo), and nickel (Ni). Beneficial mineral elements include silicon (Si) and cobalt (Co). The beneficial elements have not been deemed essential for all plants but may be essential for some. The distinction between beneficial and essential is often difficult in the case of some trace elements. Cobalt for instance is essential for nitrogen fixation in legumes. It may also inhibit ethylene formation and extend the life of cut roses. Silicon, deposited in cell walls, has been found to improve heat and drought tolerance and increase resistance to insects and fungal infections. Silicon, acting as a beneficial element, can help compensate for toxic levels of manganese, iron, phosphorus and aluminium as well as zinc deficiency. A more holistic approach to plant nutrition would not be limited to nutrients essential to survival but would include mineral elements at levels beneficial for optimum growth. With developments in analytical chemistry and the ability to eliminate contaminants in nutrient cultures, the list of essential elements may well increase in the future.

The use of soil for greenhouse production before the 1960s was common. Today a few growers still use soil in their mixes. The bulk of production is in soilless mixes. Soilless mixes must provide support, aeration, nutrient and moisture retention just as soils do, but the addition of fertilizers or nutrients are different. Many soilless mixes have calcium, magnesium, phosphorus, sulphur, nitrogen, potassium and some micronutrients incorporated as a pre-plant fertilizer. Nitrogen and potassium still must be applied to the crop during production. Difficulty in blending a homogenous mix using pre-plant fertilizers may often result in uneven crops and possible toxic or deficient levels of nutrients. Soilless mixes that require addition of micro and macronutrients applied as liquid throughout the growth of the crop, may actually give the grower more control of his crop. To achieve optimum production, the grower can adjust nutrient levels to compensate for other environmental factors during the growing season. The absorption of mineral ions is dependent on a number of factors in addition to weather conditions. These include the cation exchange capacity or CEC and the pH or relative amount of hydrogen (H+) or hydroxyl ions (OH-) of the growing medium, and the total alkalinity of the irrigation water.

Cation Exchange Capacity (CEC)

The cation exchange capacity refers to the ability of the growing medium to hold exchangeable mineral elements within its structure. These cations include ammonium, nitrogen, potassium, calcium, magnesium, iron, manganese, zinc and copper. Peat moss and mixes containing bark, sawdust and other organic materials all have some level of cation exchange capacity.

pH

The term pH refers to the alkalinity or acidity of a growing media water solution. This solution consists of mineral elements dissolved in ionic form in water. The reaction of this solution whether it is acid, neutral or alkaline will have a marked effect on the availability of mineral elements to plant roots. When there is a greater amount of hydrogen (H^+) ions, the solution will be acid (<7.0). If there is more hydroxyl (OH^-) ions, the solution will be alkaline (>7.0). A balance of hydrogen to hydroxyl ions yields a pH neutral soil (= 7.0). The range for most crops is 5.5 to 6.2 or slightly acidic. This creates the greatest average level for availability for all essential plant nutrients. Extreme fluctuations of higher or lower pH can cause deficiency or toxicity of nutrients.

17.2. The Elements for Complete Plant Nutrition

The following is a brief guideline of the role of essential and beneficial mineral nutrients that are crucial for growth. Eliminate any one of these elements, and plants will display abnormalities of growth, deficiency symptoms, or may not reproduce normally.

1. Macronutrients

Nitrogen: It is a major component of proteins, hormones, chlorophyll, vitamins and enzymes essential for plant life. Nitrogen metabolism is a major factor in stem and leaf growth (vegetative growth). Too much application can delay flowering and fruiting. Deficiencies can reduce yields, cause yellowing of the leaves and stunt growth.

Phosphorus: It is necessary for seed germination, photosynthesis, protein formation and almost all aspects of growth and metabolism in plants. It is essential for flower and fruit formation. Low pH (<4) results in phosphate being chemically locked up in organic soils. Deficiency symptoms are purple stems and leaves; maturity and growth are retarded. Yields of fruit and flowers are poor. Premature drop of fruits and flowers may often occur. Phosphorus must be applied close to the plant's roots in order for the plant to utilize it. Large applications of phosphorus without adequate levels of zinc can cause a zinc deficiency.

Potassium: It is necessary for formation of sugars, starches, carbohydrates, protein synthesis and cell division in roots and other parts of the plant. It helps to adjust water balance, improves stem rigidity and cold hardiness, enhances flavor and color on fruit and vegetable crops, increases the oil content of fruits and is important for leafy crops. Deficiencies result in low yields, mottled, spotted or curled leaves, scorched or burned look to leaves.

Sulphur: It is a structural component of amino acids, proteins, vitamins and enzymes and is essential to produce chlorophyll. It imparts flavor to many vegetables. Deficiencies show as light green leaves. Sulphur is readily lost by leaching from soils and should be applied with a nutrient formula. Some water supplies may contain sulphur.

Magnesium: It is a critical structural component of the chlorophyll molecule and is necessary for functioning of plant enzymes to produce carbohydrates, sugars and fats. It is used for fruit and nut formation and essential for germination of seeds. Deficient plants appear chlorotic, show yellowing between veins of older leaves; leaves may droop. Magnesium is leached by watering and must be supplied. It can be applied as a foliar spray to correct deficiencies.

Calcium: It activates enzymes, is a structural component of cell walls, influences water movement in cells and is necessary for cell growth and division. Some plants must have calcium to take up nitrogen and other minerals. Calcium is easily leached. Calcium, once deposited in plant tissue, is immobile (non-translocatable) so there must be a constant supply for growth. Deficiency causes stunting of new growth in stems, flowers and roots. Symptoms range from distorted new growth to black spots on leaves and fruit. Yellow leaf margins may also appear.

2. Micronutrients

Iron: Iron is necessary for many enzyme functions and as a catalyst for the synthesis of chlorophyll. It is essential for the young growing parts of plants. Deficiencies are pale leaf color of young leaves followed by yellowing of leaves and large veins. Iron is lost by leaching and is held in the lower portions of the soil structure. Under conditions of high pH (alkaline), iron is rendered unavailable to plants. When soils are alkaline, iron may be abundant but unavailable. Applications of an acid nutrient formula containing iron chelates, held in soluble form, should correct the problem.

Manganese: Is involved in enzyme activity for photosynthesis, respiration, and nitrogen metabolism. Deficiency in young leaves may show a network of green veins on a light green background similar to an iron deficiency. In the advanced stages, the light green parts become white, and leaves are shed. Brownish, black, or greyish spots may appear next to the veins. In neutral or alkaline soils plants often show deficiency symptoms. In highly acid soils, manganese may be available to the extent that it results in toxicity.

Boron: It is necessary for cell wall formation, membrane integrity, calcium uptake and may aid in the translocation of sugars. Boron affects at least 16 functions in plants. These functions include flowering, pollen germination, fruiting, cell division, water relationships and the movement of hormones. Boron must be available throughout the life of the plant. It is not translocated and is easily leached from soils. Deficiencies kill terminal buds leaving a rosette effect on the plant. Leaves are thick, curled and brittle. Fruits, tubers and roots are discolored, cracked and flecked with brown spots.

Zinc: It is a component of enzymes or a functional cofactor of a large number of enzymes including auxins (plant growth hormones). It is essential to carbohydrate metabolism, protein synthesis and internodal elongation (stem growth). Deficient plants have mottled leaves with irregular chlorotic areas. Zinc deficiency leads to iron deficiency causing similar symptoms. Deficiency occurs on eroded soils and is least available at a pH range of 5.5 - 7.0. Lowering the pH can render zinc more available to the point of toxicity.

Copper: Copper is concentrated in roots of plants and plays a part in nitrogen metabolism. It is a component of several enzymes and may be part of the enzyme systems that use carbohydrates and proteins. Deficiencies cause die back of the shoot tips, and terminal leaves develop brown spots. Copper is bound tightly in organic matter and may be deficient in highly organic soils. It is not readily lost from soil but may often be unavailable. Too much copper can cause toxicity.

Molybdenum: It is a structural component of the enzyme that reduces nitrates to ammonia. Without it, the synthesis of proteins is blocked and plant growth ceases. Root nodule (nitrogen fixing) bacteria also require it. Seeds may not form completely, and nitrogen deficiency may occur if plants are lacking molybdenum. Deficiency signs are pale green leaves with rolled or cupped margins.

Chlorine: It is involved in osmosis (movement of water or solutes in cells), the ionic balance necessary for plants to take up mineral elements and in photosynthesis. Deficiency symptoms include wilting, stubby roots, chlorosis (yellowing) and bronzing. Odours in some plants may be decreased. Chloride, the ionic form of chlorine used by plants, is usually found in soluble forms and is lost by leaching. Some plants may show signs of toxicity if levels are too high.

Nickel: It has just recently won the status as an essential trace element for plants. It is required for the enzyme urease to break down urea to liberate the nitrogen into a useable form for plants. Nickel is required for iron absorption. Seeds need nickel in order to germinate. Plants grown without additional nickel will gradually reach a deficient level at about the time they mature and begin reproductive growth. If nickel is deficient, plants may fail to produce viable seeds.

Sodium: It is involved in osmotic (water movement) and ionic balance in plants.

Cobalt: Cobalt is required for nitrogen fixation in legumes and in root nodules of nonlegumes. The demand for cobalt is much higher for nitrogen fixation than for ammonium nutrition. Deficient levels could result in nitrogen deficiency symptoms.

Silicon: It is found as a component of cell wall. Plants receiving soluble silicon produce stronger, tougher cell wall making them a mechanical barrier to piercing and sucking insects. This significantly enhances plant heat and drought tolerance. Foliar sprays of silicon have also shown benefits reducing populations of aphids on field crops. Tests have also found that silicon can be deposited by the plants at the site of infection by fungus to combat the penetration of the cell walls by the attacking fungus. Improved leaf erectness, stem strength and prevention or depression of iron and manganese toxicity have all been noted as effects from silicon. Silicon has not been determined essential for all plants but may be beneficial for many.

17.3. Biological Functions, Nutrient Deficiencies and Sources

Deficiencies of nutrients in plants have various visual symptoms that are usually similar regardless of the species. The most common deficiency symptom is reduced growth, which is difficult to detect and diagnose at a glance. Other visual symptoms usually involve changes in colouration following a specific pattern, such as from the leaf tip down the midrib towards the base of the leaf or from the leaf margin toward the midrib, or between the veins of the leaf. Such symptoms may appear in new leaves or old leaves, indicating the phloem-mobility of the deficient nutrient and the ability of the plant to translocate existing stocks of the deficient nutrient. In many cases, internodal distances will shorten as well.

It may be remembered that plants are autotrophs, gathering solar energy for fixation of atmospheric carbon into energy-rich compounds consumed by heterotrophs, which gather the energy-rich plants for consumption and completion of their (and our) life cycles. In addition to energy, plants also harvest the mineral nutrients required by animal life. For the most part, the essential elements required by higher animals are similar to those of plants. However, animals require sodium (only beneficial in plants), selenium (beneficial to only a small group of Se-hyper accumulating plants, iodine (essential only to certain marine alga), silicon (beneficial to a number of plant species), and cobalt (essential to N_2-fixing Rhizobium symbiotically associated with leguminous plants). Yet other trace elements essential to animal and human nutrition, but apparently neither essential nor beneficial for plants, are chromium, lithium, fluorine and vanadium (although some propose that chromium and vanadium).

1. Nitrogen

Biological function of N

- Protein synthesis (one or more N per amino acid)
- Base pairs for RNA/DNA
- Prosthetic groups for protein (ex.: heme group of chlorophyll)
- Hormones (ABA, cytokinins)
- Metal uptake (phytosiderophores) and transport in xylem & phloem (ex: Cu with amines)
- Osmoregulation

- Chemical defenses, alkaloids, misc. biochemicals (ex: mescaline, cocaine, morphine, nicotine, caffeine, quinine)

Deficiency

Mild N deficiency will restrict plant growth, but often in a subtle manner that can only be assessed by comparison to plants grown with an adequate N supply. Moderate N deficiency will cause leaves to be light green or yellowish. Severe symptoms include necrosis (tissue death) starting at the tips of older leaves, with the tissue death developing a V-pattern down the midrib toward the base of the leaf.

Nitrogen Fertilizers

i. **Anhydrous ammonia (NH_3):** First among N fertilizers in terms of synthesis path is anhydrous ammonia, since it is the product of the Haber-Bosch process. It contains 82% N and is a volatile gas at atmospheric pressure and ambient temperature but can be easily liquified with pressure and refrigeration. Agronomic use of anhydrous ammonia therefore, requires proper storage infrastructure. Application must be subsurface (to avoid immediate volatilization), preferably into a soil that is not too dry so that the ammonia gas can dissolve into the soil water.

ii. **Aqua ammonia, ammonium hydroxide (NH_4OH):** Ammonia gas is dissolved in water to give a 33% solution (saturated). Ammonia is still volatile if not contained. Bulk of fertilizer weight is water, not N.

iii. **Ammonium nitrate (NH_4NO_3):** Solid, white powder usually pelletized and is conditioned to improve handling, 33 or 34-0-0. Potentially explosive because of presence of both reducing and oxidizing agents (NH_4 and NO_3) in same compound.

iv. **Urea [$CO(NH_2)_2$]:** White powder, usually pelletized, 45 or 46-0-0. First organic compound ever synthesized inorganically. Soil micro-organisms cause urea to *hydrolyze* to ammonium carbonate, which may then spontaneously reform as (volatile) ammonia gas and ammonium bicarbonate if pH remains high, by the following reaction:

$$CO(NH_2)_2 \longrightarrow 2\ NH_4+ + CO_32- \leftrightarrow NH_3 + NH_4+ + HCO_3-$$

Incorporation of urea into the soil is done so that the pH buffer capacity reduces the potential volatilization loss of ammonia which is essential for efficient use of urea. It is the most common fertilizer in many regions with infrastructure inadequate to transport and store liquified ammonia. Urea is produced industrially by high temp/high pressure reaction of ammonia and CO2:

$2\,NH_3 + CO_2 \longrightarrow CO(NH_2)_2$

v. **Ammonium sulphate, [$(NH_4)_2SO_4$]:** 21-0-0, It is produced either by addition of ammonia to sulphuric acid using virgin materials or as by product of coke industry in which the flue gases containing ammonia from the organic N in coal are reacted with sulphuric acid solutions.

vi. **Calcium nitrate, [$Ca\,(NO_3)_2$]:** 15-0-0. Not particularly popular in U.S. as fertilizer except for nutrient solutions/greenhouses, but more popular in Europe, perhaps because of its suitability for soils in danger of acidification.

vii. **Potassium nitrate, (KNO_3):** 14-0-44. Commonly used in greenhouses and sometimes for high value crops. Produced by reaction of KCl with nitric acid, followed by volatilization of HCl:

$KCl + HNO_3 \leftrightarrow KNO_3 + HCl(g)$

viii. **Ammonium bicarbonate, (NH4HCO3):** 17-0-0. Relatively unstable N fertilizer source that spontaneously reforms as ammonia, carbon dioxide, and water, causing major volatilization losses:

$NH_4HCO_3 \leftrightarrow NH_3 + CO_2 + H_2O$

Overall N fertilizer recovery rates for this source are low ~30%, but this is the major synthetic N source used in China.

ix. **Ammonium phosphates:** MAP, monoammonium phosphate, $NH_4H_2PO_4$, typically 12-52-0, and DAP, diammonium phosphate, $(NH_4)_2HPO_4$, typically 18-46-0. Formulated by reacting various quantities of ammonia with phosphoric acid. More often regarded as P fertilizer than N fertilizer since fertilizer application of DAP to achieve desired N rates exceed usual P rates, in most cases.

Slow-release N fertilizers

These fertilizers are high value-added products, that is a relatively cheap material (fertilizer N), is remanufactured into a more expensive product for a specific purpose fertilizer availability timed to more closely match the gradual and continual plant nutrient needs, also minimizing sudden leaching of excess fertilizer. These products are typically used for high-value agricultural needs such as turf and greenhouses, rather than row crops. The basic strategy is to either make the N insoluble or trap it with some sort of barrier.

Urea form fertilizers are manufactured by the reaction of formaldehyde, CH_2O, with urea, $CO(NH_2)_2$, to produce compounds such as methylene diurea:

Another technique in the manufacture of controlled-release fertilizers is coating the fertilizer. Sulphur-coated urea (SCU) typically consists of 30-40% N (as urea),

10- 30% S, 2-3% conditioner, and 2-3% sealant. Soil microorganisms oxidize the sulphur to sulphate, breaching the coating and allowing the urea to be released. Other coatings are formulated with micropores and cracks to permit slow diffusion of the fertilizer out of the fertilizer granule. Yet another group of techniques "trap" the N fertilizer in a matrix, either a gel or fritted solid or a cation exchanger such as zeolites or vermiculite, to slow the release of N.

2. Phosphorus

Biological functions

Notable in the ubiquitous ATP/ADP energy transport and storage compounds. Additionally, sugar phosphates form the "rails" of the nucleic acids DNA and RNA (which N-containing bases forming the "rungs"). Phospholipids are an important constituent of membrane chemistry and phosphoproteins are essential for life functions.

Deficiency symptoms

Results of P deficiency are spread more or less evenly around the plant, usually with no glaring visual deficiency symptom except for stunted growth and late maturity. Grassy species, including corn, will show reddening of leaves if P is severely deficient.

The forms of P in soil are: soluble, mineral, and organic.

Phosphatic fertilizers

Rock Phosphate: Source of all mineral P for fertilizers. Ore grade contains 6 to 15% P. Mined from deposits in Florida, Tennessee, North Carolina, Morocco, and elsewhere. Very insoluble. Effectively used as fertilizer material only on very acid soils or if acidulated or partially acidulated.

Single Superphosphate: Basically the product of the reaction of rock phosphate, usually a fluorapatite, with sulphuric acid according to the reaction:

$$2\ Ca_5(PO_4)_3F + 7\ H_2SO_4 \rightarrow 3\ Ca(H_2PO_4)_2 + 7\ CaSO_4 + 2\ HF$$

producing a mixture of monocalcium phosphate, a water-soluble phosphate, and gypsum, along with the volatile, and very nasty, hydrofluoric acid gas as a by-product. The commercial product contains between 7 to 9.5% P, or 16 to 22% P_2O_5.

Phosphoric Acid, (H_3PO_4): Produced by the same process as single superphosphate, but using more sulphuric acid.

$$2\ Ca_5(PO_4)_3F + 10\ H_2SO_4 \rightarrow 6\ H_3PO_4 + 10\ CaSO_4 + 2\ HF$$

Phosphoric acid is then separated mechanically from the gypsum for use. Phosphoric acid is rarely used directly as a fertilizer except for purposes of fertigation and commercial hydroponics, particularly when the source water contains (calcium) bicarbonate alkalinity (=high pH) that would otherwise cause phosphate precipitation. This material is, however, the acid used to manufacture the ammonium phosphates and the next fertilizer. Pure phosphoric acid is a solid that melts at 42°C. Phosphoric acid of commerce is typically ~23%P, or 53-54% 16 to 22% P_2O_5.

Triple Superphosphate: The active ingredient is calcium monophosphate, as in the single superphosphate, but the product carries less gypsum content because it was made with phosphoric acid instead of sulphuric acid:

$$2\,Ca_5(PO_4)_3F + 12\,H_3PO_4 + 9\,H_2O \rightarrow 9\,Ca(H_2PO_4)_2 + 2\,HF$$

This material contains 17 to 23% P or 44 to 52% P_2O_5. It was the most popular P fertilizer until surpassed in usage by ammonium phosphates. Much of the gypsum generated as a by-product is diverted into construction.

Ammonium Phosphates: MAP, monoammonium phosphate, $NH_4H_2PO_4$, typically 12-52-0, and DAP, diammonium phosphate, $(NH_4)_2HPO_4$, typically 20-50-0. Formulated by reacting various quantities of ammonia with phosphoric acid. More often regarded as P fertilizer than N fertilizer since fertilizer application of DAP to achieve desired N rates exceeds usual P rates, in most cases. Completely water-soluble P.

Superphosphoric Acid: A phosphoric acid mixture containing a substantial proportion of one or more polyphosphoric acids (pyrophosphoric acid, $H_4P_2O_7$; tripyrophosphoric acid, $H_5P_3O_{10}$; tetrapyrophosphoric acid, $H_6P_4O_{13}$. Manufactured by "roasting" phosphoric acid to remove some water, thereby polymerizing two or more phosphates. Typically 30 to 36% P, or 68 to 83% P_2O_5

Ammonium Polyphosphates, APP: Product of reacting polyphosphoric acid, with ammonia to form the final product, with a typical fertilizer composition of 10- 34-0, 10% N and 15% P. The product is very water-soluble, and has the capability to chelate some micronutrients, such as Fe and Zn.

3. Potassium

Biological functions

Potassium is associated with the movement of water, nutrients and carbohydrates in plant tissue. It's involved with enzyme activation within the plant, which affects protein, starch and adenosine triphosphate (ATP) production. The production of ATP can regulate the rate of photosynthesis. Potassium also helps regulate the opening and closing of the stomata, which regulates the exchange of water vapor, oxygen and carbon dioxide. If K is deficient or not supplied in adequate amounts, it stunts plant growth and reduces yield.

Deficiency symptoms

Potassium-deficient plants are easily distinguished by their tendency to wilt on dry, sunny days. The overall appearance of the plant is wilted or drooping. Deficient plants will have a stocky appearance with short internodes. Younger leaves' growth is inhibited, and they have small leaf blades. Leaves may also be dark to bluish-green, have a bronze metallic shine, or have a wavy appearance. In some species older leaves show blotchy chlorosis. In monocots such as maize, leaves may have inverted V-shaped chlorosis.

Potassium fertilizers

Potassium Chloride, (KCl): This single K source amounts to ~90% of all K fertilizers consumed in the U.S. For the most part, it is mined as sylvite ore (KCl, which is the potassium analog of halite, or rock salt, NaCl) mixed with NaCl, beneficiated to remove some of the contaminants, and sold. These products are often red because of residual contaminants which, although not water-soluble, are not harmful for soil application. Some KCl is the result of crystallization from brine, either from solution mining of KCl ore or precipitation from hypersaline lakes (e.g., Searles Lake and Dead Sea), often with carnallite $KClMgCl_2$ as an intermediate precipitate. The final form of KCl from crystallization processes is white crystals. Typical fertilizer analysis of KCl is 0-0-60 (=52% K), where ideal composition is 63%).

Potassium Sulphate, (K_2SO_4): About 5% of the U.S. market, this fertilizer is used particularly for horticultural crops in which chloride uptake is a problem, as in tobacco. 0-0-50.

Potassium Magnesium Sulphate, (K_2SO_4 + 2 $MgSO_4$): This is the primary K fertilizer produced by German and French mines from langbeinite ore and historically was called "double manure salts." The analysis is 0-0-18, but it also contains the secondary nutrients S (23%) and Mg (11%).

Potassium Nitrate, (KNO_3): 13-0-44 (less the 14-0-47 of pure salt because of conditioners added against hygroscopic moisture). Formulated from KCl and nitric acid. Expensive for agronomic use, used for horticulture and greenhouses.

4. Calcium

Biological functions

Calcium is present in the cytoplasm at levels that would indicate that it is a micronutrient, ~0.1 μM, in order to prevent interference with the high levels of P. Nonetheless, small fluctuations in Ca levels in the cytoplasm are part of signalling mechanisms for environmental stress. Ca pumps are directed out of the cytoplasm, either to vacuoles, where it may be precipitated as calcium oxalate, or across the plasmalemma. The function of the majority of plant Ca is structural, in the cell walls of shoots and roots.

Deficiency symptoms

Symptoms of calcium deficiency first appear on younger leaves and tissues, growth is inhibited, and plants have a bushy appearance. The youngest leaves are usually small and misshapen with brown chlorotic spots developing along the margins, which spread to eventually unite in the center of the leaves. Veins are also brown, making a typical feature of Ca-deficient plants the dark veins of completely necrotic leaves. Leaves also may be crinkled and torn. The growth of root tips is inhibited in Ca-deficient plants.

Calcium fertilizers

- Calcium carbonate, $CaCO_3$, or calcium magnesium carbonate, $CaMg(CO_3)_2$, as agricultural limestone are the primary sources of added Ca to soils, although they are used primarily or additionally to neutralize soil acidity.
- Gypsum, $CaSO_4.2H_2O$. No remediation of soil acidity should be expected from gypsum, but Ca will become available.
- Calcium nitrate, $Ca(NO_3)_2$, contains 19% Ca.
- Chelated calcium, usually as CaEDTA or Ca forms of hydrolysed protein. If applied, often foliar application.
- Single superphosphate (18-21% Ca) and triple superphosphate (12-14% Ca)

5. Magnesium

Biological function

- The most familiar function of Mg, as the central ion in chlorophyll molecule, accounts for <25% of total plant Mg
- Mg in the cytoplasm is related to enzyme activation: for example, the substrate for ATPases is MgATP.
- Additional Mg is present in the cell wall

Deficiency Symptoms

Magnesium deficiencies in soils are relatively rare as far as plants are concerned, but animals fed on Mg-deficient forage may develop hypomagesameia (grass tetany).

"Excess" Mg is known almost only in soils derived from serpentine or olivene, both Mg-rich rocks, in which the exchangeable Ca/Mg ratio is ~1/5 instead of the more common 4/1 to 1/1; such soils often also contain toxic, high levels of bioavailable metals, among them Ni, which together with high Mg limit the distribution of "normal" vegetation, leading to "serpentine barrens".

Fertilizers

- Dolomitic agricultural limestone, $CaMg(CO_3)_2$, contains up to 13% Mg. Even calcitic limestone will contain some Mg as Mg substitution for Ca (magnesian calcite).
- Potassium magnesium sulphate, $K_2SO_4 \cdot 2MgSO_4$
- Magnesium sulphate, $Mg(SO_4) \cdot 7H_2O$, Epsum salt
- Magnesium nitrate, $Mg(NO_3)_2$, often applied as foliar spray.

6. Sulphur

Biological functions

- Amino acids: cysteine, methionine. organic-N/organic-S in plants typically has a 30-40:1 molar ratio
- coenzymes and prosthetic groups
- glutathione: antioxidant and precursor to phytochelatins
- alliins in allium (with allinase produce pungency in onions) and glycosinolates in brassica (produces isothiocynates) for chemical defenses and S storage.

Deficiency Symptoms

Deficiency symptoms are yellowing of whole plant, moderately phloem mobile. Response to S fertilizer in many crops and regions. With the reduction of sulphur at the smokestacks, agriculture may be losing the only benefit of acid rain.

Sulphur fertilizers

- Potassium magnesium sulphate, $K_2SO_4.2MgSO_4$: 22%S
- Sulphur-coated urea: 10-20% S
- Single Superphosphate (13.9% S as gypsum; only 1% S in triple superphosphate)
- Ammonium sulphate, $(NH_4)_2SO_4$, 24% S
- Elemental Sulphur, S (100%S): oxidizes to sulphuric acid, used also as acidifying agent

Ammonium thiosulphate (ATS), $(NH_4)_2S_2O_3$, 26% S, 12% N Also $Al_2(SO_4)_3$, and FeSO4, also used as acidifying agents

17.4 Fertigation

Fertigation is a method of fertilizer application in which fertilizer is incorporated within the irrigation water by the drip system. In this system fertilizer solution is distributed evenly in irrigation. The availability of nutrients is very high therefore the efficiency is more. In this method liquid fertilizer as well as water soluble fertilizers are used. By this method, fertilizer use efficiency is increased from 80 to 90 per cent. (Table 17.1)

Table 17.1: Fertilizer efficiencies of various application methods

Nutrient	Fertilizer use efficiency (%)	
	Soil application	Fertigation
Nitrogen	30-50	95
Phosphorous	20	45
Potassium	50	80

Advantages of fertigation

- Nutrients and water are supplied near the active root zone through fertigation which results in greater absorption by the crops.
- As water and fertilizer are supplied evenly to all the crops through fertigation there is possibility for getting 25-50 per cent higher yield.
- Fertilizer use efficiency through fertigation ranges between 80-90 per cent, which helps to save a minimum of 25 per cent of nutrients.
- By this way, along with less amount of water and saving of fertilizer, time, labour and energy use is also reduced substantially.

Table 17.2: Water saving, yield and profit under drip and drip fertigation systems

Crops	Water Saving (%)	Yield (t/ha)			Profit (Rs/ha)		
		Conventional	Drip	Drip+ Fertgn	Conventional	Drip	Drip + Fertgn
Banana	35	26	30	37	81000	98000	120000
Sugarcane	29	120	160	207	30000	47000	68000
Tomato	32	45	56	65	56000	77000	95000

Fertilizer used in fertigation

- Urea, potash and highly water soluble fertilizers are available for applying through fertigation.

- Application of super phosphorus through fertigation must be avoided as it makes precipitation of phosphate salts. Thus phosphoric acid is more suitable for fertigation as it is available in liquid form.
- Special fertilisers like mono ammonium phosphate (Nitrogen and Phosphorus), poly feed (Nitrogen, Phosphorus and Potassium), Multi K (Nitrogen and Potassium), Potassium sulphate (Potassium and Sulphur) are highly suitable for fertigation0 as they are highly soluble in water. Fe, Mn, Zn, Cu, B, Mo are also supplied along with special fertilisers.

Table 17.3: Fertilizers commonly used in fertigation

Name	$N - P_2O_5 - K_2O$ content	Solubility (g/l) at 20 C
Ammonium nitrate	34-0-0	1830
Ammonium sulphate	21-0-0	760
Urea	46-0-0	1100
Monoammonium phosphate	12-61-0	282
Diammonium phosphate	18-46-0	575
Potassium chloride	0-0-60	347
Potassium nitrate	13-0-44	316
Potassium sulphate	0-0-50	110
Monopotassium phosphate	0-52-34	230
Phosphoric acid	0-52-0	457

N fertigation

Urea is well suited for injection in micro irrigation system. It is highly soluble and dissolves in non-ionic form, so that it does not react with other substances in the water. Also urea does not cause precipitation problems. Urea, ammonium nitrate, ammonium sulphate, calcium ammonium sulphate, calcium ammonium nitrate are used as nitrogenous fertilizers in drip fertigation.

P fertigation

Application of phosphorus to irrigation water may cause precipitation of phosphate salts. Phosphoric acid and mono ammonium phosphate appears to be more suitable for fertigation.

K fertigation

Application of K fertilizer does not cause any precipitation of salts. Potassium nitrate, Potassium chloride, Potassium sulphate and mono potassium phosphate are used in drip fertigation.

Micro nutrients

Fe, Mn, Zn, Cu, B, Mo could be used as micro nutrients in drip fertigation.

18

Hints for Achieving Higher Yield of Field Crops

With the sharp development of urban and industrial growth, the area under agriculture is declining year by year. On the other hand, human and animal population is increasing at a very fast rate. So, to meet the growing demand of food and fodders, a vertical increase in yield is required. Crop productivity per unit area can be enhanced with the manipulations of some agronomist practices such as sowing the crops at appropriate time, selection of suitable field for raising a crop, adoption of best planting techniques, use of appropriate seed rates, adoption of proper irrigation schedule, proper and need based application of fertilizers, adoption of zero till technology, green manuring, mulching, addition of well decomposed farm yard manure, timely and proper weed management, selection of recommended crops and crop varieties, proper control of insects and pests, adoption of marketing tips etc.

18.1. Proper Adoption of Agronomic Practices

For getting higher yield, adopt each agronomic practice in such a way that a healthy crop is raised with judicious use of inputs which are discussed as below:

1. Selection of crops and crop varieties

All soils are not suitable for growing all crops but every crop has a specific requirement for a specific soil type. For instance groundnut requires loamy sand (light) soil as these soils facilitates more pegging and pod development. When groundnut is grown on heavy (clayey) soils, its yields are less because these soils resist pegging and pod development. Moreover, nutrient requirement of this crop are less (being leguminous crop) which can be met in light soils (loamy sand). On heavy soil vegetative growth will be high but reproductive growth will be less. Apart from this crop, arhar, bajra, guara, barley, gram etc. are best suited for light textured soils. Crops like maize, cotton, sugarcane, potato, pulses etc. are best suited on medium textured (sandy loam) soils. These crops can neither tolerate to severe water stress nor to excessive water as these crops die due to excessive stagnation of water. Rice crop requires heavy soils (clayey) as it has excessive requirement for water. Rice especially direct seeded crop sown on light soils shows severe deficiency of iron and other micro nutrients. Hence, cultivation of rice (direct and transplanted) should only be done on medium to heavy textured soils. Wheat cultivation can be done on any type of soil; however, on light soils its fertilizer requirements are more. Farmers should not cultivate any crop which is new to their areas without the consent of agriculture experts as this crop may not adapt to the area and secondly it may have marketing problems.

Always cultivate varieties which are recommended of the crops by the State Agricultural Universities because unrecommended varieties may be more liable to attack by insect pests and diseases or may have less yield potential. The most important point is that select crop varieties as per time of sowing. Varieties varies for their duration for maturity and can be classified as :

i. Long duration varieties

ii. Medium duration varieties

iii. Short duration varieties

The yields of long duration varieties are highest and must be sown during the start of recommended time of sowing. A farmer should not stick to one variety only as due to seasonal variations, its performance may not be better in a particular year may be due to more attack of insect pests and diseases. The entire area of farm should be brought under 2 to 4 or even more varieties in order to avoid risk of seasonal variations.

When the field is vacated late, adopt short duration varieties and do not go for long duration varieties. Higher yields can be achieved by selecting recommended varieties as per their time of sowing.

2. Tillage requirement

These are two main objectives of tillage. (1) to control the weeds; (II) To pulverize the soil. However, excessive tillage can destroy soil structure. In general depending upon the soil type and weed intensity, 3-5 ploughing are sufficient in order to prepare fine seed bed. The concept of zero tillage is gaining popularity these days as there is lot of saving in energy and time of the farmer and hence cost of cultivation can be reduced and net profit can be improved. A contact and non-residual herbicide like Gramoxone (Paraquat) or Sweep Power (Glufosinate Ammonium) at 2.0 to 2.5 lit/ha depending upon stage and intensity of weeds should be sprayed few days before sowing the crop and then sow the crop direct with the tractor drill i.e. without tillage. The weed intensity is less in the fields under zero till system because seeds lying in lower soil layer do not get chance to germinate. The disadvantage of this technique is that perennial weeds like *Cynodon dactylon* (Khabbal/Bermuda grass), *Sorghum halepense* (Baru), *Cyperus rotundus* (Dila/

Motha) etc. dominates in zero till fields as the vegetative parts of these weeds remains in the viable form because these are not uprooted for many years. The zero till sowing must be done with care because many factors like depth of seed placement, adequate moisture at sowing govern germination and hence crop stand.

With continuous tillering operations, especially with puddling a, hard pan develops beneath the soil layers resulting in poor infiltration rates. Under such conditions, deep cheesling is required to save performance of different tillage systems on wheat yield is presented in Table 18.1 as reported by Walia *et al* (2005). They reported that with deep tillage (with inverted plough) significantly higher wheat yields were obtained as compared to zero tillage treatments during first year and these differences were narrowed down during second and third year of experimentation, as yield differences between different treatments were found to be non-significant.

Table 18.1: Grain yield of wheat as influenced by tillage treatments

Tillage treatments for 3 years			Grain yield (kg/ha)			
1st year	2nd year	3rd year	1st year	2nd year	3rd year	Mean
Inverted	Zero	Zero	3009	3316	3470	3265
Inverted	Normal	Normal	3144	3061	3450	3218
Zero	Zero	Zero	2655	2590	3240	2828
Zero	Normal	Normal	2409	3197	3370	2992
LSD (p=0.05)			248	326	NS	245

3. Date of sowing

Date of sowing is the key agronomic practice governing crop yield. A timely sown crop gives higher grain yield as compared to late sown wheat with similar inputs i.e. seed, fertilizers, irrigation, pesticides etc. So, the economics of the farm can be easily increased by managing timely sowing. It has been seen at farmer's level that fields are lying vacant but sowings are delayed without any valid reason i.e. just due to the laziness of the farmer. Early/timely sown crop develops good growth parameters i.e. tillering, leaf area, stout stem, well developed reproductive parts because optimum time is available with the crop plants for completion of vegetative and reproductive stages. On the other hand, late sown crops hurriedly complete vegetative and reproductive stages resulting in poor development of growth parameters and hence poor yield. It is non-monetary input for increasing crop yields. On the basis of field experiments, it has been proved that adequate yield reductions occur with delay in sowing but a farmer has to spent nearly same money on farm inputs for timely and late sown crops. So, sowing the crop at optimum time is a big success in agriculture. Sometimes fields are not vacated in time by the previous crops for sowing of succeeding crop. For example, sowing of wheat is delayed after cotton, basmati rice, potato, sugarcane etc. and under such situations go for zero tillage by applying pre-sowing irrigation before harvest of previous crops and select crop variety recommended for late sowings.

Another indirect benefit of early sowing is that some problematic weeds can be easily controlled by adopting this technique. For instance *Phalaris minor* (Sitti/ Gullidanda) is a troublesome weed of wheat in rice-wheat rotation, can be controlled by early sowing of wheat (among the recommended time) because at that time suitable temperature required for germination of this weed is not available. At this prevailing temperature especially in Punjab conditions, wheat can germinate and *Phalaris minor* seeds germinate in less number. Data presented in Table 18.2 (Virk *et al*, 2003) indicated that a significant decrease in grain yield of wheat was obtained by sowing wheat on November 10 and November 25 as compared to October 25 as well as November 25 sown crop yielded significantly less than November 10 sown crop during both years. Data also indicated that dry matter of *Phalaris minor* was significantly higher in November 10 sown crop (as this period is ecologically most suitable for *P. minor* seeds to germinate) as compared to early sown crop (October 25) and late sown crop (November 25). With delayed sowing more seeds of *P. minor* germinates prior to crop sowing. Early sown crop yielded more due to less intensity of *P. minor* and more availability of time for the crop to nature.

Table 18.2: Effect of sowing date on dry matter accumulation by *Phalaris minor* and grain yield of wheat

Date of sowing	Dry matter of *P.minor**		Grain yield (q/ha)		
	I	II	I	II	Mean
October 25	5.0(24.2)	5.1 (25.1)	36.0	40.9	38.5
November 10	5.8(32.6)	5.7 (34.7)	31.8	34.9	33.4
November 25	5.2(25.8)	4.8(23.9)	29.8	32.0	30.9
LSD (P=0.05)	0.17	0.23	1.6	1.3	

*Figures in parentheses are original values and figures without parenthesis are transformed and statistically analysed values

4. Crop rotation

Crop rotation have lot of benefits in agriculture.

i. Maintains soil fertility
ii. Maintains physical conditions of soil
iii. Helpful in non-monetary control of weeds
iv. Avoids excessive depletion of ground water
v. Nutrients from deeper layers can be exploited by sowing deep rooted crops
vi. Less build-up of soil borne diseases and insect pests
vii. Deficiency of micro-nutrients can be avoided
viii. Improvement in crop productivity

ix. Parasitic weeds can be effectively controlled by growing non-host specific crop in the infested field

x. It is an eco-friendly approach

Growing of same crop year after year on the same field have lot of problems resulting in declining trend in yield with passage of time. For crop sustainability, crop rotation is an ideal approach. For maintaining soil fertility as well as natural resources always add legume crop in non-legume cropping system. The area under the crops such as rice which exhausts natural resources especially ground water, should be kept under control. Eco-friendly crops such as legumes and oil seeds must be included in the cropping system. Do not prefer non-legume crop rotation such as rice-wheat as these crops exhausts the field for same type of nutrients resulting in deficiency of nutrient elements especially micro-nutrients such as zinc, iron, manganese, copper etc. Another point of consideration is that deep rooted and shallow rooted crops should be kept in a cropping system to improve soil fertility.

Rice-wheat cropping system is very exhaustive system resulting in significant reduction in available N, P and K content (Table 18.3) in soil as compared to the cropping system in which summer moong or potato was included (Anonymous, 2008). Soil analysis was done seven years of completion of each cropping system on fixed plots.

Table 18.3: Soil fertility status after 7 cropping cycles (0-15 cm depth)

Cropping Systems	Available nutrients (kg/ha)		
	N	P	K
Rice-wheat	218.1	32.1	128.6
Maize-Wheat-Summer Moong	232.8	46.2	142.6
Maize-Potato-Summer Moong	240.0	45.2	149.9
Maize-Potato-Onion	252.1	48.9	160.8
Summer Groundnut-Potato-Pearlmillet (Fodder)	229.6	45.8	136.7
Initial status	185.9	43.7	131.9

Another advantage of crop rotation is that management of weeds becomes very simple with the adoption of this manipulation. *Phalaris minor* (*Gullidanda/Sitti*) is a troublesome weed of wheat in rice-wheat cropping system. Rotation of wheat with berseem or potato or rapeseed and mustards helps in elimination/reduction in the population of this problematic weed. The experimental results reported by Singh (2002) indicated that growing of berseem for three years helps in eliminating seed bank of *Phalaris minor* from soil (Table 18.4) as grown up weeds of this weed are cut with berseem and no seed is produced by *P. minor*. Similarly inclusion of potato in the cropping system also helps uprooting of *Phalaris minor* plants during earthing up or at the time of digging of potato resulting in poor seed bank in soil. So, inclusion of berseem or potato in cropping system, helps to eliminate *P. minor* within three years.

Table 18.4: Effect of crop rotation on dry matter and seed bank of *Phalaris minor*

Crop Rotation	Dry matter of *P. minor (q/ ha)* (at harvest of *Rabi* crops)	Seeds of *P. minor* (no/ m^2)*(After harvest of *Rabi* crops)
Rice-Wheat (Herbicide)	2.09	1.80
Rice-Wheat (Control)	45.53	10.30
Rice-Potato-Wheat	0.0	0.0
Rice-Potato-Sunflower	0.0	0.0
Rice-Berseem	0.0	0.0
Rice-Gobhi Sarson	0.13	0.50

* 0-7.5 cm soil depth

5. Plant population

An ideal plant stand is a key to success of all field crops. A gappy stand (poor population) may result in less yield with same inputs. Moreover, weed population is more in the field with less crop stand because crop is unable to provide proper smothering potential on the weeds. So, crop with normal stand had always fewer problems of weeds. To attain proper crop stand, use of viable seed at recommended seed rates is very important. So, before sowing any crop, check the viability of seeds and do not use seed with poor viability. The seed should be free from weed seeds, broken seeds, diseased seeds and foreign bodies. The growth and yield attributes will be on the higher side in crop with normal stand. However, use of excessive seed rates especially in non-tillering crops like maize, cotton etc. is harmful as it results in poor development of reproductive parts. So, always use viable seed at recommended rates for attaining higher yields.

6. Planting pattern

Planting pattern is another agronomic technique by which crop yields can be increased with the use of same or even less inputs. The objective of appropriate planting techniques is to cover the field space uniformly for best utilization of solar energy and other inputs required for plant growth. Closer spacing, cross sowing, bed planting are few important planting techniques which can be exploited for higher yield. In wheat crop, the seed rate for closer (15 cm) sowing and cross sowing (22.5 cm x 22.5 cm) is as recommended for this crop with normal planting techniques. In cross sowing half the seed rate is used in one direction and half in another direction. Kler and Bains (1985) reported significantly higher PARI (%) in bi-direction (cross) sowing as compared to unidirection sowing when recorded at milk and hard dough stage. Also significantly higher wheat yield was recorded in bidirectional sowing than unidirection sowing due to more photosynthetic efficiency of former treatment.

Bed planting is another technique with lot of advantages such as :

i. No crop mortality due to excessive rainfall or irrigation
ii. No crop injury due to shallow water table (waterlogged soils)
iii. Less lodging due to speedy winds
iv. Less infestation of weeds
v. Mechanical interculture is possible
vi. Less attack of insect pests and diseases
vii. Good quality produce can be obtained
viii. Less seed rate is required
ix. Less dose of nitrogen and other elements
x. High saving in irrigation water

Bed planting technique can be adopted in wheat, rapeseed and mustard, gram, maize, moong, mash, soybean, arhar, sugarcane and many more crops. The technique is very helpful for the crops which loose germination with rainfall especially on medium to heavy textured soils such as soybean, moong, mash, maize etc. So, the risk of germination in flat sowing technique during rainy season crops can be covered with bed planting technique. Growing of intercrops in bed planted crops has lot of scope. Few examples are : (I) Intercropping of mentha (in furrows) in bed planted wheat crop is very common in Punjab and (II) Intercropping of summer moong, summer mash in spring crop of sugarcane and gobhi sarson, wheat and many vegetables in autumn crop of sugarcane for high productivity.

7. Fertilizer application

Fertilizer is the costlier input in agriculture and its judicious use not only improves yield but also reduces cost of cultivation resulting in higher returns. The recommendation of fertilizers especially N, P and K is different for different crops depending upon the response of a particular crop to a particular nutrient element, soil type, irrigation facilities available, environmental conditions etc. So, recommended dose of macro- nutrients (N, P and K) and need based use of sulphur and micro-nutrients must be done for higher yields. Time of fertilizer application is very important and delay in their applications may give poor response. It has been seen that most of the farmers make injudicious use of fertilizers i.e. they used very high dose of fertilizers especially nitrogen which results in crop lodging and more attack of insect pests and disease due to more succulency of the crop. Hence, always use recommended dose of fertilizers at recommended time in order to make agriculture more economical.

Soil testing is very important before sowing of each crop and as per test report, fertilizer should be added in order to make judicious use of inputs. Soil samples from plough layer (0-15 cm depth) be taken from 5 to 7 places from the field at random and mix the soil and a composite sample be analysed from a reliable laboratory.

Apart from quantity of fertilizer, their method of application is very important. Side placement of fertilizer is very effective than broadcast method and as far as possible drill the fertilizer for higher efficiency. Drilling of fertilizer is also very helpful in reducing weed growth as in drilling method, fertilizer is placed in the vicinity of crop seeds, hence crop plants will remove fertilizer at higher rates than weed plants. Broadcasting of fertilizer provides equal chances for crop and weed plants to rob the fertilizer. It was reported that significantly less population of *Phalaris minor* in placement method (2.5 cm below seeds of wheat) as compared to broadcast method of fertilizer application (Table 18.5). A significant increase in grain yield of wheat was recorded in placement method of fertilizer application.

Table 18.5. Effect of fertilizer application methods in population of *P. minor* and grain yield of wheat

Treatments	*P.minor* plants (No./sq.m)	Grain yield (q/ha)
Broadcast	1.9	48.7
Placement (2.5 cm below seed)	1.6	54.7
L.S.D. (p=0.05%)	0.1	2.45

8. Addition of organic manures

Farm yard manures (FYM), compost, green manuring and crop residues are the important sources of organic manures. Few advantages of organic manures are:

i. Provides all (macro and micro) nutrients for plant growth

ii. Improves physical and chemical conditions of the soil

iii. Improve water retention capacity of soil

iv. Makes the soil porous which helps better plant growth

v. Helps in modernization of soil temperature

vi. Improves quality of farm produce

vii. Improve microbial growth in soil.

viii. Improves crop yield as some crops like maize, potato, turmeric, vegetables are high responsive to organic manures.

Farm yard manure is very common organic manure and is used widely by the farmers. Well decomposed FYM should be used. On the other hand, use of undecomposed (raw) FYM may lead to more weed problems and more attack by termites. Generally fodders mixed with mature weed plants are fed to animals and the weed seeds come in the intact form along with cow dung which is not composted in proper pits. These weed seeds in the viable form are added in the field along with FYM where they create big problem. So, always use well decomposed FYM for better results and never use raw FYM.

Green manuring is another source of organic manures. Few crops like Sun hemp/ Saan (*Crotalaria juncea*), Dhaincha (*Sesbania aculeata)* and cowpea (*Vigna unguiculata* L.) are best suited for green manuring due to quick growth and leguminous in nature.

Anyone of these can be sown when a period of 40 to 50 days is available between harvest of one crop and sowing of next crop. These crops are quick growing and less input responsive crop, acquires considerable biomass within a short period of time. Bury these crops in soil just before puddling rice and 7-10 days before sowing other crops like maize, potato, wheat etc. Bury these crops before flowering as mature crop will take more time for decomposition.

Addition of crop residues is another option with the farmers to enrich soil. Generally straw of few crops like wheat, maize etc. is very economical and cannot be spared for incorporation. However, rice straw can be incorporated in soil as it is poor quality fodder. Either incorporate rice straw in soil after combine harvesting by running disc harrow twice in the field or adopt zero till technology and rice straw can be left on soil surface as surface mulch. However, either remove or spread the loose straw uniformly in the field before sowing of wheat crop. For better germination sow wheat at proper soil moisture and at proper depth and zero till is successful provided germination is ensured. Weed population is generally less in zero till wheat as weed seeds do not germinate due to non-availability of eight or if germinate they are in less number due to the presence of thick layer of rice straw. In case of pulses like moong and mash, incorporate plant material into the soil after removing the pods.

9. Adoption of intensive cropping system

The area under cultivation is decreasing day by day due to fast development in urbanization, so the only option to increase food production is adoption of intensive cropping systems. Due to development in mechanization, rice-wheat cropping system is widely adopted. As both crops are non-legumes, so depletion of soil is high and to maintain soil fertility, non-legume crop(s) are to be added in the prevalent cropping system. Rice had adverse effect on soil properties and in order to maintain soil health, alternate cropping systems must be adopted. Five cropping systems viz. Rice- Wheat, Maize-Wheat-Moongbean, Maize-Potato-Moongbean, Maize-Potato-Onion and Groundnut-Potato-Pearl-millet (Fodder) were compared by Walia *et al.* (2010) for seven years. Experimental results indicated that highest rice equivalent yield and hence highest profit was recorded in Maize-Potato-Onion which was followed by Groundnut-Potato-Pearlmillet (Fodder), Maize-Potato-Moong-bean, Maize-Wheat- Moongbean and all these systems were found better than the prevailing Rice-Wheat cropping system. Also irrigation water applied (cm) was lowest in Maize- Wheat-Moongbean followed by Groundnut-Potato-Pearlmillet (Fodder), Maize- Potato-Onion and Rice-Wheat. Also water productivity was highest in Maize-Potato- Pearlmillet (Fodder) and lowest in Rice-Wheat cropping system (Table 18.6).

Table 18.6. Rice equivalent yield, irrigation water applied and water productivity of different cropping systems

Cropping Systems	Rice equivalent yield (q/ha)	Irrigation water applied (cm)	Water productivity (kg grain/ m^3 irrigation
Rice-Wheat	132.4	212	0.625
Maize-Wheat-Moongbean	161.8	92	1.759
Maize-Potato-Moongbean	191.0	105	1.819
Maize-Potato-Onion	278.6	130	2.138
Groundnut-Potato-Pearlmillet (Fodder)	233.0	103	1.262

10. Irrigation techniques

Water is the most valuable natural resource required for plant growth. Due to excessive and injudicious use of water, the water table is going down which is very serious concern and big challenge in future. Generally farmers believe in flood irrigation due to free supply of electricity especially in Punjab. The area under rice is increasing and due to its high water requirements, water table is going down year by year. So, there is a need to use irrigation water more judiciously which is only possible through drip or sprinkler irrigation technologies. It is advisable to select low water requiring crops and watering should be done with drip irrigation which results in lot of saving of water as well as in improving water use efficiency.

Bed and ridge planting techniques are very helpful in saving irrigation as only furrows are irrigated which requires less time per unit area. Direct seeding of rice also requires less water than transplanted rice as it does not need puddling which requires huge amount of water. Irrigation water can be saved by replacing coarse and fine rice with basmati (super fine varieties) rice as water requirements of latter group of rice is low. So, there is a great need to economise irrigation water for the welfare of States in general and India in particular.

11. Proper weed management

Weeds compete with the crops for all essential inputs like nutrients, soil moisture, light, space etc. Weeds must be controlled before the start of critical period of crop weed competition in order to attain higher crop yields. Weed flora varies with type of crop, soil type, available irrigation facilities, tillage techniques, environmental conditions etc. In general grass type weeds are more competitive than broad leaved weeds. Crop weed association is very common and those weeds having ecological similar requirements as of crop, develop association with that particular crop. For instance *Echinochloa crusgalli* (Swank/Maddal) is associated with rice and *Phalaris minor* (Gullidanda/ Sitti) with wheat in rice-wheat system due to their similar ecological requirements. Few weeds such as *Trianthema portulacastrum* (Itsit/Chupati) germinates in flushes and start producing seed during very young age, are very difficult to control.

Crop losses due to weeds may vary with type of weeds and their densities. In general 20 to 50 per cent losses due to weeds in field crops are very common. These losses are very less in smothering crop like rapeseed and mustards and high in less smothering crops like mentha and turmeric. Walia (2014) reported that with effective control of weeds, 283.8 per cent oil yield in menthe can be increased as compared to unweeded control (Table 18.7). This yield increase in chemical method is 118.2 per cent in maize, 102.4 per cent in transplanted rice and 83.7 per cent in wheat infested with *Phalaris minor*.

Table 18.7. Grain yield of important field crops as influenced by different treatments

Methods of weed control	Wheat (q/ha) infested with		Rice (q/ha)	Maize (q/ha)	Sugarcane (q/ha)	Groundnut (q/ha)	Mentha oil (kg/ha)
	P. minor	Wild oats					
Use of herbicide	45.0 (83.7)	41.0 (92.2)	60.1 (102.4)	31.2 (118.2)	520 (77.5)	18.1 (52.11)	68.7 (283.7)
Mechanical Control	34.4 (40.4)	37.3 (77.6)	55.7 (87.5)	28.0 (95.8)	413 (41.0)	16.2 (36.1)	53.1 (196.7)
Control (Unweeded)	24.5	21.0	29.7	14.3	293	11.9	17.9

Figures in parentheses are percentage of yield increase over unweeded control

Weeds can be controlled with the adoption of preventive, mechanical, cultural, chemical, biological and integrated methods of weed control. We should not depend upon a single method and the adoption of integrated approach is the best weed management method. Preventive, cultural and biological methods have limited scope. Mechanical method of weed control is losing its importance due to high wages of farm labour and its uncertainty in timely availability. Chemical method is very favourable method among farmers. More than 90 per cent area under wheat and rice is treated with one or the other herbicide under Punjab conditions. However, excessive and indiscriminate use of herbicides is harmful not only on present crop but also on following crops due to high residual effects of some herbicide groups. These herbicides also pollute the environment.

On the other side, herbicides are very useful tool in agriculture provided they are used judiciously. Always use recommended dose of recommended herbicides at recommended time and method. Do not use unrecommended herbicides as well as unrecommended brands of recommended herbicides.

Always spray the herbicides in straight strips/bands and do not move the nozzle to and fro. For pre-plant or pre-emergence herbicides, use 500 to 750 litres of water/ ha and for post-emergence applications use 200 to 250 litre/ha of water for effective weed control. Always use flat fan or flood zet nozzle for herbicide use and never use cone type nozzle. The height of nozzle should be 50 cm from ground/crop canopy. Do not spray on windy days. Also during hot months of May and June, always spray either in morning or evening hours and do not spray during noon. Do calibration before spraying the herbicide and use same nozzle and depute the same person who have done calibration for spraying in the field. Check the nozzle before use by spraying plain water on a pucca floor and if wetting is not uniform then change the nozzle and when wetting is uniform then nozzle is alright. Before each filling, stir the spray fluid thoroughly. Knapsack sprayer is very common spraying equipment for herbicides. Tractor mounted sprayers can also be used especially for pre-plant or pre-emergence herbicides. Always purchase herbicides from a reliable shop keeper after obtaining receipt. The above listed points must be kept in mind while using herbicides for attaining better control of weeds and hence higher yield.

12. Plant protection measures

The losses by insect pests and diseases are also high but less than that by weeds. Always treat the seed with recommended fungicide in order to control root borne diseases. For the control of termites, seeds can also be treated with suitable insecticide just before sowing. Later on spray of recommended fungicide or insecticide should be done as per the disease organism/type of insect appearing on the crop. Always use recommended pesticides as per need and do not use these agrochemicals in excess. Spray only when population of insects exceeds thrash hold level. After use, store the left over pesticides (fungicides and insecticides) on separate shelves after proper levelling and away from herbicides. All the pesticides should be kept away from the approach of children.

13. Precision farming

Precision or site-specific crop management refers to a management system of production agriculture, using diverse technologies to increase field productivity and protect the environment. Under precision agriculture, however, inputs are applied in each part of the field according to its unique set of conditions. Moreover, when to apply, how to apply, how much to apply, kind of inputs in relation to water, nutrient, pesticides etc., the residual effect of nutrient and crop-residue and left over water on the succeeding crop and their behaviour with the environment in time and space are studied from a very close angle so that resources wastages may be reduced to minimum possible. Normally farmers follow one uniform practice of water, nutrient and pesticides application at their farm. While in this concept the variation observed within the field itself is to be taken care of. Even a field is to be visualized critically and assurance of balanced supply in desired amount in each nook and corner is to be maintained, led to sustainable yield levels in different cropping systems.

With precision farming, cost on inputs can be reduced while net profit is increased. Application of nitrogen with leaf colour charts in important crops like rice, wheat, maize etc. helps in saving nitrogen as it is applied as per the requirement of crop. In this technique, the colour of leaf is matched with colour chart and nitrogen dose is applied as per actual requirement of plant. Laser land leveling is another approach for saving irrigation water. Generally fields are not in proper level and hence requiring variable depth of irrigation water at variable sites of the field, thus, requiring huge amount of water for first irrigation. Moreover at lower spots of field, crop may be damaged due to excessive water. So, with proper laser land leveling technique, not only water is saved but a uniform healthy crop can be grown. Input cost on irrigation water and nutrients can also be saved with the adoption of bed planting system of crops. Tensiometer can be used in transplanted paddy to save irrigation water. Also transplanted puddle rice do not need water stagnation upto maturity and after 3 weeks of transplanting apply irrigation two days after the ponded water percolates in the field.

14. Cultivation of high value crops

Medicinal and aromatic crops, vegetables, floriculture are high value crops which improves the economic condition of the farmer. The decision to cultivate high value crops should only be taken after studying the market. Also their cultivation should be started on small scale and when good marketing response is achieved, the area can be increased slowly and slowly. There are lot of medicinal and aromatic crops which can be grown successfully in India. However, South India is very ideal for the cultivation of all spices and in North India cultivation of Mentha, turmeric, funnel, menthe etc. can be done depending upon the marketing demand. These crops are not becoming popular due to fluctuations in the market. Cultivation of vegetables is highly profitable provided they are produced as per demand in the market. Also cultivation of off season vegetables under protected conditions is highly profitable. Food processing and storage has a lot of scope in vegetable crops in order to cover glut in the market. In floriculture, cultivations of marigold and gladiolus is very common and highly economical.

15. Integrated farming

The various components of farming system are cropping system, dairy farming, poultry farming, pisciculture, fruit and vegetable growing, floriculture, agroforestry, agri-horticulture, silviculture, rabbitary, ruckery, emu rearing, vermiculture, biogas production, agro-based small scale industry and milk-based household industry.

The integrated farming system brings prosperity to the farmers by judicious combinations of various available enterprises. It also involves recycling of farm resources resulting in optimization of farm input use. The farming system approach provides flow of money to the farmer round the year by disposal of milk, eggs, honey, fish, mushroom, fruits, vegetables, etc. Also with the effective recycling of bi-products of the corresponding enterprise as manure will result in improving the physico- chemical and biological properties of soil.

It also involves the integration of minor occupations viz. growing of flowers and vegetables, nurseries under polythene cover or in small poly houses, establishment of fruit and forest plant nurseries. The farm produce can be used for value addition products such as vegetable pickles, murabas, tomato sauces, garlic/onion powder, turmeric powder etc. Fruit preparations like jelly, jam, amla murabba, apple murabba etc. are effective tool to make farming system approach more realistic and self-dependent.

16. Marketing

Both quantity and quality of farm produce, plays very important role for increasing farm income. The uncleaned crop produce having infestation of weed seeds will fetch low price in the market. So, always grow clean crop free of weed plants especially thorny ones. For example, the seeds of *Rumex spinosus* have big thorns and their presence reduces market value of the wheat produce. Similarly, presence of *Convolvulus arvensis* (Lehli/Hirankhuri) in seed crop of wheat leads to rejection of the produce by seed certification agencies as the seeds of this weed cannot be separated out from wheat grains by any mean due to their akin size. The produce with high moisture content is rejected in the market, so always harvest crop at proper moisture (in grain) or dry the produce in the sun before bringing to the market.

18.2. Conclusion

In agriculture, concepts of timely operations is very important in order to make this profession more profitable. The expenditure on inputs will remain more or less the same under optimum time and delayed operations. Proper sowing time along with optimum crop stand is very important for getting high yields. Judicious and need based use of inputs like fertilizers, irrigations, pesticides etc. should be done and their indiscriminate use must be avoided. Do-not stick to one rotation and go for intensive intercropping system and high value crops. So, timely agronomic operations along with recommended practices are the key to successful farming for attaining higher yields.

19

Crop Management in Problematic Soils

With the current food production level of 275.11 million tonnes, India has to step up food grain production by next decade to feed the burgeoning population (Report Ministry of Agriculture). India with 2.3% of the global geographical area, supports about 16.8% human population, 20% of the world's livestock. However, per capita availability of land for producing agricultural commodities has declined from 0.34 ha in 1951 to about 0.23 ha in 1980 and 0.15 ha in 0.15 ha during 2015 and it is expected to decline further to about 0.09 ha by 2050 A.D. With the possibility of horizontal expansion or putting more land under cultivation being remote, further augmentation in yield in Indian agriculture will be forced more and more into marginal or problematic soils accompany about 57% geographical area with increasing competition for good quality lands. Marginal and problematic soils with proper management have a good potential for increasing food production to meet the expanding needs of the country. Crop production under the following five special situations has been described:

1. Salt affected soils
2. Acidic soils
3. Waterlogged soils
4. Eroded soils

19.1. Salt Affected Soils (Alkali and Saline Soils)

Salt affected soils cover an area of about 8.6 million hectares in India; almost half are in canal commands. To meet out the food needs of an increasing population which is estimated 1.4 billion by the year 2025 and 1.7 billion by the year 2050 A.C, India would have to increase its grain production to higher levels. Assured irrigation, increase fertilizer use, high yielding varieties and plant protection measures have already helped appreciably to increase yield per unit area. However considering the limitations for further increase in production through these methods and shortage of arable land, agriculture will be pushed more and more in problematic soils. These soils are mostly found in States of Uttar Pradesh, Haryana, Punjab, Madhya Pradesh, Bihar and Andhra Pradesh. Based on exchangeable Sodium, pH, quantum and nature of soils, these soils have been classified into two broad groups' alkali and saline soils.

A. Alkali Soils

Alkali soils, also known as sodic or solonetz soils, are those in which pH exceeds 8.5, exchangeable sodium percentage (ESP) in more than 15. The electrical conductivity of the saturated paste (EC) is less than 4 dSm^{-1}. These soils are dominated by CO_3 and HCO_3 of Na, contain 2-4% amorphous $CaCO_3$ in the surface soil and a hard $CaCO_3$ *kankar* pan in the sub-soil. Due to high pH and high ESP, alkali soils develop poor physical properties. The predominant soluble salts are carbonates and bicarbonates of sodium and, therefore, these soils are also termed as sodic soils. Locally these soils are called bara, rakkar or kallar soils. During the dry season, salts of sodium accumulate at the surface and show up as white or brown salt crust. The brown colour is due to the presence of dispersed organic compounds. Micro-relief of the area and geo-chemical mobility of the salts play a major role in the formation of alkali soils. The dispersed clay particles block the soil pores resulting in very low infilteration rate, crusting and hardening of surface soil upon drying. Sodication also takes place through the use of high RSC waters, saline irrigation waters low in $Cl:SO_4$ ratio and contribution from shallow brackish ground water. These soils are highly dispersed and have poor water and air permeability affecting plant growth. The ground water quality is generally good but in some cases, it may contain high RSC.

Factors Affecting Plant Growth in Alkali Soils

Plant growth, yield/or quality of crop produce in alkali soils is affected by anyone or combination of the following factors:

i. Poor water and air permeability as a result of high dispersion of soil aggregates and clay particles.

ii. Excess of Na which is toxic *per se* to the plants and causes imbalance due to antagonistic effect on K and Ca nutrition.

iii. Difficulties in preparing seedbed as all the soil does not come to proper moisture condition at the same time, resulting in clod formation and poor tilth of the soil.

iv. Low permeability of water due to poor conductance from the lower to the upper soil layers and restricted effective root zone.

v. Hard crust on the surface layer which greatly hinders the seedling emergence and reduces the germination percentage and thus, plant population.

vi. Deficiency of Ca since nearby all the soluble and exchangeable Ca is precipitated as insoluble $CaCO_3$.

vii. Toxic concentration of HCO^{-}_{3} and CO^{2-}_{3} ions.

viii. Continuous loss of fertile soil due to wind and water erosion.

ix. Low activity of useful microbes due to high pH and excess exchangeable Na.

x. Decreased solubility and availability of micronutrients like zinc (Zn), iron (Fe) and manganese (Mn) due to high pH, $CaCO_3$ and soluble CO_3^{2-} and HCO_3^- irons.

xi. Presence of hard $CaCO_3$ layer which is a physical barrier for the movement of water, salts and vertical root growth of crops and trees.

xii. Increased solubility and accumulation of certain elements like fluorine (F), selenium (Se) and molybdenum (Ma) to toxic level in plants, that may affect crop yield and/or health of the animals feeding on such crops. Increased concentration of fluorine can cause severe problems of bone diseases in animals. Selenium toxicity in the form of *Degnala* disease, which affects skin and hair, is quite common in cattle fed on fodder and straw of crops raised on alkali soil.

Depending upon the amount of exchangeable Na, pH, nature and stage of crop growth, environmental conditions and the overall management levels under field conditions, plant growth is adversely affected due to combination of the above factors.

B. Saline Soils

These soils contain excessive quantities of soluble salts predominantly chlorides and sulphates of sodium, calcium and magnesium because of which plant growth is affected. These salts show up on the surface as a white crust which becomes particularly conspicuous with the movement of dissolved salts from the lower soil depths to the surface and their accumulation after the evaporation of water. In saline soils EC exceeds 4 dSm^{-1} and pH is less than 8.5 and the ESP is usually less than 15. These soils are in a flocculated state and hence a fairly permeable. The plant growth in such soils is affected by the osmotic effects of salts, the toxicity of the soluble ions like sodium, chloride and boron and lesser availability of essential nutrients due to competitions. High salt concentration prevents germination of seeds. Where seeds may germinate, young plants either make a retarded growth or die after germination. Hence, the growth of crops is irregular and patchy. Different crops respond differently to soil salinity. Few crops resist to salinity while others are very sensitive and important examples are as follows.

Crop Groups Based on Response to Salt Stress

Sensitive group		Resistant group	
Highly sensitive	Medium sensitive	Medium tolerant	Highly tolerant
Lentil	Radish	Spinach	Barley
Mash	Cowpea	Sugarcane	Rice (Transplanted)
Chickpea	Broadbean	Raya	Cotton
Beans	Vetch	Rice (direct sowing)	Sugarbeet
Peas	Cabbage	Wheat	Turnip
Carrot	Cauliflower	Pearlmillet	Tobacco
Onion	Cucumber	Oats	Safflower
Lemon	Gourds	Alfalfa	Taramira
Orange	Tomato	Blue panic-grass	Date palm
Grapes	Sweet potato	Para grass	Mesquite
Plum	Millet	Sudan grass	
Pear	Maize	Guara	
Apple	Berseem	Pome granate, Desi Babul (*Acacia*)	

Recommended cropping sequences for saline soils

- Pearlmillet-Barley
- Pearlmillet-Wheat
- Pearlmillet-Mustard
- Sorghum-Wheat/Barley
- Sorghum-Mustard

Cultural Practices

In saline soils salts often accumulates in few cm of soil during non-crop period. When high water table exist may result in excessive salt accumulation in root zone particularly in arid and semi-arid regions. Under these conditions both germination and yield are adversely affected. Successful crop rising in saline soils can be achieved by:

- A heavy pre-sowing irrigation to leach the accumulated salts from the root zone is very useful and essential to improve germination and early growth. Whenever good quality water available, pre-sowing irrigation should be given.
- Deep tillage and inversion have been found to reduce sensitivity hazards.
- Higher seed rate and closer spacing are advisable to counter these effects. For cotton, pearl millet, sorghum, wheat, barley, mustard about 25% higher seed rate over that recommended or normal soils ensure good crop stand. In

case of transplanted crops, number of seedlings per hill should be increased to compensate any loss in germination.

- Chemical treatment of seed and seedling with Ca salt increases tolerance of wheat in saline soils.
- Furrow planting may help in obtaining better crop stand and yield in saline conditions because salt tend to accumulate on the ridges under furrow method of irrigation.

Soil Fertility Management in Saline Soils

The depth and frequency of irrigation depends upon:

- Proliferation and depth of root penetration
- Capacity of soil to store and transmit water
- Nature of plant responses to soil water stress

In saline soils, the evapotranspiration's needs of crops are modified by presence of soluble salts as the amount of salts increases the water in the soil becomes less available to the plants. The soil may appear quite moist. This is because osmotic pressure of soil solution increases with the increase in ion concentration or salt concentration and the plants are unable to extract water as readily as the crops can extract water from relatively normal soil.

In saline soils, crops tend to have poor crop stand and transfer less water, therefore, light and frequent irrigation with one or two deep irrigation in the crop season should be applied for keeping the osmotic potential at low level.

Leaching and Drainage

Leaching involves removing excess salts from top soil layer to reduce root zone salinity to a desired level. The efficiency with which salts can be leached per unit of applied water depends upon the characteristics of soil profile.

- Depth of applied water
- Method of water application
- Absorption and exchange reaction due to leaching
- The initial wetness of soil

A leaching study showed that to leach as predetermined fraction of salt from the profile, water require continuous ponding in the order: Peat – Clay – Sandy loam.

Nutrient management techniques to increase nutrient use efficiency in saline and alkali soils can be used as under:

- In saline soils the fertilizer having low salt index should be used as career of nitrogen, phosphorus and potassium, e.g. ammonium sulphate nitrate, calcium ammonium nitrate, superphosphate, muriate of potash.
- In alkali soils, the fertilizer having high salt index should be used as carrier of nitrogen, phosphorus and potassium, for example calcium ammonium nitrate, ammonium sulphate, ammonium sulphate nitrate, mono and di-ammonium phosphates, muriate of potash, potassium sulphate.

Method of Application

- Urea and cation micronutrients applied as foliar application
- Nitrogenous fertilizers place at a soil depth of 10-15 cm
- Phosphate fertilizers should be band placed
- Heavier application of nitrogenous fertilizers in one time should be avoided

The response of fertilizers have been noted only upto moderate level of salinity or alkalinity but at higher levels the response is negated.

Sodic soils are deficient in

- Calcium, nitrogen, zinc
- Organic matter
- Marginal in iron
- These soils are put to cultivation after application of amendment, organic and chemical ones (gypsum or iron pyrites)
- Need not be dressed with Ca separately
- Soils reclaimed by iron pyrites do not exhibit Fe deficiency too in crop production
- These soils do not respond to phosphorus during initial 3-5 years of reclamation due to high amount of native available phosphorus
- Response to K is also not reported due to presence of K bearing minerals and large contribution from non-exchangeable K (above 90%) to total uptake by crop.

Alkali soils, therefore, require special management of N and Zn which is entirely different from their counterpart normal soils.

Form and Method of Nitrogen Application

- Among different forms, ammonium sulphate is found most suitable form of N for all crops at high pH due to residual acidity.

- At low pH, urea and ammonium sulphate were at par perhaps due to soil buffering
- In rice pre-transplanting submergence for one week further curtailed NH_3 volatilization by about 10%
- Increasing levels of nitrogen before puddling and one week of pre-transplanting submergence increased grain and straw yields and N uptake of rice as salts leach down
- Basal dressing of 1/2 or1/3 nitrogen before puddling of partially reclaimed sodic soil increased grain yield of rice

Nitrogen management

There is an arbitrary recommendation of applying 25-30% more N for different crops in sodic soils than normal soil with a view to compensate N loss from alkali soil. However, more the N applied more could be its loss from soil probably due to poor crop stand. Toxic effect of NO_2 accumulation in plants due to reduced activity of NO_3 reductases which becomes toxic to plants.

Phosphorus management

Alkali soils do not respond to phosphorus application during first 3-5 years of reclamation due to high available P status of soil on account of reaction between insoluble calcium phosphate and sodium carbonate to cause soluble sodium phosphate.

Rice crop responded to 30 kg P_2O_5/ha from 4th year onwards due to depletion of native P through leaching, fixation and uninterrupted crop mining. Application of phosphorus to either or both in rice-wheat cropping system increased the grain yield of rice and wheat only after 5 and 11 years of cropping, respectively. Probably rice with a shallow root system fails to make use of higher P status of soil depths unlike wheat which is able to tap P from lower depths. Positive response to applied P has been observed up to 120 kg P_2O_5/ha in case of berseem crop.

Potassium management

Predominance of micaceous minerals in the soils of many arid and semi-arid regions, Na-K exchange in biotite and dissolution of muscoutic structural units release large amount of available K in alkaline sodium soil environment. Crop response to applied K fertilizers has not been observed in alkali soils of the Indo-Gangetic plains.

Zinc management

Among micronutrients, alkali soils have acute deficiency of Zinc and iron is observed in rice. Basal dressing to both rice and wheat grown in sequence in alkali soils at 40 and 20 kg $ZnSO_4$/ha respectively is recommended. Spray of 0.5% aqueous solution of Zn sulphate (1000 l/ha) twice after 20 days of transplanting/ sowing of crops at an interval of 10 days was superior to its soil application with respect to yields and uptake and recovery of applied zinc. Soil applied of zinc sulphate raised Zn soil status. Continuous application of Zn for 3 years led to build up to Zn in various forms in the soil.

Integrated Nutrient Management

Incorporation of any sort of organic material helps build up of humus and hence, fertility status of soil improves by way of:

- Reducing soil pH
- Nitrogen loss through acid production
- Immobilization of nature and applied nitrogen, respectively.
- Green manuring of leguminous crops before rice is beneficial
- Alternatively press mud or water hyacinth (aquatic weed) or FYM manure or any other potential source of nutrients particularly nitrogen, may be utilized to supplement fertilizer nitrogen in alkali soils for sustainable crop production with simultaneous improvement in soil.
- Though integrated use of water hyacinth or press mud or FYM in alkali soil, fertilizer nitrogen may be saved up to 50% of the recommended dose with an additional advantage of soil improvement.
- Besides, organic sources also had residual effect on subsequent wheat crop and raised wheat yield by more than 20%
- Use of organics also brought down soil pH and ESP

19.2 Acidic Soils

These soils are characterized by low pH levels generally less than 5.5.

Reclamation of Acid Soils

The acidic soil management should aim at improving the production potential of soil by addition of amendments or by agronomic manipulation to obtain maximum productivity. Application of lime as an amendment to neutralize the exchangeable Al^{3+} is quite effective for amelioration of sub-soil acidity, application of gypsum ($CaSO_4$) is more suitable than lime which may be due to non-phytotoxicity of

$A1SO_4$, sulphate component and water solubility of gypsum. The acid tolerant plant species and varieties which have Aluminium tolerant mechanisms can be grown in acidic soils. The rice has good tolerance to acidity as flooding of rice raises pH of field to almost neutrality. Short duration rice varieties grown in uplands tolerate acidity. Minor millets (*Panicum miliare*) and finger millet (*Eleusine coracana*) are tolerant to acidity and their crops show meagre response to liming. Gram, lentil, groundnut, sorghum, field peas and maize show medium response to liming. Pigeonpea, cotton and soybean are sensitive to soil acidity and these crops respond well to liming. Fodder crops like dinanath grass, napier bajra hybrid, guinea grass, maize and Setaria can be grown under agroforestry system in acid soil conditions. The grasses viz., napier bajra, dinanath grass, Setaria, *Paspalum notatum*, Kikuji grass and guinea grass can be successfully grown under acidic soils. Among cereals/legumes, lucerne, cowpea, maize, jowar are important in acidic soils. Important tree/shrubs for acid soils are *Sesbania, grandiflora, Leucaena leucecephala* spp.

Acid and Acid Sulphate Soils

Characterizerd by low pH generally less than 5.5. In acid soils growth is affected by excessive amount of dissolved ions of Al, Fe, Mn and Cu and become toxic. Acid soils are available on low lying coastal lands which are generally rich in pyrite and on oxidation form H_2SO_4. Acid sulphate soils have $pH < 4$. The crops in these soils suffer mainly because of Al toxicity and nutrient deficiency specially P.

Application of $CaCO_3$ is recommended according to time requirement of soil. It varies with the factors like:

(i) pH of the soil

(ii) Clay content

(iii) Cation exchange capacity

(iv) Climate

For reclamation of acid sulphate soil water management is the key to soil management. Root development is always restricted in acid sulphate soils unless water is available for irrigation or to maintain a high water table artificially as flooding alleviates activities; rice is the best choice to be grown on acid sulphate soils in warm region. Rice yield can be as good as of non-acid soils if recommended dose of lime as well as fertilizers are applied. The benefits of lime and phosphorus applications may be due to increases supply of nitrogen by stimulation of ammonification and microbial fixation of nitrogen process.

Application Method for Reclamation in Acid Sulphate Soils

Liming material should be evenly broadcast and mix in soil. A few weeks in advance to sowing of a crop for completely exchange process. Proper moisture of soil enhance the exchange process generally lime should be applied after every 5 years. However, the soil should be periodically get checked for pH.

For reclamation of acid sulphate soils water management is the key to soil management. Root development is always restricted in acid sulphate soils unless water is available for irrigation or to maintain a high water table artificially, as flooding eliminates acidity. Rice is the best choice to be grown on acid sulphate soils in warm regions.

Total reclamation can be slow and expensive but reclamation is comparatively easier or less costly on soils which do not have resources of pyrites or high value of exchangeable aluminium. On such soils moderate application of lime will significantly reduce soluble aluminium levels and enable crops to respond to conventional fertilizer application without necessarily raising pH above 4.5.

Agronomic Techniques

Many of principal crop and vegetation are sensitive to acid soil and grown on these soils get injury. Rice, millets, oats, tea and potato are crops which can tolerate to some extent to soil acidity on low pH soil. Among these crops rice is well adapted to acid sulphate soils because it grows under flooded conditions which depresses Al toxicity and increases availability of phosphate. Rice yield can be as good as of non-acids soils if soil is submerged a long as possible before transplanting and recommended dose of lime and MnO_2 and NPK are applied. The benefits of lime and phosphorus application may be due to increased supply of N by stimulation of ammonification and microbial fixation of nitrogen processes.

Nutrient Management in Acid Soils

Nitrogen

- The availability of nitrogen in acid soils usually increases on liming by increases the pH.
- The effect of lime on mineralization of soil nitrogen was not so pronounced in green manured soil as in untreated (no green manure)
- In a field experiment, average release of 12 ppm of NH_4-N from a soil was recorded, the pH value was raised to 7.2 after 37 days of lime application
- The calcium ammonium nitrate was as better source for growing wheat in acid soils as it contains 8.1% Ca.

Phosphorus

Management of phosphorus in acid soils depends upon:

- Extent of acidity
- Status of sulphur
- Crop duration

Evidence of greater availability of phosphorus as rendered by liming of acid soils has been observed and found significant increase in the phosphorus content of the leaves of plants in number of crops. This effect of lime diminished in the presence of P fertilizers suggesting that liming helped in the mobilization of nature soil phosphorus.

The effect of continuous manuring for about ten years with acidifying fertilizers e.g. sulphate of ammonia, superphosphate etc. in the acid red loam soil of Kanke (Ranchi) has been quite evident from pH, available phosphorus potential and lime potential.

Potassium and Calcium

Reported higher concentration of Ca in soybean, mungbean (green gram) and barley plants grown after liming in the acid red soils indicating greater availability to the crops.

- The availability of K on the other hand, in general, was found to decrease
- It was also reported that no definite effect of lime on the availability of soil K
- Liming increased or decreased the exchangeable K in soil only slightly
- Calcium contents of plants also increased or decreased the exchangeable K in soil only slightly
- Calcium content of plants, however, increased considerably due to liming

Micronutrient management

- Crop response to various micronutrients in acid soils is low to medium with very few cases of high response.
- Response of micronutrients such as Fe, Mn, Zn and Cu are very limited in acid soils
- However, the behaviour of zinc in submerged low and acid soils is quite different in which rice responds to zinc application

19.3. Waterlogged Soils

The soils with very shallow water table are termed as water logged soils.

Waterlogged law: The working group on problem identification in 1 m area adopted the following norms for identification of waterlogged area:

i	Waterlogged area	Water table within 2 m from the low surface.
ii	Adopted area for waterlogging	Water table between 2-3 m from the low surface.
iii	Safe area	Water table below 3 m from the low surface.

For water logged areas, the following norms are supported for the classification of different categories.

(i) Water-table within 2 m from soil surface during pre-monsoon (April/May).

(ii) Water-table within 1 m from soil surface during post-monsoon (October-November).

Under these two conditions, the area may be treated as waterlogged which adversely affect crop yield and suitable drainage measures be adopted to remove excess water. The areas where the water-table may be deep presently but may rise under conditions of frequent irrigations, heavily rainfall and inadequate drainage are termed as critical and are potential areas for water-logging. It is estimated that about 6.0 m ha land suffers from surface flooding (3.4 m ha) and high water-table (2.6 m ha). The continuous availability of water in the root zone lowers aeration which in return results into poor growth and nutrient uptake.

Critical and potential areas for waterlogging

Critical area for water logging has water table between 2 and 3 meter from soil surface during pre-monsoon and are between 1 and 2 m during post monsoon period. Waterlogging may develop for a short period if suitable measures are adopted which are located specific and these may comprise drainage system, land development and scientific irrigation management. During monsoon, irrigated area with water table between 3-5 m during pre-monsoon may be considered as potential areas for waterlogging.

Reclamation of waterlogged and salt affected lands

Waterlogging can be controlled by adopting suitable drainage measures and through scientific methods of water management. Saline lands are reclaimed by addition of chemical amendments i.e. gypsum ($CaSO_4.2\ H_2O$). Sub-surface drainage is ideal for leaking of excess salts and reaction products after addition of chemical amendments from saline soils. Acid soils are reclaimed by adding lime to soil.

Classification of Drainage System

There are two major types of drainage system; surface and sub-surface. In addition to these, two ground water control systems; by pumping is called vertical drainage and reclamation efforts of saline and waterlogged lands by planting certain trees with high evapo-transpiration rate and resistance to advance soil climatic conditions is called bio drainage.

A. Surface drainage

It is adopted to remove excess water from surface of crop land, temporary surface water disposal is done through a network of surface drains, linking crop land with main outlet.

Cultivation of tolerant species: In water logged soils growing of excess water tolerant forage crops viz., dicanthium, guinea grass, iseilema, maize, napier-bajra hybrid, paragrass, setaria and signal grass etc. are beneficial for increasing productivity. Tree species suitable for waterlogged saline soils are *Tamarix articulate, Prosopis juliflora, Acacia farnesiana, Pithecellobium dulce* and *Parfinsonia aculeata* etc. whereas *Eucalyptus, Casuarina* and *Populus* trees can be grown in saline waterlogged areas near canal commands. Among shrubs and under shrubs *Salicornia, Italoxylon, Salsola, Tamarix* and *Suaeda* are important for waterlogged soils.

Methods of Field Drainage

The accomplishment of surface drainage of agricultural lands includes land farming, low smoothening, low grading or leveling, bedding system, open ditches and levels.

- **Land farming:** Land farming is a process of changing natural topography with the aim to control movement of water contour from land surface. This practice includes land smoothening, low grading or leveling and shallow field drains crossable by farm machinery.
- **Land smoothening:** In field lands where there are only minor differences in field levels, good surface drainage can be achieved by land smoothening.
- **Land grading and leveling**: For efficient drainage, the grade in the dissection of drains can be 0.05 to 0.5%, however, preferable slope is 0.1%.
- **Bedding system:** This is suitable for par hydraulic conductivity soils having slopes upto 1.5%. The levelled surface of field is converted into sloping area by repeated ploughings and the beds are always kept in parallel to the greatest slope of the field such that deep furrow serves as a field drain. The length of bed depends on field conditions and it may vary from 100 to 300 m. The width of bed depends upon slope, land use, soil permeability, farm machinery to be used and the type of farming operations.

- **Open ditches:** It includes 3 systems i.e. parallel field drain, random system and parallel open ditch system.
- **Parallel field drain system:** This system is used when surface is uniform with few noticeable ridges or depressions. The channels are spaced further apart than in a bedding system and there have greater capacity than the dead furrows.
- **Random system:** This system is most suitable for areas having deep depressions and it is not technically feasible to fill depressions. The slope of channel can be as flat as 10:1 where depth of drains is more than 60 cm or 4:1, of the field are formed parallel to drains.
- **Parallel open ditch system:** This not only provides surface drainage to an area but also helps to control the ground water-table.
- **Levels:** These are commonly known as embankments and these are used in conjunction with drainage works to exclude run-off from uplands entering the areas. This method is commonly used in flow plains and coastal areas.

B. Sub-surface drainage

Two types of man-made systems of sub-surface are: (i) horizontal (ii) vertical. The horizontal drainage involves engineering structures laid horizontally parallel to ground and vertical drainage is mainly achieved by pumping out ground water through tube wells. These structures are laid horizontally and works like tube well.

- **Bio-drainage:** The bio drainage system involves growing certain categories of plants which draw their main water supply directly from ground water or the capillary fringe just below it. The tree species belong to this group are *Eucalyptus*, Poplar, *Tamarix muskit* and *Acacia* spp. Similarly, non-woody plants like bushes, sedges and grasses are called as phreatophytes. Two species of *Eucalyptus* which are commonly used as bio-drainage are *Eucalyptus tereticornis, Eucalyptus camaldulensis*. These trees transpire luxuriantly when ground water contact has been made. These tree species usually use much water than required for normal growth conditions.
- **Vertical drainage:** It refers to removal of excess water by means of a hole either under pressure or through pumping. The success of vertical drainage system depends upon the presence of favourable unconfined aquifers within 50 m depth below soil surface to permit pumping of groundwater on sustained basis. Good to moderate quality groundwater can be utilized for irrigation purposes.
- **Horizontal drainage:** A horizontal drainage system consists of the laterals, collectors and main drains. The water entering the laterals flows into the collectors, which carry it to the main drain.

- **Drainage:** Subsurface drainage provides an efficient means for lowering water table to prevent secondary salinization. In addition to initial cost of installing subsurface drainage and problem of disposal of drainage water could assume serious preparation in areas where no natural outlet like sea is available.

Waterlogged soil

Water logging refers to the presence of excess water or water above saturation in or near to the root zone or standing water above ground for any substantial period of time. An agricultural land is said to water logged when soil pores in the root zone get saturated with water.

The Government of India in 1942 establish a land reclamation to tackle with this problem and set up land reclamation, Irrigation and Power Research Institute at Amritsar during 1947 (LRIPRI) where different means and ways to reclaim the soils are being evolved.

In India this problem is more acute in Indo-Gangetic plains mainly due to :

- Flat topography
- Frequent flooding during monsoon
- Canal irrigation in these area

Effect of Water Logging

Excess amount of water in the root zone interfere with the biological processes. The main effects are:

i. Reduced aeration and reduced O_2 supply to roots

ii. Lack of O_2 supply to roots results in reduced respiration of the protoplasm of root cells resulting in death of roots.

iii. Reduced absorption of water and nutrients (P, K, Ca, Mg)

iv. CO_2 is accumulated in no aerated soils and produce toxic and inhibitory effect on roots (CO_3, HCO_3 and H_2CO_3)

v. Root growth is affected due to reduction in cell division and cell elongation

vi. Loss of nutrients – The continuous percolation and draining of water may lead to leaching and removal of nutrients particularly N in sufficient quantity. On the other hand due to reduced conditions there may be reduction of NO_3 to gaseous form which is lost to atmosphere.

$NO\text{-}_3 > NO\text{-}_2 > N_2O\text{-} > N_2$

vii. Metabolic processes of plants are affected (ATP)

viii. Decomposition of O.M. is slowed down (the end products are – CO_2, H+, Methane, H_2S)

ix. pH is changed – In acidic soils it increases and in salty soils it decreases

x. Plant growth is reduced due to welting and leaf fall

xi. Water logging give rise to soil salinity and alkalinity conditions due to excess evaporation resulting in salt deposition to the surface layer.

Causes of water logging

i. High rainfall which recharges the water table results in rising the water table.

ii. Surface flow of rain water from upper regions to lower regions also results in prolonged inundation of lower system.

iii. Seepage from rivers and canal system. In Punjab water table has risen very faster in some areas due to canal irrigation.

iv. Intensive-cropping and growing of high water requiring crops like paddy is also a source of water logging.

v. Blockage of natural drains due to construction of roads, rails, canals etc. Such type of construction has resulted blockage of natural water courses which otherwise were helpful to drain away the excess water.

Control Measures

1. Mechanical measures
2. Agronomic measures

1. Mechanical Measures

i. **Drainage:** The most effective answer to water logging is a properly designed drainage system. In fact, the importance of drainage is so well recognized in advance countries that it is always considered integral part of any irrigation scheme.

In West Germany for example no irrigation project is taken up without a corresponding drainage project. In this country, too, instructions have already been issued that irrigation project estimate should include the cost of drainage work.

ii. **Field drains**: In water logged areas or prone to water logging field, drains to drain individual field will be necessary. But as their construction and maintenance is to be done by farmers, the later may need technical advice, with regard to such matters as alignment depth and grade of drains. Drains may be deep enough to keep the ground water at a level which would not be harmful to standing crops. In Maharashtra excavation of drains to a permeable sub stratum where it was available at reasonable depth, has been successful. In case of soils with low permeability, sub-surface drains may be required.

iii. **Borrow pits:** Water generally stagnates in the borrower pits formed as a result of excavation of canals, and road and rails embankments. This is not desirable in an area prone to water logging. The commission recommends that the indiscriminate excavation of borrow pits should not be allowed. If they are unavoidable, they should be so excavated that water will not stagnate but flow into the near drainage system.

iv. **Use of ground water for irrigation:** Adequate provision for surface and ground water use should be made in the planning for new projects, because it helps to keep water logging under control. Farmers should also be encouraged to dig well and sink shallow tube wells. Construction of large number of shallow tube wells by farmers in Punjab has helped to kept the water table down. The commission recommends that necessary technical and financial assistance should be given to the farmers for the construction of wells and shallow tube wells.

v. **Lining of canals:** Lining of canals is of great importance in areas which on the introduction of irrigation are liable to water logging and salinity. If lining is done at the initial stage it will cost much less than if done later.

2. Agronomic Measures

In areas susceptible to water logging, it will be desirable to introduce only those crops which need light irrigation. Crops need heavy irrigation will have to be limited.

i. **Water management:** Land behaviour of groundwater levels after the introduction of irrigation has to be studied. The depth of sub-soil water table has to be observed in selected open wells distributed over with a greater concentration in low lying areas. The observations are generally made twice a year once in June before the monsoon and second time in October after the monsoon. Our objective is to know the variations in the water level from year to year.

ii. **Planting method:** Ridge planting technique should be adopted in water logged areas. This makes the plants less affected due to shallow water table or stagnation of water in the fields.

This method includes:

- Narrow ridge and furrow system in which ridges are mini-bunds furrows act as a drain for excess water.
- Broad bed and furrow method: In which broad beds of different width (1-2 m) are made and a furrow is provided after each bed for draining excess water. The crops are raised on beds.

Fertilizer Management

There are four types of losses of nitrogenous fertilizers:

i. **Immobilization :** Broadcast urea is more susceptible to this loss by Algae (15-40%)

ii. **Ammonia Volatilization :** These types of losses are low in submerged soils

iii. **Nitrification and denitrification :** In aerobic surface soil layer the ammonia is oxidized to NO_3 which diffuses into anaerobic layer below and is lost by denitrification.

iv. **Leaching and runoff :** Broadcasted urea may be subjected to leaching and runoff losses from soils. In light soils with high percolation rate, urea loss due to leaching is more.

Management of Nutrient Loss

i. **Deep placement:** It reduces losses through nitrification, denitrification, urea super granules can be placed deeply.

ii. **Slow release fertilizers:** To delay the availability of N to the plant until it has strong root system which can compete with the mechanism of losses and biological immobilization.

iii. **Split application**: The concentration of N in flooded water may be reduced by split application of N fertilizer resulting in reduced leaching or runoff losses. The application may be made with the need of the crop.

iv. Higher doses of N (25% than normal) to compensate losses.

v. Ammonical or amide forms of N fertilizer should be preferred as these are less subjected to denitrification losses.

19.4. Eroded Soils

Soil erosion occurs mainly through the forces of wind and water.

Wind Erosion Mechanism and Processes

- **Saltation:** The major portion of soil carried by the wind moves in a series of short bounces is referred as saltation. It carries fine particles of 0.1 to 0.5 mm in diameter. Saltation is caused by direct pressure of wind on soil particles and their collision with other particles. The soil particles are pushed by the wind and are placed almost vertically in the first stage of saltation. Some of grains size only as short distance and some leaf's 30 cm or higher depending upon wind velocity and size from the ground.
- **Suspension:** Very fine soil particles which are less than 0.1 mm diameter are carried into suspension. There are kicked up into air by action of particles. The movement of fine dust particles in suspension is governed by characteristic movement of wind. The suspended material is carried to long distance from its original location. Therefore, it is a complete loss of soil particles to the eroded area.
- **Surface creep:** The soil particles larger than 0.5 mm in diameter but smaller than 1.0 mm are along the surface comes in impact of particles in saltation to form a surface creep. The soil moved in saltation and surface creep usually remains within the eroded area, when the exception winds are blowing from different dissections.

Soil and Water Conservation on Agricultural Land

- **Agronomic measures**: Agronomic measures help to intercept rain water and reduce the splash effect; improve better intake of water by the soil by improving the organic matter and soil structure; reduce run-off losses through use of contour cultivation, close and dense crops, strip cropping, mixed cropping and mulches.
- **Contour farming:** It is a simple conservation practice of farming operation, viz. ploughing, cultivation, planting across the slope, keeping the same level as far as possible. The ridges, furrows and the rows of plants are placed across the slope. These form a continual series of miniature reservoir and act as barriers are small and individually, but their number is large so their cumulative effect is more in reducing run-off, soil erosion and losses of plant nutrient.
- **Cover management:** Legume crops are erosion resistant; provide better soil cover and better protection to land against erosion than ordinary wide-row erosion permitting crops. Cowpea, green gram, black gram, groundnut, soybean, sunhemp and dhaincha are good cover crops and effectively reduce soil erosion. Nearness of crop canopy to soil surface and extent of crop cover determine erosion control. Other factors like inoculation with Rhizobium or fertilization with single superphosphate prove beneficial.

- **Mulching:** The surface mulches are useful for protecting the soil against blowing and beating action of rainfall, reducing evaporation losses, increasing water infiltration, checks weeds growth, improves soil structure and finally increases crop yield. Mulching is the most effective measure of soil and water conservation but the mulching material is non-available due to its competition for annual fodder poses limitation in its adoptions which invites for strategies of *in-situ* mulching or utilization of weed biomass for mulching. Growing of sunhemp green manure in 60 cm wide contour strips between maize rows and recycling after 30-days as soil mulch reduces erosion and cheeks weed growth.
- **Conservation tillage:** Low intensity tillage favours consolidation of soil and imparts erosion resistance through better consolidation, structure, infiltration, pose distribution and depression storage.
- **Intercropping:** The line sowing of mixed crops give arise to intercropping practice. Intercropping of sorghum and pigeon-pea in 1:1 ratio has proved sustainable system in semi-arid regions of south-eastern Rajasthan.
- **Strip cropping:** Contour strip cropping is the growing of erosion permitting crops in strips of suitable widths across the slopes on contours, alternate to the strip of soil protecting and erosion resistant crops. The strip cropping shortens the slope length, hinders movements of run-off, help in desilt it and increases the absorption of run-off water. Dense foliage cover of erosion resistant crops viz., groundnut, moth bean and horse gram, prevents beating action of rain drops.
- **Mixed cropping:** Indian farmers have been practicing mixed cropping for better land cover and acting as protection against beating action of rain drops, soil erosion and crop assurance under adverse climatic situations. This is practiced in semi-arid regions to provide as insurance against complete failure of the given crop under adverse climatic conditions.
- **Bench terracing:** It consists of a series of platforms having suitable vertical drops along contours or on suitably graded lines across the general slope of land. Bench terracing is usually practiced on slopes ranging from 6 to 33%. The vertical drop varies from 60 to 180 cm, depending upon the slope and soil conditions. The material excavated from the upper part of terrace is used for filling the lower parts.
- **Grassed waterways:** Grass waterway is a safe disposal structure for higher granting run-off. It protects the land against rills and gullies. Waterway is constructed according to a proper design and vegetative cover is established to save the channel section. The best suited grasses are *Panicum repens, Brachiaria mutica, Cynodon plectostachyus, Cynodon dactylon* and *Paspalum notatum*.

- **Water erosion:** Soil erosion caused by water can be distinguished into (i) sheet erosion (ii) rill erosion (iii) gully erosion.
- **Sheet erosion:** The erosion in which soil metrix is lost but remains undetected for long period and removes a thin veneer of soil uniformly from large areas during every rain, even producing a run-off is sheet erosion. It can be detected from muddy colour of run-off arising from the fields.
- **Rill erosion:** When sheet erosion is allowed to continue unchecked, the silt laden run-off forms a well-defined, minute finger shaped grooves over the entire field is known as rill erosion. The rills formed don't pose any hindrance to farm machinery. The rills vanish after cultivation but reappear with ensuing rains.
- **Gully erosion:** When rill erosion is neglected, the tiny grooves develop into wider and deeper channels known as gully erosion. The gullies obstruct the farm operations. These gullies tend to deeper and widen with every heavy rainfall. These gullies cut the entire field into small fragments and in course of time, make them unfit for cultivation.

Water erosion mechanism and processes

The soil erosion caused by rainfall is due to impact of energy from

(i) falling rain drops obliquely or vertically from above, and

(ii) surface flow more or less horizontally along the ground surface. The main role of falling rain drop is to detach the soil particles (called splash) and that of the surface flow (outride the rills and gullies) is to transport the soil, which gets accelerated on sloping topography.

Causes of erosion

i. Removal of vegetation from soil

ii. Extension of cultivation beyond steep slopes

iii. Permanent agriculture is possible on certain slopping lands but not without adoption of approval soil conservation practices

iv. Shifting cultivation: Clearance of rich forest soil. Forests are uprooted and cleared. Areas are cultivated for 2-3 years and then abandon for 5-15 years and again for cultivation. Common in Tribal Community called Jhooming.

Factors effects soil erosion by water

- Precipitation : Intensity and amount
- Degree and length of slope

- Type of soil e.g. fine soils are more susceptible to erosion
- Nature of ground cover and land use

Effect of erosion

- Soil is lost and its fertility depleted
- Most of water run-off are utilized to sea
- Flood water become more frequently damage both for life and property

19.5. Management of Problematic Soils

Characterization of Problem Soils

The process of weathering, especially in arid and semi-arid regions releases large amount of soluble salts. These salts keep on accumulating at some depth in the soil profiles depending upon rainfall, soil topography and other factor to ionic mobility of salts causing spatial and temporal variation in salt distribution in soil profile.

I. Alkali Soils

- Also called sodic soils
- PH exceeds 8.5
- ESP > 15
- There is conspicuous presence of CO_3 and HCO_3
- Na among the soluble salts
- EC ranges from very low to very high

Effects

Alkali soils develop poor physical properties due to high pH and high ESP. Poor infiltration rate due to dispersed clay particles block the soil pores and so cause crusting and hardening of surface soil upon drying. Water and air movement as well as root growth remain restricted. Soil physical property could deteriorate even at lower value than this.

Reclamation Technique

The main objective of reclamation in alkali soil is to reduce exchange Na content to safe level for optimum plant growth. The reclamation can be achieved either by applying chemical or creating favourable biological activity in alkali soils.

1. Chemical Reclamation

The exchange Na is commonly replaced by Ca through ions like H and Mg. Ca in exchanged reaction is derived from either added Ca amendment or art of Ca present in the nature $CaCO_3$ of soil which is mobilized by using acid or acid forwards.

Chemical used

Gypsum ($CaSO_4.2H_2O$) is most commonly used @ 10-15 ton/ha containing about 70% ($CaSO_4.2H_2O$). About half of actual gypsum requirement is sufficient to reclaim 15 cm surface soil to start growing with rice as the 1st crop.

(a) **Phyrites (FeS_2):** also used in recent years on large scale in some part of country. Biological oxidation of pyrite is carried out by *Thiobaccilus ferroxidins* bacteria only under low pH conditions. But in case of oxidation of element sulphur by *Thiobaccilus thioxidens* very slow in alkali soils. The acid produced through oxidation reaction with $CaCO_3$ of soil to release Ca for exchange with Na on the clay complex. Due to this problem efficiency of pyrite remain between 70-80% as compared with gypsum at recommended dose of application.

(b) **Mineral acids:** Sulphuric acid and Nitric acid and other by products of gypsum from industries manufacturer are used. Citric acid, phosphoric acid and oxalic acids are also used.

(c) **Continuous cropping:** A good amount of biomass of plant parts and root portion is added every year, it will help to increase biological activity.

2. Biological Reclamation

The use of gypsum is generally recommended and it is a costly input. However, the alternative method of biological reclamation seems to be one such alternative. Alkali soils of India and Pakistan contain high level of $CaCO_3$ C is insoluble $CaCO_3$ can be made soluble by adding acid or acid forwards. However, it could be best done by indicating biological activities either through addition of organic matter or cultivation of salt tolerant crops.

Choice of crop and cropping system

During *kharif* season in sodic soils of Sutlej Ganga Plains rice is favoured as first crop because it can tolerate standing water and medium ESP. During *rabi* season only shallow rooted crops like wheat, barley, berseem, shaftal could be grown in the initial years.

Since 0-15 cm surface depth is considered for reclamation of alkali soils. When rice is taken in crop sequence the reclamation of alkali soils is enhanced and the pH of surface comes down to about 9 after three years and it became possible to grow less tolerant and deep rooted crops like Jowar, bajra, oilseeds, cotton, sugarcane, mustard, rapeseed, sunflower, linseed, groundnut are semi-tolerant. It was observed that oilseed crop like rapeseed, mustard and toria can also be grown as first crop followed by jowar and bajra for fodder in *kharif* in highly deteriorated alkali soils with application of 50% growth regulator as gypsum.

Agronomic Techniques for Successful Crop Raising

I. Poor germination stand in alkali soils can be recovered by following special techniques

a) Increased seed rate

b) Restoring spacing

c) Increased number of plants/hill in case of transplanted crops

II. Soil fertility management

The use of manures and fertilizers bring about improvement in chemical and physical properties of soil. Increased crop growth from the addition of manures tends to promote the soil aggregation, reduces crust formation and increased biotic activity.

Application of FYM and other organic manures hasten the reclamation process of sodic soils through solubilization of natural Ca as a result of action of organic acids produced during decomposition of humus. Among green manure plants Dhaincha has been found to be more successful in alkali soils.

III. Irrigation water management

On alkali soils water accumulates following rain or heavy irrigation. It remains on soil surface for long period until it evaporates and percolates. During the drying process particularly in a period of high evaporative atmospheric demand during summer months, the soil surface dries up quickly but there is no change in the water content below 15 cm soil depth. This indicates a very poor water upward flux to replenish the water loss taking place at surface.

19.6. Soil Conservation Practices

The aim of soil conservation is to obtain greatest possible permanent benefit from land. This means using the land and at same time maintaining or improving its productive capacity. Following soil conservation practices should be followed.

1. Agronomic measures
2. Mechanical measures

1. Agronomic measures

These help to intercept rain, reduce splash effect, help to obtain better intake of water. Improve infiltration rate of soil by improving organic matter content and soil structure help to retard or reduce over land runoff through use of contour cultivation, mulches, dense growing crop, strip cropping, mixed cropping.

i. **Contour cultivation** – Crops grow across slope keeping same level as far as possible. Ridges and rows are also placed across slope which reduces runoff, soil erosion and loss of plant nutrients as compared with up and down cultivation. Contour farming conserve soil fertility and increases crop yields. Where slope is 2.2% on alluvial soil and they conserve 11.3 kg N, 11.7 kg P, 44 kg K, 398.1 kg CaO_2, 180.1 kg Mg_2O in one season. Much less power is required to be exerted by man, animal and machines in across slope instead of up and down slope. Less wear and tear of implements and less time is required. Bunding is done along contour to reduce length of slope. All the cultural operations are done parallel to these bunds.

ii. **Mulching:** Surface mulching is used to prevent soil from blowing and being passed away to reduce evaporation. Mulching increase infiltration rate, control weeds, improve soil structure, ultimately increase yields.

iii. **Interculture:** Kills weed and produce 5-6 cm thick soil layer which acts as mulch and help to reduce evaporation from soil surface. It also breaks the surface crust.

iv. **Cover crop:** Growing of crops which provide maximum cover and reduce run off and soil loss which as legumes. Cultivated legumes in general form dense cover and hence protect better cultivated land against erosion than other crops. The crops are cowpea, moong, urdbean, moth, groundnut, sunhemp.

v. **Strip cropping:** Contour strip with growing of soil exposing and erosion permitting crops in strips at suitable width across slope on contours. Alternatively with strip of soil protecting and erosion resistant crops. It reduces length of slope, check movement of run-off water, help to increases absorption of rain water by soil. Dense foliage of erosion resistant crops prevents the rain from beating soil surface directly Groundnut and moth are most suitable crops for checking erosion. But seed rate of legumes need to be tripled. The most effective width of contour strip cereals like jowar, bajra is 21.6m and for legumes 7.2 m.

vi. **Permanent or temporary buffer strip cropping:** Strip are established to take case of critical i.e. steep or lightly eroded slopes in fields under contour strip cropping. These strips do not form part of rotation practiced in normal check erosion. Generally planted with perennial legume, grasses, shrubs on permanent or temporary basis.

vii. **Mixed cropping:** Small land holding is responsible for lack of interest in strip cropping. Some of objective of mix cropping are

(a) Better and continuous cover of land;

(b) Good protecting against rain beating action

(c) Compete protection against soil erosion

(d) Insurance of one or more crop.

Roots of various spp. in mixed cropping system feed at different depth in soil. The line sowing of mixed crop gives rise to practice of intercropping. For instance mixed cropping of wheat and gram are more assured under rain fed conditions in kandi areas of Punjab.

2. Mechanical Measures

These are also called engineering measures and are adopted to supplement agronomic practices when the latter alone are not adequately effective. These include basin listing, sub-soiling, contour bunding, graded bunding and bench terracing on steep slope.

Main objective

I. To increase time of water stagnation by reducing run-off and thereby providing an opportunity for infiltration of water.

II. To divide a long slope into several short ones so as to reduce velocity of run-off, thus prevent erosion.

Few important mechanical measures for conservation of soil are as follows:

1. **Basin listing:** Consist in making of small interrupted basin along contour with a special implement called Basin Lister. This helps to retain rain water as it falls and an especially effective on retentive soil having mild slope.
2. **Sub soiling:** Consisting breaking with a sub-soiler to conserve more rain water by improving physical conditions of soil. This enhances greater moisture infiltration into soil, reduces both runoff and soil erosion. Subsoil work at depth of 30-60 cm and at a spacing of 90-180 cm.
3. **Contour bunding:** Consists in making a comparatively narrow based bunds at intervals across the slope of land. It is an important measure that conserve soil water in arid and semi-arid areas with high infiltration and permeability. Commonly adopted on agriculture land, up to slope of about 6%. Spacing between bunds should not exceed 150 cm for vertical drop or 67.5 cm for horizontal spacing whichever is less.

4. **Graded bunding or channel terraces:** It is used in areas receiving rainfall 780 cm/year but irrespective of soil texture. In clay soil graded bunding has to be used even for areas having less than 80 cm annual rainfall.

5. **Bench terracing:** consists of series of platform having suitable vertical drops along contour or on suitable graded line across generals slope of land. The vertical drop may vary from 60-180 cm depending upon slope and soil conditions. Bench terraces can be practiced on undulating land and intensive farming areas. Bench terracing helps to retain soil moisture, manure and fertilizer that facilitate application of irrigation if available. In rainfed areas terracing is usually practiced on slopes ranging from 6-33%. It may have to be used on gentle slope, if irrigation is to be applied to crop.

6. **Wind break:** It is a narrow belt of trees planted in one or a few rows in a field or in one or more sites of a farm to promote protection from wind.

7. **Shelter belt:** Comprises of wide belt of trees planted in several rows at right angles to prevailing wind direction. Wind break and shelter belt are not applied in India on significant scale in systematic pattern.

Practices used to control wind damage/erosion

- Crop residue may be left on field until planting time or stubble mulch tillage use to retain or protected amount of residue on surface of soil.
- A mulch such as manure @ 15 tonne /ha is applied just before or immediately after planting a crop.
- Keep tillage practices to avoid the pulverizing the small lots
- The soil surface is not smooth but left rough
- Tillage and planting machines are used which prepares 2-7 inches ridges in height
- Cultivating and planting operations are done at right angle or across from direction of prevailing winds
- Strip cropping used, strip are designed so that vegetative land may furnish protection to cultivated area.
- Field wind breaks of trees or shrubs are planted in rows at right angle to prevailing winds. These will provide effective control for 10 times their height on the ground (soil surface).

20

Weed Management With Judicious Use of Herbicides

Herbicides are the chemicals used for killing or suppressing the growth of weed plants from field / horticulture / plantation crops as well as non-cropped areas/ water bodies etc. The herbicide is said to be an ideal herbicide when it is very safe to the crop on which it is applied and on the other hand, it must control all associated weeds effectively. Also, ideal herbicides should leave no residual effects on the succeeding crop(s). Herbicides are of two types i.e.

(i) Selective herbicides and

(ii) Non-selective herbicides.

A selective herbicide is that which is very safe to crop plants but unsafe to weed plants e.g. atrazine in maize, clodinafop in wheat etc. On the other hand, non-selective herbicides are those which kill all green plants including crop and weed plants which come in their contact e.g. paraquat and glufosinate ammonium. However, a non-selective herbicide can be used as a selective herbicide by applying directed application only on weeds with the use of spray hood. Avoid the contact of non-selective especially translocated herbicides glufosinate ammonium with the crop at the time of spray in order to avoid crop losses. Glufosinate ammonium is non-selective, very popularly used for controlling annual and perennial weeds from non-cropped lands as blanket application and from wider row crops as directed application only on weeds. It is a translocated type of herbicide and its contact with the crop must be avoided. On the other hand, paraquat (Gramoxone) is a non-translated herbicide and its limited contact especially with lower leaves leads to no loss of the crop. However, its contact with growing points of crop plants is harmful.

20.1. Importance of Herbicides

Herbicides are the agrochemicals used widely in agricultural crops for the control of weeds as pre or post – emergence, which have both beneficial and harmful effects. Many farmers use herbicides wrongly, for example using over/under dose, wrong time and method of application, use of un-recommended herbicides, repeated use of herbicides etc. which may lead to huge crop losses. Judicious use of herbicides control weeds very effectively and prove very safe not only to the present crop but also to the succeeding crops. An ideal herbicide is that which provides very effective control of weeds and should leave no residues in crop produce on which it is applied and must be safe for the following (succeeding) crops.

In crop production, it is not possible to eliminate the use of herbicides because chemical weed control is the only option available with the farmers for controlling weeds from the field crops as it is very economical method as compared to other methods. The scope of mechanical weeding is decreasing day by day due to high wages of farm labour and their non – availability when required. Chemical weeding has become a more popular method of weed control in field crops as compared to other methods as it eliminates early crop weed competition, controls both inter-row and intra-row weeds very effectively, less influenced by environmental factors, very economical and quick method of weed control etc. Herbicides are a very useful tool for improving crop produce provided these are used judiciously. Indiscriminate/wrong use of these agrochemicals (herbicides) may lead to many harmful effects such as environmental pollution, residues in crop produce as well as in soil, crop toxicity, human health problems, herbicide drift and run-off, un-uniform control of weeds etc.

However, herbicides are much safer as compared to other pesticides like insecticides, fungicides, rodenticides etc. and on the other hand losses due to weeds for agricultural crops are much more than insects and disease organisms. The use of herbicide is very popular especially in rice and wheat for the control of grassy and broadleaf weeds. In Punjab State about 90 to 95 per cent area each under rice and wheat is treated with one or the other herbicide, which may be due to the presence of morphologically similar weeds in these crops, which are very difficult to control with mechanical methods of weed control. Hence farmers prefer chemical weeding, as it is a more effective and economical method of weed control. Apart from these crops, herbicide use is also very common in maize, cotton, sugarcane, oilseeds, pulses, fodder crops, vegetables etc.

Few non-selective herbicides such as paraquat and glufosinate ammonium are very common in non-cropped areas as well as in horticultural and plantation crops. For annual weeds, use paraquat whereas for joint control of annual and perennial weeds, use glufosinate ammonium. Paraquat is non-translated and glufosinate ammonium is translocated non - selective herbicide. For perennial weed control, use low dose so that leaves are not killed quickly and herbicide may be translocated to the root stock in higher amounts. The spray can be repeated if need be i.e. if weeds regenerate or on left over weeds. For the control of *Cyperus rotundus,* repeated spray of 2, 4-D is very effective. The objective for perennial weed control is to kill the root stock with any translocated herbicide and foliage must be killed due to starvation. So, judicious use of herbicide may lead to good results, which are not possible with other methods of weed control.

Agriculture cannot be made successful/economical without the use of herbicides but these agrochemicals must be used judiciously so that their benefits can be harvested and their harmful aspects could be avoided. Adoption of following points/tips will not only help in effective kill of weeds but will lead to more safety of crop plants without any soil or environmental pollution.

20.2. Tips For Judicious use of Herbicides

Following points must be taken into consideration in order to obtain judicious use of applied herbicides.

1. Identification of weed flora

Herbicides are meant for killing the specific weed(s) and all spectrum of weed kill is different for different herbicides. So, before planning herbicidal control in a particular crop, it is very important to know the weed flora present in the field so that a selective and effective herbicide is selected and can be purchased for spraying. Even the history of previous weed flora (of last year) must be known especially for pre-emergence applications. For instance, farmers of Punjab State invariably use pre-emergence (to weeds) herbicides such as butachlor, anilofos, pretilachlor, pyrazosulfuron etc. in the fields which do not have specific (target) weeds such as *Echinochloa crusgalli* (Swank/Maddal), *Ischaemum rugosum* (wrinkle grass) etc. for which these are effective. Hence, it is non-judicious use of herbicides, which must be avoided. In case sedges and grassy weeds are present in rice crop, select bispyribac which is applied as post-emergence.

Similarly, in wheat non-judicious use of herbicides is also on the increase, as farmers do not select specific herbicide as per present weed flora at the time of spray. For instance, 2, 4-D is commonly used for the control of all types of broadleaf weeds from wheat whereas few hardy weeds (difficult to control) are not killed with 2, 4-D. Select metsulfuron when Kandyali Palak (*Rumex spinosus*) is present and for control of Buttonweed (*Malva neglecta*) effectively, use carfentrazone. These herbicides also control other broadleaf weeds from wheat crops and these are substitutes for 2, 4-D. So, select herbicide as per weed flora. Similarly, if wild oats (*Avena ludoviciana*) and Gullidanda (*Phalaris minor*) are present, use clodinafop, pinoxaden or fenoxaprop-p-ethyl. Prefer to select mesosulfuron + iodosulfuron 3.6 WDG or sulfosulfuron + metsulfuron 75 WG or fenoxaprop 8% + metribuzin 14% or metribuzin + clodinafop in fields of wheat crop having joint infestations of grassy and broadleaf weeds.

So, knowledge regarding the type of weeds infesting the crop is very important and herbicides should be selected according to type/dominance of weed flora. It is not advisable to use herbicides every year and its use can be omitted if weeds are not present or they are in low population. Other methods of weed control like mechanical weeding or cultural control measures can be adopted if weed pressure is low. Only need based use of herbicides should be done i.e. when it is not possible to control weeds infesting the crop with any other weed control method.

2. Selection of herbicides

Lot of herbicides and brands of herbicides are available in the market. The use of un-recommended herbicides or un-recommended brands of recommended herbicides may lead to many serious problems such as poor or no control of weeds, toxicity on crop / complete crop failure or many times toxicity on succeeding crops is also observed. Many Punjab farmers used un-recommended brands of sulfosulfuron or unrecommended chemicals for the control of *Phalaris minor* from wheat, which resulted in 70 to 100% damage of wheat crop. Also, due to severe residual toxicity especially due to toxic metabolites of sulfosulfuron in soil, these farmers were unable to grow any catch crops such as summer moong/ summer mash/cucurbits/fodders like maize, sorghum, bajra etc. Poor selectivity (tolerance) of wheat crop to unapproved brands of sulfosulfuron may be due to poor quality or incompatibility of used surface-active agents (surfactants) with herbicide i.e. sulfosulfuron. Also, few farmers use metribuzin (un-recommended herbicide) for the control of *Phalaris minor* in wheat and suffered a great loss due to its low selectivity margin as slight overdosing may kill wheat crop.

So, after identifying the weed flora of the crop/field, suitable herbicide should be selected which is duly recommended by State Agricultural Universities. The use of un-recommended herbicides as well as un-recommended brands of recommended herbicides may lead to severe problems such as poor control of weeds as well as crop toxicity. So, for safe (judicious) use of herbicide, use only recommended herbicides along with their approved brands. Never select the wrong herbicide and always select the herbicide as per the weed flora present in the crop. Do not adopt one policy (one herbicide) for your entire farm but selection of herbicide may vary from one field to another field depending upon weed flora and nature of succeeding crop.

Apart from above points, the nature of succeeding crops and type of the crop growing in the adjoining fields should be kept in mind. For instance, do not use sulfosulfuron but go for other recommended herbicides like clodinafop, fenoxaprop-p-ethyl or pinoxadin for the control of *Phalaris minor* in wheat fields, which are to be succeeded, by maize, sorghum, bajra etc in order to avoid risk of herbicide residues on succeeding crops. Similarly, the use of 2,4-D in wheat should be avoided when the adjoining field is of broadleaf crops like oil seed crops, gram, vegetables etc. During summers, do not use 2,4-D in the sugarcane field adjoining to the cotton field as cotton is very sensitive to 2,4-D as even traces of 2,4-D can damage the cotton crop very badly. Apart from wind drift, vapour drift is also very important in 2,4-D which may damage the distant crops (sensitive) especially during hot days so judicious use of herbicides will lead to no harmful effects.

3. Tips for herbicide purchase

After selection of a particular herbicide, selection of a reliable dealer in your nearby town/city is very important. In case no reliable dealer (shopkeeper) is present in your village/town, then you must visit the adjoining big city to purchase herbicide. The most important point is to obtain a receipt inclusive of the batch number from the shopkeeper/dealer. If any pesticide dealer hesitates to give you a receipt, it may be assumed that he is selling poor quality products. Never buy herbicide as per the advice of the dealer but ask for a particular herbicide as per your situation. Also check the expiry date of herbicide before payment and never purchase the expired packet/container. Damage to crop, if any, will not be owned by the dealer/shopkeeper in the absence of receipt. So, always purchase recommended herbicide along with approved brands from a trustworthy shopkeeper only after obtaining proper receipt.

4. Calibration of spray pump

Calibration is the procedure to have an estimate regarding the volume of water required for spraying a unit area. The pump, nozzle and operator who has done calibration should not be changed at the time of spray of herbicide in the field. For pre-emergence application of herbicides, do calibration on bare soil and for post-emergence herbicides, it must be done in the crop field, which is to be treated with the herbicide. Never dissolve herbicide in a particular (estimated) volume of water without calibration, which may lead to either scarcity, or surplus of spray fluid. Weeds will be killed only when these are exposed to a minimum lethal dose of herbicide which is only possible with calibration. As per calibrated amount for one acre, the spray fluid can be made in bulk or as stock solution i.e. total calibrated amount divided by fifteen (the capacity of knapsack sprayer). So, to obtain uniform weed kill in the field, spray uniformly over the entire field, which is only possible with calibration.

5. Selection of spray pump and nozzle

There are many types of spray pumps (equipment's) and nozzles available in the market. For pre-plant and pre-emergence herbicide applications use tractor mounted sprayer or knapsack sprayer fitted with flood zet nozzle which have high discharge rates (Fig 20.1). On the other hand, for post-emergence applications use either power sprayers or knapsack sprayer fitted with flat fan or flood zet nozzle of small orifice (hole). More quantity of spray fluid (> 200 lit./acre) is required for more efficacy of pre-plant or pre-emergence herbicides. Higher volume of water will enable the herbicide molecules to enter into soil water stream. For post-emergence herbicides, always use less spray volume (100 to 120 lit/acre) so that there is less bounce off of spray droplets from leaves of weed plant and more retention of spray fluid is ensured. High spray volume will produce bigger droplets, which will not retain on foliage but will fall on ground quickly after spray. More retention of spray droplets is required for more absorption of herbicide in the target plants. Do not use defective nozzle with un-uniform spraying pattern. To check the nozzle, spray plain water on a pucca floor and if the wetting is uniform, the nozzle is all right and if wetting is not uniform then either replace spray tip or nozzle. For effective control of weeds uniform spray of herbicide is very essential which depends only on type of pump, nozzle and patience of worker doing spray.

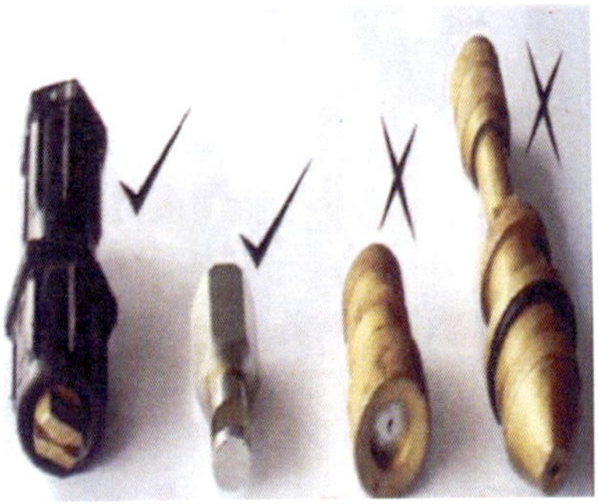

Flat fan Flood zet Single action Triple action
Types of nozzles

Knap sack sprayer

Fig. 20.1: Types of nozzles and spray with knap sack sprayer

6. Selection of time of herbicide spray

Pre-plant herbicides should be applied just before sowing the crop on the prepared field and there should be minimum gap between land preparation and herbicide application as well as between herbicide application and sowing of crop for better results. Also, pre-emergence herbicides must be applied immediately after sowing the crop so that there is good soil moisture for better herbicide results. Another very important point is that during hot months, do not spray during noon hours but always spray pre-emergence herbicides either in the morning or in the evening hours. Many post-emergence herbicides (for example 2,4-D) have no direct effect with soil moisture but foliage of weeds should be fully open so that there is more retention, adsorption, penetration and translocation of herbicide. So, do not spray post-emergence herbicides on wilted foliage. On the other hand, avoid spray during excessive foggy hours of the day (morning) during winters, as retention of spray fluid on foliage will be less due to the presence of more water on the foliage. Do not spray on windy days which may lead to not only un-uniform application of herbicide but drift of herbicide will also be high which can damage the adjoining sensitive crops. Another very important point is that always spray on actively growing weeds so that there is maximum uptake and transport of herbicide in the target plants. For instance, herbicides used in wheat for control of *Phalaris minor* are more effective on fertilized fields (especially nitrogen) than on non-fertilized fields, which may be due to more uptake due to fast metabolic activities under the former conditions.

The last but not the least important point regarding time of herbicide application is that these must be sprayed as per recommended growth stage of crop and weeds. Application of herbicide before the recommended time may lead to crop toxicity whereas its delayed application may lead to poor weed control. For instance, wheat can tolerate 2,4-D when applied at maximum tillering stage but it's wrong application (before or after the optimum stage) may lead to many deformities in wheat plant such as tubular leaves, branching of the ears, swelling of wheat ears, spikelet's without grains etc. In transplanted paddy, there should be minimum gap between puddling and transplanting as well as between transplanting and herbicide application for better weed control efficiency. For effective control of perennial weeds, time of application is very important. Always spray on actively growing weeds and do not spray on grown up weeds due to their poor metabolic activities. Cut the foliage of grown up weeds and for better results spray post-emergence herbicides when new growth appears.

7. Dose of herbicide

Depending upon crop tolerance, nature of soil, climatic conditions and type of weeds present in the field, herbicides are recommended at different doses. The content of soil organic matter also decides the dose of pre-emergence herbicide to be applied. Fields with high organic matter content need a higher dose of herbicide due to more adsorption capacity of such fields. Herbicides must be used at the recommended doses as per State Agricultural University. The dose of herbicide is recommended to fulfill the following two criteria.

(I) Weeds must receive lethal doses (minimum dose required for killing the target weeds).

(II) Crops must tolerate even double the recommended dose of herbicide.

The use of lower dose may lead to poor weed control as well as early development of resistance in weeds and on the other hand use of higher dose may lead to crop toxicity as well as residual problems in the present as well as in the succeeding crops. Under dose of herbicide may lead to poor weed control and hence more losses in grain yield apart from increasing soil weed seed bank status which leads to more weed problems during coming years.

So, for proper and sustainable weed control, use herbicides at the recommended rates. Also the phenomenon of resistance in weeds is delayed when herbicides are used at proper rates, whereas under/over dose of herbicide may lead to early development of resistance in weeds. Proper dose of herbicide will provide better weed control, no soil residual problems on succeeding crops, no residues in crop produces as well as any environmental pollution. Dose of herbicide is also recommended depending upon soil type and organic matter content. For instance, use of atrazine at 1.0 kg/ha on heavy textured soil and at 0.625 kg/ha on light textured soil for weed control in maize is recommended.

8. Method of herbicide application

Out of different methods of herbicide application, only two methods viz. spray and broadcast are commonly used. Broadcast method of herbicide application is only recommended in puddled transplanted paddy after mixing with sand or fertilizers in standing water. Ponding of water for 5 to 7 days after the application of herbicide, improves efficacy. So, uniform mixing and uniform spread of pre-emergence to weed herbicides is required for proper weed control in transplanted rice. Delay in transplanting after puddling and delay in herbicide application after transplanting seedlings in the field, may lead to poor weed control. Ponding of water integrates the effect of applied herbicides.

In all field crops except transplanted rice, herbicides are applied through sprays i.e. recommended dose of herbicide is mixed in recommended (calibrated) volume of water and this spray fluid is sprayed uniformly over a unit area. The most important point in proper weed management is that spray must be as uniform as possible. Do not use contaminated/dirty water for preparing spray fluid, as it will choke the nozzle frequently while spraying. So always use clean water for spray fluid and spray in strips with slight overlapping (Figure 20.2). Do not move the nozzle to and fro while spraying. The height of nozzle from ground surface or crop canopy must be 50 cm. It will be appreciable to divide the field into small plots equivalent to the number of spray pumps (total spray fluid in litres divided by 15, capacity of pump) for uniform spray. Prefer to use multi boom nozzles (2 to 10) for uniformity of spray. Always use a flat fan or flood zet nozzle for spraying herbicides and don't use cone type nozzles. The height of nozzle from ground surface or crop canopy should be 50 cm and it should be neither more nor less for uniform application of herbicide. Spray it with a regular pressure of 1.5 to 2.0 bars and use pressure regulator valves for uniform spray.

Few farmers broadcast even foliar uptake herbicides (translocated) after mixing with sand and as a result no control of weeds is attained so this practice must be avoided. The pre-plant herbicides are volatile in nature and these must be incorporated in top soil immediately after spray in order to avoid vapour losses. Leaving the field without incorporation or delayed sowing of the crop after spray will not only reduces the concentration of herbicide in the field but will lead to more environmental pollution. For better results, herbicides must be applied as per recommendations, for example, metsulfuron/ azimsulfuron/ ethoxysulfuron/ bensulfuron-methyl are recommended as post-emergence for the control of broadleaf weeds and sedges from transplanted rice. These herbicides must be applied through spray after draining excess water from the field so that leaves of these target weeds may come in contact with the herbicide solution. Few farmers, spray these herbicides in the ponded field and no/poor control of weeds is attained because weed plants don't receive lethal dose required for killing. To conclude, uniform spray of herbicides leads to effective control of weeds and on the other hand un-uniform spray not only leads to poor weed control but crop toxicity can also be seen.

Wrong methods

Right methods

Fig. 20.2: Right and wrong methods of herbicide application

9. Spray volume

As discussed earlier too, use high spray volume (>200 lit/acre) for better efficacy of pre-plant or pre-emergence herbicides. So, spray all pre-emergence herbicides just after sowing on moist fields for better results. The efficacy of pendimethalin is directly correlated with soil moisture and spray volume. Poor results are obtained with delayed application, as these herbicides are effective only on emerging weed seedlings not on emerged seedlings. For foliar applied herbicides use less water (100-120 lit/acre) so that there is long contact of spray droplets with the foliage due to more retention of smaller droplets on the leaf surface. It will improve retention, absorption and translocation of herbicide into the foliage and hence there will be better kill of weed plants. For postemergence applications, use nozzle with fine droplet (flat fan) and for pre-emergence herbicides, use nozzle with coarse droplets (flood zet). The spray volume has an indirect relation with pressure of the pump i.e. more pump pressure less will be the discharge rate i.e. amount of spray fluid used. For instance, the discharge rate of backpack power sprayer is much less than that of knap-sack sprayer. The use of a pressure regulator valve is very important for uniform application of herbicide. The herbicide solution must be as fresh as possible and do not store spray fluid till the next day.

10. Adopt proper spray technology

Adoption of proper spray technology is a key factor for judicious use of herbicides. Few important points for proper/judicious use of herbicides are:

- Calibration of the pump is a must to find out the estimate of water required for spraying a unit area.
- The person who had done calibration must be deputed for spray.
- For herbicide spray, use flat fan or flood zet nozzle and do not use cone type nozzle.
- Before spray, check the nozzle by spraying plain water on the pucca floor and if wetting is not uniform then either replace the tip or nozzle.
- Spray in parallel bands (strips) with slight overlapping. Do not move spray lance to and fro.
- Prefer to use a multiboom nozzle for uniform application.
- Never spray on windy days. Always spray at right angles to the direction of the wind.
- The height of the nozzle should be 50 cm from ground/foliage of the crop.
- Always use a pressure regulator valve while spraying.
- Stir spray fluid before each filling.
- If soil moisture is less, increase spray volume.
- During hot months spray either in the morning or in the evening hours.
- Discard leftover solution on waste lands.

- For precise application of herbicide, divide the field into number of plots equivalent to number of spray pumps to be used (total spray volume divided by 15; the capacity of sprayer)
- Do not eat/smoke while spraying.
- Always wear a full sleeve shirt, trousers, gas masks etc. while spraying.
- Do not spray with an empty stomach.
- Take bath after spray is over and wash clothes.
- Store left over herbicides in a separate store after proper labeling.
- Destroy the herbicide containers after spray.

11. Use herbicide in rotation

Do not stick to one herbicide group every year, as it will lead to early development of resistance as well as shift in weed flora. To avoid these serious problems, use herbicides in rotation i.e. go on changing herbicide groups every year. For instance, continuous use of isoproturon in Punjab and Haryana in wheat had led to the problem of resistance in *Phalaris minor* to this herbicide. Similarly use of clodinafop for 2 to 3 years for the control of *Phalaris minor* in wheat has led to the problem of broad leaf weeds in wheat. So, for sustainable weed management adopt rotational use of herbicides as per weed problems.

12. Select herbicide as per cropping sequence

The recommended herbicides are very safe for the crop on which these are being used. Few active chemicals (molecules) of recommended herbicides or their metabolites produced in soil after their applications are very toxic on the following/succeeding crops. Due to residual toxicity, these succeeding crops show no germination or sometimes when seeds germinate, their growth and development is poor. For instance, wheat can tolerate even double the recommended dose of sulfosulfuron but residual toxicity on succeeding crops like maize, sorghum, bajra etc have been observed even with the recommended dose of this herbicide to wheat crop. Similarly, application of atrazine in potato or late sown fodders may show residual toxicity on few succeeding crops like oats, wheat, oil seeds, berseem, cucurbits etc. So, selection of herbicide should be done by keeping in view the succeeding crop also as per the cropping system of our farm.

13. Adopt integrated weed management approach

When more than one method of weed control is used, it is called integrated weed management (IWM). This approach has many advantages like development of resistance to herbicides in weeds can be delayed, environmental pollution and residual problems can be reduced, reduction in weed seed bank of soil etc. Generally, farmers spray herbicides in their crops and later on they don't bother about the escaped weeds. However, the left over or hardy weeds should be either retreated with suitable herbicide or must be uprooted for better results.

Band application i.e. application of herbicide only on crop rows (in wider row sown crops only) is a very safe and effective method for reducing herbicide residual toxicity in soil. Atrazine at 1/3rd of the recommended dose can be sprayed on rows of maize crop and weeds growing in inter row area can be treated with non-residual contact herbicide (paraquat) or can be intercultured with hand tools or tractor mounted implements. The herbicide residues can be reduced in soil with the adoption of this technique.

In maize few grassy weeds such as *Acrachne racemose, Commelina benghalensis* and *Bracharia reptans* are not killed with the application of atrazine (recommended herbicide) and these can be very effectively controlled with the tank mix application of atrazine with pendimethalin or alachlor or trifluralin as reported by Walia *et al* (2007). Pre-emergence application of atrazine at recommended rate (1.0 kg/ha) resulted in higher dry matter accumulation by weeds and less grain yield of maize as compared to tank mix application of atrazine at lower dose (0.75 kg/ha) with pendimethalin (0.75 kg/ha) or alachlor (1.25 kg/ha) or trifluralin (1.20 kg/ha) as pre-emergences. Hence tank mix application of herbicides increased grain yield of maize due to effective control of weeds.

20.3. Conclusions

It may be concluded that herbicide use in agriculture is necessary for controlling weeds. However, their need-based use is required and avoid their indiscriminate use which is done at present. If herbicides are used judiciously i.e. as per recommended schedule, they do not leave any residues in soil, no environmental pollution is created, no residues are detected from crop produce. Also, there will be no pollution of underground water. The problems due to herbicides may be observed when their indiscriminate use is made i.e. overdosing, wrong time of application, use of un-recommended herbicides or their brands, use of herbicides in many splits etc. Due to these wrong practices not only the present crop but succeeding crops may be damaged and there will be more problems of crop residues. For exploiting the benefits of herbicides, always use recommended herbicides with recommended technology. On the other hand, sole dependence upon the herbicides for weed control is also not desirable. For sustainable weed management, integrate judicious use of herbicides with other methods of weed control such as preventive, mechanical, cultural etc.

References

Anonymous, 2022. Package and practices for cultivation of *kharif* crops, PAU, Ludhiana (Punjab)

Anonymous, 2022-23. Package and practices for cultivation of *rabi* crops, PAU, Ludhiana (Punjab)

Al-Kodmany, K. (2018). "The vertical farm: a review of developments and implications for the vertical city". *Buildings*. **8** (2): 1–24.

Ali Sheikh Amir, S. Elamathi, Singh Suryendra, Debbarma, Victor and Ghosh, Gautam. 2017. Leaf Colour Chart for Proficient Nitrogen Management in Transplanted Rice (*Oryza sativa* L.) in Eastern Uttar Pradesh, India. *Int.J.Curr.Microbiol.App.Sci.* **6**(11): 5367-5372.

Auld, B.A., Talbot, H.E. and Radburn, K.B. 1992. Host range of three isolates of *Alternaria zinniae* a potential biocontrol agent for *Xanthium. Plant Preot.* **7**: 114-116.

Balole, T. V. ; Legwaila, G. M., 2006. *Sorghum bicolor* (L.) Moench. Record from Protabase. Brink, M. & Belay, G. (Editors). PROTA (Plant Resources of Tropical Africa / Resources végétales de l'Afrique tropicale), Wageningen, Netherlands.

Bauer A and Black A L 1992. Organic carbon effects on available water capacity on three soil textural groups. *Soil Sci Soc Am J* **56**: 248-54.

Beri V, Sidhu B S and Bahl G S 2002. Crop residue management. *In: Recent Advances in Agronomy* (Singh G, Kolar J S and Sekhon H S, eds). Indian Soc Agron, IARI, New

Beri V, Sidhu B S, Bahl G S and Bhat A K 1995. Nitrogen and phosphorus transformation as affected by crop residues management practices and their influence on crop yields. *Soil Use and Management* **11**: 51-54.

Bhagat R M and Verma T S 1991. Impact of rice straw management on soil physical properties and wheat yield. *Soil Sci* **152**: 108-15.

Bhandari A L and Walia S S 2000. Effect of FYM, green manuring and crop residue incorporation on crop and soil productivity under combine harvested rice-wheat system. *Proc. International conf on managing the natural resources for sustainable crop production in the 21st century*. Feb. 14-18, 2000. IARI, New Delhi, India **4:** 19-20.

Bhat A K, Beri V and Sidhu B S 1991. Effect of long term recycling of crop residues on soil productivity. *J Indian Soc Soil Sci* **39**: 380-83.

Bhattacharya, P and Gehlot, D. 2004. Current status of regulatory mechanism in organic farming. *Fertilizer News* **49**(11): 15-31.

Bhattacharya, P. and Chakraborty, G. 2005. Current status of organic farming in India and other countries. *Journal of Fertilizers* **1**(9): 111-113.

Bowers, R.C. (1982) Commercialization of microbial biological control agents. In: Charudattan, R. and Walker, H.L. (eds) Biological Control of Weeds with Plant Pathogens. John Wiley and Sons, New York, pp. 157–173

Brown, W.L. and Darrah, L.L. 1985. Origin, Adaptation, and Types of Corn. *National Corn Handbook*-**10,** pp.1-6.

Bruce, A.A. and Louise, M. 1995**.** Constraints in the development of Bio-herbicides, *Weed Technology* **93**: 638-652.

Chand, R., Prasanna, P. A. L. and A. Singh. 2011. Farm size and productivity: Understanding the strengths of smallholders and improving their livelihoods. *Economic and Political Weekly*, **46** (26 & 27): 10-14.

Charudattan, R. and Walkar, H.Z. 1982. "In biological control of weeds with plant pathogens". John Wiley and Sons, New York. 293 p.

Charudattan, R. Devaleria, J.T. and Prange, V.J. 1990. Special problems associated with aquatic weed control. P. 287-303 in R.R. Baker and P.E. Dunn eds. New Directions in Biological Control. Alternatives for suppressing agricultural pests and diseases. Alan. R. Liss Inc. New York.

Chaudary, OP and Kashyap RK 1987. Field fertilization vis-à-vis incidence of insect- pest, complex and marketable yield in egg plants. (*Solanum melangina*). *Italino di Enotomologia* : **41** 3-20.

Chengyan Zheng, Yu Jiang, Changqing Chen, Yanni Sun, Jinfei Feng, Aixing Deng, Zhenwei Song, Weijian Zhang, 2014. The impacts of conservation agriculture on crop yield in China depend on specific practices, crops and cropping regions, *The Crop Journal,* **2**, Issue 5, Pages 289-296.

Dhiman S D, Nandal D P and Om Hari 2000. Productivity of rice (*Oryza sativa*)-wheat (*Triticum aestivum*) cropping system as affected by its residue management and fertility levels. *Indian J Agron* **45**: 1-5.

Dillon, S.L., Shapter, F.M., Henry, R.J., Cordeiro, G., Izquierdo, L., Lee, L.S. 2007. Domestication to crop improvement: Genetic resources for *Sorghum* and *Saccharum* (Andropogoneae). *Ann. Bot. (London, U.K.)* **100**:975–989.

FAO. 2013. Climate-Smart Agriculture: Sourcebook. Rome, Italy: Food and Agriculture Organization of the United Nations

FAO. 2019. FAO's work on climate change. United Nations Climate Change Conference 2019. Rome. 40 pp. www. fao.org/3/ca7126en/ca7126en.pdf.

Friedrich T, Derpsch R, Kassam AH 2012. Global overview of the spread of conservation agriculture. *Field Action Sci Rep* **6**:1–7

Ghidan AY, Al-Antary TM, Awwad AM, Ayad JY, 2018. Physiological effect of some nanomaterials on pepper (*Capsicum annuum* L.) plants. *Fresenius Environmental Bulletin.* **27**(11):7872-7878

Ghidan, Alaa and Antary, Tawfiq. 2019 "Applications of Nanotechnology in Agriculture". Applications of Nanobiotechnology, edited by Margarita Stoytcheva, Roumen Zlatev, *IntechOpen,* **10**.5772/intechopen.88390.

Ghosh S K 2000. Organic farming for sustainable development. *Kisan World* **27**(5): 14-17.

Gupta, R.K.; Kaur, J.; Kang, J.S.; Singh, H.; Kaur, S.; Sayed, S.; Gaber, A.; Hossain, A. 2022. Tillage in Combination with Rice Straw Retention in a Rice–Wheat System Improves the Productivity and Quality of Wheat Grain through Improving the Soil Physio-Chemical Properties. *Land* **11**, 1693.

Haggblade, Steven & Tembo, Gelson, 2003. "Conservation farming in Zambia:," EPTD discussion papers 108, International Food Policy Research Institute (IFPRI).

Harrington LW 2008. A brief history of conservation agriculture in Latin America, South Asia and Sub-Saharan Africa. PACA, 1st floor, NASC Complex, DPS Marg, Pusa

Hegde R N 1997. Organic farming. *Kisan World* **24**(8):43-44.

Herrmann, G.A. and Rundgren, G. 2007. Certification and Accreditation. In: *The World of Organic Agriculture: Statistics and Emerging Trends, 2007.* Willer, H. and Yussefi-Menzler, M. (Eds.).

Hobbs PR, Sayre K, Gupta R 2008. The role of conservation agriculture in sustainable agriculture. *Philos Trans R Soc* **363**:543–555

https://agritech.tnau.ac.in/agriculture/agri_nutrientmgt_fertigation.html Nutrient Management, Fertigation https://climate.nasa.gov/effects/

https://climatechange.chicago.gov/climate-impacts/climate-impacts-agriculture-and-food-supply

https://justenergy.com/blog/the-long-term-effects-of-global-warming/

https://www.agricultureinindia.net/case-studies/case-studies-on-organic-agriculture-in-india/20951 (Nitiparna R)

https://www.noaa.gov/education/resource-collections/climate/climate-change-impacts

https://www.sopa.org/statistics/soybean-production-by-state/?search_type=search_by_year&arrive=2020&starting_year_value=&ending_year_value=&submit=Search, SOPA Databank

https://www.statista.com/statistics/267271/worldwide-oilseed-production-since-2008/, Worldwide oilseed production in 2021/2022, by type

https://www.un.org/en/climatechange/what-is-climate-change

https://www.usgs.gov/faqs/what-are-long-term-effects-climate-change#:~:text=Scientists%20have%20predicted%20that%20long,resources%20in%20semi%2Darid%20regions

Huber, B., Kilcher, L. and Schmid, O. 2007. Standards and Regulations. In: *The World of Organic Agriculture: Statistics and Emerging Trends, 2007.* Willer, H. and Yussefi-Menzler, M. (Eds.).

International Federation of Organic Agriculture Movements (IFOAM), Bonn, Germany and Research Institute of Organic Agriculture (FiBL), Frick, Switzerland. pp. 56-66.

International Federation of Organic Agriculture Movements (IFOAM), Bonn, Germany and Research Institute of Organic Agriculture (FiBL), Frick, Switzerland. pp. 67-74.

Hussain Z M, Choudhury M H K and Alam Q K 2001. Effects of ecological agriculture on soil properties and arthropid diversity in rice-based cropping systems in flood plain areas in Bangladesh. *Intl Conference on Nature Farming and Ecological Balance*, ISNF, CCS HAU, Hisar, India. March 7-10.p 160.

Jat, M.L., Parvesh Chandna, Raj Gupta, S.K. Sharma and M.A. Gill. 2006. Laser Land Leveling: A Precursor Technology for Resource Conservation. Rice-Wheat Consortium Technical Bulletin Series 7. New Delhi: Rice-Wheat Consortium for the Indo-Gangetic Plains.

Jeetendra, Aryal and Jat ML, 2015. Laser land levelling: how it strikes all the right climate-smart chords, (international maize and wheat improvement center (CIMMYT))

Kenny, D.S. 1986. De vine the way it was developed: an industrialist's view, *Weed Sci.* **34** (Suppl. 1) 15-16.

Kilcher, L. 2007. How organic agriculture contributes to sustainable development. In: *The World of Organic Agriculture : Statistics and Emerging Trends, 2007.*

Kler D S, Kumar A, Chhina G S, Kaur R and Uppal R S 2001. Essentials of organic farming – a review. *Environ Ecol* **19**: 776-98.

Kler D S, Uppal R S, Kumar A and Kaur G 2002. Imbalance and maintenance of eco- system – a review. *Environ Ecol* **20**: 20-48.

Kumar A and Yadav D S 2003. Long term nutrient management for sustainability in rice- wheat cropping system. *Fert News* **48**(8): 27-28 & 31-34.

Kumar, S. and Jain, D. K. 2005. Are linkages between crops and livestock important for the sustainability of the farming system? *Asian Economic Review*, **47** (1):90-101.

Kumara D. Suresh and Palanisami K. 2010, Impact of Drip Irrigation on Farming System: Evidence from Southern India, *Agricultural Economics Research Review* **23** pp 265-272

Liebhardt W C, Andrews R W, Culik M N, Harwood R R, Janke R R, Radke J K and Rieger-Schwartz S L 1989. Crop production during conversion from conventional to low input methods. *Agron J* **81:** 150-59.

Lindwall CW, Sonntag B 2010. landscape transformed: the history of conservation tillage and direct seeding, knowledge impact in society. University of Saskatchewan, Saskatoon

Lipper, L., Thornton, P., Campbell, B. *et al.* 2014. Climate-smart agriculture for food security. *Nature Clim Change* **4**, 1068–1072. https://doi.org/10.1038/nclimate2437

MAFF (Ministry of Agriculture, Forestry and Fisheries, Japan, 2000, Climate change and Agriculture.

Mahapatra B S, Ramasubramanian T and Chowdhary H 2009. Organic farming for sustainable agriculture: Global and Indian perspective. *Indian J Agron* **54** : 178-85.

Manos, D.-P.; Xydis, G. 2019. "Hydroponics: are we moving towards that direction only because of the environment? A discussion on forecasting and a systems review". *Environmental Science and Pollution Research.* **26** (13): 12662–12672.

Marwaha, B.C. and Jat, S.L. 2004. Status and scope of organic farming in India. *Fertiliser News* **49**(11): 41-48.

Mazher Farid Iqbal, Muzzammil Hussain, Naeem Faisal, Javed Iqbal, Afif Ur Rehman, Maqsood Ahmad and Javed Akhtar Padyar, 2017. Happy seeder zero tillage equipment for sowing of wheat in standing rice stubbles, *Int. J. Adv. Res. Biol. Sci.* **4** (4): 101-105

Mishra V K and Sharma R B 1997. Effect of fertilizers alone and in combination with manure on physical properties and productivity of Entisol under rice-based cropping system. *J Indian Soc Soil Sci* **45** : 84-88.

More S D 1994. Effect of farm wastes and organic manures on soil physical properties, nutrient availability and yield of rice-wheat grown on sodic vertisol. *J Indian Soc Soil Sci* **42** : 25-58.

NAAS (National Academy of Agricultural Sciences) 2009. State of Indian Agriculture, New Delhi.

Nayyar V K and Chhibba I M 2000. Effect of green manuring on micronutrient availability in rice-wheat cropping system in north-west India. *In Long-term soil fertility experiments in rice-wheat cropping system.* Rice Wheat Consortium Paper Series 6 (Abrol I P, Bronson K F, Duxbury J M and Gupta R K eds.). pp 68-72, New Delhi, India : Rice-wheat Consortium for Indo-Gangetic Plains.

NIIR Project Consultancy Services, Date: October 23, 2018

Ohno T and Enrich M S 1977. Inhibitory effects of crop residue derived organic ligands on phosphate adsorption kinetics. *J Environ Qual* **26**: 889-95.

Palaniappan S P and Annadurai K 1999. *Organic Farming : Theory and Practice.* Scientific Publishers (India) Jodhpur, India.

Parkash Y S, Bhadoria P B S and Rakshit Amitava 2002. Comparative efficacy of organic manures on the changes in soil properties and nutrient availability in an Alfisol. *J Indian Soc Soil Sci* **50**: 217-21.

Patil P V, Chalwade P B, Solanke A S and Kulkarni V K 2003. Effect of fly ash and FYM on physico-chemical properties of vertisols. *J Soils & Crops* **13** : 59-64.

Perszewski R 2005. Ideas leading to no-till's second revolution. http://www.no-tillfarmer.com/pages/Feature-Articles---Ideas-Leading-To-No-Tills-Second-Revolution.Php. Accessed 2 June 2014

Ping Cong,Yuyi Li, Jing Wang, Zhijuan Gao, Huancheng Pang, Li Zhang,Na Liu & Jianxin Dong, 2020. Increasing straw incorporation rates improves subsoil fertility and crop yield in the Huang-Huai-Hai Plain of China, *Archives of Agronomy and Soil Science* **66**, Issue 14

Pittelkow CM, Liang X, Linquist BA, van Groenigen KJ, Lee J, Lundy ME, van Gestel N, Six J, Venterea RT, van Kessel C. 2015. Productivity limits and potentials of the principles of conservation agriculture. *Nature*. **15**;517(7534):365-8. doi: 10.1038/nature13809

Prasad, K. and Gill, M.S. Gill. 2009. Development and strategies perspective for organic farming in India. *Indian J. Agron.* **54** (2): 186-192.

Prasad, R. 1993. Development of bioherbicides for integrated weed management in forestry. In Proc. Int. Symp. *Indian Society of Weed Science*. Hissar, Nov. 18-20, 193, **I**: 269-274.

Ramteke J R, Dixit A J, Patil R A and Kanade Von 2001. Effect of organic farming on soil quality and productivity of hybrid rice. *Intl Conference on Nature Farming and Ecological Balance*, ISNF, CCS HAU, Hisar, India. March 7-10, p 161.

Ray S S and Gupta R P 2001. Effect of green manuring and tillage practices on physical properties of puddled loam soil under rice-wheat cropping system. *J Indian Soc Soil Sci* **49**: 670-78.

Rickman, J.F., 2002. Manual for laser land leveling. Rice-Wheat Consortium Technical Bulletin Series 5. New Delhi: Rice-Wheat Consortium for the Indo-Gangetic Plains.

Saritha A., Ramanjaneyulu A. V., Sainath N. and Umarani. E., 2020. Nutritional Importance and Value Addition in Maize. *Biotica Research Today* **2**(9): 974- 977.

Shah, S.K. 2011. Towards Adopting Nanotechnology in Irrigation. Micro Irrigation Systems. Karnataka, India: India Water Portal, 2011.

Sharma A.R., Jat M.L., Saharawat Y.S., Singh V.P., Singh R. 2012 Conservation agriculture for improving productivity and resource-use efficiency: prospects and research needs in Indian context, *Indian Journal of Agronomy*, **57** (IAC Special Issue), pp. 131-140

Sidhu H S, Mahal J S, Dhaliwal I S, Bector Vishal, Singh Manpreet, Sharda Ajay and Singh Thakar, 2008, Tractor Operated Laser Land Leveler, Published by: Coordinating Cell, AICRP on Farm Implements and Machinery Central Institute of Agricultural Engineering, Bhopal.

Singh M, Brar K S and Bakhetia D R C 1997. Management of sugarcane stalk borer (*Chilo auricillus*) with egg parasitoid (*Trichogramma chilonis*). Third Agric. Sci. Cong. PAU, Ludhiana, India, p. 12-15.

Singh, J.M., Jasdev Singh, Hardeep Kumar, Sukhpal Singh, Jatinder Sachdeva, Baljinder Kaur, Shruti Chopra and Prem Chand 2019, Management of Paddy Straw in Punjab: An Economic Analysis of Different Techniques, *Ind. Jr. of Agri. Econ.* **74**, No.3, July-Sept.2019

Singh R P 1993. Botanical pesticides in developing countries: Current status and future trends. pp 236-69. In: *Pesticides: their ecological impact in developing countries.* (Dhaliwal G S and Singh B eds). Commonwealth Pub., New Delhi, India.

Singh, J. P., Gangwar, B., Pandey, D. K. and Kochewad, S. A. 2011. Integrated farming system model for small farm holders of western plain zone of Uttar Pradesh. PDFSR Bulletin No. 05, pp. 58. Project Directorate for Farming Systems Research, Modipuram, Meerut, India.

Smith, R.J. Jr. 1986. Biological control of northern joint vetch (*Aeschynomene virginica*) in rice (*Oryza sativa*) and soyabeans (*Glycine max.*) – A researcher's view. *Weed Sci.* **34**: (Suppl. 1) 17: 23.

Srivastava O P 1998. Integrated nutrient management for sustained fertility of soil. *Indian J agric chem.* **31**:1-12.

Stockdale, E.A., Lampkin, N.H., Hovi, M., Keatinge, R., Lennartsson, E.K.M., McDonald, D.W., Padel, S., Tattersall, EH., Wolfe, M.S. and Watson, C.A. 2001. Agronomic and environmental implications of organic farming system. *Advances in Agronomy* **70**: 261-327.

Sur H S, Sidhu A S, Singh Rachhpal, Aggarwal G C and Sandhu K S 1993. Long term effect of green manuring on soil physical properties and production potential in green manuring – maize – wheat sequence. *Ann. Agric. Res.* **14** : 125-31.

Surekha J and Rao P Arjuna 2000. Management of fruit borer of Bhendi with organic sources of NPK and certain insecticides and its effect on Bhendi yield. *Pestology* **24**: 35-39.

TeBeest, D.O., Cisar, C.R. and Spiegel, F.W. 1992. Partial characterisation of progeny from a cross between *Colletotrichum gleosporioides f. sp. aeschynomene* and *C. gloeosporioides* from *Carva. Plant Prot.* **7**: 171.

Templeton, G.E. 1982. In biological control weeds with plant pathogens eds. R. Charudattan and R. Walker. John Willey and Sons, N.Y. pp. 29-44.

Tillman D 1998. The greening of the green revolution. *Nature London* **396**(6708): 211-12.

Tiwari A, Nema H R, Tembhare B R and Sharma S K 2000. Soil physical environment in long term fertilizer experiment on Typic Haplustorts. *JNKVV Res J* **34** : 29-33.

Uddin Rakib M and Suliaman M F 2021 "Smart Indoor fogponics farming system". IOP Conf. Ser.: *Earth Environ. Sci* 673012012.

Venkateswara Rao S, Azam K M, Moula S P and Subba Rao I V 1989. Management of chilli pod borer *Spodoptera litura* F. through regulation of fertilizers. *The Andhra agric J.* **36** : 183-85.

Walia, U.S. 2018. *Weed management* by Kalyani Publishers, Daryaganj, New Delhi.

Walia, U.S., Singh, S. and Singh, B. (2007) Integrated Approach for the control of hardy weeds in groundnut (*Arachis hypogaea L.*) *Indian J. Weed Sci.* **39** (1 and 2): 112–115.

Walia, U.S., Singh, S. and Singh, B. 2007. Integrated control of hardy weeds in maize (*zea mays L*). *Indian J. Weed Sci.* **39** (1 and 2): 17–20.

Walker, J., Mikandrow, A. and Mittar, G.D. 1991. Species of *Colletotrichum and Xanthium* (asteraceae) with comments on some taxonomic and nomenclatural problems *Weed Res.* **95**: 1175-1193.

Walker, T. S.; Bais, H. P.; Grotewold, E.; Vivanco, J. M. 2003. "Root exudation and rhizosphere biology". *Plant Physiology.* **132** (1): 44–51.

Wynen, E. 1998. Evaluating the potential contribution of organic agriculture to sustainability goals. Environment and Natural Resources Service, Research, Teaching and Training Division, Sustainable Development Department, Food and Agriculture Organization of United Nations (FAO), Rome.

Yang, X.B.A. and Tebeest, D.O. 1993. Epidemiological mechanisms of mycoherbicide effectiveness. *Phytopathology* **83**: 891-893.